ELEMENTS OF ADVANCED ENGINEERING MATHEMATICS

PETER V. O'NEIL

*The University of Alabama
at Birmingham*

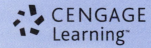
CENGAGE
Learning™

Australia • Brazil • Japan • Korea • Mexico • Singapore • Spain • United Kingdom • United States

Elements of Advanced Engineering Mathematics, 1st Edition
Peter V. O'Neil

Executive Director, Global Publishing Program:
Chris Carson

Senior Developmental Editor:
Hilda Gowans

Editorial Assistant:
Nancy Saundercook

Associate Marketing Manager:
Lauren Betsos

Media Editor:
Chris Valentine

Director, Content and Media Production:
Sharon Smith

Content Project Manager:
Jean Buttrom

Production Service:
RPK Editorial Services

Copyeditor:
Fred Dahl

Proofreader:
Harlan James

Indexer:
Shelly Gerger-Knecthl

Compositor:
Integra Software Services

Senior Art Director:
Michelle Kunkler

Internal Designer:
Carmela Pereira

Cover Designer:
Andrew Adams

Cover Images:
© Tund/Shutterstock

Permissions Account Manager, Text:
Mardell Glinski Shultz

Permissions Account Manager, Images:
Deanna Ettinger

Text and Images Permissions Researcher:
Kristiina Paul

Senior First Print Buyer:
Doug Wilke

For product information and technology assistance, contact us at
**Cengage Learning Customer & Sales Support,
1-800-354-9706.**

For permission to use material from this text or product, submit all requests online at
www.cengage.com/permissions.
Further permissions questions can be emailed to
permissionrequest@cengage.com.

Library of Congress Control Number: 2009930547
ISBN-13: 978-0-495-66818-3
ISBN-10: 0-495-66818-4

Cengage Learning
200 First Stamford Place, Suite 400
Stamford, CT 06902
USA

Cengage Learning is a leading provider of customized learning solutions with office locations around the globe, including Singapore, the United Kingdom, Australia, Mexico, Brazil, and Japan. Locate your local office at: **international.cengage.com/region.**

Cengage Learning products are represented in Canada by Nelson Education Ltd.

For your course and learning solutions, visit
www.cengage.com/ engineering.

Purchase any of our products at your local college store or at our preferred online store **www.ichapters.com.**

Printed in the United States of America
1 2 3 4 5 6 7 13 12 11 10 09

$\int \int_{\Sigma} f(x, y, z) \, d\sigma$ surface integral of f over Σ (11.5.4)

n   often denotes a unit normal vector   (11.6)

$f(x_0-)$, $f(x_0+)$ left and right limits, respectively, of $f(x)$ at x_0 (12.1)

$\mathcal{F}[f]$ or $\hat{f}$ Fourier transform of f (13.3)

$\mathcal{F}_c[f]$ or $\hat{f}_c$ Fourier cosine transform of f (13.4)

$\mathcal{F}_s[f]$ or $\hat{f}_s$ Fourier sine transform of f (13.4)

$P_n(x)$ the *nth* Legendre polynomial   (14.2)

$\Gamma(x)$ gamma function   (14.3)

$J_v(x)$ Bessel function of the first kind of order v (14.3)

$Y_n(x)$ Bessel function of the second kind of order n (14.3)

$\Delta^2 u$ Laplacian of u (17.1)

ELEMENTS OF ADVANCED ENGINEERING MATHEMATICS

Contents

PART 2   Vectors, Linear Algebra, and Systems of Linear Differential Equations   103

PART 5   Partial Differential Equations   337

Preface

This book is intended to provide students with an efficient introduction and accessibility to ordinary and partial differential equations, linear algebra, vector analysis, Fourier analysis, special functions, and eigenfunction expansions for their use as tools of inquiry and analysis in modeling and problem solving. It should also serve as preparation for further reading where this suits individual needs and interests.

Although much of this material appears in Advanced Engineering Mathematics, 6th edition, *Elements* has been completely rewritten to provide a natural flow of the material in this shorter format.

Many types of computations, such as construction of direction fields or the manipulation of Bessel functions and Legendre polynomials in writing eigenfunction expansions, require the use of software packages. A short MAPLE primer is included as Appendix B. This is designed to enable the student to quickly master the use of MAPLE for such computations. Other software packages also can be used.

In producing this book, I would like express my appreciation to the editorial leadership at Cengage Learning who helped bring it from concept to final form. In particular, thanks are due to Chris Carson, Hilda Gowans, and Rose Kernan.

PART 1

Ordinary Differential Equations

$$[f](s) = \int_0^\infty e^{-st} t^{-1/2}\, dt = 2\int_0^\infty e^{-sx^2}\, dx \ (\text{set } x = t^{1/2}) = \frac{2}{\sqrt{s}} \int_0^\infty e^{-z^2}\, dz \ (\text{set } z = x$$

$$= \sqrt{\frac{\pi}{s}}$$

$$\mathcal{L}[f](s) = \int_0^\infty e$$

$$= 2\int_0^\infty e^{-s} \, dx$$

$$= \sqrt{\frac{\pi}{s}}$$

$$](s) = \int_0^\infty e^{-st} t^{-1/2}\, dt$$

$$= 2\int_0^\infty e^{-sx^2}\, dx \ (\text{set } x = t^{1/2})$$

$$= \frac{2}{\sqrt{s}} \int_0^\infty e^{-z^2}\, dz \ (\text{set } z = x\sqrt{s})$$

$$[f](s) = \int_0^\infty e^{-st} t^{-1/2}$$

CHAPTER 1

First-Order Differential Equations

1.1   Terminology and Separable Equations

We begin with *ordinary differential equations*, which are equations that contain one or more derivatives of a function of a single variable. These model a rich variety of phenomena of interest in the sciences, engineering, economics, ecological studies and other areas.

This chapter is devoted to first-order differential equations, in which only the first derivative of the unknown function appears. As an example,

$$y' + xy = 0$$

is a first-order equation for the unknown function $y(x)$. A *solution* of a differential equation is any function satisfying the equation. It is routine to check that $y = ce^{-x^2/2}$ is a solution of $y' + xy = 0$ for any constant c.

We will develop techniques for solving several kinds of first-order equations which arise in important contexts, beginning with separable equations.

Separable Differential Equation

A differential equation is *separable* if it can be written (perhaps after some algebraic manipulation) as

$$\frac{dy}{dx} = A(x)B(y).$$

This equation, in theory, can be solved by writing it in differential form as

$$\frac{1}{B(y)}dy = A(x)dx.$$

Now the variables have been separated with all terms involving y on one side and all terms involving x on the other. Integrate the separated equation,

$$\int \frac{1}{B(y)}\, dy = \int A(x)\, dx$$

to obtain an equation for the solution $y(x)$.

EXAMPLE 1.1

To solve $y' = y^2 e^{-x}$, first write

$$\frac{dy}{dx} = y^2 e^{-x}$$

which in differential form is

$$\frac{1}{y^2}\, dy = e^{-x}\, dx.$$

The variables have been separated. Integrate:

$$\int \frac{1}{y^2}\, dy = \int e^{-x}\, dx,$$

or

$$-\frac{1}{y} = -e^{-x} + k$$

in which k is a constant of integration. Solve for y to get

$$y(x) = \frac{1}{e^{-x} - k}.$$

This is a solution of $y' = y^2 e^{-x}$ for any number k. ◆

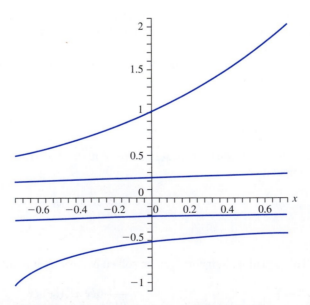

FIGURE 1.1 *Some integral curves in Example 1.1.*

This expression for $y(x)$ is called the *general solution* of this differential equation, because it contains an arbitrary constant. We obtain *particular solutions* by making specific choices for k. In Example 1.1,

$$y(x) = \frac{1}{e^{-x} - 3}, \quad y(x) = \frac{1}{e^{-x} + 3},$$

$$y(x) = \frac{1}{e^{-x} - 6}, \quad \text{and } y(x) = \frac{1}{e^{-x}} = e^x$$

are particular solutions corresponding to $k = \pm 3, 6$ and 0. Particular solutions are also called *integral curves* of the differential equation. Graphs of these integral curves are shown in Figure 1.1.

EXAMPLE 1.2

$x^2 y' = 1 + y$ is separable, since we can write it as

$$\frac{1}{1+y} \, dy = \frac{1}{x^2} \, dx.$$

The algebra of separating the variables requires that $y \neq -1$ and $x \neq 0$. Integrate to obtain

$$\ln|1 + y| = -\frac{1}{x} + k$$

with k an arbitrary constant. This equation implicitly defines the general solution. This means that we have an equation containing the solution $y(x)$, but not an explicit expression giving $y(x)$ in terms of x and k. In this example we can explicitly solve for $y(x)$. First, take the exponential of both sides of the equation to get

$$|1 + y| = e^{k - 1/x} = e^k e^{-1/x} = A e^{-1/x},$$

where we have written $A = e^k$. Since k can be any number, A can be any positive number. Now eliminate the absolute value symbol by writing

$$1 + y = \pm A e^{-1/x} = B e^{-1/x},$$

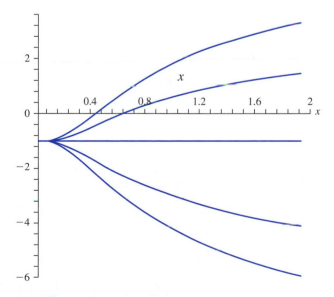

FIGURE 1.2 *Some integral curves in Example 1.2.*

where $B = \pm A$ can be any nonzero number. Then

$$y = -1 + Be^{-1/x}$$

with $B \neq 0$. Finally, observe that, if we allow $B = 0$, this expression for y gives us $y = -1$, which is also a solution of the original equation $x^2 y' = 1 + y$. We can therefore write the general solution

$$y = -1 + Be^{-1/x}$$

in which B can be any real number. Integral curves (graphs of solutions) corresponding to $B = 0, 4, 7, -5$, and -8 are shown in Figure 1.2. ♦

Each of these examples has infinitely many solutions because of the arbitrary constant in the general solution. If we specify that the solution is to satisfy a condition $y(x_0) = y_0$, with x_0 and y_0 given numbers, then we pick out the particular integral curve passing through (x_0, y_0). The differential equation, together with a condition $y(x_0) = x_0$, is called an *initial value problem*. The condition $y(x_0) = y_0$ is called an *initial condition*.

One way to solve an initial value problem is to find the general solution, then solve for the constant to find the solution satisfying the initial condition.

EXAMPLE 1.3

Solve the initial value problem

$$y' = y^2 e^{-x}; \quad y(1) = 4.$$

From Example 1.1, we know that the general solution of this differential equation is

$$y(x) = \frac{1}{e^{-x} - k}.$$

Choose k so that

$$y(1) = \frac{1}{e^{-1} - k} = 4.$$

Solve this equation for k to get

$$k = e^{-1} - \frac{1}{4}.$$

The solution of the initial value problem is

$$y(x) = \frac{1}{e^{-x} + \frac{1}{4} - e^{-1}}. \quad ♦$$

It is not always possible to find an explicit solution of a differential equation, in which y is isolated on one side of an equation and some expression in x occurs on the other side.

EXAMPLE 1.4

We will solve the initial value problem

$$y' = y\frac{(x-1)^2}{y+3}; \quad y(3) = -1.$$

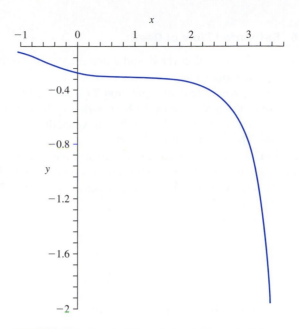

FIGURE 1.3   *Graph of the solution in Example 1.4.*

In differential form,

$$\frac{y+3}{y}\,dy = (x-1)^2\,dx$$

or

$$\left(1 + \frac{3}{y}\right) dy = (x-1)^2 dx.$$

Integrate to obtain

$$y + 3\ln|y| = \frac{1}{3}(x-1)^3 + k.$$

This equation implicitly defines the general solution. However, we cannot solve for y as an explicit expression of x.

This does not prevent us from solving the initial value problem. We need $y(3) = -1$, so put $x = 3$ and $y = -1$ into the implicitly defined general solution to get

$$-1 = \frac{1}{3}\left(2^3\right) + k.$$

Then $k = -11/3$ and the solution of the initial value problem is implicitly defined by

$$y + 3\ln|y| = \frac{1}{3}(x-1)^3 - \frac{11}{3}.$$

Part of this solution is graphed in Figure 1.3. ◆

1.1.1   Some Applications of Separable Equations

Separable differential equations arise in many contexts. We will discuss two of these.

EXAMPLE 1.5 Estimated Time of Death

A homicide victim is discovered, and a detective from the forensics laboratory is summoned to estimate the time of death.

The strategy is to find an expression $T(t)$ for the body's temperature at time t, taking into account the fact that after death the body will cool by radiating heat energy into the room. $T(t)$ can then be used to estimate the last time at which the victim was alive and had a "normal" body temperature. This was the time of death.

To do this, some information is needed. First, suppose we know that the body is located in a room that is kept at a constant 68 degrees Fahrenheit. For some time after death, the body will radiate heat into the cooler room. Assume, for want of better information, that the victim's temperature was 98.6 degrees.

By Newton's law of cooling, the body radiates heat energy into the room at a rate proportional to the difference between the room and body temperature. If $T(t)$ is the body's temperature at time t, then for some constant of proportionality k,

$$\frac{dT}{dt} = k[T(t) - 68].$$

This is a separable differential equation, since

$$\frac{1}{T - 68}\, dT = k\, dt.$$

Integrate to obtain

$$\ln|T - 68| = kt + c.$$

To solve for T, take the exponential of both sides of this equation to get

$$|T - 68| = e^{kt+c} = Ae^{kt}$$

where $A = e^c$. Then

$$T - 68 = \pm Ae^{kt} = Be^{kt},$$

so

$$T(t) = 68 + Be^{kt}.$$

Now the constants k and B must be determined, and this requires two pieces of information. Suppose the detective arrived at 9:40 p.m. and immediately measured the body temperature, obtaining 94.4 degrees. Let 9:40 be time zero for convenience. Then

$$T(0) = 94.4 = 68 + B,$$

so $B = 26.4$. So far

$$T(t) = 68 + 26.4e^{kt}.$$

To determine k, we need another measurement. The lieutenant takes the body temperature again at 11:00 p.m. and finds it to be 89.2 degrees. Since 11:00 is 80 minutes after 9:40, this means that

$$T(80) = 89.2 = 68 + 26.4e^{80k}.$$

Then

$$e^{80k} = \frac{21.2}{26.4}$$

so

$$80k = \ln\left(\frac{21.2}{26.4}\right).$$

Then

$$k = \frac{1}{80} \ln\left(\frac{21.2}{26.4}\right).$$

The temperature function is now completely known:

$$T(t) = 68 + 26.4e^{\ln(21.2/26.4)t/80}.$$

The time of death was the last time at which the body temperature was 98.6 degrees (just before it began to cool). Solve for t so that

$$T(t) = 98.6 = 68 + 26.4e^{\ln(21.2/26.4)t/80}.$$

This gives us

$$\frac{30.6}{26.4} = e^{\ln(21.2/26.4)t/80}.$$

Take the logarithm of this equation to obtain

$$\ln\left(\frac{30.6}{26.4}\right) = \frac{t}{80}\ln\left(\frac{21.2}{26.4}\right).$$

According to this model, the time of death was

$$t = \frac{80\ln(30.6/26.4)}{\ln(21.2/26.4)},$$

which is approximately -53.8 minutes. Death occurred approximately 53.8 minutes before (because of the negative sign) the first measurement at 9:40 p.m., which was chosen as time zero in the model. This puts the murder at about 8:46 p.m.

This is an estimate, because an educated guess was made of the body's temperature before death. It is also impossible to keep the room at exactly 68 degrees. However, this model is robust in the sense that small changes in the body's normal temperature and in the constant temperature of the room yield small changes in the estimated time of death. The student can check this by trying a different normal temperature for the body, say 99.3 degrees, to see how much this changes the estimated time of death. ◆

EXAMPLE 1.6 Radioactive Decay and Carbon Dating

In radioactive decay, mass is lost by its conversion to energy, which is radiated away. It has been observed that at any time t the rate of change of the mass $m(t)$ of a radioactive element is proportional to the mass itself. This means that for some constant of proportionality k that is unique to the element,

$$\frac{dm}{dt} = km.$$

Here k must be negative, because the mass is decreasing with time.

This differential equation for m is separable. Write it as

$$\frac{1}{m}\,dm = k\,dt.$$

A routine integration yields

$$\ln|m| = kt + c.$$

Since mass is positive, $|m| = m$, so

$$m(t) = e^{kt+c} = Ae^{kt}$$

in which A can be any positive number. Any radioactive element has its mass change according to a rule of this form, and this reveals an important characteristic of radioactive decay. Suppose at some time τ there are M grams. Look for h so that, at the later time $\tau + h$, exactly half of the mass has radiated away. This would mean that

$$m(\tau + h) = \frac{M}{2} = Ae^{k(\tau+h)} = Ae^{k\tau}e^{kh}.$$

But $Ae^{k\tau} = M$, so the last equation becomes

$$\frac{M}{2} = Me^{kh}.$$

Then

$$e^{kh} = \frac{1}{2}.$$

Take the logarithm of this equation to solve for h, obtaining

$$h = \frac{1}{k}\ln\left(\frac{1}{2}\right) = -\frac{1}{k}\ln(2).$$

Notice that h, the time it takes for half of the mass to convert to energy, depends only on the number k and not on the mass itself, or the time at which we started measuring the loss. If we measure the mass of a radioactive element at any time (say in years), then again h years later exactly half of this mass will have radiated away. This number h is called the *half-life* of the element. The constants h and k are both uniquely tied to the particular element and to each other by $h = -(1/k)\ln(2)$.

Now look at the numbers A and k in the expression $m(t) = Ae^{kt}$. k is tied to the element's half-life. The meaning of A is made clear by observing that

$$m(0) = Ae^0 = A.$$

A is the mass that is present at some time designated for convenience as time zero (think of this as when we start the clock). A is the *initial mass*, usually denoted m_0. Then

$$m(t) = m_0 e^{kt}.$$

It is sometimes convenient to write this expression in terms of the half-life h. Since $k = -(1/h)\ln(2)$, then

$$m(t) = m_0 e^{kt} = m_0 e^{-\ln(2)t/h}. \tag{1.1}$$

This expression is the basis for an important technique used to estimate the ages of certain ancient artifacts. The Earth's upper atmosphere is bombarded by high-energy cosmic rays, producing large numbers of neutrons which collide with nitrogen and converting some of it into radioactive carbon-14 or ^{14}C. This has a half-life $h = 5,730$ years. Over the geologically short time in which life has evolved on Earth, the ratio of ^{14}C to regular carbon in the atmosphere has remained approximately constant. This means that the rate at which a living plant or animal ingests ^{14}C is about the same now as in the past. When a living organism dies, it ceases its intake of ^{14}C, which then begins to decay. By measuring the ratio of ^{14}C to carbon in an artifact, we can estimate the amount of this decay, and hence the time it took, giving an estimate of the last time the organism lived. This method of estimating the age of an artifact is called *carbon dating*. Since an artifact may have been contaminated by exposure to other living organisms, this is a sensitive process. However, when applied rigorously and combined with other tests and information, carbon dating has proved a valuable tool in historical and archeological studies.

If we put $h = 5730$ into equation (1.1) with $m_0 = 1$, we get

$$m(t) = e^{-\ln(2)t/5730} \approx e^{-0.000120968t}.$$

As a specific example, suppose we have a piece of fossilized wood and measurements show that the ratio of ^{14}C to carbon is .37 of the current ratio. To calibrate our clock, say the wood died at time zero. If T is the time it would take for one gram of the radioactive carbon to decay to .37 of one gram, then T satisfies the equation

$$0.37 = e^{-0.000120968T}$$

from which we obtain

$$T = -\frac{\ln(0.37)}{0.000120968} \approx 8,219$$

years. This is approximately the age of the wood. ◆

Examples 1.5 and 1.6 illustrate a process called *mathematical modeling*. We use information about some process of interest to write differential equations, and perhaps initial conditions, for the quantity we want to know and then attempt to solve for this quantity.

SECTION 1.1 PROBLEMS

In each of Problems 1 through 6, determine whether $y = \varphi(x)$ is a solution of the differential equation. C is constant wherever it appears.

1. $2yy' = 1; \varphi(x) = \sqrt{x-1}$ for $x > 1$
2. $y' + y = 0; \varphi(x) = Ce^{-x}$
3. $y' = -\frac{2y+e^x}{2x}$ for $x > 0; \varphi(x) = \frac{C-e^x}{2x}$
4. $y' = \frac{2xy}{2-x^2}$ for $x \neq \pm\sqrt{2}; \varphi(x) = \frac{C}{x^2-2}$
5. $xy' = x - y; \varphi(x) = \frac{x^2-3}{2x}$ for $x \neq 0$
6. $y' + y = 1; \varphi(x) = 1 + Ce^{-x}$

In each of Problems 7 through 16, determine if the differential equation is separable. If it is, find the general solution (perhaps implicitly defined). If it is not, do not attempt a solution.

7. $3y' = 4x/y^2$
8. $y + xy' = 0$
9. $\cos(y)y' = \sin(x+y)$
10. $e^{x+y}y' = 3x$
11. $xy' + y = y^2$
12. $y' = \frac{(x+1)^2 - 2y}{2y}$
13. $x\sin(y)y' = \cos(y)$
14. $\frac{x}{y}y' = \frac{2y^2+1}{x+1}$
15. $y + y' = e^x - \sin(y)$
16. $[\cos(x+y) + \sin(x-y)]y' = \cos(2x)$

In each of Problems 17 through 21, solve the initial value problem.

17. $xy^2y' = y + 1; y(3e^2) = 2$
18. $y' = 3x^2(y+2); y(2) = 8$
19. $\ln(y^x)y' = 3x^2y; y(2) = e^3$
20. $2yy' = e^{x-y^2}; y(4) = -2$
21. $yy' = 2x\sec(3y); y(2/3) = \pi/3$

22. An object having a temperature of 90° Fahrenheit is placed in an environment kept at 60°. Ten minutes later, the object has cooled to 88°. What will be the temperature of the object after it has been in this environment for 20 minutes? How long will it take for the object to cool to 65°?

23. A thermometer is carried outside a house whose ambient temperature is 70° Fahrenheit. After 5 minutes, the thermometer reads 60°, and 15 minutes after this, 50.4°. What is the outside temperature (which is assumed to be constant)?

24. A radioactive element has a half-life of $\ln(2)$ weeks. If e^3 tons are present at a given time, how much will be left three weeks later?

25. The half-life of Uranium-238 is approximately $4.5(10^9)$ years. How much of a 10-kilogram block of $U - 238$ will be present one billion years from now?

26. Given that 12 grams of a radioactive element decay to 9.1 grams in 4 minutes, what is the half-life of this element?

27. Evaluate

$$\int_0^\infty e^{-t^2 - 9/t^2}\,dt.$$

Hint: Let

$$I(x) = \int_0^\infty e^{-t^2 - (x/t)^2}\,dt.$$

Calculate $I'(x)$ and find a differential equation for $I(x)$. Use the standard integral $\int_0^\infty e^{-t^2}\,dt = \sqrt{\pi}/2$ to determine $I(0)$ and use this initial condition to solve for $I(x)$. Finally, evaluate $I(3)$.

1.2   Linear Equations

> A first-order differential equation is *linear* if it has the form
>
> $$y' + p(x)y = q(x)$$
>
> for some functions $p(x)$ and $q(x)$.

There is an approach to solving a linear equation. Let

$$g(x) = e^{\int p(x)dx}$$

and notice that

$$g'(x) = p(x)e^{\int p(x)dx} = p(x)g(x). \qquad (1.2)$$

Multiply $y' + p(x)y = q(x)$ by $g(x)$ to obtain

$$g(x)y' + p(x)g(x)y = q(x)g(x).$$

In view of equation (1.2) this is

$$g(x)y' + g'(x)y = q(x)g(x).$$

Now we see the point to multiplying the differential equation by $g(x)$. The left side of the new equation is the derivative of $g(x)y$. The differential equation has become

$$\frac{d}{dx}(g(x)y) = q(x)g(x),$$

which we can integrate to obtain

$$g(x)y = \int q(x)g(x)dx + c.$$

If $g(x) \neq 0$, we can solve this equation explicitly for y:

$$y(x) = \frac{1}{g(x)}\int q(x)g(x)dx + \frac{c}{g(x)}.$$

This is the general solution, with arbitrary constant c.

We do not recommend memorizing this formula for $y(x)$. Just remember to multiply the linear differential equation by $e^{\int p(x)dx}$ and write one side as the derivative of a product. Because of this, $e^{\int p(x)dx}$ is called an *integrating factor* for the differential equation.

EXAMPLE 1.7

The equation $y' + y = x$ is linear with $p(x) = 1$ and $q(x) = x$. An integrating factor is

$$e^{\int p(x)dx} = e^{\int dx} = e^x.$$

Multiply the differential equation by e^x to get

$$e^x y' + e^x y = xe^x.$$

This is

$$(ye^x)' = xe^x,$$

which we integrate to obtain

$$ye^x = \int xe^x dx = xe^x - x + c.$$

Finally, solve for y by multiplying this equation by e^{-x}:

$$y = x - xe^{-x} + ce^{-x}.$$

This is the general solution. ◆

EXAMPLE 1.8

Solve the initial value problem

$$y' = 3x^2 - \frac{y}{x}; \quad y(1) = 5.$$

This differential equation is not in linear form, but we can write it as

$$y' + \frac{1}{x}y = 3x^2,$$

which is linear. An integrating factor is

$$e^{\int (1/x)dx} = e^{\ln(x)} = x$$

for $x > 0$. Multiply the differential equation by x to obtain

$$xy' + y = 3x^3$$

or

$$(xy)' = 3x^3.$$

Integrate to obtain

$$xy = \frac{3}{4}x^4 + c.$$

Solve for y to write the general solution

$$y = \frac{3}{4}x^3 + \frac{c}{x}$$

for $x > 0$. For the initial condition, we need

$$y(1) = \frac{3}{4} + c = 5.$$

Then $c = 17/4$ and the solution of the initial value problem is

$$y = \frac{3}{4}x^3 + \frac{17}{4x}. \quad \blacklozenge$$

Solving a linear differential equation may lead to integrals we cannot evaluate in elementary form. For example, an integrating factor for $y' + xy = 2$ is $e^{\int x\,dx} = e^{x^2/2}$. Multiplication by this integrating factor results in

$$(ye^{x^2/2})' = 2e^{x^2/2}.$$

Integrating this equation requires that we evaluate $\int 2e^{x^2/2}\,dx$, which cannot be done in elementary form. This does not mean that we cannot solve this differential equation. However, something beyond the simple technique discussed here must be invoked.

Here is an application of linear equations to a mixing problem.

EXAMPLE 1.9   A Mixing Problem

We want to determine how much of a given substance is present in a container in which various substances are being added, mixed, and drained out. This is a *mixing problem* and is encountered in the chemical industry, ocean currents, and other areas.

As an example, suppose a tank contains 200 gallons of brine (salt mixed with water), in which 100 pounds of salt are dissolved. A mixture consisting of 1/8 pound of salt per gallon is pumped into the tank at a rate of 3 gallons per minute, and the mixture is continuously stirred. Brine is also allowed to empty out of the tank at the same rate of 3 gallons per minute (see Figure 1.4). How much salt is in the tank at any time?

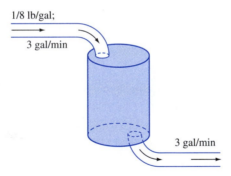

FIGURE 1.4   *Storage tank in Example 1.9.*

Let $Q(t)$ be the amount of salt in the tank at time t. The rate of change of $Q(t)$ with respect to time must equal the rate at which salt is pumped in minus the rate at which it is pumped out:

$$\frac{dQ}{dt} = (\text{rate in}) - (\text{rate out})$$

$$= \left(\frac{1}{8}\frac{\text{pounds}}{\text{gallon}}\right)\left(3\frac{\text{gallons}}{\text{minute}}\right) - \left(\frac{Q(t)}{200}\frac{\text{pounds}}{\text{gallon}}\right)\left(3\frac{\text{gallons}}{\text{minute}}\right)$$

$$= \frac{3}{8} - \frac{3}{200}Q(t).$$

This is the linear equation

$$Q'(t) + \frac{3}{200}Q = \frac{3}{8}.$$

An integrating factor is $e^{\int (3/200)dt} = e^{3t/200}$. Multiply the differential equation by this to get

$$(Qe^{3t/200})' = \frac{3}{8}e^{3t/200}.$$

Integrate to obtain

$$Qe^{3t/200} = \frac{3}{8}\frac{200}{3}e^{3t/200} + c.$$

Then

$$Q(t) = 25 + ce^{-3t/200}.$$

Now use the initial condition:

$$Q(0) = 100 = 25 + c$$

so $c = 75$ and

$$Q(t) = 25 + 75e^{-3t/200}.$$

From this, $Q(t) \to 25$ as $t \to \infty$. This is the *steady-state* value of $Q(t)$. The term $75e^{-3t/200}$ is called the *transient* part of the solution, because it decays to zero as t increases. $Q(t)$ is the sum of a steady-state part and a transient part. This type of decomposition of a solution is found in many settings. For example, the current in a circuit is often a sum of a steady-state term and a transient term.

The initial ratio of salt to brine in the tank is 100 pounds per 200 gallons, or 1/2 pound per gallon. Since the mixture pumped in has a constant ratio of 1/8 pound per gallon, we expect the brine mixture to dilute toward the incoming ratio with a terminal amount of salt in the tank of 1/8 pound per gallon times 200 gallons. This leads to the expectation that, in the long term, the amount of salt in the tank should approach 25 pounds, as the model verifies. ♦

SECTION 1.2   **PROBLEMS**

In each of Problems 1 through 5, find the general solution.

1. $y' - \frac{3}{x}y = 2x^2$
2. $y' + y = \frac{1}{2}(e^x - e^{-x})$
3. $y' + 2y = x$
4. $y' + \sec(x)y = \cos(x)$
5. $y' - 2y = -8x^2$

In each of Problems 6 through 10, solve the initial value problem.

6. $y' + 3y = 5e^{2x} - 6; y(0) = 2$
7. $y' + \frac{1}{x-2}y = 3x; y(3) = 4$
8. $y' - y = 2e^{4x}; y(0) = -3$
9. $y' + \frac{2}{x+1}y = 3; y(0) = 5$
10. $y' + \frac{5y}{9x} = 3x^3 + x; y(-1) = 4$

11. Find all functions with the property that the $y-$intercept of the tangent to the graph at (x, y) is $2x^2$.

12. A 500-gallon tank initially contains 50 gallons of brine solution in which 28 pounds of salt have been dissolved. Beginning at time zero, brine containing 2 pounds of salt per gallon is added at the rate of 3 gallons per minute, and the mixture is poured out of the tank at the rate of 2 gallons per minute. How much salt is in the tank when it contains 100 gallons of brine? *Hint*: The amount of brine in the tank at time t is $50 + t$.

13. Two tanks are connected as in Figure 1.5. Tank 1 initially contains 20 pounds of salt dissolved in 100 gallons of brine. Tank 2 initially contains 150 gallons of brine in which 90 pounds of salt are dissolved. At time zero, a brine solution containing 1/2 pound of salt per gallon is added to tank 1 at the rate of 5 gallons per minute. Tank 1 has an output that discharges

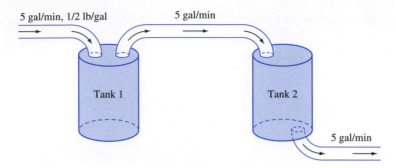

FIGURE 1.5   *Storage tank in Problem 13, Section 1.2.*

brine into tank 2 at the rate of 5 gallons per minute, and tank 2 also has an output of 5 gallons per minute. Determine the amount of salt in each tank at any time. Also determine when the concentration of salt in tank 2 is a minimum and how much salt is in the tank at that time. *Hint*: Solve for the amount of salt in tank 1 at time t and use this solution to help determine the amount in tank 2.

1.3   Exact Equations

A differential equation $M(x, y) + N(x, y)y' = 0$ can be written in differential form as

$$M(x, y)dx + N(x, y)dy = 0. \tag{1.3}$$

Sometimes this differential form is the key to writing a general solution. Recall that the differential of a function $\varphi(x, y)$ of two variables is

$$d\varphi = \frac{\partial \varphi}{\partial x}dx + \frac{\partial \varphi}{\partial y}dy. \tag{1.4}$$

If we can find a function $\varphi(x, y)$ such that

$$\frac{\partial \varphi}{\partial x} = M(x, y) \text{ and } \frac{\partial \varphi}{\partial y} = N(x, y). \tag{1.5}$$

then the differential equation $Mdx + Ndy = 0$ is just

$$M(x, y)dx + N(x, y)dy = d\varphi = 0.$$

But if $d\varphi = 0$, then $\varphi(x, y) = $ constant. The equation

$$\varphi(x, y) = c$$

with c as an arbitrary constant implicitly defines the general solution of $M + Ny' = 0$.

EXAMPLE 1.10

We will use these ideas to solve

$$\frac{dy}{dx} = \frac{2x - e^x \sin(y)}{e^x \cos(y) + 1}.$$

This equation is neither separable nor linear. Write it in the form of equation (1.3) as

$$M(x, y)dx + N(x, y)dy = (e^x \sin(y) - 2x)dx + (e^x \cos(y) + 1)dy = 0.$$

Now let $\varphi(x, y) = e^x \sin(y) + y - x^2$. Then

$$\frac{\partial \varphi}{\partial x} = e^x \sin(y) - 2x = M(x, y) \text{ and } \frac{\partial \varphi}{\partial y} = e^x \cos(y) + 1 = N(x, y),$$

so equations (1.5) are satisfied. The differential equation becomes just $d\varphi = 0$ with the general solution defined implicitly by

$$\varphi(x, y) = e^x \sin(y) + y - x^2 = c.$$

To verify that this equation does indeed implicitly define the solution of the differential equation, differentiate it implicitly with respect to x to get

$$e^x \sin(y) + e^x \cos(y) y' + y' - 2x = 0$$

and solve this for y' to get

$$y' = \frac{2x - e^x \sin(y)}{e^x \cos(y) + 1},$$

which is the original differential equation. ♦

Example 1.10 illustrates an idea. If we can find $\varphi(x, y)$ satisfying equations (1.5), we have the general solution implicitly defined by the equation $\varphi(x, y) = c$. Finding such a $\varphi(x, y)$ is usually the hard part of applying this method.

EXAMPLE 1.11

Consider

$$\frac{dy}{dx} = -\frac{2xy^3 + 2}{3x^2 y^2 + 8e^{4y}}.$$

This is neither linear nor separable. Write

$$(2xy^3 + 2)dx + (3x^2 y^2 + 8e^{4y})dy = 0$$

so

$$M(x, y) = 2xy^3 + 2 \text{ and } N(x, y) = 3x^2 y^2 + 8e^{4y}.$$

For equations (1.5), we need $\varphi(x, y)$ such that

$$\frac{\partial \varphi}{\partial x} = 2xy^3 + 2 \text{ and } \frac{\partial \varphi}{\partial y} = 3x^2 y^2 + 8e^{4y}.$$

Choose either of these equations and integrate it. If we choose the first equation, then integrate with respect to x:

$$\varphi(x, y) = \int \frac{\partial \varphi}{\partial x} dx$$

$$= \int (2xy^3 + 2)dx = x^2 y^3 + 2x + g(y).$$

In this integration, we are reversing a partial derivative with respect to x, so y is treated as a constant. This means that the constant of integration may also involve y, hence it is called $g(y)$. Now we know $\varphi(x, y)$ to within this unknown function $g(y)$. To determine $g(y)$, use the fact that we know what $\partial \varphi / \partial y$ must be:

$$\frac{\partial \varphi}{\partial y} = 3x^2 y^2 + 8e^{4y} = 3x^2 y^2 + g'(y).$$

This means that $g'(y) = 8e^{4y}$, so $g(y) = 2e^{4y}$. This fills in the missing piece, and

$$\varphi(x, y) = x^2y^3 + 2x + 2e^{4y}.$$

The general solution of the differential equation is implicitly defined by

$$x^2y^3 + 2x + 2e^{4y} = c,$$

in which c is an arbitrary constant. In this example, it is not clear how to solve for y explicitly in terms of x. ◆

A function $\varphi(x, y)$ satisfying equations (1.5) is called a *potential function* for the differential equation $M + Ny' = 0$. If we can find a potential function $\varphi(x, y)$, we have at least the implicit expression $\varphi(x, y) = c$ for the general solution. The method of Example 1.11 may produce a potential function if the integrations can be carried out. However, it may also be the case that a potential function does not exist.

EXAMPLE 1.12

The equation $y' + y = 0$ is separable and linear, and the general solution is $y(x) = ce^{-x}$. However, to make a point, try to find a potential function. In differential form,

$$ydx + dy = 0.$$

Here $M(x, y) = y$ and $N(x, y) = 1$. A potential function φ would have to satisfy

$$\frac{\partial \varphi}{\partial x} = y \text{ and } \frac{\partial \varphi}{\partial y} = 1.$$

If we integrate the first of these equations with respect to x, we get

$$\varphi(x, y) = \int y\, dx = xy + g(y).$$

But then we need

$$\frac{\partial \varphi}{\partial y} = 1 = x + g'(y),$$

so $g'(y) = 1 - x$, which is impossible if g is a function of y only. ◆

We call a differential equation $M + Ny' = 0$ *exact* if it has a potential function. Otherwise it is *not exact*.

There is a simple test to determine whether $M + Ny' = 0$ is exact for (x, y) in some rectangle R of the plane.

THEOREM 1.1   *Test for Exactness*

Suppose M and N are continuous for all (x, y) in some rectangle R in the (x, y) plane. Then $M + Ny' = 0$ is exact on R if and only if

$$\frac{\partial N}{\partial x} = \frac{\partial M}{\partial y}$$

for (x,y) in R. ◆

In the case of $y\,dx + dy = 0$, $M(x, y) = y$ and $N(x, y) = 1$. So

$$\frac{\partial M}{\partial y} = 1 \text{ and } \frac{\partial N}{\partial x} = 0.$$

Theorem 1.1 tells us that this differential equation is not exact. We saw this in Example 1.12.

EXAMPLE 1.13

We will solve the initial value problem

$$(\cos(x) - 2xy) + (e^y - x^2)y' = 0; \quad y(1) = 4.$$

In differential form,

$$(\cos(x) - 2xy)\,dx + (e^y - x^2)\,dy = 0,$$

with

$$M(x, y) = \cos(x) - 2xy \text{ and } N(x, y) = e^y - x^2.$$

Compute

$$\frac{\partial M}{\partial y} = -2x = \frac{\partial N}{\partial x}$$

for all (x, y). By Theorem 1.1, the differential equation is exact for all (x, y). A potential function $\varphi(x, y)$ must satisfy

$$\frac{\partial \varphi}{\partial x} = \cos(x) - 2xy \text{ and } \frac{\partial \varphi}{\partial y} = e^y - x^2.$$

Choose one of these to integrate. If we begin with the second, then integrate with respect to y:

$$\varphi(x, y) = \int (e^y - x^2)dy = e^y - x^2y + h(x).$$

The "constant of integration" is $h(x)$, because x is held fixed in a partial derivative with respect to y. Now we know $\varphi(x, y)$ to within $h(x)$. Next we need

$$\frac{\partial \varphi}{\partial x} = \cos(x) - 2xy = -2xy + h'(x).$$

This requires that

$$h'(x) = \cos(x),$$

so $h(x) = \sin(x)$. A potential function is

$$\varphi(x, y) = e^y - x^2y + \sin(x).$$

The general solution is implicitly defined by

$$e^y - x^2y + \sin(x) = c.$$

For the initial condition, choose c so that $y(1) = 4$. We need

$$e^4 - 4 + \sin(1) = c.$$

The solution of the initial value problem is implicitly defined by

$$e^y - x^2y + \sin(x) = e^4 - 4 + \sin(1). \; \blacklozenge$$

In each of Problems 1 through 5, test the differential equation for exactness. If it is exact (on some region of the plane), find a potential function and the general solution (perhaps implicitly defined). If it is not exact anywhere, do not attempt a solution.

1. $2y^2 + ye^{xy} + (4xy + xe^{xy} + 2y)y' = 0$

2. $4xy + 2x + (2x^2 + 3y^2)y' = 0$

3. $4xy + 2x^2y + (2x^2 + 3y^2)y' = 0$

4. $2\cos(x+y) - 2x\sin(x+y) - 2x\sin(x+y)y' = 0$

5. $1/x + y + (3y^2 + x)y' = 0$

In each of Problems 6 and 7, determine α so that the equation is exact. Obtain the general solution of the exact equation.

6. $3x^2 + xy^\alpha - x^2y^{\alpha-1}y' = 0$

7. $2xy^3 - 3y - (3x + \alpha x^2 y^2 - 2\alpha y)y' = 0$

In each of Problems 8 through 11, determine if the differential equation is exact in some rectangle containing the point where the initial condition is given. If it is exact, solve the initial value problem. If not, do not attempt a solution.

8. $2y - y^2\sec^2(xy^2) + (2x - 2xy\sec^2(xy^2))y' = 0$;
   $y(1) = 2$

9. $3y^4 - 1 + 12xy^3y' = 0$; $y(1) = 2$

10. $1 + e^{y/x} - \frac{y}{x}e^{y/x} + e^{y/x}y' = 0$; $y(1) = -5$

11. $x\cos(2y - x) - \sin(2y - x) - 2x\cos(2y - x)y' = 0$;
    $y(\pi/12) = \pi/8$

12. $e^y + (xe^y - 1)y' = 0$; $y(5) = 0$

1.4   Additional Applications

This section is devoted to some elementary applications of first-order differential equations. We will need Newton's second law of motion, which states that the sum of the external forces acting on an object is equal to the derivative, with respect to time, of the product of the mass and the acceleration. When the mass is constant, $dm/dt = 0$ and Newton's law reduces to the familiar $F = ma$.

Terminal Velocity   An object is falling under the influence of gravity in a medium such as water, air, or oil. We want to analyze the motion.

Let $v(t)$ be the velocity of the object at time t. Gravity pulls the object downward, while the medium retards the downward motion. Experiment has shown that this retarding force is proportional in magnitude to the square of the velocity. Let m be the mass of the object, g the usual constant acceleration due to gravity, and α the constant of proportionality in the retarding force of the medium. Choose downward as the positive direction (this is arbitrary). Let F be the magnitude of the total external force acting on the object. By Newton's law,

$$F = mg - \alpha v^2 = m\frac{dv}{dt}.$$

Suppose that at time zero the object is dropped, not thrown downward, so $v(0) = 0$. We now have an initial value problem for the velocity:

$$v' = g - \frac{\alpha}{m}v^2; \; v(0) = 0.$$

The differential equation is separable:

$$\frac{1}{g - (\alpha/m)v^2}dv = dt.$$

Integrate to get

$$\sqrt{\frac{m}{\alpha g}}\tanh^{-1}\left(\sqrt{\frac{\alpha}{mg}}v\right) = t + c.$$

This equation involves the inverse of the hyperbolic tangent function, which is defined by

$$\tanh(x) = \frac{e^{2x} - 1}{e^{2x} + 1}.$$

Solving for $v(t)$, we obtain

$$v(t) = \sqrt{\frac{mg}{\alpha}} \tanh\left(\sqrt{\frac{\alpha g}{m}}(t + c)\right).$$

Now use the initial condition:

$$v(0) = \sqrt{\frac{mg}{\alpha}} \tanh\left(c\sqrt{\frac{\alpha g}{m}}\right) = 0.$$

Since $\tanh(w) = 0$ only for $w = 0$, this requires that $c = 0$ and the solution for the velocity is

$$v(t) = \sqrt{\frac{mg}{\alpha}} \tanh\left(\sqrt{\frac{\alpha g}{m}}t\right).$$

This expression yields an interesting and perhaps nonintuitive conclusion. As $t \to \infty$, $\tanh(\sqrt{\alpha g/m}\,t) \to 1$. This means that

$$\lim_{t \to \infty} v(t) = \sqrt{\frac{mg}{\alpha}}.$$

An object falling under the influence of gravity through a retarding medium will not increase in velocity indefinitely, even given enough distance. It will instead settle eventually into an essentially constant velocity fall, approximating velocity $\sqrt{mg/\alpha}$ more closely as t increases. This limiting value is called the *terminal velocity* of the object. Skydivers commonly experience this phenomenon.

Sliding Motion on an Inclined Plane   A block weighing 96 pounds is released from rest at the top of an inclined plane of slope length 50 feet and makes an angle of $\pi/6$ radians with the horizontal. Assume a coefficient of friction of $\mu = \sqrt{3}/4$. Assume also that air resistance acts to retard the block's descent down the ramp with a force of magnitude equal to one half the block's velocity. We want to determine the velocity $v(t)$ of the block.

Figure 1.6 shows the forces acting on the block. Gravity acts downward with magnitude $mg\sin(u)$, which is $96\sin(\pi/6)$ or 48 pounds. Here $mg = 96$ is the weight (as distinguished from mass) of the block. The drag due to friction acts in the reverse direction and, in pounds, is given by

$$-\mu N = -\mu mg \cos(u) = -\frac{\sqrt{3}}{4}(96)\cos(\pi/6) = -36.$$

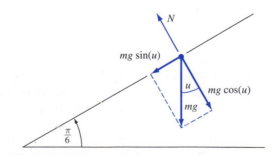

FIGURE 1.6 *Forces acting on the sliding block.*

The drag force due to air resistance is $-v/2$. The total external force acting on the block has magnitude

$$F = 48 - 36 - \frac{1}{2}v = 12 - \frac{1}{2}v.$$

Since the block weighs 96 pounds, its mass is $96/32 = 3$ slugs. From Newton's second law,

$$3\frac{dv}{dt} = 12 - \frac{1}{2}v.$$

This is the linear equation

$$v' + \frac{1}{6}v = 4.$$

Compute the integrating factor $e^{\int (1/6)dt} = e^{t/6}$. Multiply the differential equation by $e^{t/6}$ to obtain

$$v'e^{t/6} + \frac{1}{6}ve^{t/6} = 4e^{t/6}$$

or

$$(ve^{t/6})' = 4e^{t/6}.$$

Integrate to get

$$ve^{t/6} = 24e^{t/6} + c$$

so

$$v(t) = 24 + ce^{-t/6}.$$

Assuming that the block starts from rest at time zero, then $v(0) = 0 = 24 + c$. So $c = -24$ and

$$v(t) = 24\left(1 - e^{-t/6}\right).$$

This gives the block's velocity at any time. We can also determine its position. Let $x(t)$ be the position of the block at time t measured from the top. Since $v(t) = x'(t)$ and $x(0) = 0$, then

$$x(t) = \int_0^t v(\tau)\, d\tau = \int_0^t 24\left(1 - e^{-\tau/6}\right) d\tau$$

$$= 24t + 144\left(e^{-t/6} - 1\right).$$

Suppose we want to know when the block reaches the bottom of the ramp. This occurs at a time T such that $x(T) = 50$. Thus, we must solve for T in

$$24T + 144\left(e^{-T/6} - 1\right) = 50.$$

This equation cannot be solved algebraically for T, but a computer approximation yields $T \approx 5.8$ seconds.

Notice that

$$\lim_{t \to \infty} v(t) = 24.$$

Of course, this limit is irrelevant in this setting, since the block reaches the end of the ramp in about 5.8 seconds. However, if the ramp were long enough, the block would approach arbitrarily close to 24 feet per second in velocity. For practical purposes, on a sufficiently long ramp, the block will appear to settle into a constant velocity slide. This is similar to the terminal velocity experienced by objects falling in a retarding medium.

Orthogonal Trajectories   Two curves intersecting at a point P are *orthogonal* if their tangents are perpendicular at P. This occurs when the slope of one curve at P is the negative reciprocal of the slope of the other curve at P.

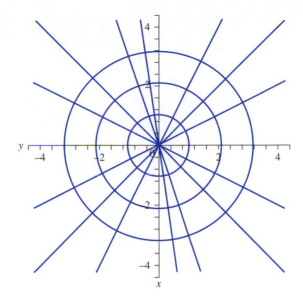

FIGURE 1.7   *Two families of orthogonal trajectories.*

Suppose we have two sets (or families) of curves, $\mathcal{F}$ and $\mathcal{G}$. We say that $\mathcal{F}$ is a set of *orthogonal trajectories* of $\mathcal{G}$ if, whenever a curve of $\mathcal{F}$ intersects a curve of $\mathcal{G}$, these curves are orthogonal at the point of intersection. When $\mathcal{F}$ is a family of orthogonal trajectories of $\mathcal{G}$, then $\mathcal{G}$ is also a family of orthogonal trajectories of $\mathcal{F}$.

For example, let $\mathcal{F}$ consist of all circles about the origin and $\mathcal{G}$ of all straight lines through the origin. Figure 1.7 shows some curves of these families. These families are orthogonal trajectories of each other. Wherever one of the lines intersects one of the circles, the line is orthogonal to the tangent to the circle there.

Given a family $\mathcal{F}$ of curves, suppose we want to find the family $\mathcal{G}$ of orthogonal trajectories of $\mathcal{F}$. Here is a strategy to do this. The curves of $\mathcal{F}$ are assumed to be graphs of a equation $F(x, y, k) = 0$ with different choices of k giving different curves. Think of these curves as integral curves of some differential equation $y' = f(x, y)$. The curves in the set of orthogonal trajectories are then integral curves of the differential equation $y' = -1/f(x, y)$ with the negative reciprocal ensuring that curves of one family are orthogonal to curves of the other family at points of intersection. The idea is to produce the differential equation $y' = f(x, y)$ from $\mathcal{F}$, then solve the equation $y' = -1/f(x, y)$.

EXAMPLE 1.14

Let $\mathcal{F}$ consist of curves that are graphs of

$$F(x, y, k) = y - kx^2 = 0.$$

These are parabolas through the origin. We want the family of orthogonal trajectories. First obtain the differential equation of $\mathcal{F}$. From $y - kx^2 = 0$ we can write $k = y/x^2$. Differentiate $y - kx^2 = 0$ to get $y' = 2kx$. Substitute for k in this derivative to get

$$y' - 2kx = 0 = y' - 2\left(\frac{y}{x^2}\right)x = 0.$$

This gives us

$$y' = 2\left(\frac{y}{x}\right) = f(x, y)$$

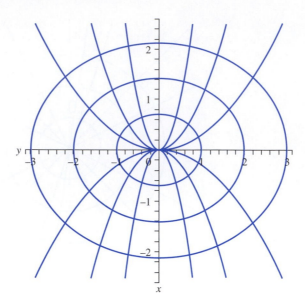

FIGURE 1.8 *Families of orthogonal trajectories in Example 1.14.*

as the differential equation of $\mathcal{F}$. This means that $\mathcal{F}$ is the family of integral curves of $y' = 2y/x$. The differential equation of the family of orthogonal trajectories is therefore

$$y' = -\frac{1}{f(x, y)} = -\frac{x}{2y}.$$

This is a separable equation that can be written as

$$2y\, dy = -x\, dx$$

with a general solution of

$$y^2 + \frac{1}{2}x^2 = c.$$

These curves are ellipses, and they make up the family $\mathcal{G}$ of orthogonal trajectories of $\mathcal{F}$. Figure 1.8 shows some of the ellipses in $\mathcal{G}$ and the parabolas in $\mathcal{F}$. ♦

SECTION 1.4   PROBLEMS

1. A 10-pound ballast bag is dropped from a hot air balloon which is at an altitude of 342 feet and ascending at 4 feet per second. Assuming that air resistance is not a factor, determine the maximum height reached by the bag, how long it remains aloft, and the speed with which it eventually strikes the ground.

2. A 48-pound box is given an initial push of 16 feet per second down an inclined plane that has a gradient of $\frac{7}{24}$. If there is a coefficient of friction of $\frac{1}{3}$ between the box and the plane and a force of air resistance equal in magnitude to $\frac{3}{2}$ of the velocity of the box, determine how far the box will travel down the plane before coming to rest.

3. A skydiver and her equipment together weigh 192 pounds. Before the parachute is opened, there is an air drag force equal in magnitude to six times her velocity. Four seconds after stepping from the plane, the skydiver opens the parachute, producing a drag equal to three times the square of the velocity. Determine the velocity and how far the skydiver has fallen at time t. What is the terminal velocity?

4. Archimedes' principle of buoyancy states that an object submerged in a fluid is buoyed up by a force equal to the weight of the fluid that is displaced by the object. A rectangular box $1 \times 2 \times 3$ feet and weighing 384 pounds is dropped into a 100-foot-deep freshwater lake. The box begins to sink with a drag due to the water having magnitude equal to $\frac{1}{2}$ the velocity. Calculate the terminal velocity of the box. Will the box have achieved a velocity of 10 feet per second by the time it reaches the bottom? Assume that the density of water is 62.5 pounds per cubic foot.

5. Suppose the box in Problem 4 cracks open upon hitting the bottom of the lake, and 32 pounds of its contents spill out. Approximate the velocity with which the box surfaces.

6. The acceleration due to gravity inside the earth is proportional to the distance from the center of the earth. An object is dropped from the surface of the earth into a hole extending straight through the planet's center. Calculate the speed the object achieves by the time it reaches the center.

7. A particle starts from rest at the highest point of a vertical circle and slides under only the influence of gravity along a chord to another point on the circle. Show that the time taken is independent of the choice of the terminal point. What is this common time?

In each of Problems 8 through 12, find the family of orthogonal trajectories of the given family of curves. If software is available, graph some curves of both families.

8. $x + 2y = k$

9. $2x^2 - 3y = k$

10. $x^2 + 2y^2 = k$

11. $y = kx^2 + 1$

12. $y = e^{kx}$

1.5 Existence and Uniqueness Questions

There are initial value problems having no solution. One example is

$$y' = 2\sqrt{y}; \quad y(0) = -1.$$

The differential equation has general solution $y = (x + c)^2$, but there is no real number c such that $y(0) = c^2 = -1$.

An initial value problem may also have more than one solution. In particular, the initial value problem

$$y' = 2\sqrt{y}; \quad y(1) = 0$$

has the zero solution $y = \varphi(x) = 0$ for all x. But it also has the solution

$$y = \psi(x) = \begin{cases} 0 & \text{for } x \leq 1 \\ (x-1)^2 & \text{for } x \geq 1. \end{cases}$$

Because existence and/or uniqueness can fail for even apparently simple initial value problems, we look for conditions that are sufficient to guarantee both existence and uniqueness of a solution. Here is one such result.

THEOREM 1.2 *Existence and Uniqueness*

Let $f(x, y)$ and $\partial f/\partial y$ be continuous for all (x, y) in a rectangle R centered at (x_0, y_0). Then there is a positive number h such that the initial value problem

$$y' = f(x, y); \quad y(x_0) = y_0$$

has a unique solution defined at least for $x_0 - h < x < x_0 + h$. ◆

The theorem gives no control over h, hence it may guarantee a unique solution only on a small interval about x_0.

EXAMPLE 1.15

The problem

$$y' = e^{x^2 y} - \cos(x - y); \quad y(1) = 7$$

has a unique solution on some interval $(1 - h, 1 + h)$, because $f(x, y) = e^{x^2 y} - \cos(x - y)$ and $\partial f / \partial y = x^2 e^{x^2 y} - \sin(x - y)$ are continuous for all (x, y), hence on any rectangle centered at $(1, 7)$. The theorem does not give us any control over the size of h. ◆

EXAMPLE 1.16

The initial value problem

$$y' = y^2; \quad y(0) = n,$$

in which n is a positive integer, has the solution

$$y(x) = -\frac{1}{x - \frac{1}{n}}.$$

This solution is defined only for $-1/n < x < 1/n$, hence on smaller intervals about $x_0 = 0$, as n is chosen larger. ◆

For this reason, Theorem 1.2 is called a *local result*, giving a conclusion about a solution only on a perhaps very small interval about the given point x_0.

SECTION 1.5   PROBLEMS

In each of Problems 1 through 4, use Theorem 1.2 to show that the initial value problem has a unique solution in some interval about the number x_0 at which the initial condition is specified. Assume routine facts about continuity of standard functions of two variables.

1. $y' = \sin(xy); \quad y(\pi/2) = 1$

2. $y' = \ln|x - y|; \quad y(3) = \pi$

3. $y' = x^2 - y^2 + 8x/y; \quad y(3) = -1$

4. $y' = \cos(e^{xy}); \quad y(0) = -4$

5. Consider the initial value problem

$$|y'| = 2y; \quad y(x_0) = y_0,$$

in which x_0 is any number.

(a) Assuming that $y_0 > 0$, find two solutions.

(b) Explain why the conclusion of part (a) does not violate Theorem 1.2.

1.6   Direction Fields

For an initial value problem

$$y' = f(x, y), y(x_0) = y_0,$$

it may be impossible to write a solution in a form from which we can conveniently draw conclusions. For example, the problem

$$y' - \sin(x)y = 4; \quad y(0) = 2$$

has the solution

$$y(x) = 4e^{-\cos(x)} \int_0^x e^{-\cos(\xi)} d\xi + 2e^{1-\cos(x)}.$$

The behavior and properties of this solution are not immediate.

One way to obtain a qualitative picture of a solution is to use the direction field of the differential equation $y' = f(x, y)$, where $f(x, y)$ is given for (x, y) in some region of the plane. From the differential equation itself, we know the slope of the integral curve passing through (x, y). This slope is $f(x, y)$.

Form a rectangular grid of points (x_i, y_j) in the plane, or perhaps over some region of the plane where $f(x, y)$ is defined. Through each grid point, draw a short line segment having slope $f(x_i, y_j)$. These line segments are called *lineal elements*. The lineal element through (x_i, y_j) is tangent to the integral curve through this point, and the collection of all the lineal elements is called a *direction field* for the differential equation $y' = f(x, y)$. If enough lineal elements are drawn, they trace out the shapes of integral curves of $y' = f(x, y)$, just as short tangent segments drawn along a curve give an outline of the shape of the curve. The direction field therefore provides a picture of how integral curves behave in the region over which the grid has been placed.

If we think of the integral curves of $y' = f(x, y)$ as the trajectories of moving particles of a fluid, then the direction field is a flow pattern of this fluid.

EXAMPLE 1.17

The differential equation

$$y' = y^2.$$

has $f(x, y) = y^2$. In this example, we know the general solution explicitly:

$$y = -\frac{1}{x + k}$$

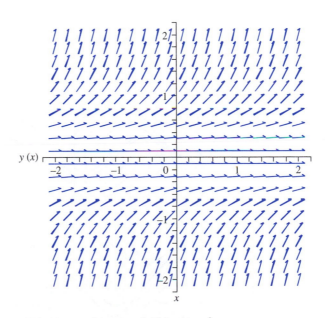

FIGURE 1.9 *Direction field for $y' = y^2$.*

in which k is an arbitrary constant. Figure 1.9 shows a direction field for this differential equation for $-2 \le x \le 2$ and $-2 \le y \le 2$. Figure 1.10 shows a direction field together with four solution

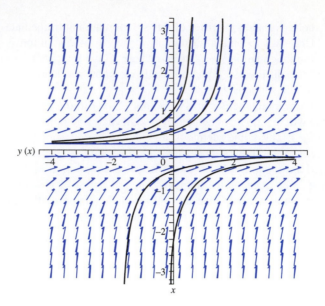

FIGURE 1.10 *Integral curves in the direction field for* $y' = y^2$.

curves corresponding to $y(0) = -2$, $y(0) = -1/2$, $y(0) = 1/2$, and $y(0) = 1$. These solution curves follow the flow of the tangent line segments making up the direction field. ◆

EXAMPLE 1.18

The differential equation

$$y' = \sin(xy)$$

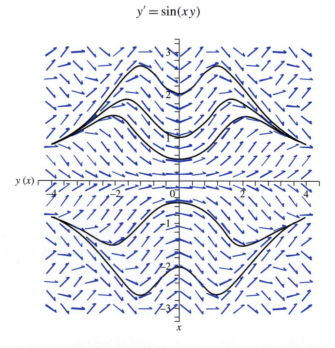

FIGURE 1.11 *Direction field and some integral curves for* $y' = \sin(xy)$.

has no solution that can be written as a finite algebraic combination of elementary functions. Figure 1.11 shows a direction field for this equation, together with five solution curves corresponding to $y(0) = -2$, $y(0) = -1/2$, $y(0) = 1/2$, $y(0) = 1$, and $y(0) = 2$. Integral curves can be drawn to fit the flow of the lineal elements of the direction field. ◆

It is not practical to draw direction fields by hand. There are several mathematical computation software packages that will carry out this function and will also sketch integral curves through specified points.

SECTION 1.6   *PROBLEMS*

In each of Problems 1 through 6, draw a direction field for the differential equation. Use it to sketch the integral curve passing through the given initial value. These problems require use of a software package containing direction field routines.

1. $y' = \sin(y)$; $y(1) = \pi/2$

2. $y' = x\cos(2x) - y$; $y(1) = 0$

3. $y' = y\sin(x) - 3x^2$; $y(0) = 1$

4. $y' = e^x - y$; $y(-2) = 1$

5. $y' - y\cos(x) = 1 - x^2$; $y(2) = 2$

6. $y' = 2y + 3$; $y(0) = 1$

1.7   Numerical Approximation of Solutions

Suppose we have an initial value problem

$$y' = f(x, y); \quad y(x_0) = y_0.$$

Choose a positive integer n and a number h. A numerical method is an algorithm for successively approximating the value of the solution at points

$$x_0 + h, x_0 + 2h, \ldots, x_0 + nh.$$

n is the number of points at which approximate values are calculated (the number of iterations of the method), and h is a (usually small) number called the *step size*. This is the distance between successive points.

We will begin with Euler's method, which is conceptually simple and displays some of the ideas encountered in numerical methods. This will serve as a bridge to a Runge-Kutta method, which is much more accurate and more commonly used. In this discussion, it will be convenient to denote the exact solution value at $x_k = x_0 + kh$ as $y(x_k)$ (an unknown number) and the approximated value at x_k as y_k. Thus, $y_k \approx y(x_k)$.

We are given $y(x_0) = y_0$, so we know this solution value exactly. Calculate $f(x_0, y_0)$ and draw the line having this slope through (x_0, y_0). This line is tangent to the solution at (x_0, y_0). Move along this tangent line to a point (x_1, y_1) where $x_1 = x_0 + h$. This number y_1 is the Euler approximation to $y(x_1)$ (see Figure 1.12). We have some hope that this is a "good" approximation for h "small" because a tangent line at a point fits the curve closely near that point. Note that (x_1, y_1) is probably not on the integral curve through (x_0, y_0), but is on the tangent to this curve at (x_0, y_0).

Next, compute $f(x_1, y_1)$. This is the slope of the tangent to the graph of the solution passing through (x_1, y_1). Draw the line through (x_1, y_1) having this slope, and move along this line to (x_2, y_2) where $x_2 = x_0 + 2h$. This determines a number y_2, which we take as the Euler approximation to $y(x_2)$. (Figure 1.12 again).

Continue in this way. Once we have reached (x_k, y_k), draw the line through this point having slope $f(x_k, y_k)$ and move along this line to (x_{k+1}, y_{k+1}). Take y_{k+1} as the Euler approximation to $y(x_{k+1})$.

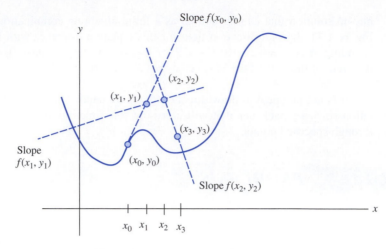

FIGURE 1.12   *The Euler approximation scheme.*

This is the idea of the method. It is sensitive to how much $f(x, y)$ changes if x and y are varied by a small amount. The method also tends to accumulate error, since we use an approximation y_k to make the next approximation y_{k+1}. Following segments of lines is conceptually simple but is not as accurate as some other methods.

We will derive an expression for the approximate value y_k at x_k. From Figure 1.12,

$$y_1 = y_0 + f(x_0, y_0)(x_1 - x_0).$$

At the next step,

$$y_2 = y_1 + f(x_1, y_1)(x_2 - x_1).$$

After the approximation y_k has been computed, the next approximate value is

$$y_{k+1} = y_k + f(x_k, y_k)(x_{k+1} - x_k).$$

Since each $x_{k+1} - x_k = h$, the method can be summarized as follows.

Euler's Method

Define y_{k+1} in terms of x_k and y_k by

$$y_{k+1} = y_k + hf(x_k, y_k)$$

for $k = 0, 1, 2, \ldots, n - 1$.

EXAMPLE 1.19

Consider

$$y' = x\sqrt{y}; \quad y(2) = 4.$$

This problem, with separable differential equation, is easily solved exactly:

$$y(x) = \left(1 + \frac{x^2}{4}\right)^2.$$

TABLE 1.1	*Euler's Method Applied to* $y' = x\sqrt{y}$; $y(2) = 4$	
x	$y(x)$	**Euler approximation of** $y(x)$
2.0	4	4
2.05	4.205062891	4.200000000
2.1	4.42050650	4.410062491
2.15	4.646719141	4.630564053
2.2	4.88410000	4.861890566
2.25	5.133056641	5.104437213
2.3	5.394006250	5.358608481
2.35	5.667475321	5.624818168
2.4	5.953600000	6.903489382
2.45	6.253125391	6.195054550
2.5	6.566406250	6.499955415

We will apply Euler's method and use the exact solution to gauge the accuracy. Use $h = 0.05$ and $n = 10$. Then $x_0 = 2$ and $x_{10} = 2 + (10)(0.05) = 2.5$, so we are computing approximate values on $[2, 2.5]$. These values are computed by

$$y_{k+1} = y_k + (.05)x_k\sqrt{y_k} \text{ for } k = 0, 1, 2, \ldots, 9.$$

Table 1.1 gives the Euler approximate values together with values computed from the exact solution. Although the approximation appears quite good on $[2, 2.5]$, the approximate values are increasingly less than the actual solution values as x moves away from 2. ◆

It can be shown that the error in Euler's method is proportional to h. For this reason, Euler's method is called a *first-order method*. We can increase the accuracy in an Euler approximation by choosing a smaller h at the cost of more computing time.

Modified Euler Method

There is a variation of Euler's method defined by setting

$$y_{k+1} = y_k + hf(x_k + h/2, y_k + hf(x_k, y_k)/2).$$

This approximation scheme is called the *modified Euler method*.

There are many approximation schemes that are better than the Euler and modified Euler methods. We will mention one, a Runge-Kutta (RK) method that is a fourth-order method in the sense that the error is proportional to h^4. To get some sense of how the method is obtained, begin with the fact that

$$y(x_{k+1}) - y(x_k) = \int_{x_k}^{x_{k+1}} y'(x)\,dx = \int_{x_k}^{x_k+h} y'(x)\,dx,$$

$$y(x_{k+1}) = y(x_k) + \int_{x_k}^{x_{k+1}} y'(x)\,dx.$$

Now recall Simpson's method for approximating an integral:

$$\int_a^{a+h} f(x)\,dx \approx \frac{h}{6}[f(a) + 4f(a + h/2) + f(a + h)].$$

This gives us

$$y(x_{k+1}) \approx y(x_k) + \frac{h}{6}[y'(x_k) + 2y'(x_k + h/2) + 2y'(x_k + h/2) + y'(x_k + h)], \qquad (1.6)$$

in which we have written the term $4y'(x_k + h/2)$ as a sum of two terms with a view to making the following approximations.

1. Approximate $y'(x_k) \approx f(x_k, y_k) = W_{k1}$.
   This is the slope of the solution at x_k using Euler's method.
2. Approximate $y'(x_k + h/2) \approx f(x_k + h/2, y_k + hW_{k1}/2) = W_{k2}$.
   This is an approximation of the slope at $x_k + h/2$ using Euler's method.
3. Approximate $y'(x_k + h/2) \approx f(x_k + h/2, y_k + hW_{k2}/2) = W_{k3}$.
   This is a modified Euler method for the slope at $x_k + h/2$.
4. Approximate $y'(x_{k+1}) \approx f(x_{k+1}, y_k + hW_{k3}) = W_{k4}$.
   This is also an Euler approximation at x_{k+1}.

The Runge-Kutta method (or RK method) consists of using these approximations in equation (1.6).

Runge-Kutta Method

Define y_{k+1} by

$$y_{k+1} = y_k + \frac{h}{6}(W_{k1} + 2W_{k2} + 2W_{k3} + W_{k4}).$$

There is actually an entire class of approximation schemes called Runge-Kutta methods, and we have given just one of these.

EXAMPLE 1.20

Consider

$$y' = \frac{1}{y}\cos(x + y); \; y(0) = 1.$$

Let $h = 0.1$, and compute

$$W_{k1} = \frac{1}{y_k}\cos(x_k + y_k),$$

$$W_{k2} = \frac{1}{y_k + .05W_{k1}}\cos(x_k + y_k + .05 + .05W_{k1}),$$

$$W_{k3} = \frac{1}{y_k + .05W_{k2}}\cos(x_k + y_k + .05 + .05W_{k2}),$$

$$W_{k4} = \frac{1}{y_k + .01W_{k3}}\cos(x_k + y_k + .1 + .1W_{k3}),$$

and

$$y_{k+1} = y_k + \frac{0.1}{6}(W_{k1} + 2W_{k2} + 2W_{k3} + W_{k4}).$$

TABLE 1.2	Runge-Kutta Method Applied to $y' = (1/y)\cos(x + y);\ y(0) = 1$

x	y_k
0.0	1
0.1	1.046496334
0.2	1.079334533
0.3	1.00247716
0.4	1.110582298
0.5	1.111396688
0.6	1.103521442
0.7	1.087597013
0.8	1.064096638
0.9	1.033337679
1.0	0.9954821879

Table 1.2 gives approximate values of $y(x)$ for $n = 10$. ◆

SECTION 1.7   PROBLEMS

In each of Problems 1 through 6, use Euler's method to generate approximate numerical values of the solution using $h = 0.05$ and ten iterations ($n = 10$). In each of Problems 1 through 5, compare approximate solution values with exact solution values.

1. $y' = y\sin(x);\ y(0) = 1$

2. $y' = x + y;\ y(1) = -3$

3. $y' = 3xy;\ y(0) = 5$

4. $y' = 2 - x;\ y(0) = 1$

5. $y' = y - \cos(x);\ y(1) = -2$

6. $y' = x - y^2;\ y(0) = 4$

In each of Problems 7 through 11, use the Runge-Kutta method to obtain approximate values of the solution at selected points, using $h = 0.2$ and $n = 10$. In Problems 8 and 11, solve the problem exactly to compare exact solution values with the approximated values.

7. $y' = \sin(x + y);\ y(0) = 2$

8. $y' = y - x^2;\ y(1) = -4$

9. $y' = \cos(y) + e^{-x};\ y(0) = 1$

10. $y' = y^3 - 2xy;\ y(3) = 2$

11. $y' = -y + e^{-x};\ y(0) = 4$

CHAPTER 2

Linear
Second-Order
Equations

A second-order differential equation is one containing a second derivative but no higher derivative. We will focus on linear second-order equations, which have many important applications.

2.1 Theory of the Linear Second-Order Equation

This section lays the foundations for writing solutions of the second-order linear differential equation

$$y'' + p(x)y' + q(x)y = f(x). \tag{2.1}$$

Generally, we assume that p and q are continuous, at least on the interval where we seek solutions. The function f is called a *forcing function* for the differential equation, and in some applications, it can have finitely many jump discontinuities.

In treating equation (2.1), it is useful to consider two cases.

The Homogeneous Case
Equation (2.1) is *homogeneous* if $f(x)$ is identically zero. In this case the differential equation is

$$y'' + p(x)y' + q(x)y = 0. \tag{2.2}$$

The most significant fact about solutions of equation (2.2) is that sums of solutions and constant multiples of solutions are again solutions.

THEOREM 2.1

If y_1 and y_2 are solutions of equation (2.2) and c_1 and c_2 are numbers, then $c_1 y_1 + c_2 y_2$ is also a solution. ◆

35

In this statement, c_1 and c_2 can be real or complex numbers. The conclusion is easily verified by substituting $c_1y_1 + c_2y_2$ into the differential equation.

EXAMPLE 2.1

It is routine to check that $\sin(2x)$ and $\cos(2x)$ are solutions of $y'' + 4y = 0$. Therefore

$$c_1 \sin(2x) + c_2 \cos(2x)$$

is also a solution for any numbers c_1 and c_2. ◆

We call $c_1y_1 + c_2y_2$ a *linear combination* of y_1 and y_2. This linear combination is uninteresting if one of the functions is already a constant multiple of the other. Indeed, if $y_2 = ky_1$, then

$$c_1y_1 + c_2y_2 = c_1y_1 + kc_2y_1$$
$$= [c_1 + kc_2]y_1,$$

and this is again just a constant multiple of y_1. We therefore usually consider linear combinations $c_1y_1 + c_2y_2$ only when y_1 and y_2 are not constant multiples of each other. In this case, we say that y_1 and y_2 are *linearly independent*. If one function is a constant multiple of the other, we say that these functions are *linearly dependent*. In Example 2.1, $\sin(2x)$ and $\cos(2x)$ are linearly independent solutions, because these functions are not constant multiples of each other. The importance of this is that any linear combination $c_1 \sin(2x) + c_2 \cos(2x)$ with c_1 and c_2 nonzero is a new solution, not just a constant multiple of $\sin(2x)$ or $\cos(2x)$.

In the second-order linear case, two linearly independent solutions are enough to specify all possible solutions.

THEOREM 2.2

Let y_1 and y_2 be linearly independent solutions of equation (2.2) on some interval, or perhaps for all real x. Then every solution is a linear combination of y_1 and y_2. ◆

This means that the expression $c_1y_1 + c_2y_2$ contains every possible solution of equation (2.2) when y_1 and y_2 are linearly independent. This leads us to call $y(x) = c_1y_1(x) + c_2y_2(x)$ the *general solution* of equation (2.2) when y_1 and y_2 are linearly independent solutions. In this case, y_1 and y_2 are said to form a *fundamental set of solutions*.

EXAMPLE 2.2

e^x and e^{2x} are linearly independent, are solutions of $y'' - 3y' + 2y = 0$, and form a fundamental set of solutions. The general solution is

$$y(x) = c_1e^x + c_2e^{2x}. \quad ◆$$

Theorem 2.2 provides a strategy for finding the general solution of equation (2.2).

Strategy for Writing the General Solution of $y'' + p(x)y' + q(x)y = 0$

Find two linearly independent solutions y_1 and y_2. The general solution is $y(x) = c_1 y_1(x) + c_2 y_2(x)$.

This does not tell us how to find such y_1 and y_2. We will deal with this in Section 2.2.

The Nonhomogeneous Case

We now want to know what the general solution of equation (2.1) looks like when $f(x)$ is nonzero at least for some x. In this case, the differential equation is *nonhomogeneous*.

The main difference between the homogeneous and nonhomogeneous cases is that, in the latter, sums and constant multiples of solutions need not be solutions.

EXAMPLE 2.3

$\sin(2x) + 2x$ and $\cos(2x) + 2x$ are solutions of the nonhomogeneous equation $y'' + 4y = 8x$. However, if we substitute the sum of these solutions, $\sin(2x) + \cos(2x) + 4x$, into the differential equation, we find that this sum is not a solution. Further, if we multiply one of these solutions by 2, taking, say, $2\sin(2x) + 4x$, we find that this is not a solution either. ◆

However, given any two solutions Y_1 and Y_2 of equation (2.1), we find that their *difference* $Y_1 - Y_2$ is a solution, not of the nonhomogeneous equation, but of the *associated homogeneous equation* (2.2). To see this, substitute $Y_1 - Y_2$ into equation (2.1):

$$(Y_1 - Y_2)'' + p(x)(Y_1 - Y_2)' + q(x)(Y_1 - Y_2)$$
$$= [Y_1'' + p(x)Y_1' + q(x)Y_1] - [Y_2'' + p(x)Y_2' + q(x)Y_2]$$
$$= f(x) - f(x) = 0.$$

Notice that the general solution of the associated homogeneous equation (2.2) has the form $c_1 y_1 + c_2 y_2$, where y_1 and y_2 are linearly independent solutions of the homogeneous equation (2.2). Since $Y_1 - Y_2$ is a solution of this homogeneous equation, then for some numbers c_1 and c_2:

$$Y_1 - Y_2 = c_1 y_1 + c_2 y_2,$$

which means that

$$Y_1 = c_1 y_1 + c_2 y_2 + Y_2.$$

But Y_1 is *any* solution of equation (2.1). This means that we can obtain any solution by appropriately choosing the constants c_1 and c_2 in $c_1 y_1 + c_2 y_2 + Y_2$.

We will summarize this discussion as a general conclusion, in which we use Y_p (for *particular solution*) instead of Y_2 of the discussion.

THEOREM 2.3

Let Y_p be any solution of equation (2.1). Let y_1 and y_2 be linearly independent solutions of equation (2.2). Then the expression

$$c_1 y_1 + c_2 y_2 + Y_p$$

contains every solution of equation (2.1). ◆

For this reason, we call $c_1 y_1 + c_2 y_2 + Y_p$ the *general solution* of equation (2.1).

Theorem 2.3 suggests a strategy for finding all solutions of the nonhomogeneous equation (2.1).

---

1. Find two linearly independent solutions y_1 and y_2 of the associated homogeneous equation $y'' + p(x)y' + q(x)y = 0$.
2. Find *any* particular solution Y_p of the nonhomogeneous equation $y'' + p(x)y' + q(x)y = f(x)$.
3. The general solution of $y'' + p(x)y' + q(x)y = f(x)$ is

$$y(x) = c_1 y_1(x) + c_2 y_2(x) + Y_p(x),$$

in which c_1 and c_2 can be any real numbers.

EXAMPLE 2.4

We will find the general solution of

$$y'' + 4y = 8x.$$

We know from Example 2.1 that the general solution of $y'' + 4y = 0$ is $c_1 \sin(2x) + c_2 \cos(2x)$. Substitution verifies that $Y_p(x) = 2x$ is a particular solution of the nonhomogeneous equation. Therefore the general solution of $y'' + 4y = 8x$ is

$$y = c_1 \sin(2x) + c_2 \sin(2x) + 2x.$$

This expression contains every solution of the given nonhomogeneous equation, simply by choosing different values of the constants c_1 and c_2.   ◆

In carrying out this strategy, the hard part is usually finding a particular solution Y_p, which was simply given in Example 2.4. We will develop methods for finding a particular solution Y_p in Section 2.3.

2.1.1   The Initial Value Problem

The general solution of equation (2.1) contains two arbitrary constants. Therefore, to specify a unique solution, two pieces of information must be given. These are usually the values the solution and its first derivative must take at a particular point x_0.

In general, then, the initial value problem for the linear second-order differential equation has the form

$$y'' + p(x)y' + q(x)y = f(x);\ y(x_0) = A,\ y'(x_0) = B,$$

in which x_0, A, and B are given. This means that the solution is to pass through a given point (x_0, A) with slope B.

EXAMPLE 2.5

We will solve the initial value problem

$$y'' + 4y = 8x;\ y(\pi) = 1,\ y'(\pi) = -6.$$

From Example 2.4, the differential equation has general solution

$$y(x) = c_1 \sin(2x) + c_2 \cos(2x) + 2x.$$

We need

$$y(\pi) = c_2 \cos(2\pi) + 2\pi = c_2 + 2\pi = 1$$

so $c_2 = 1 - 2\pi$. Next, we need

$$y'(\pi) = 2c_1 \cos(2\pi) - 2c_2 \sin(2\pi) + 2 = 2c_1 + 2 = -6$$

so $c_1 = -4$. The unique solution of the initial value problem is

$$y(x) = -4 \sin(2x) + (1 - 2\pi) \cos(2x) + 2x. \quad \blacklozenge$$

We now have strategies for solving equations (2.1) and (2.2) and the initial value problem. We must be able to find two linearly independent solutions of the homogeneous equation and any one particular solution of the nonhomogeneous equation. We will now develop important cases in which we can carry out these steps.

SECTION 2.1   *PROBLEMS*

In each of Problems 1 through 5, verify that y_1 and y_2 are solutions of the homogeneous differential equation, write the general solution, and solve the initial value problem.

1. $y'' + 36y = 0$; $y(0) = -5$, $y'(0) = 2$
   $y_1(x) = \sin(6x)$, $y_2(x) = \cos(6x)$

2. $y'' - 16y = 0$; $y(0) = 12$, $y'(0) = 3$
   $y_1(x) = e^{4x}$, $y_2(x) = e^{-4x}$

3. $y'' + 3y' + 2y = 0$; $y(0) = -3$, $y'(0) = -1$
   $y_1(x) = e^{-2x}$, $y_2(x) = e^{-x}$

4. $y'' - 6y' + 13y = 0$; $y(0) = -1$, $y'(0) = 1$
   $y_1(x) = e^{3x} \cos(2x)$, $y_2(x) = e^{3x} \sin(2x)$

5. $y'' - 2y' + 2y = 0$; $y(0) = 6$, $y'(0) = 1$
   $y_1(x) = e^x \cos(x)$, $y_2(x) = e^x \sin(x)$

In Problems 6 through 10, use the results of Problems 1 through 5, respectively, and the given particular solution Y_p to write the general solution of the nonhomogeneous equation.

6. $y'' + 36y = x - 1$, $Y_p(x) = (x - 1)/36$

7. $y'' - 16y = 4x^2$; $Y_p(x) = -x^2/4 + 1/2$

8. $y'' + 3y' + 2y = 15$; $Y_p(x) = 15/2$

9. $y'' - 6y' + 13y = -e^x$; $Y_p(x) = -8e^x$

10. $y'' - 2y' + 2y = -5x^2$; $Y_p(x) = -5x^2/2 - 5x - 4$

2.2   The Constant Coefficient Homogeneous Equation

We will solve the constant coefficient linear homogeneous equation

$$y'' + ay' + by = 0, \tag{2.3}$$

in which a and b are constants (numbers). A method suggests itself if we read the differential equation like a sentence. We want a function y such that the second derivative, plus a constant multiple of the first derivative, plus a constant multiple of the function itself, is equal to zero for all x. This behavior suggests an exponential function $e^{\lambda x}$, because derivatives of $e^{\lambda x}$ are all constant multiples of $e^{\lambda x}$. We therefore try to find λ so that $e^{\lambda x}$ is a solution.

Substitute $e^{\lambda x}$ into equation (2.3) to get

$$\lambda^2 e^{\lambda x} + a\lambda e^{\lambda x} + b e^{\lambda x} = 0.$$

Since $e^{\lambda x}$ is never zero, the exponential factor cancels and we are left with a quadratic equation for λ:

$$\lambda^2 + a\lambda + b = 0. \tag{2.4}$$

Equation (2.4) is the *characteristic equation* of the differential equation (2.3). It has roots

$$\frac{1}{2}(-a \pm \sqrt{a^2 - 4b}),$$

leading to three cases.

Case 1: Real, Distinct Roots

This occurs when $a^2 - 4b > 0$. The distinct roots are

$$\lambda_1 = \frac{1}{2}(-a + \sqrt{a^2 - 4b}) \text{ and } \lambda_2 = \frac{1}{2}(-a - \sqrt{a^2 - 4b}).$$

$e^{\lambda_1 x}$ and $e^{\lambda_2 x}$ are linearly independent solutions solutions, and in this case the general solution of equation (2.3) is

$$y = c_1 e^{\lambda_1 x} + c_2 e^{\lambda_2 x}.$$

EXAMPLE 2.6

The characteristic equation of

$$y'' - y' - 6y = 0$$

is

$$\lambda^2 - \lambda - 6 = 0.$$

This can be read directly from the coefficients of the differential equation. This quadratic equation has distinct real roots 3 and -2. The general solution is

$$y = c_1 e^{3x} + c_2 e^{-2x}. \quad \blacklozenge$$

Case 2: Repeated Roots

This occurs when $a^2 - 4b = 0$, and the root of the characteristic equation is $\lambda = -a/2$. One solution of the differential equation is $e^{-ax/2}$.

We need a second, linearly independent solution. We will invoke a method called *reduction of order*, which will produce a second solution if we already have one solution. Attempt a second solution $y(x) = u(x)e^{-ax/2}$. Compute

$$y' = u'e^{-ax/2} - \frac{a}{2}ue^{-ax/2}$$

and

$$y'' = u''e^{-ax/2} - au'e^{-ax/2} + \frac{a^2}{4}ue^{-ax/2}.$$

Substitute these into equation (2.3) to get

$$u''e^{-ax/2} - au'e^{-ax/2} + \frac{a^2}{4}ue^{-ax/2}$$

$$+ au'e^{-ax/2} - a\frac{a}{2}ue^{-ax/2} + bue^{-ax/2}$$

$$= e^{-ax/2}\left[u'' - \left(b - \frac{a^2}{4}\right)\right] = 0.$$

Since $b - a^2/4 = 0$ in this case, and since $e^{-ax/2}$ never vanishes, this equation reduces to

$$u'' = 0.$$

This has solutions $u(x) = cx + d$, with c and d as arbitrary constants. Therefore, any function $y = (cx + d)e^{-ax/2}$ is also a solution of equation (2.3) in this case. Since we need only one solution that is linearly independent from $e^{-ax/2}$, choose $c = 1$ and $d = 0$ to get the second solution $xe^{-ax/2}$. The general solution in this repeated roots case is

$$y = c_1 e^{-ax/2} + c_2 x e^{-ax/2}.$$

This is often written $y = e^{-ax/2}(c_1 + c_2 x)$.

It is not necessary to repeat this derivation every time we encounter the repeated root case. Simply write one solution $e^{-ax/2}$, and the second (linearly independent) solution is $xe^{-ax/2}$.

EXAMPLE 2.7

We will solve $y'' + 8y' + 16y = 0$. The characteristic equation is

$$\lambda^2 + 8\lambda + 16 = 0$$

with repeated root $\lambda = -4$. The general solution is

$$y = c_1 e^{-4x} + c_2 x e^{-4x}. \quad \blacklozenge$$

Case 3: Complex Roots

The characteristic equation has complex roots when $a^2 - 4b < 0$. Because the characteristic equation has real coefficients, these roots appear as complex conjugates, $\alpha + i\beta$ and $\alpha - i\beta$, in which α can be zero but β is nonzero. Now the general solution is

$$y = c_1 e^{(\alpha+i\beta)x} + c_2 e^{(\alpha-i\beta)x}.$$

or

$$y = e^{\alpha x}\left(c_1 e^{i\beta x} + c_2 e^{-i\beta x}\right). \tag{2.5}$$

This is correct, but it is sometimes convenient (for example, for graphing purposes) to have a solution that involves only real quantities. We can find such a solution using an observation by the eighteenth century Swiss mathematician Leonhard Euler, who showed that, for any real number β,

$$e^{i\beta x} = \cos(\beta x) + i\sin(\beta x).$$

By replacing x with $-x$, we also have

$$e^{-i\beta x} = \cos(\beta x) - i\sin(\beta x).$$

Then

$$c_1 e^{i\beta x} + c_2 e^{-i\beta x} = c_1 e^{\alpha x}[\cos\beta x + i\sin(\beta x)] + c_2 e^{\alpha x}[\cos(\beta x) - i\sin(\beta x)]$$

$$= e^{\alpha x}[(c_1 + c_2)\cos(\beta x) + i(c_1 - c_2)\sin(\beta x)]. \tag{2.6}$$

This is a solution for any real or complex numbers c_1 and c_2. In particular, if we choose $c_1 = c_2 = 1/2$, we obtain the solution $e^{\alpha x}\cos(\beta x)$. And if we choose $c_1 = -c_2 = 1/2i$, we obtain $e^{\alpha x}\sin(\beta x)$.

Since these are two linearly independent solutions, and any two linearly independent solutions form a fundamental set, then we can also write the general solution in this case as

$$y(x) = c_1 e^{\alpha x} \cos(\beta x) + c_2 e^{\alpha x} \sin(\beta x). \tag{2.7}$$

This formulation has the advantage of involving only real functions.

In practice, we do not repeat the rationale for the general solution (2.7) each time we encounter Case 3. Simply write the general solution (2.7), when $\alpha \pm i\beta$ are the roots of the characteristic equation.

EXAMPLE 2.8

Solve $y'' + 2y' + 3y = 0$. The characteristic equation is

$$\lambda^2 + 2\lambda + 3 = 0$$

with complex conjugate roots $-1 \pm i\sqrt{2}$. With $\alpha = -1$ and $\beta = \sqrt{2}$, the general solution is

$$y = c_1 e^{-x} \cos(\sqrt{2}x) + c_2 e^{-x} \sin(\sqrt{2}x). \quad \blacklozenge$$

EXAMPLE 2.9

Solve $y'' + 36y = 0$. The characteristic equation is

$$\lambda^2 + 36 = 0$$

with complex roots $\lambda = \pm 6i$. Now $\alpha = 0$ and $\beta = 6$, so the general solution is

$$y(x) = c_1 \cos(6x) + c_2 \sin(6x). \quad \blacklozenge$$

We are now able to solve the constant coefficient homogeneous equation

$$y'' + ay' + by = 0$$

in all cases. Here is a summary:

Let λ_1 and λ_2 be the roots of the characteristic equation

$$\lambda^2 + a\lambda + b = 0.$$

Then:

1. If λ_1 and λ_2 are real and distinct,

$$y(x) = c_1 e^{\lambda_1 x} + c_2 e^{\lambda_2 x}.$$

2. If $\lambda_1 = \lambda_2$,

$$y(x) = c_1 e^{\lambda_1 x} + c_2 x e^{\lambda_1 x}.$$

3. If the roots are complex $\alpha \pm i\beta$,

$$y(x) = c_1 e^{\alpha x} \cos(\beta x) + c_2 e^{\alpha x} \sin(\beta x).$$

   PROBLEMS

In each of Problems 1 through 10, write the general solution.

1. $y'' - y' - 6y = 0$
2. $y'' - 2y' + 10y = 0$
3. $y'' + 6y' + 9y = 0$
4. $y'' - 3y' = 0$
5. $y'' + 10y' + 26y = 0$
6. $y'' + 6y' - 40y = 0$
7. $y'' + 3y' + 18y = 0$
8. $y'' + 16y' + 64y = 0$
9. $y'' - 14y' + 49y = 0$
10. $y'' - 6y' + 7y = 0$

In each of Problems 11 through 20, solve the initial value problem.

11. $y'' + 3y' = 0; \; y(0) = 3, \; y'(0) = 6$
12. $y'' + 2y' - 3y = 0; \; y(0) = 6, \; y'(0) = -2$
13. $y'' - 2y' + y = 0; \; y(1) = y'(1) = 0$
14. $y'' - 4y' + 4y = 0; \; y(0) = 3, \; y'(0) = 5$
15. $y'' + y' - 12y = 0; \; y(2) = 2, \; y'(2) = -1$
16. $y'' - 2y' - 5y = 0; \; y(0) = 0, \; y'(0) = 3$
17. $y'' - 2y' + y = 0; \; y(1) = 12, \; y'(1) = -5$
18. $y'' - 5y' + 12y = 0; \; y(2) = 0, \; y'(2) = -4$
19. $y'' - y' + 4y = 0; \; y(-2) = 1, \; y'(-2) = 3$
20. $y'' + y' - y = 0; \; y(-4) = 7, \; y'(-4) = 1$

2.3   Solutions of the Nonhomogeneous Equation

From Theorem 2.3, a key to solving the nonhomogeneous linear equation (2.1) is to find a particular solution Y_p. We will present two methods for doing this.

2.3.1   The Method of Variation of Parameters

We want to find a particular solution Y_p of $y'' + p(x)y' + q(x)y = f(x)$.

Suppose we know two linearly independent solutions y_1 and y_2 of the associated homogeneous equation. One strategy, called the *method of variation of parameters* is to look for functions u_1 and u_2 so that

$$Y_p(x) = u_1(x)y_1(x) + u_2(x)y_2(x).$$

To see how to choose u_1 and u_2, substitute Y_p into the differential equation. We must compute two derivatives. First,

$$Y_p' = u_1 y_1' + u_2 y_2' + u_1' y_1 + u_2' y_2.$$

Simplify this derivative by imposing the condition that

$$u_1' y_1 + u_2' y_2 = 0. \tag{2.8}$$

Now,

$$Y_p' = u_1 y_1' + u_2 y_2'$$

so

$$Y_p'' = u_1' y_1' + u_2' y_2' + u_1 y_1'' + u_2 y_2''.$$

Substitute Y_p into the differential equation to get

$$u_1' y_1' + u_2' y_2' + u_1 y_1'' + u_2 y_2''$$
$$+ p(x)(u_1 y_1' + u_2 y_2') + q(x)(u_1 y_1 + u_2 y_2) = f(x).$$

Rearrange terms to write

$$u_1[y_1'' + p(x)y_1' + q(x)y_1]$$
$$+ u_2[y_2'' + p(x)y_2' + q(x)y_2]$$
$$+ u_1'y_1' + u_2'y_2' = f(x).$$

The two terms in square brackets are zero because y_1 and y_2 are solutions of $y'' + p(x)y' + q(x)y = 0$. The last equation therefore reduces to

$$u_1'y_1' + u_2'y_2' = f(x). \tag{2.9}$$

Solve equations (2.8) and (2.9) for u_1' and u_2' to get

$$u_1'(x) = -\frac{y_2(x)f(x)}{W(x)} \text{ and } u_2'(x) = \frac{y_1(x)f(x)}{W(x)} \tag{2.10}$$

where

$$W(x) = \begin{vmatrix} y_1(x) & y_2(x) \\ y_1'(x) & y_2'(x) \end{vmatrix} = y_1(x)y_2'(x) - y_2(x)y_1'(x).$$

$W(x)$ is called the *Wronskian* of y_1 and y_2. It can be shown that $W(x) \neq 0$ if y_1 and y_2 are linearly independent, which is assumed to be the case here. Integrate equations (2.10) to obtain

$$u_1(x) = -\int \frac{y_2(x)f(x)}{W(x)} dx \text{ and } u_2(x) = \int \frac{y_1(x)f(x)}{W(x)} dx. \tag{2.11}$$

Once we have u_1 and u_2, we have a particular solution $Y_p = u_1 y_1 + u_2 y_2$, and the general solution of $y'' + p(x)y' + q(x)y = f(x)$, giving

$$y = c_1 y_1 + c_2 y_2 + Y_p.$$

EXAMPLE 2.10

Find the general solution of

$$y'' + 4y = \sec(x)$$

for $-\pi/4 < x < \pi/4$.

The characteristic equation of $y'' + 4y = 0$ is $\lambda^2 + 4 = 0$ with complex roots $\lambda = \pm 2i$. Two linearly independent solutions of the associated homogeneous equation $y'' + 4y = 0$ are

$$y_1(x) = \cos(2x) \text{ and } y_2(x) = \sin(2x).$$

Now look for a particular solution of the nonhomogeneous equation. First compute the Wronskian

$$W(x) = \begin{vmatrix} \cos(2x) & \sin(2x) \\ -2\sin(2x) & 2\cos(2x) \end{vmatrix} = 2(\cos^2(x) + \sin^2(x)) = 2.$$

Use equations (2.11) with $f(x) = \sec(x)$ to obtain

$$u_1(x) = -\int \frac{\sin(2x)\sec(x)}{2} dx$$

$$= -\int \frac{2\sin(x)\cos(x)\sec(x)}{2} dx$$

$$= -\int \frac{\sin(x)\cos(x)}{\cos(x)} dx$$

$$= -\int \sin(x) dx = \cos(x)$$

and

$$u_2(x) = \int \frac{\cos(2x)\sec(x)}{2}\, dx$$

$$= \int \frac{(2\cos^2(x) - 1)}{2\cos(x)}\, dx$$

$$= \int \left(\cos(x) - \frac{1}{2}\sec(x) \right) dx$$

$$= \sin(x) - \frac{1}{2}\ln|\sec(x) + \tan(x)|.$$

This gives us the particular solution

$$Y_p(x) = u_1(x)y_1(x) + u_2(x)y_2(x)$$

$$= \cos(x)\cos(2x) + \left(\sin(x) - \frac{1}{2}\ln|\sec(x) + \tan(x)| \right)\sin(2x).$$

The general solution of $y'' + 4y' = \sec(x)$ is

$$y(x) = c_1\cos(2x) + c_2\sin(2x)$$

$$+ \cos(x)\cos(2x) + \left(\sin(x) - \frac{1}{2}\ln|\sec(x) + \tan(x)| \right)\sin(2x). \quad \blacklozenge$$

2.3.2 The Method of Undetermined Coefficients

We will discuss a second method for finding a particular solution of the nonhomogeneous equation, which, however, applies only to the constant coefficient case

$$y'' + ay' + by = f(x).$$

The idea behind the *method of undetermined coefficients* is that sometimes we can guess a general form for $Y_p(x)$ from the appearance of $f(x)$.

EXAMPLE 2.11

We will find the general solution of $y'' - 4y = 8x^2 - 2x$.

It is routine to find the general solution $c_1 e^{2x} + c_2 e^{-2x}$ of the associated homogeneous equation. We need a particular solution $Y_p(x)$ of the given nonhomogeneous equation.

Because $f(x) = 8x^2 - 2x$ is a polynomial, and derivatives of polynomials are polynomials, it is reasonable to think that there might be a polynomial solution. Further, no such solution can include a power of x higher than 2. If $Y_p(x)$ had an x^3 term, this term would be retained by the $-4y$ term of the differential equation, and $y'' - 4y$ has no such term.

This informal reasoning suggests that we try a particular solution $Y_p(x) = Ax^2 + Bx + C$. Compute $y'(x) = 2Ax + B$ and $y''(x) = 2A$ and substitute into these the differential equation to get

$$2A - 4(Ax^2 + Bx + C) = 8x^2 - 2x.$$

Write this as

$$(-4A - 8)x^2 + (-4B + 2)x + (2A - 4C) = 0.$$

This second-degree polynomial must be zero for all x if Y_p is to be a solution. But a second-degree polynomial has only two roots, unless all of its coefficients are zero. Therefore,

$$-4A - 8 = 0$$
$$-4B + 2 = 0$$
$$2A - 4C = 0.$$

Solve these to get $A = -2$, $B = 1/2$, and $C = -1$. This gives us the particular solution

$$Y_p(x) = -2x^2 + \frac{1}{2}x - 1.$$

The general solution is

$$y(x) = c_1 e^{2x} + c_2 e^{-2x} - 2x^2 + \frac{1}{2}x - 1. \quad \blacklozenge$$

EXAMPLE 2.12

Find the general solution of $y'' + 2y' - 3y = 4e^{2x}$.

The general solution of $y'' + 2y' - 3y = 0$ is $c_1 e^{-3x} + c_2 e^x$.

Now look for a particular solution. Because derivatives of e^{2x} are constant multiples of e^{2x}, we suspect that a constant multiple of e^{2x} might serve. Try $Y_p(x) = Ae^{2x}$. Substitute this into the differential equation to get

$$4Ae^{2x} + 4Ae^{2x} - 3Ae^{2x} = 5Ae^{2x} = 4e^{2x}.$$

This works if $5A = 4$, so $A = 4/5$. A particular solution is $Y_p(x) = 4e^{2x}/5$.

The general solution is

$$y(x) = c_1 e^{-3x} + c_2 e^x + \frac{4}{5}e^{2x}. \quad \blacklozenge$$

EXAMPLE 2.13

Find the general solution of $y'' - 5y' + 6y = -3\sin(2x)$.

The general solution of $y'' - 5y' + 6y = 0$ is $c_1 e^{3x} + c_2 e^{2x}$.

We need a particular solution Y_p of the nonhomogeneous equation. Derivatives of $\sin(2x)$ are constant multiples of $\sin(2x)$ or $\cos(2x)$. Derivatives of $\cos(2x)$ are also constant multiples of $\sin(2x)$ or $\cos(2x)$. This suggests that we try a particular solution $Y_p(x) = A\cos(2x) + B\sin(2x)$. Notice that we include both $\sin(2x)$ and $\cos(2x)$ in this first attempt, even though $f(x)$ just has a $\sin(2x)$ term. Compute

$$Y_p'(x) = -2A\sin(2x) + 2B\cos(2x) \text{ and } Y_p''(x) = -4A\cos(2x) - 4B\sin(2x).$$

Substitute these into the differential equation to get

$$-4A\cos(2x) - 4B\sin(2x) - 5[-2A\sin(2x) + 2B\cos(2x)]$$
$$+ 6[A\cos(2x) + B\sin(2x)] = -3\sin(2x).$$

Rearrange terms to write

$$[2B + 10A + 3]\sin(2x) = [-2A + 10B]\cos(2x).$$

But $\sin(2x)$ and $\cos(2x)$ are not constant multiples of each other, unless these constants are zero. Therefore,

$$2B + 10A + 3 = 0 = -2A + 10B.$$

Solve these to get $A = -15/52$ and $B = -3/52$. A particular solution is

$$Y_p(x) = -\frac{15}{52}\cos(2x) - \frac{3}{52}\sin(2x).$$

The general solution is

$$y(x) = c_1 e^{3x} + c_2 e^{2x} - \frac{15}{52}\cos(2x) - \frac{3}{52}\sin(2x). \quad \blacklozenge$$

The method of undetermined coefficients has a trap built into it. Consider the following.

EXAMPLE 2.14

Find a particular solution of $y'' + 2y' - 3y = 8e^x$.

Reasoning as before, try $Y_p(x) = Ae^x$. Substitute this into the differential equation to obtain

$$Ae^x + 2Ae^x - 3Ae^x = 8e^x.$$

But then $8e^x = 0$, which is a contradiction. $\quad \blacklozenge$

The problem here is that e^x is also a solution of the associated homogeneous equation, so the left side will vanish when Ae^x is substituted into $y'' + 2y' - 3y = 8e^x$.

Whenever a term of a proposed $Y_p(x)$ is a solution of the associated homogeneous equation, multiply this proposed solution by x. If this results in another solution of the associated homogeneous equation, multiply it by x again.

EXAMPLE 2.15

Revisit Example 2.14. Since $f(x) = 8e^x$, our first impulse was to try $Y_p(x) = Ae^x$. But this is a solution of the associated homogeneous equation, so multiply by x and try $Y_p(x) = Axe^x$. Now

$$Y_p' = Ae^x + Axe^x \text{ and } Y_p'' = 2Ae^x + Axe^x.$$

Substitute these into the differential equation to get

$$2Ae^x + Axe^x + 2(Ae^x + Axe^x) - 3Axe^x = 8e^x.$$

This reduces to $4Ae^x = 8e^x$, so $A = 2$, yielding the particular solution $Y_p(x) = 2xe^x$. The general solution is

$$y(x) = c_1 e^{-3x} + c_2 e^x + 2xe^x. \quad \blacklozenge$$

EXAMPLE 2.16

Solve $y'' - 6y' + 9y = 5e^{3x}$.

The associated homogeneous equation has characteristic equation $(\lambda - 3)^2 = 0$ with repeated roots $\lambda = 3$. The general solution of this associated homogeneous equation is $c_1 e^{3x} + c_2 x e^{3x}$.

For a particular solution, we might first try $Y_p(x) = Ae^{3x}$, but this is a solution of the homogeneous equation. Multiply by x and try $Y_p(x) = Axe^{3x}$. This is also a solution of the homogeneous equation, so multiply by x again and try $Y_p(x) = Ax^2 e^{3x}$. If this is substituted into the differential equation we obtain $A = 5/2$, so a particular solution is $Y_p(x) = 5x^2 e^{3x}/2$. The general solution is

$$y = c_1 e^{3x} + c_2 x e^{3x} + \frac{5}{2} x^2 e^{3x}. \quad \blacklozenge$$

The method of undetermined coefficients is limited by our ability to "guess" a particular solution from the form of $f(x)$. Table 2.1 provides a list of functions to try in the initial attempt at $Y_p(x)$. In this list, $P(x)$ is a given polynomial of degree n, $Q(x)$ and $R(x)$ are polynomials of degree n with undetermined coefficients for which we must solve, and c and β are constants.

If any term on the left is a solution of the associated homogeneous equation, multiply the proposed solution on the right by x. If this is again a solution of the associated homogeneous equation, multiply by x again.

2.3.3   The Principle of Superposition

Suppose we want to find a particular solution of

$$y'' + p(x)y' + q(x)y = f_1(x) + f_2(x) + \cdots + f_N(x).$$

It is routine to check that, if Y_j is a solution of

$$y'' + p(x)y' + q(x)y = f_j(x),$$

then $Y_1 + Y_2 + \cdots + Y_N$ is a particular solution of the original differential equation.

TABLE 2.1	*Functions to Try for $Y_p(x)$ in the Method of Undetermined Coefficients*
$f(x)$	$Y_p(x)$
$P(x)$	$Q(x)$
Ae^{cx}	Re^{cx}
$A\cos(\beta x)$	$C\cos(\beta x) + D\sin(\beta x)$
$A\sin(\beta x)$	$C\cos(\beta x) + D\sin(\beta x)$
$P(x)e^{cx}$	$Q(x)e^{cx}$
$P(x)\cos(\beta x)$	$Q(x)\cos(\beta x) + R(x)\sin(\beta x)$
$P(x)\sin(\beta x)$	$Q(x)\cos(\beta x) + R(x)\sin(\beta x)$
$P(x)e^{cx}\cos(\beta x)$	$Q(x)e^{cx}\cos(\beta x) + R(x)e^{cx}\sin(\beta x)$
$P(x)e^{cx}\sin(\beta x)$	$Q(x)e^{cx}\cos(\beta x) + R(x)e^{cx}\sin(\beta x)$

EXAMPLE 2.17

Find a particular solution of

$$y'' + 4y = x + 2e^{-2x}.$$

To find a particular solution, consider two problems:

> *Problem* 1: $y'' + 4y = x$
> *Problem* 2: $y'' + 4y = 2e^{-2x}$

Using undetermined coefficients, we find a particular solution $Y_{p_1}(x) = x/4$ of Problem 1, and a particular solution $Y_{p_2}(x) = e^{-2x}/4$ of Problem 2. A particular solution of the given differential equation is

$$Y_p(x) = \frac{1}{4}x + \frac{1}{4}e^{-2x}.$$

Using this, the general solution is

$$y(x) = c_1 \cos(2x) + c_2 \sin(2x) + \frac{1}{4}\left(x + e^{-2x}\right). \quad \blacklozenge$$

SECTION 2.3   PROBLEMS

In each of Problems 1 through 6, find the general solution using the method of variation of parameters for a particular solution.

1. $y'' + y = \tan(x)$
2. $y'' - 4y' + 3y = 2\cos(x+3)$
3. $y'' + 9y = 12\sec(3x)$
4. $y'' - 2y' - 3y = 2\sin^2(x)$
5. $y'' - 3y' + 2y = \cos(e^{-x})$
6. $y'' - 5y' + 6y = 8\sin^2(4x)$

In each of Problems 7 through 16, find the general solution using the method of undetermined coefficients for a particular solution.

7. $y'' - y' - 2y = 2x^2 + 5$
8. $y'' - y' - 6y = 8e^{2x}$
9. $y'' - 2y' + 10y = 20x^2 + 2x - 8$
10. $y'' - 4y' + 5y = 21e^{2x}$

11. $y'' - 6y' + 8y = 3e^x$
12. $y'' + 6y' + 9y = 9\cos(3x)$
13. $y'' - 3y' + 2y = 10\sin(x)$
14. $y'' - 4y' = 8x^2 + 2e^{3x}$
15. $y'' - 4y' + 13y = 3e^{2x} - 5e^{3x}$
16. $y'' - 2y' + y = 3x + 25\sin(3x)$

In each of Problems 17 through 24, solve the initial value problem.

17. $y'' - 4y = -7e^{2x} + x$; $y(0) = 1$, $y'(0) = 3$
18. $y'' + 4y' = 8 + 34\cos(x)$; $y(0) = 3$, $y'(0) = 2$
19. $y'' + 8y' + 12y = e^{-x} + 7$; $y(0) = 1$, $y'(0) = 0$
20. $y'' - 3y' = 2e^{2x}\sin(x)$; $y(0) = 1$, $y'(0) = 2$
21. $y'' - 2y' - 8y = 10e^{-x} + 8e^{2x}$; $y(0) = 1$, $y'(0) = 4$
22. $y'' - y' + y = 1$; $y(1) = 4$, $y'(1) = -2$
23. $y'' - y = 5\sin^2(x)$; $y(0) = 2$, $y'(0) = -4$
24. $y'' + y = \tan(x)$; $y(0) = 4$, $y'(0) = 3$

2.4   Spring Motion

A spring suspended vertically and allowed to come to rest has a *natural length L*. An object (bob) of mass m is attached at the lower end, pulling the spring d units past its natural length. The bob

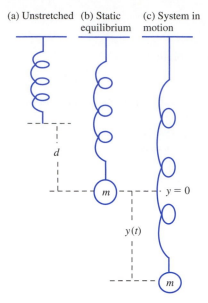

FIGURE 2.1 *Spring at natural and equilibrium lengths, and in motion.*

comes to rest in its *equilibrium position*. The bob is then displaced vertically a distance y_0 units and released from rest or with some initial velocity. We want to construct a model allowing us to analyze the motion of the bob.

Let $y(t)$ be the displacement of the bob from the equilibrium position at time t (Figure 2.1) and take this equilibrium position to be $y = 0$. Down is chosen as the positive direction (an arbitrary choice). Now consider the forces acting on the bob. Gravity pulls it downward with a force of magnitude mg. By Hooke's law, the spring exerts a force ky on the object, where k is the *spring constant*, a number quantifying the "stiffness" of the spring. At the equilibrium position, the force of the spring is $-kd$, which is negative because it acts upward. If the object is pulled downward a distance y from this position, an additional force $-ky$ is exerted on it. The total force due to the spring is therefore $-kd - ky$. The total force due to gravity and the spring is $mg - kd - ky$. Since at the equilibrium point this force is zero, then $mg = kd$. The net force acting on the object due to gravity and the spring is therefore just $-ky$.

There are forces tending to retard or damp out the motion. These include air resistance or perhaps viscosity of a medium in which the object is suspended. A standard assumption, verified by experiment, is that the retarding forces have magnitude proportional to the velocity y'. Then, for some constant c called the *damping constant*, the retarding forces equal cy'. The total force acting on the bob due to gravity, damping, and the spring itself is $-ky - cy'$.

Finally, there may be a driving force $f(t)$ acting on the bob. In this case, the total external force is

$$F = -ky - cy' + f(t).$$

Assuming that the mass is constant, Newton's second law of motion gives us

$$my'' = -ky - cy' + f(t),$$

or

$$y'' + \frac{c}{m}y' + \frac{k}{m}y = \frac{1}{m}f(t). \tag{2.12}$$

This is the *spring equation*. Solutions give the displacement of the bob as a function of time and enable us to analyze the motion under various conditions.

2.4.1   Unforced Motion

The motion is unforced if $f(t) = 0$. Now the spring equation is homogeneous and the characteristic equation has roots

$$\lambda = -\frac{c}{2m} \pm \frac{1}{2m}\sqrt{c^2 - 4km}.$$

As we might expect, the solution for the displacement, and hence the motion of the bob, depends on the mass, the amount of damping, and the stiffness of the spring.

Case 1: $c^2 - 4km > 0$

Now the roots of the characteristic equation are real and distinct,

$$\lambda_1 = -\frac{c}{2m} + \frac{1}{2m}\sqrt{c^2 - 4km} \text{ and } \lambda_2 = -\frac{c}{2m} - \frac{1}{2m}\sqrt{c^2 - 4km}.$$

The general solution of the spring equation in this case is

$$y(t) = c_1 e^{\lambda_1 t} + c_2 e^{\lambda_2 t}.$$

Clearly, $\lambda_2 < 0$. Since m and k are positive, $c^2 - 4km < c^2$ so $\sqrt{c^2 - 4km} < c$ and $\lambda_1 < 0$ also. Therefore,

$$\lim_{t \to \infty} y(t) = 0$$

regardless of initial conditions. In the case that $c^2 - 4km > 0$, the motion simply decays to zero as time increases. This case is called *overdamping*.

EXAMPLE 2.18   Overdamping

Let $c = 6, k = 5$, and $m = 1$. Now the general solution is $y(t) = c_1 e^{-t} + c_2 e^{-5t}$. Suppose the bob was initially drawn upward 4 feet from equilibrium and released downward with a speed of 2 feet per second. Then $y(0) = -4$ and $y'(0) = 2$, and we obtain

$$y(t) = \frac{1}{2} e^{-t} \left(-9 + e^{-4t} \right).$$

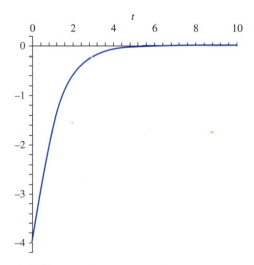

FIGURE 2.2   *Overdamped, unforced motion in Example 2.18.*

Figure 2.2 is a graph of this solution. Since $-9 + e^{-4t} < 0$ for $t > 0$, then $y(t) < 0$ and the bob always remains above the equilibrium point. Its velocity $y'(t) = e^{-t}(9 - 5e^{-4t})/2$ decreases to zero as t increases, so the bob moves downward toward equilibrium with decreasing velocity, approaching arbitrarily close to but never reaching this position and never coming completely to rest. ◆

Case 2: $c^2 - 4kmc = 0$
In this case, the general solution of the spring equation is

$$y(t) = (c_1 + c_2 t)e^{-ct/2m}.$$

This case is called *critical damping*. While $y(t) \to 0$ as $t \to \infty$, as with overdamping, now the bob can pass through the critical point, as the following example shows.

EXAMPLE 2.19   Critical Damping

Let $c = 2$ and $k = m = 1$. Then $y(t) = (c_1 + c_2 t)e^{-t}$. Suppose the bob is initially pulled up 4 feet above the equilibrium position and then pushed downward with a speed of 5 feet per second. Then $y(0) = -4$ and $y'(0) = 5$, so

$$y(t) = (-4 + t)e^{-t}.$$

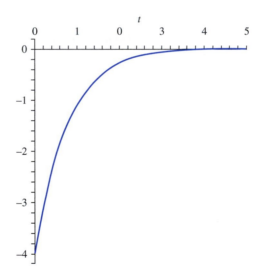

FIGURE 2.3 *Critically   damped,   unforced motion in Example 2.19.*

Since $y(4) = 0$, the bob reaches the equilibrium 4 seconds after it was released and then passes through it. In fact, $y(t)$ reaches its maximum when $t = 5$ seconds, and this maximum value is $y(5) = e^{-5}$, which is about 0.007 units below the equilibrium point. The velocity $y'(t) = (5 - t)e^{-t}$ is negative for $t > 5$, so the bob's velocity decreases after the 5-second point. Since $y(t) \to 0$ as $t \to \infty$, the bob moves with decreasing velocity back toward the equilibrium point as time increases. Figure 2.3 is a graph of this displacement function. ◆

In general, when critical damping occurs, the bob either passes through the equilibrium point exactly once, as in Example 2.19, or never reaches it at all, depending on the initial conditions.

Case 3: $c^2 - 4km < 0$

Here the spring constant and mass of the bob are sufficiently large that $c^2 < 4km$ and damping is less dominant. This is called *underdamping*. The general underdamped solution has the form

$$y(t) = e^{-ct/2m}[c_1 \cos(\beta t) + c_2 \sin(\beta t)]$$

in which

$$\beta = \frac{1}{2m}\sqrt{4km - c^2}.$$

Since c and m are positive, $y(t) \to 0$ as $t \to \infty$, as in the other two cases. This is not surprising in the absence of an external driving force. However, with underdamping, the motion is oscillatory because of the sine and cosine terms in the displacement function. The motion is not periodic, however, because of the exponential factor $e^{-ct/2m}$ which causes the amplitudes of the oscillations to decay as time increases.

EXAMPLE 2.20 Underdamping

Let $c = k = 2$ and $m = 1$. The general solution is

$$y(t) = e^{-t}[c_1 \cos(t) + c_2 \sin(t)].$$

Suppose the bob is driven downward from a point 3 feet above equilibrium with an initial speed of 2 feet per second. Then $y(0) = -3$ and $y'(0) = 2$, and the solution is

$$y(t) = -e^{-t}(3\cos(t) + \sin(t)).$$

The behavior of this solution is more easily visualized if we write it in *phase angle form*. Choose C and δ so that

$$3\cos(t) + \sin(t) = C\cos(t + \delta).$$

For this we need

$$3\cos(t) + \sin(t) = C\cos(t)\cos(\delta) - C\sin(t)\sin(\delta).$$

Then

$$C\cos(\delta) = 3 \text{ and } C\sin(\delta) = -1.$$

Then

$$\frac{C\sin(\delta)}{C\cos(\delta)} = \tan(\delta) = -\frac{1}{3},$$

so

$$\delta = \arctan\left(-\frac{1}{3}\right) = -\arctan\left(\frac{1}{3}\right).$$

To solve for C, write

$$C^2\cos^2(\delta) + C^2\sin^2(\delta) = C^2 = 3^2 + 1 = 10.$$

Then $C = \sqrt{10}$ and the phase angle form of the solution is

$$y(t) = \sqrt{10}e^{-t}\cos(t - \arctan(1/3)).$$

The graph is a cosine curve with decaying amplitude squeezed between graphs of $y = \sqrt{10}e^{-t}$ and $y = -\sqrt{10}e^{-t}$. Figure 2.4 shows $y(t)$ and these two exponential functions as reference curves.

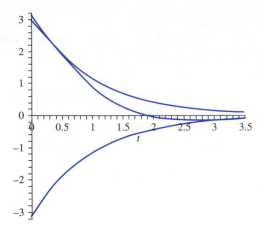

FIGURE 2.4   *Underdamped, unforced motion in Example 2.20.*

The bob passes back and forth through the equilibrium point as t increases. Specifically, it passes through the equilibrium point exactly when $y(t) = 0$, which occurs at times

$$t = \arctan\left(\frac{1}{3}\right) + \frac{2n+1}{2}\pi$$

for $n = 0, 1, 2, \cdots$. ◆

Next we will pursue the effect of a driving force on the motion of the bob.

2.4.2 Forced Motion

Different driving forces will result in different motion. We will analyze the case of a periodic driving force $f(t) = A\cos(\omega t)$. Now the spring equation (2.12) is

$$y'' + \frac{c}{m}y' + \frac{k}{m}y = \frac{A}{m}\cos(\omega t). \tag{2.13}$$

We have solved the associated homogeneous equation in all cases on c, k, and m. For the general solution of equation (2.13), we need only a particular solution. Application of the method of undetermined coefficients yields the particular solution

$$Y_p(t) = \frac{mA(k - m\omega^2)}{(k - m\omega^2)^2 + \omega^2 c^2}\cos(\omega t)$$

$$+ \frac{A\omega c}{(k - m\omega^2)^2 + \omega^2 c^2}\sin(\omega t).$$

It is customary to denote $\omega_0 = \sqrt{k/m}$ to write

$$Y_p(t) = \frac{mA(\omega_0^2 - \omega^2)}{m^2(\omega_0^2 - \omega^2)^2 + \omega^2 c^2}\cos(\omega t)$$

$$+ \frac{A\omega c}{m^2(\omega_0^2 - \omega^2)^2 + \omega^2 c^2}\sin(\omega t).$$

We will analyze some specific cases to get some insight into the motion with this forcing function.

Case 1: Overdamped Forced Motion

Overdamping occurs when $c^2 - 4km > 0$. Suppose $c = 6, k = 5, m = 1, A = 6\sqrt{5}$, and $\omega = \sqrt{5}$. If the bob is released from rest from the equilibrium position, then $y(t)$ satisfies the initial value problem

$$y'' + 6y' + 5y = 6\sqrt{5}\cos(\sqrt{5}t); \ y(0) = y'(0) = 0.$$

The solution is

$$y(t) = \frac{\sqrt{5}}{4}(-e^{-t} + e^{-5t}) + \sin(\sqrt{5}t),$$

a graph of which is shown in Figure 2.5. As time increases, the exponential terms decay to zero, and the displacement behaves increasingly like $\sin(\sqrt{5}t)$, oscillating up and down through the equilibrium point with approximate period $2\pi/\sqrt{5}$. Contrast this with the overdamped motion without the forcing function, in which the bob began above the equilibrium point and moved with decreasing speed down toward it, but never reached it.

Case 2: Critically Damped Forced Motion

Let $c = 2, m = k = 1, \omega = 1$, and $A = 2$. Assume that the bob is released from rest from the equilibrium point. The initial value problem modeling the motion is

$$y'' + 2y' + y = 2\cos(t); \ y(0) = y'(0) = 0$$

with a solution of

$$y(t) = -te^{-t} + \sin(t).$$

Figure 2.6 is a graph of this solution, which is critically damped forced motion. As t increases, the term with the exponential factor decays, although not as fast as in the overdamping case where there is no factor of t. Nevertheless, after sufficient time, the motion settles into nearly (but not exactly because $-te^{-t}$ is never zero for $t > 0$) a sinusoidal motion back and forth through the equilibrium point.

Case 3: Underdamped Forced Motion

Let $c = k = 2, m = 1, \omega = \sqrt{2}$, and $A = 2\sqrt{2}$, so $c^2 - 4km < 0$. Suppose the bob is released from rest at the equilibrium position. The initial value problem is

$$y'' + 2y' + 2y = 2\sqrt{2}\cos(\sqrt{2}t); \ y(0) = y'(0) = 0$$

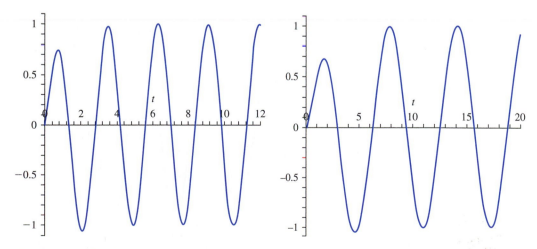

FIGURE 2.5 *Overdamped, forced motion.*   **FIGURE 2.6** *Critically damped, forced motion.*

with solution

$$y(t) = -\sqrt{2}e^{-t}\sin(t) + \sin(\sqrt{2}t).$$

This is an example of underdamped forced motion. Unlike overdamping and critical damping, the exponential term e^{-t} has a trigonometric factor $\sin(t)$. Figure 2.7 is a graph of this solution. As time increases, the term $-\sqrt{2}e^{-t}\sin(t)$ becomes less influential, and the motion settles nearly into an oscillation back and forth through the equilibrium point with period nearly $2\pi/\sqrt{2}$.

2.4.3 Resonance

In the absence of damping, an important phenomenon called resonance can occur. Suppose $c = 0$ but that there is still a driving force $A\cos(\omega t)$. Now the spring equation (2.12) is

$$y'' + \frac{k}{m}y = \frac{A}{m}\cos(\omega t).$$

From the particular solution Y_p found in Section 2.4.2 with $c = 0$, we find that this spring equation has general solution

$$y(t) = c_1\cos(\omega t) + c_2\sin(\omega t) + \frac{A}{m(\omega_0^2 - \omega^2)}\cos(\omega t)$$

in which $\omega_0 = \sqrt{k/m}$. This number is called the *natural frequency* of the spring system and it is a function of the stiffness of the spring and the mass of the bob, while ω is the *input frequency* and is contained in the driving force. This general solution assumes that the natural and input frequencies are different. Of course, the closer we choose the natural and input frequencies, the larger the amplitude of the $\cos(\omega t)$ term in the solution.

Resonance occurs when the natural and input frequencies are the same. Now the differential equation is

$$y'' + \frac{k}{m}y = \frac{A}{m}\cos(\omega_0 t). \tag{2.14}$$

The solution derived for the case that $\omega \neq \omega_0$ does not apply to equation (2.14). To find the general solution in the present case with $\omega = \omega_0$, first find the general solution of the associated homogeneous equation

$$y'' + \frac{k}{m}y = 0.$$

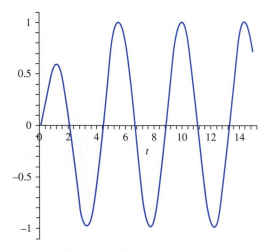

FIGURE 2.7 *Underdamped, forced motion.*

This has general solution

$$y_h(t) = c_1 \cos(\omega_0 t) + c_2 \sin(\omega_0 t).$$

Now we need a particular solution of equation (2.14). To use the method of undetermined coefficients, we might try a function of the form $a \cos(\omega_0 t) + b \sin(\omega_0 t)$. However, these are solutions of the associated homogeneous equation, so instead try

$$Y_p(t) = at \cos(\omega_0 t) + bt \sin(\omega_0 t).$$

Substitute $Y_p(t)$ into equation (2.14) to get

$$-2a\omega_0 \sin(\omega_0 t) + 2b \cos(\omega_0 t) = \frac{A}{m} \cos(\omega_0 t).$$

Thus, choose

$$a = 0 \text{ and } 2b\omega_0 = \frac{A}{m}.$$

This gives us the particular solution

$$Y_p(t) = \frac{A}{2m\omega_0} t \sin(\omega_0 t).$$

The general solution is

$$y(t) = c_1 \cos(\omega_0 t) + c_2 \sin(\omega_0 t) + \frac{A}{2m\omega_0} t \sin(\omega_0).$$

This solution differs from that in the case $\omega \neq \omega_0$ in the factor of t in the particular solution. Because of this, solutions increase in amplitude as t increases. This phenomenon is called *resonance*.

As an example, suppose $c_1 = c_2 = \omega_0 = 1$ and $A/2m = 1$. Now the solution is

$$y(t) = \cos(t) + \sin(t) + t \sin(t).$$

Figure 2.8 displays the increasing amplitude of the oscillations with time.

While there is always some damping in the real world, if the damping constant is close to zero compared to other factors and if the natural and input frequencies are (nearly) equal, then oscillations can build up to a sufficiently large amplitude to cause resonance-like behavior.

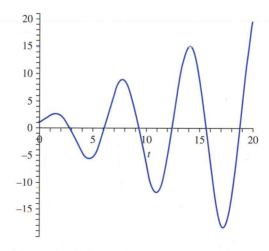

FIGURE 2.8 *Resonance.*

This caused the collapse of the Broughton Bridge near Manchester, England, in 1831 when a column of soldiers marching across maintained a cadence (input frequency) that happened to closely match the natural frequency of the material of the bridge. More recently, the Tacoma Narrows Bridge in the state of Washington experienced increasing oscillations driven by high winds, causing the concrete roadbed to oscillate in sensational fashion until it collapsed into the Puget Sound. This occurred on November 7, 1940. At one point, one side of the roadbed was about twenty-eight feet above the other as it thrashed about. Local news crews were on hand, and motion pictures of the collapse are available in many engineering and science schools and on the Web.

2.4.4   Beats

In the absence of damping, an oscillatory driving force can also cause a phenomenon called beats. Suppose $\omega \neq \omega_0$ and consider

$$y'' + \omega_0^2 y = \frac{A}{m}\cos(\omega t).$$

Assume that the object is released from rest from the equilibrium position, so $y(0) = y'(0) = 0$. The solution is

$$y(t) = \frac{A}{m(\omega_0^2 - \omega^2)}[\cos(\omega t) - \cos(\omega_0 t)].$$

The behavior of this solution reveals itself more clearly if we write it as

$$y(t) = \frac{2A}{m(\omega_0^2 - \omega^2)}\sin\left(\frac{1}{2}(\omega_0 + \omega)t\right)\sin\left(\frac{1}{2}(\omega_0 - \omega)t\right).$$

This formulation reveals a periodic variation of amplitude in the solution, depending on the relative sizes of $\omega_0 + \omega$ and $\omega_0 - \omega$. This periodic variation is called a *beat*. As an example, suppose $\omega_0 + \omega = 5$ and $\omega_0 - \omega = 1/2$, and that the constants are chosen so that $2A/m(\omega_0^2 - \omega^2) = 1$. Now the displacement function is

$$y(t) = \sin\left(\frac{5t}{2}\right)\sin\left(\frac{t}{4}\right).$$

Figure 2.9 is a graph of this solution.

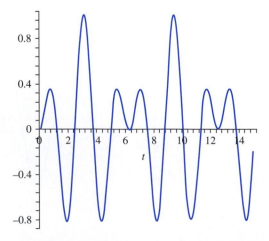

FIGURE 2.9   *Beats.*

2.4.5   Analogy with an Electrical Circuit

In a resistance R, inductance L, and capacitance C (RLC) circuit with electromotive force $E(t)$, Kirchhoff's laws yield a differential equation for the current:

$$Li'(t) + Ri(t) + \frac{1}{C}q(t) = E(t).$$

Since $i = q'$, this is a second-order differential equation for the charge:

$$q'' + \frac{R}{L}q' + \frac{1}{LC}q = \frac{1}{L}E.$$

Assuming that the resistance, inductance, and capacitance are constant, this equation is exactly analogous to the spring equation with a driving force, which has the form

$$y'' + \frac{c}{m}y' + \frac{k}{m}y = \frac{1}{m}f(t).$$

This means that solutions of the spring equation immediately translate into solutions of the circuit equation with the following correspondences:

$$\text{displacement function } y(t) \Longleftrightarrow \text{ charge } q(t)$$

$$\text{velocity } y'(t) \Longleftrightarrow \text{ current } i(t)$$

$$\text{driving force } f(t) \Longleftrightarrow \text{ electromotive force } E(t)$$

$$\text{mass } m \Longleftrightarrow \text{ inductance } L$$

$$\text{damping constant } c \Longleftrightarrow \text{ resistance } R$$

$$\text{spring modulus } k \Longleftrightarrow \text{ reciprocal } 1/C \text{ of the capacitance.}$$

SECTION 2.4   PROBLEMS

1. This problem gauges the relative effects of initial position and velocity on the motion in the unforced overdamped case. Solve the initial value problems

   $$y'' + 4y' + 2y = 0; \ y(0) = 5, \ y'(0) = 0$$

   and

   $$y'' + 4y' + 2y = 0; \ y(0) = 0, \ y'(0) = 5.$$

   Graph the solutions on the same set of axes.

2. Repeat the experiment of Problem 1, except now use the critically damped unforced equation

   $$y'' + 4y' + 4y = 0.$$

3. Repeat the experiment of Problem 1 for the underdamped unforced equation

   $$y'' + 2y' + 5y = 0.$$

Problems 4 through 9 explore the effects of changing the initial position or initial velocity on the motion of the object. In each, use the same set of axes to graph the solution of the initial value problem for the given values of A and observe the effect that these changes cause in the solution.

4. $y'' + 4y' + 2y = 0; \ y(0) = A, \ y'(0) = 0$; A has values $1, 3, 6, 10, -4$ and -7.

5. $y'' + 4y' + 2y = 0; \ y(0) = 0, \ y'(0) = A$; A has values $1, 3, 6, 10, -4$ and -7.

6. $y'' + 4y' + 4y = 0; \ y(0) = A, \ y'(0) = 0$; A has values $1, 3, 6, 10, -4$ and -7.

7. $y'' + 4y' + 4y = 0; \ y(0) = 0, \ y'(0) = A$; A has values $1, 3, 6, 10, -4$ and -7.

8. $y'' + 2y' + 5y = 0; \ y(0) = A, \ y'(0) = 0$; A has values $1, 3, 6, 10, -4$ and -7.

9. $y'' + 2y' + 5y = 0; \ y(0) = 0, \ y'(0) = A$; A has values $1, 3, 6, 10, -4$ and -7.

10. An object having a mass of 1 gram is attached to the lower end of a spring having a modulus of 29 dynes per centimeter. The mass is, in turn, adhered to a dashpot that imposes a damping force of $10v$ dynes, where $v(t)$ is the velocity of the mass at time t in centimeters per second. Determine the motion of the mass if it is pulled down 3 centimeters from equilibrium and then struck upward with a blow sufficient to impart a velocity of 1 centimeter per second. Graph the solution. Solve the problem when the initial velocity is, in turn, 2, 4, 7, and 12 centimeters per second. Graph these solutions on the same axes to visualize the influence of the initial velocity on the motion.

11. How many times can the mass pass through the equilibrium point in overdamped motion? What condition can be placed on the initial displacement to ensure that it never passes through equilibrium?

12. How many times can the mass pass through equilibrium in critical damping? What condition can be placed on $y(0)$ to ensure that the mass never passes through the equilibrium point? How does the initial velocity influence whether the mass passes through the equilibrium point?

13. In underdamped unforced motion, what effect does the damping constant have on the frequency of the oscillations?

14. Suppose $y(0) = y'(0) \neq 0$. Determine the maximum displacement of the mass in critically damped unforced motion. Show that the time at which this maximum occurs is independent of the initial displacement.

15. Consider overdamped forced motion governed by

$$y'' + 6y' + 2y = 4\cos(3t).$$

(a) Find the solution satisfying $y(0) = 6$, $y' = 0$.

(b) Find the solution satisfying $y(0) = 0$, $y'(0) = 6$.

Graph these solutions on the same set of axes to compare the effects of initial displacement and velocity on the motion.

16. Carry out the program of Problem 15 for the critically damped forced system governed by

$$y'' + 4y' + 4y = 4\cos(3t).$$

17. Carry out the program of Problem 15 for the underdamped forced system governed by

$$y'' + y' + 3y = 4\cos(3t).$$

CHAPTER 3

The Laplace Transform

3.1   Definition and Notation

The Laplace transform is an important tool for solving certain kinds of initial value problems, particularly those involving discontinuous forcing functions, as occur frequently in areas such as electrical engineering.

We will see that the Laplace transform converts an initial value problem to an algebra problem, leading us to attempt the following procedure:

initial value problem $\Longrightarrow$ algebra problem

$\Longrightarrow$ solution of the algebra problem

$\Longrightarrow$ solution of the initial value problem.

This is an effective strategy, because some algebra problems are easier to solve than initial value problems. We begin in this section with the definition and elementary properties of the transform.

Laplace Transform

The *Laplace transform* of a function f is a function $\mathcal{L}[f]$ defined by

$$\mathcal{L}[f](s) = \int_0^\infty e^{-st} f(t)\,dt.$$

The integration is with respect to t and defines a function of the new variable s for all s such that this integral converges.

Because the symbol $\mathcal{L}[f](s)$ may be awkward to write in computations, we will make the following convention. We will use a lowercase letter for a function put into the transform and the corresponding uppercase letter for the transformed function. In this way,

$$\mathcal{L}[f]=F,\ \mathcal{L}[g]=G,\ \text{and } \mathcal{L}[h]=H$$

and so on. If we include the variable, these would be written

$$\mathcal{L}[f](s)=F(s),\ \mathcal{L}[g](s)=G(s),\ \text{and } \mathcal{L}[h](s)=H(s)$$

It is also customary to use t (for time) as the variable of the input function and s for the variable of the transformed function.

EXAMPLE 3.1

Let a be any real number and $f(t)=e^{at}$. The Laplace transform of f is the function defined by

$$\mathcal{L}[f](s)=\int_0^\infty e^{-st}e^{at}\,dt$$

$$=\int_0^\infty e^{(a-s)t}\,dt=\lim_{k\to\infty}\int_0^k e^{(a-s)t}\,dt$$

$$=\lim_{k\to\infty}\left[\frac{1}{a-s}e^{(a-s)t}\right]_0^k$$

$$=-\frac{1}{a-s}=\frac{1}{s-a}$$

provided that $s>a$. The Laplace transform of $f(t)=e^{at}$ can be denoted $F(s)=1/(s-a)$ for $s>a$. ◆

We rarely determine a Laplace transform by integration. Table 3.1 is a short table of Laplace transforms of familiar functions, and much longer tables are available. In the table, n denotes a nonnegative integer, and a and b are constants. Reading from Table 3.1, if $f(t)=\sin(3t)$ then by entry (6)

$$F(s)=\frac{3}{s^2+9},$$

TABLE 3.1   *Laplace Transforms of Selected Functions*

$f(t)$	$F(s)$	$f(t)$	$F(s)$
(1) 1	$\frac{1}{s}$	(8) $t\sin(at)$	$\frac{2as}{(s^2+a^2)^2}$
(2) t^n	$\frac{n!}{s^{n+1}}$	(9) $t\cos(at)$	$\frac{s^2-a^2}{(s^2+a^2)^2}$
(3) e^{at}	$\frac{1}{s-a}$	(10) $e^{at}\sin(bt)$	$\frac{b}{(s-a)^2+b^2}$
(4) $t^n e^{at}$	$\frac{n!}{(s-a)^{n+1}}$	(11) $e^{at}\cos(bt)$	$\frac{s-a}{(s-a)^2+b^2}$
(5) $e^{at}-e^{bt}$	$\frac{a-b}{(s-a)(s-b)}$	(12) $\sinh(at)$	$\frac{a}{s^2-a^2}$
(6) $\sin(at)$	$\frac{a}{s^2+a^2}$	(13) $\cosh(at)$	$\frac{s}{s^2-a^2}$
(7) $\cos(at)$	$\frac{s}{s^2+a^2}$	(14) $\delta(t-a)$	e^{-as}

and if $k(t) = e^{2t} \cos(5t)$, then by entry (11)

$$K(s) = \frac{s-2}{(s-2)^2 + 25}.$$

There are also software routines for transforming functions.

The Laplace transform is *linear*, which means that the transform of a sum is the sum of the transforms, and constants factor through the transform:

$$\mathcal{L}[f+g] = F + G \text{ and } \mathcal{L}[cf] = cF$$

for all s such that $F(s)$ and $G(s)$ are both defined and for any number c.

Given $F(s)$, we sometimes need to find $f(t)$ such that $\mathcal{L}[f] = F$. This is the reverse process of computing the transform of f, and we refer to it as taking an *inverse Laplace transform*. This inverse transform is denoted $\mathcal{L}^{-1}$, and

$$\mathcal{L}^{-1}[F] = f \text{ exactly when } \mathcal{L}[f] = F.$$

For example, the inverse Laplace transform of $1/(s-a)$ is e^{at}.

If we use Table 3.1 to find an inverse transform, read from the right column to the left column. For example, using the table and the linearity of the Laplace transform, we can read that

$$\mathcal{L}^{-1}\left[\frac{3}{s^2+16} - \frac{7}{s-5} \right] = \frac{3}{4} \sin(4t) - 7e^{5t}.$$

The inverse Laplace transform $\mathcal{L}^{-1}$ is linear because $\mathcal{L}$ is. This means that

$$\mathcal{L}^{-1}[F+G] = \mathcal{L}^{-1}[F] + \mathcal{L}^{-1}[G] = f + g$$

and for any number c

$$\mathcal{L}^{-1}[cF] = c\mathcal{L}^{-1}[F] = cf.$$

SECTION 3.1   PROBLEMS

In each of Problems 1 through 5, use Table 3.1 to determine the Laplace transform of the function.

1. $f(t) = 3t \cos(2t)$

2. $g(t) = e^{-4t} \sin(8t)$

3. $h(t) = 14t - \sin(7t)$

4. $w(t) = \cos(3t) - \cos(7t)$

5. $k(t) = -5t^2 e^{-4t} + \sin(3t)$

In each of Problems 6 through 10, use Table 3.1 to determine the inverse Laplace transform of the function.

6. $R(s) = \dfrac{7}{s^2 - 9}$

7. $Q(s) = \dfrac{s}{s^2 + 64}$

8. $G(s) = \dfrac{5}{s^2 + 12} - \dfrac{4s}{s^2 + 8}$

9. $P(s) = \dfrac{1}{s+42} - \dfrac{1}{(s+3)^4}$

10. $F(s) = \dfrac{-5s}{(s^2 + 1)^2}$

3.2   Solution of Initial Value Problems

To apply the Laplace transform to the solution of an initial value problem, we must be able to transform a derivative. This involves the concept of a piecewise continuous function.

This means that f has (at most) finitely many discontinuities on the interval, and these are all jump discontinuities. The function graphed in Figure 3.1 has jump discontinuities at t_0 and t_1. The magnitude of a jump discontinuity is the width of the gap in the graph there. In Figure 3.1, the magnitude of the jump at t_1 is

$$|\lim_{t \to t_1+} f(t) - \lim_{t \to t_1-} f(t)|.$$

By contrast, let

$$g(t) = \begin{cases} 1/t & \text{for } 0 < t \leq 1 \\ 0 & \text{for } t = 0. \end{cases}$$

Then g is continuous on $(0, 1]$ but is not piecewise continuous on $[0, 1]$ because $\lim_{t \to 0+} g(t) = \infty$.

THEOREM 3.1 *Transform of a Derivative*

Let f be continuous for $t \geq 0$ and suppose f' is piecewise continuous on $[0, k]$ for every $k > 0$. Suppose also that $\lim_{k \to \infty} e^{-sk} f(k) = 0$ if $s > 0$. Then,

$$\mathcal{L}[f'](s) = sF(s) - f(0). \tag{3.1}$$

♦

This states that the transform of $f'(t)$ is s times the transform of $f(t)$, minus $f(0)$, the original function evaluated at $t = 0$. This can be proven by integration by parts.

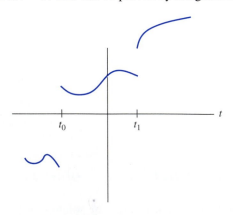

FIGURE 3.1 *Typical jump discontinuities.*

If f has a jump discontinuity at 0, as occurs if f is an electromotive force that is switched on at time zero, then the conclusion of the theorem must be amended to read

$$\mathcal{L}[f'[s] = sF(s) - f(0+)$$

where $f(0+) = \lim_{t \to 0+} f(t)$.

There is an extension of Theorem 3.1 to higher derivatives. If n is a positive integer, let $f^{(n)}$ denote the nth derivative of f.

THEOREM 3.2 *Transform of a Higher Derivative*

Let $f, f', \ldots, f^{(n-1)}$ be continuous for $t > 0$ and suppose $f^{(n)}$ is piecewise continuous on $[0, k]$ for every $k > 0$. Suppose also that

$$\lim_{k \to \infty} e^{-sk} f^{(j)}(k) = 0$$

for $s > 0$ and $j = 1, 2, \ldots, n - 1$. Then

$$\mathcal{L}[f^{(n)}](s) = s^n F(s) - s^{n-1} f(0) - s^{n-2} f'(0) - \cdots - s f^{(n-2)}(0) - f^{(n-1)}(0). \tag{3.2}$$

♦

The second derivative case $n = 2$ occurs sufficiently often that we will record the formula separately for this case:

$$\mathcal{L}[f''](s) = s^2 F(s) - s f(0) - f'(0). \tag{3.3}$$

We are now prepared to use the Laplace transform to solve some initial value problems.

EXAMPLE 3.2

We will solve $y' - 4y = 1$; $y(0) = 1$.

We already know how to solve this problem, but we will apply the Laplace transform to illustrate the technique. Take the transform of the differential equation, using the linearity of $\mathcal{L}$ and equation (3.1) to write

$$\mathcal{L}[y' - 4y](s) = \mathcal{L}[y'](s) - 4\mathcal{L}[y](s)$$

$$= (sY(s) - y(0)) - 4Y(s) = \mathcal{L}[1](s).$$

Insert the initial data $y(0) = 1$ and use Table 3.1 to find that $\mathcal{L}[1](s) = 1/s$. Then

$$(s - 4)Y(s) - 1 = \frac{1}{s}.$$

There is no derivative in this equation! $\mathcal{L}$ has converted the differential equation into an algebraic equation involving s and $Y(s)$, the transform of the unknown function $y(t)$. Solve for $Y(s)$ to obtain

$$Y(s) = \frac{1}{s - 4} + \frac{1}{s(s - 4)}.$$

This is the transform of the solution of the initial value problem. The solution is $y(t)$, which we obtain by applying the inverse transform

$$y = \mathcal{L}^{-1}[Y] = \mathcal{L}^{-1}\left[\frac{1}{s - 4}\right] + \mathcal{L}^{-1}\left[\frac{1}{s(s - 4)}\right]$$

$$= e^{4t} + \frac{1}{4}(e^{4t} - 1).$$

Here we used entry (3) of Table 3.1 with $a = 4$ and entry (5) with $a = 0$ and $b = 4$. The solution is

$$y(t) = e^{4t} + \frac{1}{4}(e^{4t} - 1) = \frac{5}{4}e^{4t} - \frac{1}{4}. \quad \blacklozenge$$

EXAMPLE 3.3

Solve

$$y'' + 4y' + 3y = e^t; \, y(0) = 0, \, y'(0) = 2.$$

Using the linearity of $\mathcal{L}$ and equations (3.1) and (3.3), we obtain

$$\mathcal{L}[y''] + 4\mathcal{L}[y'] + 3\mathcal{L}[y]$$
$$= [s^2Y - sy(0) - y'(0)] + 4[sY - y(0)] + 3Y$$
$$= [s^2Y - 2] + 4sY + 3Y$$
$$= \mathcal{L}[e^t] = \frac{1}{s-1}.$$

Then

$$(s^2 + 4s + 3)Y(s) - 2 = \frac{1}{s-1}.$$

Solve for Y to get

$$Y(s) = \frac{2s - 1}{(s - 1)(s^2 + 4s + 3)}$$
$$= \frac{2s - 1}{(s - 1)(s + 1)(s + 3)}.$$

To read the inverse transform from Table 3.1, use a partial decomposition to write the quotient on the right as a sum of simpler quotients. We will carry out the algebra of this decomposition. First write

$$\frac{2s - 1}{(s - 1)(s + 1)(s + 3)} = \frac{A}{s - 1} + \frac{B}{s + 1} + \frac{C}{s + 3}.$$

To solve for the constants, observe that if we added the fractions on the right, the numerator would have to equal the numerator $2s - 1$ of the fraction on the left. Therefore,

$$A(s + 1)(s + 3) + B(s - 1)(s + 3) + C(s - 1)(s + 1) = 2s - 1.$$

We can solve for A, B, and C by inserting values of s into this equation. Put $s = 1$ to get $8A = 1$, so $A = 1/8$. Put $s = -1$ to get $-4B = -3$, so $B = 3/4$. Put $s = -3$ to get $8C = -7$, so $C = -7/8$. Then

$$Y(s) = \frac{1}{8}\frac{1}{s - 1} + \frac{3}{4}\frac{1}{s + 1} - \frac{7}{8}\frac{1}{s + 3}.$$

Invert this to obtain the solution

$$y(t) = \frac{1}{8}e^t + \frac{3}{4}e^{-t} - \frac{7}{8}e^{-3t}. \quad \blacklozenge$$

Partial fraction decompositions are frequently used with the Laplace transform. The appendix at the end of this chapter reviews the algebra of this technique.

Notice that the transform method does not first solve for the general solution and then solve for the constants to satisfy the initial conditions. Equations (3.1), (3.2), and (3.3) insert the initial conditions directly into an algebraic equation for the transform of the unknown function. Still, we could have solved the problem of Example 3.3 by methods from Chapter 2. In the next section, we will see the real power of the transform.

PROBLEMS

In each of Problems 1 through 10, use the Laplace transform to solve the initial value problem.

1. $y' + 4y = 1$; $y(0) = -3$

2. $y' - 9y = t$; $y(0) = 5$

3. $y' + 4y = \cos(t)$; $y(0) = 0$

4. $y' + 2y = e^{-t}$; $y(0) = 1$

5. $y' - 2y = 1 - t$; $y(0) = 4$

6. $y'' + y = 1$; $y(0) = 6$, $y'(0) = 0$

7. $y'' - 4y' + 4y = \cos(t)$; $y(0) = 1$, $y'(0) = -1$

8. $y'' + 9y = t^2$; $y(0) = y'(0) = 0$

9. $y'' + 16y = 1 + t$; $y(0) = -2$, $y'(0) = 1$

10. $y'' - 5y' + 6y = e^{-t}$; $y(0) = 0$, $y'(0) = 2$

3.3 Shifting and the Heaviside Function

In this section, we introduce techniques which will enable us to solve initial value problems and other problems having discontinuous forcing functions.

3.3.1 The First Shifting Theorem

We will show that the Laplace transform of $e^{at} f(t)$ is the transform of $f(t)$ shifted a units to the right. This shift is achieved by replacing s by $s - a$ in $F(s)$ to obtain $F(s - a)$.

THEOREM 3.3 *First Shifting Theorem*

For any number a,

$$\mathcal{L}[e^{at} f(t)](s) = F(s - a). \tag{3.4}$$

This conclusion is also called *shifting in the s variable*. The proof is a straightforward appeal to the definition:

$$\mathcal{L}[e^{at} f(t)](s) = \int_0^\infty e^{-st} e^{at} f(t) dt$$

$$= \int_0^\infty e^{-(s-a)} f(t) dt = F(s - a). \ \blacklozenge$$

EXAMPLE 3.4

We know from Table 3.1 that $\mathcal{L}[\cos(bt)] = s/(s^2 + b^2) = F(s)$. For the transform of $e^{at} \cos(bt)$, replace s with $s - a$ to get

$$\mathcal{L}[e^{at} \cos(bt)](s) = \frac{s - a}{(s - a)^2 + b^2}. \ \blacklozenge$$

EXAMPLE 3.5

Since $\mathcal{L}[t^3] = 6/s^4$, then

$$\mathcal{L}[t^3 e^{7t}](s) = \frac{6}{(s-7)^4}.$$

Every formula for the Laplace transform of a function is also a formula for the inverse Laplace transform of a function. The inverse version of the first shifting theorem is

$$\mathcal{L}^{-1}[F(s-a)] = e^{at} f(t). \tag{3.5}$$

♦

EXAMPLE 3.6

Compute

$$\mathcal{L}^{-1}\left[\frac{4}{s^2 + 4s + 20}\right].$$

The idea is to manipulate the given function of s to the form $F(s-a)$ for some F and a. Then we can apply the inverse form of the shifting theorem, equation (3.5). Complete the square in the denominator to write

$$\frac{4}{s^2 + 4s + 20} = \frac{4}{(s+2)^2 + 16} = F(s+2)$$

if

$$F(s) = \frac{4}{s^2 + 16}.$$

From Table 3.1 $F(s)$ has inverse $f(t) = \sin(4t)$. By equation (3.5)

$$\mathcal{L}^{-1}\left[\frac{4}{s^2 + 4s + 20}\right]$$
$$= \mathcal{L}^{-1}[F(s+2)]$$
$$= e^{-2t} f(t) = e^{-2t} \sin(4t). \quad ♦$$

EXAMPLE 3.7

Compute

$$\mathcal{L}^{-1}\left[\frac{3s-1}{s^2 - 6s + 2}\right].$$

Follow the strategy of Example 3.6. Manipulate $F(s)$ to a function of $s-a$ for some a:

$$\frac{3s-1}{s^2 - 6s + 2} = \frac{3s-1}{(s-3)^2 - 7}$$
$$= \frac{3(s-3) + 8}{(s-3)^2 - 7}$$
$$= \frac{3(s-3)}{(s-3)^2 - 7} + \frac{8}{(s-3)^2 - 7}$$
$$= G(s-3) + K(s-3)$$

where

$$G(s) = \frac{3s}{s^2 - 7} \text{ and } K(s) = \frac{8}{s^2 - 7}.$$

By equation (3.5),

$$\mathcal{L}^{-1}\left[\frac{3s - 1}{s^2 - 6s + 2}\right] = \mathcal{L}^{-1}[G(s - 3)] + \mathcal{L}^{-1}[K(s - 3)]$$

$$= e^{3t}\mathcal{L}^{-1}[G(s)] + e^{3t}\mathcal{L}^{-1}[K(s)]$$

$$= e^{3t}\mathcal{L}^{-1}\left[\frac{3s}{s^2 - 7}\right] + e^{3t}\mathcal{L}^{-1}\left[\frac{8}{s^2 - 7}\right]$$

$$= 3e^{3t}\cosh(\sqrt{7}t) + \frac{8}{\sqrt{7}}e^{3t}\sinh(\sqrt{7}t). \quad \blacklozenge$$

3.3.2 The Heaviside Function and Pulses

Heaviside Function

Functions having jump discontinuities are efficiently treated by using the *unit step function* or *Heaviside function H* defined by

$$H(t) = \begin{cases} 0 & \text{for } t < 0 \\ 1 & \text{for } t \geq 0. \end{cases}$$

H is graphed in Figure 3.2. We will also use the *shifted Heaviside function* $H(t - a)$ of Figure 3.3, in which $a > 0$. This is the Heaviside function shifted a units to the right:

$$H(t - a) = \begin{cases} 0 & \text{for } t < a \\ 1 & \text{for } t \geq a. \end{cases}$$

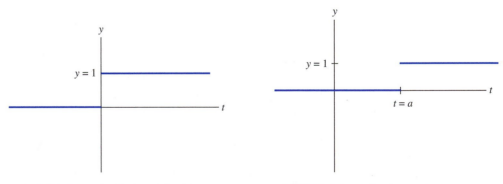

FIGURE 3.2 *The Heaviside function.*   **FIGURE 3.3** *Shifted Heaviside function.*

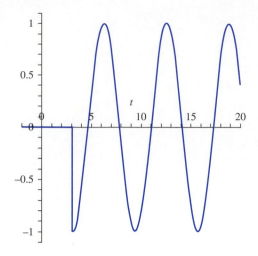

FIGURE 3.4 $H(t - \pi)\cos(t)$.

$H(t - a)$ can be used to turn a signal (function) off until time $t = a$, then turn it on. In particular,

$$H(t - a)g(t) = \begin{cases} 0 & \text{for } t < a \\ g(t) & \text{for } t \geq a. \end{cases}$$

To illustrate, Figure 3.4 shows $H(t - \pi)\cos(t)$. This is the familiar cosine function for $t \geq \pi$ but is turned off (equals 0) for $t < \pi$. Multiplying a function $f(t)$ by $H(t - a)$ leaves the graph of $f(t)$ unchanged for $t \geq a$ but replaces it by 0 for $t < a$.

We can also use the Heaviside function to define a *pulse*. If $a < b$, then

$$H(t - a) - H(t - b) = \begin{cases} 0 & \text{for } t < a \\ 1 & \text{for } a \leq t < b \\ 0 & \text{for } t \geq b. \end{cases}$$

Figure 3.5 shows the pulse $H(t - a) - H(t - b)$ with $a < b$. Pulses are used to turn a signal off until time $t = a$, then turn it on until time $t = b$, after which it is switched off again. Figure 3.6

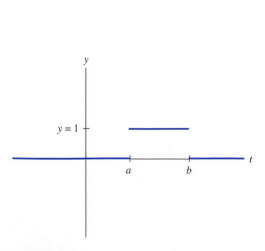

FIGURE 3.5 *A pulse $H(t - a) - H(t - b)$.*

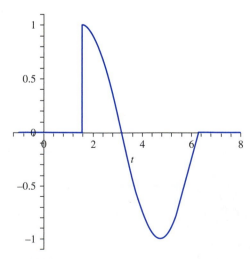

FIGURE 3.6 $(H(t - \pi/2) - H(t - 2\pi))$ $\sin(t)$.

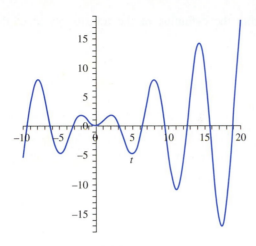

FIGURE 3.7 $t \sin(t)$.

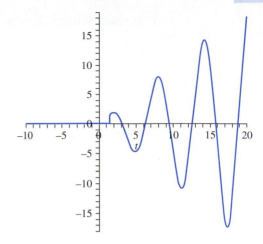

FIGURE 3.8 $H(t - 3/2) \sin(t)$.

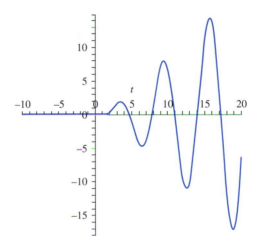

FIGURE 3.9 $H(t - 3/2)(t - 3/2) \sin(t - 3/2)$.

shows this effect for $[H(t - \pi/2) - H(t - 2\pi)] \sin(t)$, which is zero before time $t = \pi/2$ and after time $t = 2\pi$, and equals $\sin(t)$ between these times.

It is important to understand the difference between $g(t)$, $H(t - a)g(t)$ and $H(t - a)g(t - a)$. Figures 3.7, 3.8, and 3.9 show, respectively, graphs of $t \sin(t)$, $H(t - 3/2)t \sin(t)$ and $H(t - 3/2)(t - 3/2) \sin(t - 3/2)$. $H(t - 3/2)t \sin(t)$ is zero until time $3/2$ and then equals $t \sin(t)$, while $H(t - 3/2)(t - 3/2) \sin(t - 3/2)$ is zero until time $3/2$ and then is the graph of $t \sin(t)$ shifted $3/2$ units to the right.

Using the Heaviside function, we can state the second shifting theorem, which is also called shifting in the t variable.

THEOREM 3.4 *Second Shifting Theorem*

$$\mathcal{L}[H(t - a) f(t - a)](s) = e^{-as} F(s). \tag{3.6}$$

♦

This result follows directly from the definition of the transform and of the Heaviside function.

EXAMPLE 3.8

Suppose we want $\mathcal{L}[H(t-a)]$. Write

$$H(t-a) = H(t-a)f(t-a)$$

with $f(t) = 1$ for all t. Since $F(s) = \mathcal{L}[1](s) = 1/s$, then by the second shifting theorem,

$$\mathcal{L}[H(t-a)](s) = e^{-as}F(s) = \frac{1}{s}e^{-as}. \quad \blacklozenge$$

EXAMPLE 3.9

Compute $\mathcal{L}[g]$ where

$$g(t) = \begin{cases} 0 & \text{for } t < 2 \\ t^2 + 1 & \text{for } t \geq 2. \end{cases}$$

To apply the second shifting theorem, we must write $g(t)$ as a function, or perhaps sum of functions, of the form $f(t-2)H(t-2)$. To do this, first write $t^2 + 1$ as a function of $t - 2$:

$$t^2 + 1 = (t-2+2)^2 + 1 = (t-2)^2 + 4(t-2) + 5.$$

Then

$$g(t) = H(t-2)(t^2+1)$$
$$= H(t-2)(t-2)^2 + 4H(t-2)(t-2) + 5H(t-2).$$

Now apply the second shifting theorem to each term on the right:

$$\mathcal{L}[g] = \mathcal{L}[H(t-2)(t-2)^2] + 4\mathcal{L}[H(t-2)(t-2)] + 5\mathcal{L}[H(t-2)]$$
$$= e^{-2s}\mathcal{L}[t^2] + 4e^{-2s}\mathcal{L}[t] + 5e^{-2s}\mathcal{L}[1]$$
$$= e^{-2s}\left[\frac{2}{s^3} + \frac{4}{s^2} + \frac{5}{s}\right]. \quad \blacklozenge$$

As usual, any formula for $\mathcal{L}$ can be read as a formula for $\mathcal{L}^{-1}$. The inverse version of the second shifting theorem is

$$\mathcal{L}^{-1}[e^{-as}F(s)](t) = H(t-a)f(t-a). \tag{3.7}$$

This enables us to compute the inverse transform of a known transformed function that is multiplied by an exponential e^{-as}.

EXAMPLE 3.10

Compute

$$\mathcal{L}^{-1}\left[\frac{se^{-3s}}{s^2+4}\right].$$

The presence of e^{-3s} suggests the use of equation (3.7). From Table 3.1 we read that

$$\mathcal{L}^{-1}\left[\frac{s}{s^2+4}\right]=f(t)=\cos(2t).$$

Then

$$\mathcal{L}^{-1}\left[\frac{se^{-3s}}{s^2+4}\right](t)=H(t-3)\cos(2(t-3)).\ \blacklozenge$$

EXAMPLE 3.11

Solve the initial value problem

$$y''+4y=f(t);\ y(0)=y'(0)=0$$

where

$$f(t)=\begin{cases}0 & \text{for } t<3 \\ t & \text{for } t\geq 3.\end{cases}$$

First, apply $\mathcal{L}$ to the differential equation using equations (3.1) and (3.3):

$$\mathcal{L}[y'']+4\mathcal{L}[y]=[s^2-sy(0)-y'(0)]Y(s)+4Y(s)$$
$$=s^2Y(s)+4Y(s)=(s^2+4)Y(s)=\mathcal{L}[f].$$

To compute $\mathcal{L}[f]$, use the second shifting theorem. Since $f(t)=H(t-3)t$, we can write

$$\mathcal{L}[f]=\mathcal{L}[H(t-3)t]$$
$$=\mathcal{L}[H(t-3)(t-3+3)]$$
$$=\mathcal{L}[H(t-3)(t-3)]+3\mathcal{L}[H(t-3)]$$
$$=\frac{e^{-3s}}{s^2}+\frac{3e^{-3s}}{s}.$$

In summary, we have

$$(s^2+4)Y(s)=\frac{1}{s^2}e^{-3s}+\frac{3}{s}e^{-3s}=\frac{3s+1}{s^2}e^{-3s}.$$

The transform of the solution is therefore

$$Y(s)=\frac{3s+1}{s^2(s^2+4)}e^{-3s}.$$

The solution is the inverse transform of $Y(s)$. To take this inverse, use a partial fractions decomposition, writing

$$\frac{3s+1}{s^2(s^2+4)}=\frac{A}{s}+\frac{B}{s^2}+\frac{Cs+D}{s^2+4}.$$

After solving for A,B,C, and D, we obtain

$$Y(s)=\frac{3s+1}{s^2(s^2+4)}e^{-3s}$$
$$=\frac{3}{4}\frac{1}{s}e^{-3s}-\frac{3}{4}\frac{s}{s^2+4}e^{-3s}+\frac{1}{4}\frac{1}{s^2}e^{-3s}-\frac{1}{4}\frac{1}{s^2+4}e^{-3s}.$$

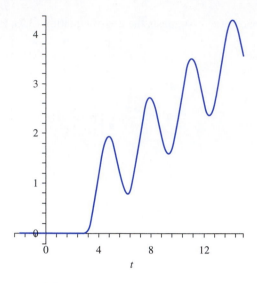

FIGURE 3.10   *Graph of the solution in Example 3.11.*

Now apply the second shifting theorem to write the solution

$$y(t) = \frac{3}{4}H(t-3) - \frac{3}{4}H(t-3)\cos(2(t-3))$$

$$+ \frac{1}{4}H(t-3)(t-3) - \frac{1}{8}H(t-3)\sin(2(t-3)).$$

This solution is 0 until time $t = 3$. Since $H(t-3) = 1$ for $t \geq 3$, then for these times,

$$y(t) = \frac{3}{4} - \frac{3}{4}\cos(2(t-3)) + \frac{1}{4}(t-3)$$

$$- \frac{1}{8}\sin(2(t-3)).$$

In summary,

$$y(t) = \begin{cases} 0 & \text{for } t < 3 \\ \frac{1}{8}[2t - 6\cos(2(t-3)) - \sin(2(t-3))] & \text{for } t \geq 3. \end{cases}$$

Figure 3.10 shows part of the graph of this solution.   ◆

EXAMPLE 3.12

Sometimes we encounter a function having several jump discontinuities. Here is an example of writing such a function in terms of step functions. Let

$$f(t) = \begin{cases} 0 & \text{for } t < 2 \\ t - 1 & \text{for } 2 \leq t < 3 \\ -4 & \text{for } t \geq 3. \end{cases}$$

Figure 3.11 shows a graph of f. There are jump discontinuities of magnitude 1 at $t = 2$ and magnitude 6 at $t = 3$.

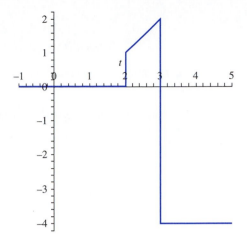

FIGURE 3.11 $f(t)$ *in Example 3.12.*

Think of $f(t)$ as consisting of two nonzero parts: the part that is $t - 1$ for $2 \le t < 3$ and the part that is -4 for $t \ge 3$. We want to turn on $t - 1$ at time $t = 2$ and turn it off at time $t = 3$, then turn -4 on at time $t = 3$ and leave it on.

The first effect is achieved by multiplying $t - 1$ by the pulse $H(t - 2) - H(t - 3)$. The second is achieved by multiplying -4 by $H(t - 3)$. Thus, write

$$f(t) = [H(t - 2) - H(t - 3)](t - 1) - 4H(t - 3). \quad \blacklozenge$$

EXAMPLE 3.13

Suppose the capacitor in the circuit of Figure 3.12 initially has a charge of zero, and that there is no initial current. At time $t = 2$ seconds, the switch is thrown from position B to A, held there for 1 second, and then switched back to B. We want the output voltage E_{out} on the capacitor.

From the circuit, write

$$E(t) = 10[H(t - 2) - H(t - 3)].$$

By Kirchhoff's voltage law,

$$Ri(t) + \frac{1}{C}q(t) = E(t),$$

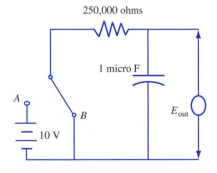

FIGURE 3.12 *The circuit of Example 3.13.*

or

$$250,000q'(t) + 10^6 q(t) = E(t).$$

We want to solve for $q(t)$ subject to the condition $q(0) = 0$. Take the Laplace transform of the differential equation to get

$$250,000[s\,Q(s) - q(0)] + 10^6 Q(s) = \mathcal{L}[E(t)].$$

Now

$$\mathcal{L}[E(t)](s) = 10\mathcal{L}[H(t-2)](s) - 10\mathcal{L}[(t-3)](s)$$

$$= \frac{10}{s}e^{-2s} - \frac{10}{s}e^{-3s}.$$

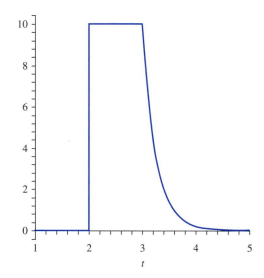

FIGURE 3.13 $E_{\text{out}}(t)$ *in Example 3.13.*

Now we have an equation for Q:

$$2.5(10^5)s\,Q(s) + 10^6 Q(s) = \frac{10}{s}e^{-2s} - \frac{10}{s}e^{-3s}.$$

Then

$$Q(s) = 4(10^{-5})\frac{1}{s(s+4)}e^{-2s} - 4(10^{-5})\frac{1}{s(s+4)}e^{-3s}.$$

Use a partial fractions decomposition to write

$$Q(s) = 10^{-5}\left[\frac{1}{s}e^{-2s} - \frac{1}{s+4}e^{-2s}\right] - 10^{-5}\left[\frac{1}{s}e^{-3s} - \frac{1}{s+4}e^{-3s}\right].$$

Apply the second shifting theorem to obtain

$$q(t) = 10^{-5}H(t-2)[1 - e^{-4(t-2)}] - 10^{-5}H(t-3)[1 - e^{-4(t-3)}].$$

Finally, the output voltage is $E_{\text{out}}(t) = 10^6 q(t)$. Figure 3.13 shows a graph of $E_{\text{out}}(t)$. ◆

SECTION 3.3   *PROBLEMS*

In each of Problems 1 through 12, find the Laplace transform of the function.

1. $(t^3 - 3t + 2)e^{-2t}$

2. $e^{-3t}(t - 2)$

3. $f(t) = \begin{cases} 1 & \text{for } 0 \le t < 7 \\ \cos(t) & \text{for } t \ge 7 \end{cases}$

4. $e^{-4t}(t - \cos(t))$

5. $f(t) = \begin{cases} t & \text{for } 0 \le t < 3 \\ 1 - 3t & \text{for } t \ge 3 \end{cases}$

6. $f(t) = \begin{cases} 2t - \sin(t) & \text{for } 0 \le t < \pi \\ 0 & \text{for } t \ge \pi \end{cases}$

7. $e^{-t}(1 - t^2 + \sin(t))$

8. $f(t) = \begin{cases} t^2 & \text{for } 0 \le t < 2 \\ 1 - t - 3t^2 & \text{for } t \ge 2 \end{cases}$

9. $f(t) = \begin{cases} \cos(t) & \text{for } 0 \le t < 2\pi \\ 2 - \sin(t) & \text{for } t \ge 2\pi \end{cases}$

10. $f(t) = \begin{cases} -4 & \text{for } 0 \le t < 1 \\ 0 & \text{for } 1 \le t < 3 \\ e^{-t} & \text{for } t \ge 3 \end{cases}$

11. $te^{-t}\cos(3t)$

12. $e^t(1 - \cosh(t))$

In each of Problems 13 through 20, find the inverse Laplace transform.

13. $\dfrac{1}{s^2 - 4s + 5}$

14. $\dfrac{1}{s^2 + 4s + 12}$

15. $\dfrac{e^{-2s}}{s^2 + 9}$

16. e^{-5s}/s^3

17. $\dfrac{1}{s^2 + 6s + 7}$

18. $\dfrac{3}{s + 2}e^{-4s}$

19. $\dfrac{s + 2}{s^2 + 6s + 1}$

20. $\dfrac{s - 4}{s^2 - 8s + 10}$

In each of Problems 21 through 26, solve the initial value problem.

21. $y'' + 4y = f(t); \ y(0) = 1, \ y'(0) = 0$, with

$$f(t) = \begin{cases} 0 & \text{for } 0 \le t < 4 \\ 3 & \text{for } t \ge 4 \end{cases}$$

22. $y'' - 2y' - 3y = f(t); \ y(0) = 1, \ y'(0) = 0$, with

$$f(t) = \begin{cases} 0 & \text{for } 0 \le t < 4 \\ 12 & \text{for } t \ge 4 \end{cases}$$

23. $y''' - 8y = g(t); \ y(0) = y'(0) = y''(0) = 0$, with

$$g(t) = \begin{cases} 0 & \text{for } 0 \le t < 6 \\ 2 & \text{for } t \ge 6 \end{cases}$$

24. $y'' + 5y' + 6y = f(t); \ y(0) = y'(0) = 0$, with

$$f(t) = \begin{cases} -2 & \text{for } 0 \le t < 3 \\ 0 & \text{for } t \ge 3 \end{cases}$$

25. $y''' - y'' + 4y' - 4y = 0; \ y(0) = y'(0) = 0, \ y''(0) = 1$, with

$$f(t) = \begin{cases} 1 & \text{for } 0 \le t < 5 \\ 2 & \text{for } t \ge 5 \end{cases}$$

26. $y'' - 4y' + 4y = f(t); \ y(0) = -2, \ y'(0) = 1$, with

$$f(t) = \begin{cases} t & \text{for } 0 \le t < 3 \\ t + 2 & \text{for } t \ge 3 \end{cases}$$

27. Determine the output voltage in the circuit of Figure 3.12, assuming that at time zero the capacitor is charged to a potential of 5 volts and the switch is opened at time zero and closed 5 seconds later. Graph this output.

28. Determine the output voltage in the RL circuit of Figure 3.14 if the current is initially zero and

$$E(t) = \begin{cases} 0 & \text{for } 0 \le t < 5 \\ 2 & \text{for } t \ge 5. \end{cases}$$

Graph this output function.

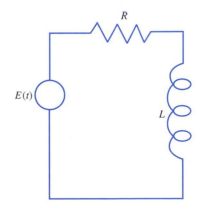

FIGURE 3.14 *The RL circuit of Problem 34.*

29. Solve for the current in the RL circuit of Problem 28 if the current is initially zero and

$$E(t) = \begin{cases} k & \text{for } 0 \leq t < 5 \\ 0 & \text{for } t \geq 5. \end{cases}$$

30. Determine $\mathcal{L}[e^{-2t}\int_0^t e^{2w}\cos(3w)\,dw]$. *Hint*: Use the first shifting theorem.

3.4   Convolution

If $f(t)$ and $g(t)$ are defined for $t \geq 0$, then the *convolution* $f * g$ of f with g is a function $f * g$, where

$$(f * g)(t) = \int_0^t f(t-\tau)g(\tau)d\tau$$

for $t \geq 0$ such that this integral is defined.

The rationale for this definition is that (in general) the transform of a product of functions does not equal the product of their transforms. However, the transform of a convolution is the product of the transforms. This fact is called the *convolution theorem*.

THEOREM 3.5   *The Convolution Theorem*

$$\mathcal{L}[f * g] = \mathcal{L}[f]\mathcal{L}[g].$$

Equivalently,

$$\mathcal{L}[f * g](s) = F(s)G(s).$$

The inverse transform version of the convolution theorem is

$$\mathcal{L}^{-1}[FG] = f * g. \tag{3.8}$$

This states that the inverse transform of a product of two functions $F(s)$ and $G(s)$ is the convolution $f * g$ of the inverse transforms of the functions. This fact is sometimes useful in computing an inverse transform. ◆

EXAMPLE 3.14

Compute

$$\mathcal{L}^{-1}\left[\frac{1}{s(s-4)^2}\right].$$

Certainly, we can do this by a partial fractions decomposition. To illustrate the use of the convolution, however, write

$$F(s) = \frac{1}{s} \text{ and } G(s) = \frac{1}{(s-4)^2}$$

so we are computing the inverse transform of a product. By the convolution theorem,

$$\mathcal{L}^{-1} \left[\frac{1}{s(s-4)^2} \right] = f * g,$$

where

$$f(t) = \mathcal{L}^{-1} \left[\frac{1}{s} \right] = 1$$

and

$$g(t) = \mathcal{L}^{-1} \left[\frac{1}{(s-4)^2} \right] = te^{4t}.$$

Then

$$\mathcal{L}^{-1} \left[\frac{1}{s(s-4)^2} \right] = f(t) * g(t)$$

$$= 1 * e^{4t} = \int_0^t \tau e^{4\tau} d\tau$$

$$= \frac{1}{4}te^{4t} - \frac{1}{16}e^{4t} + \frac{1}{16}. \quad \blacklozenge$$

Convolution is commutative:

$$f * g = g * f.$$

This can be proved by a simple change of variables in the integral defining the convolution.

In addition to its use in computing the inverse transform of products, convolution allows us to solve certain general initial value problems.

EXAMPLE 3.15

Solve the initial value problem

$$y'' - 2y' - 8y = f(t); \quad y(0) = 1, y'(0) = 0.$$

We want a formula for the solution that will hold for any "reasonable" forcing function f. Apply the Laplace transform to the differential equation in the usual way, obtaining

$$s^2 Y(s) - s - 2(sY(s) - 1) - 8Y(s) = F(s).$$

Then

$$(s^2 - 2s - 8)Y(s) = s - 2 + F(s).$$

Then

$$Y(s) = \frac{s-2}{s^2 - 2s - 8} + \frac{1}{s^2 - 2s - 8} F(s).$$

Factor $s^2 - 2s - 8 = (s-4)(s+2)$ and use a partial fractions decomposition to write

$$Y(s) = \frac{1}{3} \frac{1}{s-4} + \frac{2}{3} \frac{1}{s+2} + \frac{1}{6} \frac{1}{s-4} F(s) - \frac{1}{6} \frac{1}{s+2} F(s).$$

Now apply the inverse transform to obtain the solution

$$y(t) = \frac{1}{3}e^{4t} + \frac{2}{3}e^{-2t} + \frac{1}{6}e^{4t} * f(t) - \frac{1}{6}e^{-2t} * f(t)$$

which is valid for any function f for which these convolutions are defined. ◆

Convolution also enables us to solve some kinds of integral equations, which are equations in which the unknown function appears in an integral.

EXAMPLE 3.16

Solve for $f(t)$ in the integral equatiion

$$f(t) = 2t^2 + \int_0^t f(t-\tau)e^{-\tau}d\tau.$$

Recognize the integral on the right as the convolution of $f(t)$ with e^{-t}. Therefore, the integral equation has the form

$$f(t) = 2t^2 + f(t) * e^{-t}.$$

Apply the Laplace transform and the convolution theorem to this equation to get

$$F(s) = \frac{4}{s^3} + \frac{1}{s+1}F(s).$$

Then

$$F(s) = \frac{4}{s^3} + \frac{4}{s^4},$$

which we invert to obtain

$$f(t) = 2t^2 + \frac{2}{3}t^3. \quad ◆$$

SECTION 3.4   PROBLEMS

In each of Problems 1 through 8, use the convolution theorem to help compute the inverse Laplace transform of the function. Wherever they occur, a and b are positive constants.

1. $\dfrac{1}{(s^2+4)(s^2-4)}$

2. $\dfrac{1}{s^2+16}e^{-4s}$

3. $\dfrac{s}{(s^2+a^2)(s^2+b^2)}$

4. $\dfrac{s^2}{(s-3)(s^2+5)}$

5. $\dfrac{1}{s(s^2+a^2)^2}$

6. $\dfrac{1}{s^4(s-5)}$

7. $\dfrac{1}{s(s+2)}e^{-4s}$

8. $\dfrac{2}{s^2(s^2+5)}$

In each of Problems 9 through 16, use the convolution theorem to write a formula for the solution in terms of f.

9. $y'' - 5y' + 6y = f(t); y(0) = y'(0) = 0$

10. $y'' + 10y' + 24y = f(t); y(0) = 1, y'(0) = 0$

11. $y'' - 8y' + 12y = f(t); y(0) = -3, y'(0) = 2$

12. $y'' - 4y' - 5y = f(t); y(0) = 2, y'(0) = 1$

13. $y'' + 9y = f(t); y(0) = -1, y'(0) = 1$

14. $y'' - k^2y = f(t); y(0) = 2, y'(0) = -4$

15. $y^{(3)} - y'' - 4y' + 4y = f(t)$; $y(0) = y'(0) = 1$, $y''(0) = 0$

16. $y^{(4)} - 11y'' + 18y = f(t)$; $y(0) = y'(0) = y''(0) = y^{(3)}(0) = 0$

In each of Problems 17 through 23, solve the integral equation.

17. $f(t) = -1 + \int_0^t f(t - \tau)e^{-3\tau} d\tau$

18. $f(t) = -t + \int_0^t f(t - \tau) \sin(\tau) d\tau$

19. $f(t) = e^{-t} + \int_0^t f(t - \tau) d\tau$

20. $f(t) = -1 + t - 2 \int_0^t f(t - \tau) \sin(\tau) d\tau$

21. $f(t) = 3 + \int_0^t f(\tau) \cos(2(t - \tau)) d\tau$

22. $f(t) = \cos(t) + e^{-2t} \int_0^t f(\tau)e^{2\tau} d\tau$

3.5   Impulses and the Dirac Delta Function

Informally, an *impulse* is a force of extremely large magnitude applied over a very short period of time (imagine hitting your thumb with a hammer). We can model this idea as follows. First, for any positive number ϵ consider the pulse δ_ϵ defined by

$$\delta_\epsilon(t) = \frac{1}{\epsilon}[H(t - \epsilon) - H(t)].$$

This pulse, which is graphed in Figure 3.15, has a magnitude (height) of $1/\epsilon$ and a duration of ϵ. The *Dirac delta function* is thought of as a pulse of infinite magnitude over an infinitely short duration and is defined to be

$$\delta(t) = \lim_{\epsilon \to 0+} \delta_\epsilon(t).$$

This is not a function in the conventional sense but is a more general object called a *distribution*. For historical reasons, it continues to be known as the Dirac function after the Nobel laureate physicist P.A.M. Dirac. The *shifted delta function* $\delta(t - a)$ is zero except for $t = a$, where it has an infinite spike.

To take the Laplace transform of the delta function begin with

$$\delta_\epsilon(t - a) = \frac{1}{\epsilon}[H(t - a) - H(t - a - \epsilon)].$$

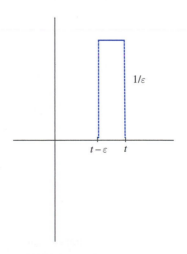

FIGURE 3.15 $\delta_\epsilon(t)$.

This has transform

$$\mathcal{L}[\delta_\epsilon(t-a)] = \frac{1}{\epsilon}\left[\frac{1}{s}e^{-as} - \frac{1}{s}e^{-(a+\epsilon)s}\right]$$
$$= \frac{e^{-as}(1-e^{-\epsilon s})}{\epsilon s}.$$

This suggests that we define

$$\mathcal{L}[\delta(t-a)] = \lim_{\epsilon \to 0+} \frac{e^{-as}(1-e^{-\epsilon s})}{\epsilon s} = e^{-as}.$$

In particular, we can choose $a = 0$ to get

$$\mathcal{L}[\delta(t)] = 1.$$

The following result is called the *filtering property* of the delta function. Suppose at time $t = a$ a signal is impacted with an impulse by mutliplying the signal by $\delta(t-a)$, and the resulting signal is then summed over all positive time by integrating it from zero to infinity. We claim that this yields exactly the value $f(a)$ of the signal at time a.

THEOREM 3.6   *Filtering Property of the Delta Function*

Let $a > 0$ and let $\int_0^\infty f(t)\,dt$ converge. Suppose also that f is continuous at a. Then

$$\int_0^\infty f(t)\delta(t-a)\,dt = f(t). \quad \blacklozenge$$

If we apply the filtering property to $f(t) = e^{-st}$, we get

$$\int_0^\infty e^{-st}\delta(t-a)\,dt = e^{-as}$$

consistent with the definition of the Laplace transform of the delta function. Now change the notation in the filtering property and write it as

$$\int_0^\infty f(\tau)\delta(t-\tau)\,d\tau = f(t),$$

and we recognize the convolution of f with δ. The last equation becomes

$$f * \delta = f.$$

The delta function therefore acts as an identity for the "product" defined by convolution.

In solving an initial value problem involving the delta function by using the Laplace transform, we proceed as we have been doing, except that now we must use the transform of the delta function.

EXAMPLE 3.17

We will solve

$$y'' + 2y' + 2y = \delta(t-3); \quad y(0) = y'(0) = 0.$$

Apply the transform to the differential equation to get

$$s^2 Y(s) + 2sY(s) + 2Y(s) = e^{-3s}$$

so

$$Y(s) = \frac{1}{s^2 + 2s + 2}e^{-3s}.$$

The solution is the inverse transform of $Y(s)$. To compute this, first write

$$Y(s) = \frac{1}{(s+1)^2 + 1}e^{-3s}.$$

Because $\mathcal{L}^{-1}[1/(s^2 + 1)] = \sin(t)$, a shift in the s–variable gives us

$$\mathcal{L}^{-1}\left[\frac{1}{(s+1)^2 + 1}\right] = e^{-t}\sin(t).$$

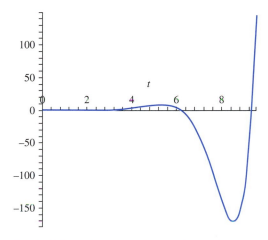

FIGURE 3.16 *Graph of the solution in Example 3.17.*

Now shift in the t–variable to obtain

$$y(t) = H(t - 3)e^{(t-3)}\sin(t - 3).$$

Figure 3.16 is a graph of this solution. ◆

Transients can be generated in a circuit during switching, and the high input voltages associated with them can create excessive current in the components, damaging the circuit. Transients can also be harmful, because they contain a broad spectrum of frequencies. Introducing a transient into a circuit can therefore force the circuit with a range of frequencies, and if one of these is near the natural frequency of the system, resonance can occur, resulting in oscillations large enough to cause damage. For this reason, engineers sometimes use a delta function to model a transient and study its effect on a circuit being designed.

EXAMPLE 3.18

Suppose the current and charge on the capacitor in the circuit of Figure 3.17 are zero at time $t = 0$. We want to describe the output voltage response to a transient modeled by $\delta(t)$.

The output voltage is $q(t)/C$ so we will determine $q(t)$. By Kirchhoff's voltage law,

$$Li' + Ri + \frac{1}{C}q = i' + 10i + 100q = \delta(t).$$

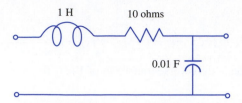

FIGURE 3.17   *Circuit of Example 3.18 with*
$E_{in}(t) = \delta(t)$.

Since $i' = q$, then

$$q'' + 10q' + 100q = \delta(t).$$

Assume the initial conditions $q(0) = q'(0) = 0$. Apply the transform to the initial value problems to get

$$s^2 Q(s) + 10s\,Q(s) + 100Q(s) = 1.$$

Then

$$Q(s) = \frac{1}{s^2 + 10s + 100} = \frac{1}{(s+5)^2 + 75}.$$

The last expression is preparation for shifting in the s–variable. Since

$$\mathcal{L}^{-1}\left[\frac{1}{s^2 + 75}\right] = \frac{1}{5\sqrt{3}}\sin(5\sqrt{3}t)$$

then

$$q(t) = \mathcal{L}^{-1}\left[\frac{1}{(s+5)^2 + 75}\right] = \frac{1}{5\sqrt{3}}e^{-5t}\sin(5\sqrt{3}t).$$

The output voltage is

$$\frac{1}{C}q(t) = 100q(t) = \frac{20}{\sqrt{3}}e^{-5t}\sin(5\sqrt{3}t).$$

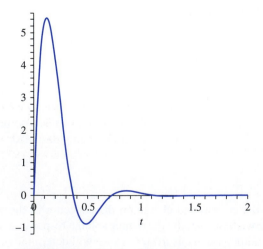

FIGURE 3.18   *Graph of the output voltage in Example 3.18.*

A graph of this output voltage is given in Figure 3.18. The circuit output displays damped oscillations at its natural frequency, even though it was not explicitly forced by oscillation of this frequency. ◆

In each of Problems 1 through 5, solve the initial value problem and graph the solution.

1. $y'' + 5y' + 6y = 3\delta(t-2) - 4\delta(t-5)$; $y(0) = y'(0) = 0$

2. $y'' - 4y' + 13y = 4\delta(t-3)$; $y(0) = y'(0) = 0$

3. $y''' + 4y'' + 5y' + 2y = 6\delta(t)$; $y(0) = y'(0) = y''(0) = 0$

4. $y'' + 16y = 12\delta(t - 5\pi/8)$; $y(0) = 3$, $y'(0) = 0$

5. $y'' + 5y' + 6y = B\delta(t)$; $y(0) = 3$, $y'(0) = 0$

6. An object of mass m is attached to the lower end of a spring of modulus k. Assume that there is no damping. Derive and solve an equation of motion for the object, assuming that at time $t = 0$ it is pushed down from the equilibrium position with an initial velocity v_0. With

what momentum does the object leave the equilibrium position?

7. Suppose, in the setting of Problem 6, the object is struck a downward blow of magnitude mv_0 at time $t = 0$. How does the position of this object compare with that of the object in Problem 6 at any positive time t?

8. A 2-pound weight is attached to the lower end of a spring, stretching it 8/3 inches. The weight is allowed to come to rest in the equilibrium position. At some later time, which we call time $t = 0$, the weight is struck a downward blow of magnitude 1/4 pound (an impulse). Assume no damping in the system. Determine the velocity with which the weight leaves the equilibrium position as well as the frequency and magnitude of the oscillations.

Appendix on Partial Fractions Decompositions

Partial fractions decomposition is an algebraic manipulation designed to write a quotient $P(x)/Q(x)$ of polynomials as a sum of simpler quotients, where simpler will be defined by the process.

Let P have degree m, let Q have degree n, and assume that $n > m$. If this is not the case, then divide Q into P. Assume that P and Q have no common roots and that Q has been completely factored into linear and/or irreducible quadratic factors. A factor is irreducible quadratic if it is second degree with complex roots, hence it cannot be factored into linear factors with real coefficients. An example of an irreducible quadratic factor is $x^2 + 4$.

The partial fractions decomposition consisting of writing $P(x)/Q(x)$ as a sum $S(x)$ of simpler quotients is as follows.

Rule 1   If $x - a$ is a factor of $Q(x)$ but $(x - a)^2$ is not, then include in $S(x)$ a term of the form

$$\frac{A}{x - a}.$$

Rule 2   If $(x - a)^k$ is a factor of $Q(x)$ with $k > 1$, but $(x - a)^{k+1}$ is not a factor, then include in $S(x)$ a sum of terms of the form

$$\frac{B_1}{x - a} + \frac{B_2}{(x - a)^2} + \cdots + \frac{B_k}{(x - a)^k}.$$

Rule 3   If $ax^2 + bx + c$ is an irreducible quadratic factor of $Q(x)$, but no higher power is a factor of $Q(x)$, then include in $S(x)$ a term of the form

$$\frac{Cx + D}{ax^2 + bx + c}.$$

Rule 4   If $(ax^2 + bx + c)^k$ is a product of irreducible factors of $Q(x)$, but $(ax^2 + bx + c)^{k+1}$ is not a factor of $Q(x)$, then include in $S(x)$ a sum of terms of the form

$$\frac{C_1 x + D_1}{ax^2 + bx + c} + \frac{C_2 x + D_2}{(ax^2 + bx + c)^2} + \cdots + \frac{C_k x + D_k}{(ax^2 + bx + c)^k}.$$

When each factor of $Q(x)$ has contributed one or more terms to $S(x)$ according to these rules, we have an expression of the form

$$\frac{P(x)}{Q(x)} = S(x)$$

with the coefficients to be determined. One way to do this is to add the terms in $S(x)$, set the numerator of the resulting quotient equal to $P(x)$, which is known, and solve for the coefficients of the terms in $S(x)$ by equating coefficients of like powers of x.

EXAMPLE 3.19

We will decompose

$$\frac{2x - 1}{x^3 + 6x^2 + 5x - 12}$$

into a sum of simpler fractions. First factor the denominator, then use Rules 1 through 4 to write the form of a partial fractions decomposition as

$$\frac{2x - 1}{x^3 + 6x^2 + 5x - 12} = \frac{2x - 1}{(x - 1)(x + 3)(x + 4)}$$

$$= \frac{A}{x - 1} + \frac{B}{x + 3} + \frac{C}{x + 4}.$$

If the fractions on the right are added, the numerator of the resulting quotient must equal $2x - 1$, which is the numerator of the original quotient. Therefore,

$$A(x + 3)(x + 4) + B(x - 1)(x + 4) + C(x - 1)(x + 3) = 2x - 1.$$

There are at least two ways we can find A, B, and C.

Method 1   Multiply the factors on the left and collect the coefficients of each power of x to write

$$A(x^2 + 7x + 12) + B(x^2 + 3x - 4) + C(x^2 + 2x - 3)$$

$$= (A + B + C)x^2 + (7A + 3B + 2C)x + (12A - 4B - 3C) = 2x - 1.$$

Equate the coefficient of each power of x on the left to the coefficient of that power of x on the right, obtaining a system of three linear equations in three unknowns:

$$A + B + C = 0 \text{ from the coefficients of } x^2$$

$$7A + 3B + 2C = 2 \text{ from the coefficients of } x$$

$$12A - 4B - 3C = -1 \text{ from the constant term.}$$

Solve these three equations, obtaining $A = 1/20$, $B = 7/4$, and $C = -9/5$. Then

$$\frac{2x-1}{x^3+6x^2+5x-12} = \frac{1}{20}\frac{1}{x-1} + \frac{7}{4}\frac{1}{x+3} - \frac{9}{5}\frac{1}{x+4}.$$

Method 2 Begin with

$$A(x+3)(x+4) + B(x-1)(x+4) + C(x-1)(x+3) = 2x - 1$$

and assign values of x that make it easy to determine A, B, and C. Put $x = 1$ to get $20A = 1$, so $A = 1/20$. Put $x = -3$ to get $-4B = -7$, so $B = 7/4$. And put $x = -4$ to get $5C = -9$, so $C = -9/5$. This yields the same result as Method 1, but in this example, Method 2 is probably easier and quicker. ◆

EXAMPLE 3.20

Decompose

$$\frac{x^2+2x+3}{(x^2+x+5)(x-2)^2}$$

into partial fractions. First, observe that $x^2 + 2x + 3$ has complex roots and so is irreducible. Thus, use the form

$$\frac{x^2+2x+3}{(x^2+x+5)(x-2)^2} = \frac{A}{x-2} + \frac{B}{(x-2)^2} + \frac{Cx+D}{x^2+x+5}.$$

If we add the fractions on the right, the numerator must equal $x^2 + 2x + 3$. Therefore,

$$A(x-2)(x^2+x+5) + B(x^2+x+5) + (Cx+D)(x-2)^2 = x^2+2x+3.$$

Second, expand the left side and collect terms to write this equation as

$$(A+C)x^3 + (-A+B-4C+D)x^2 + (3A+B+4C-4D)x - 10A + 5B + 4D$$
$$= x^2 + 2x + 3.$$

Equate coefficients of like powers of x to get

$$A + C = 0$$
$$-A + B - 4C + D = 1$$
$$3A + B + 4C - 4D = 2$$
$$-10A + 5B + 4D = 3.$$

Solve these to obtain $A = 1/11$, $B = 1$, $C = -1/11$, and $D = -3/11$. The partial fractions decomposition is

$$\frac{x^2+2x+3}{(x^2+x+5)(x-2)^2} = \frac{1}{11}\frac{1}{x-2} + \frac{1}{(x-2)^2} - \frac{1}{11}\frac{x+3}{x^2+x+5}. ◆$$

CHAPTER 4

Series Solutions

Sometimes we can solve an initial value problem explicitly. For example, the problem

$$y' + 2y = 1; \, y(0) = 3$$

has the unique solution

$$y(x) = \frac{1}{2}(1 + 5e^{-2x}).$$

This solution is in *closed form*, which means that it is a finite algebraic combination of elementary functions (such as polynomials, exponentials, sines and cosines, and the like).

We may, however, encounter problems for which there is no closed form solution. For example,

$$y'' + e^x y = x^2; \; y(0) = 4$$

has the unique solution

$$y(x) = e^{-e^x} \int_0^x \xi^2 e^{e^\xi} \, d\xi + 4e^{-e^x},$$

This solution, while explicit, has no elementary, closed form expression.

In such a case, we might try a numerical approximation. However, we also may be able to write a series solution that contains useful information. In this chapter, we will deal with two kinds of series solutions: power series (this section) and Frobenius series (the next section).

4.1   Power Series Solutions

A function f is called *analytic* at x_0 if $f(x)$ has a power series representation in some interval about x_0. That is, for some positive h,

$$f(x) = \sum_{n=0}^{\infty} a_n (x - x_0)^n$$

for $x_0 - h < x < x_0 + h$.

The numbers a_n are the coefficients in this power series representation.

It can be shown that a linear first-order equation

$$y' + p(x)y = f(x)$$

has analytic solutions about a point x_0 if p and f are analytic at x_0. A linear second-order equation

$$y'' + p(x)y' + q(x)y = f(x)$$

also has analytic solutions about x_0 if p, q and f are analytic at x_0. These results extend to higher-order linear differential equations with analytic coefficients.

We are therefore justified in seeking power series solutions of linear equations having analytic coefficients. This strategy may be carried out by substituting $y = \sum_{n=0}^{\infty} a_n (x - x_0)^n$ into the differential equation and attempting to solve for the a_n's. We will illustrate this process.

EXAMPLE 4.1

We will find power series solutions of

$$y'' + x^2 y = 0$$

expanded about $x_0 = 0$. Substitute $y = \sum_{n=0}^{\infty} a_n x^n$ into the differential equation. This will require that we compute

$$y' = \sum_{n=1}^{\infty} n a_n x^{n-1} \text{ and } y'' = \sum_{n=2}^{\infty} (n-1) n a_n x^{n-2}.$$

Notice that the series for y' begins at $n = 1$, because the derivative of the constant term a_0 is zero. Similarly, the series for y'' begins at $n = 2$.

Substitute the power series into the differential equation to obtain

$$\sum_{n=2}^{\infty} (n-1) n a_n x^{n-2} + x^2 \sum_{n=0}^{\infty} a_n x^n = 0,$$

or

$$\sum_{n=2}^{\infty} n(n-1) a_n x^{n-2} + \sum_{n=0}^{\infty} a_n x^{n+2} = 0. \tag{4.1}$$

We will shift indices in each summation so that the power of x in both summations is the same, allowing us to combine terms from both summations. One way to do this is to write

$$\sum_{n=2}^{\infty} (n-1) n a_n x^{n-2} = \sum_{n=0}^{\infty} (n+2)(n+1) a_{n+2} x^n$$

and

$$\sum_{n=0}^{\infty} a_n x^{n+2} = \sum_{n=2}^{\infty} a_{n-2} x^n.$$

Now equation (4.1) is

$$\sum_{n=0}^{\infty}(n+2)(n+1)a_{n+2}x^n + \sum_{n=2}^{\infty}a_{n-2}x^n = 0.$$

We can combine the terms for $n \geq 2$ in one summation. This requires that we write the $n = 0$ and $n = 1$ terms in the last equation separately, or else we lose terms:

$$2a_2x^0 + 2(3)a_3x + \sum_{n=2}^{\infty}[(n+2)(n+1)a_{n+2} + a_{n-2}]x^n = 0.$$

The left side can be zero for all x in some interval $(-h, h)$ only if the coefficient of each power of x is zero:

$$a_2 = a_3 = 0$$

and

$$(n+2)(n+1)a_{n+2} + a_{n-2} = 0 \text{ for } n \geq 2.$$

The last equation gives us

$$a_{n+2} = -\frac{1}{(n+2)(n+1)}a_{n-2} \text{ for } n = 2, 3, \cdots. \tag{4.2}$$

This is a *recurrence relation* for the coefficients of the series solution. It gives us a_4 in terms of a_0, a_5 in terms of a_1, and so on. Recurrence relations always give a coefficient in terms of one or more previous coefficients, allowing us to generate as many terms of the series solution as we want. To illustrate, use $n = 2$ in equation (4.2) to obtain

$$a_4 = -\frac{1}{(4)(3)}a_0 = -\frac{1}{12}a_0.$$

With $n = 3$,

$$a_5 = -\frac{1}{(5)(4)}a_1 = -\frac{1}{20}a_1.$$

In turn, we obtain

$$a_6 = -\frac{1}{(6)(5)}a_2 = 0$$

because $a_2 = 0$;

$$a_7 = -\frac{1}{(7)(6)}a_3 = 0$$

because $a_3 = 0$;

$$a_8 = -\frac{1}{(8)(7)}a_4 = \frac{1}{(56)(12)}a_0 = \frac{1}{672}a_0,$$

$$a_9 = -\frac{1}{(9)(8)}a_5 = \frac{1}{(72)(20)}a_1 = \frac{1}{1440}a_1,$$

and so on. Thus far, we have the first few terms of the series solution about 0:

$$y(x) = a_0 + a_1 x + 0x^2 + 0x^3 - \frac{1}{12}a_0 x^4$$

$$- \frac{1}{20}a_1 x^5 + 0x^6 + 0x^7 + \frac{1}{672}x^8 + \frac{1}{1440}x^9 + \cdots$$

$$= a_0 \left(1 - \frac{1}{12}x^4 + \frac{1}{672}x^8 + \cdots \right)$$

$$+ a_1 \left(x - \frac{1}{20}x^5 + \frac{1}{1440}x^9 + \cdots \right).$$

This is actually the general solution, since a_0 and a_1 are arbitrary constants. Since $a_0 = y(0)$ and $a_1 = y'(0)$, a unique solution is determined by specifying these two constants. ♦

Sometimes we must expand one or more coefficients of the differential equation in a power series to apply the method.

EXAMPLE 4.2

Solve

$$y'' + xy' - y = e^{3x}.$$

Again, we will look for a series solution about zero. Substitute

$$y(x) = \sum_{n=0}^{\infty} a_n x^n \text{ and } e^{3x} = \sum_{n=0}^{\infty} \frac{3^n}{n!}x^n$$

to obtain

$$\sum_{n=2}^{\infty} n(n-1)a_n x^{n-2} + \sum_{n=1}^{\infty} na_n x^n - \sum_{n=0}^{\infty} a_n x^n = \sum_{n=0}^{\infty} \frac{3^n}{n!}x^n.$$

Shift indices in the first summation to write this equation as

$$\sum_{n=0}^{\infty} (n+2)(n+1)a_{n+2}x^n + \sum_{n=1}^{\infty} na_n x^n - \sum_{n=0}^{\infty} a_n x^n = \sum_{n=0}^{\infty} \frac{3^n}{n!}x^n.$$

Collect terms from $n=1$ into one summation and write $n=0$ terms separately to obtain

$$2a_2 - a_0 + \sum_{n=1}^{\infty}[(n+2)(n+1)a_{n+2} + (n-1)a_n]x^n = 1 + \sum_{n=1}^{\infty}\frac{3^n}{n!}x^n.$$

Equate coefficients of like powers of x on both sides of this equation to obtain

$$2a_2 - a_0 = 1$$

and

$$(n+2)(n+1)a_{n+2} + (n-1)a_n = \frac{3^n}{n!}.$$

Then

$$a_2 = \frac{1}{2}(1 + a_0)$$

and

$$a_{n+2} = \frac{\dfrac{3^n}{n!} - (n-1)a_n}{(n+2)(n+1)}.$$

This is the recurrence relationship for the current problem, and we can use it to generate as many coefficients in the series solution as we want. The first few terms are

$$y(x) = a_0 + a_1 x + \frac{1}{2}(1 + a_0)x^2 + \frac{1}{2}x^3$$

$$+ \left(\frac{1}{3} - \frac{a_0}{24}\right)x^4 + \frac{7}{40}x^5 + \frac{1}{30}\left(\frac{57}{24} + \frac{a_0}{8}\right)x^6 + \cdots. \quad \blacklozenge$$

SECTION 4.1   *PROBLEMS*

In each of Problems 1 through 10, find the recurrence relation and use it to generate the first five terms of a power series solution about 0.

1. $y' - xy = 1 - x$

2. $y' - x^3 y = 4$

3. $y' + (1 - x^2)y = x$

4. $y'' + 2y' + xy = 0$

5. $y'' - xy' + y = 3$

6. $y'' + xy' + xy = 0$

7. $y'' - x^2 y' + 2y = x$

8. $y'' + xy = \cos(x)$

9. $y'' + (1 - x)y' + 2y = 1 - x^2$

10. $y'' + xy' = 1 - e^x$

4.2   Frobenius Solutions

We will focus on the differential equation

$$P(x)y'' + Q(x)y' + R(x)y = F(x). \tag{4.3}$$

If $P(x) \neq 0$ on some interval, then we can divide by $P(x)$ to obtain the standard form

$$y'' + p(x)y' + q(x)y = f(x). \tag{4.4}$$

If $P(x_0) = 0$, we call x_0 a *singular point* of equation (4.3). This singular point is *regular* if

$$(x - x_0)\frac{Q(x)}{P(x)} \quad \text{and} \quad (x - x_0)^2 \frac{R(x)}{P(x)}$$

are analytic at x_0. A singular point that is not regular is an *irregular singular point*.

EXAMPLE 4.3

$$x^3(x-2)^2 y'' + 5(x+2)(x-2)y' + 3x^2 y = 0$$

has singular points at 0 and 2. Now

$$(x-0)\frac{Q(x)}{P(x)} = \frac{5x(x+2)(x-2)}{x^3(x-2)^2} = \frac{5}{x^2}\left(\frac{x+2}{x-2}\right)$$

is not analytic (or even defined) at 0, so 0 is an irregular singular point. But

$$(x-2)\frac{Q(x)}{P(x)} = \frac{5(x+2)}{x^3}$$

and

$$(x-2)^2 \frac{R(x)}{P(x)} = \frac{3}{x}$$

are both analytic at 2, so 2 is a regular singular point of this differential equation.   ◆

We will not treat the case of an irregular singular point. If equation (4.3) has a regular singular point at x_0, there may be no power series solution about x_0, but there will be a *Frobenius series* solution, which has the form

$$y(x) = \sum_{n=0}^{\infty} c_n (x - x_0)^{n+r}$$

with $c_0 \neq 0$. We must solve for the coefficients c_n and a number r to make this series a solution. We will look at an example to get some feeling for how this might proceed, and then examine the method more critically.

EXAMPLE 4.4

Zero is a regular singular point of

$$x^2 y'' + 5xy' + (x+4)y = 0.$$

Substitute $y = \sum_{n=0}^{\infty} c_n x^{n+r}$ to obtain

$$\sum_{n=0}^{\infty}(n+r)(n+r-1)c_n x^{n+r-2} + \sum_{n=0}^{\infty} 5(n+r)c_n x^{n+r}$$

$$+ \sum_{n=0}^{\infty} c_n x^{n+r+1} + \sum_{n=0}^{\infty} 4c_n x^{n+r} = 0.$$

Notice that the $n=0$ term in the proposed series solution is $c_0 x^r$, which is not constant if $c_0 \neq 0$, so the series for the derivatives begins with $n=0$, unlike what we saw with power series. Shift indices in the third summation to write this equation as

$$\sum_{n=0}^{\infty}(n+r)(n+r-1)c_n x^{n+r-2} + \sum_{n=0}^{\infty} 5(n+r)c_n x^{n+r}$$

$$+ \sum_{n=1}^{\infty} c_{n-1} x^{n+r} + \sum_{n=0}^{\infty} 4c_n x^{n+r} = 0.$$

Now we can combine terms to write

$$[r(r-1)+5r+4]c_0 x^r$$

$$+\sum_{n=1}^{\infty}[(n+r)(n+r-1)c_n+5(n+r)c_n+c_{n-1}+4c_n]x^{n+r}=0.$$

This will be satisfied if the coefficient of each power of x is zero. From the x^r term (assuming that we can have $c_0 \neq 0$), this gives us

$$r(r-1)+5r+4=0.$$

This is called the *indicial equation* and is used to solve for r, obtaining the repeated root $r=-2$. From the coefficient of x^{n+r} in the series, we obtain

$$(n+r)(n+r-1)c_n+5(n+r)c_n+c_{n-1}+4c_n=0,$$

or with $r=-2$,

$$(n-2)(n-3)c_n+5(n-2)c_n+c_{n-1}+4c_n=0.$$

From this, we obtain the recurrence relation

$$c_n=-\frac{1}{(n-2)(n-3)+5(n-2)+4}c_{n-1} \text{ for } n=1,2,\cdots.$$

This simplifies to

$$c_n=-\frac{1}{n^2}c_{n-1} \text{ for } n=1,2,\cdots.$$

Solve for some coefficients:

$$c_1=-c_0$$

$$c_2=-\frac{1}{4}c_1=\frac{1}{4}c_0=\frac{1}{2^2}c_0$$

$$c_3=-\frac{1}{9}c_2=-\frac{1}{(2\cdot3)^2}c_0$$

$$c_4=-\frac{1}{16}c_3=\frac{1}{(2\cdot3\cdot4)^2}c_0$$

and so on. In general, we see that

$$c_n=(-1)^n\frac{1}{(n!)^2}c_0$$

for $n=1,2,3,\cdots$. We have found the Frobenius solution

$$y(x)=c_0\left[x^{-2}-x^{-1}+\frac{1}{4}-\frac{1}{36}x+\frac{1}{576}x^2+\cdots\right]$$

$$=c_0\sum_{n-0}^{\infty}(-1)^n\frac{1}{(n!)^2}x^{n-2}$$

for $x \neq 0$. One can check that this series converges for all nonzero x. ◆

Usually, we cannot expect the recurrence equation for c_n to have such a simple form.

The example shows that an equation with a regular singular point may have only one Frobenius series solution about that point. A second, linearly independent solution is needed. The following theorem tells us how to produce two linearly independent solutions. For convenience, the statement is posed in terms of $x_0 = 0$.

THEOREM 4.1

Suppose 0 is a regular singular point of

$$P(x)y'' + Q(x)y' + R(x)y = 0.$$

Then

(1) The differential equation has a Frobenius solution

$$y(x) = \sum_{n=0}^{\infty} c_n x^{n+r}$$

with $c_0 \neq 0$. This series converges at least in some interval $(0, h)$ or $(-h, 0)$.

Suppose that the indicial equation has real roots r_1 and r_2 with $r_1 \geq r_2$. Then the following conclusions hold.

(2) If $r_1 - r_2$ is not a positive integer, then there are two linearly independent Frobenius solutions

$$y_1(x) = \sum_{n=0}^{\infty} c_n x^{n+r_1} \text{ and } y_2(x) = \sum_{n=0}^{\infty} c_n^* x^{n+r_2}$$

with $c_0 \neq 0$ and $c_0^* \neq 0$. These solutions are valid at least in an interval $(0, h)$ or $(-h, 0)$.

(3) If $r_1 - r_2 = 0$, then there is a Frobenius solution

$$y_1(x) = \sum_{n=0}^{\infty} c_n x^{n+r_1}$$

with $c_0 \neq 0$, and there is a second solution

$$y_2(x) = y_1 \ln(x) + \sum_{n=1}^{\infty} c_n^* x^{n+r_1}.$$

These solutions are linearly independent on some interval $(0, h)$.

(4) If $r_1 - r_2$ is a positive integer, then there is a Frobenius solution

$$y_1(x) = \sum_{n=0}^{\infty} c_n x^{n+r_1}.$$

with $c_0 \neq 0$, and there is a second solution

$$y_2(x) = k y_1(x) \ln(x) + \sum_{n=0}^{\infty} c_n^* x^{n+r_2}$$

with $c_0^* \neq 0$. y_1 and y_2 are linearly independent solutions on some interval $(0, h)$. ◆

The *method of Frobenius* consists of using the Frobenius series and Theorem 4.1 to solve equation (4.3) in some interval $(-h, h)$, $(0, h)$, or $(-h, 0)$, where we are assuming that the regular singular point of interest is at $x = 0$. Proceed as follows:

1. Substitute $y = \sum_{n=0}^{\infty} c_n x^{n+r}$ into the differential equation and solve for the roots r_1 and r_2 of the indicial equation for r. This yields a Frobenius solution.

2. These roots tell us which case of Theorem 4.1 applies, and this in turn, provides a template for a second solution, which is linearly independent from the first. Once we know what this second solution looks like, we can substitute its general form into the differential equation and solve for the coefficients and, in Case 4, the constant k.

EXAMPLE 4.5 Case 2 of Theorem 4.1 and Bessel Functions of the First Kind

We will illustrate Case 2 of Theorem 4.1 with an important example. The differential equation

$$x^2 y'' + xy' + (x^2 - v^2)y = 0 \tag{4.5}$$

is called *Bessel's equation of order v*, with v being any nonnegative real number. Here "order" refers to the value of the nonnegative parameter v, since Bessel's equation is second order.

Because 0 is a regular singular point, there is the Frobenius solution

$$y = \sum_{n=0}^{\infty} c_n x^{n+r}.$$

Substitute this into Bessel's equation to obtain, after some rearrangement,

$$[r(r-1) + r - v^2]c_0 x^r + [r(r+1) + (r+1) - v^2]c_1 x^{r+1}$$

$$+ \sum_{n=2}^{\infty} [[(n+r)(n+r-1) + (n+r) - v^2]c_n + c_{n-2}]x^{n+r} = 0. \tag{4.6}$$

Set the coefficient of each power of x equal to 0. Since we require in this method that $c_0 \neq 0$, we obtain the indicial equation

$$r^2 - v^2 = 0$$

from the coefficient of x^r. The roots are $r = \pm v$.

Set $r = v$ in the coefficient of x^{r+1} in equation (4.6) to obtain

$$(2v + 1)c_1 = 0,$$

hence $c_1 = 0$.

From the coefficient of x^{n+r} in equation (4.6), we have

$$[(n+r)(n+r-1) + (n+r) - v^2]c_n + c_{n-2}$$

for $n = 2, 3, \cdots$. Since $r = v$, this reduces to

$$c_n = -\frac{1}{n(n+2v)}c_{n-2}$$

for $n = 2, 3, \cdots$. Since $c_1 = 0$, then

$$c_3 = c_5 = c_{\text{odd}} = 0.$$

For the even-indexed coefficients, write

$$
c_{2n} = -\frac{1}{2n(2n+2v)}c_{2n-2} = -\frac{1}{2^2 n(n+v)}c_{2n-2}
$$

$$
= -\frac{1}{2^2 n(n+v)}\frac{-1}{2(n-1)[2(n-1)+2v]}c_{2n-4}
$$

$$
= \frac{1}{2^4 n(n-1)(n+v)(n+v-1)}c_{2n-4}
$$

$$
= \cdots = \frac{(-1)^n}{2^{2n}n(n-1)\cdots(2)(1)(n+v)(n-1+v)\cdots(1+v)}c_0
$$

$$
= \frac{-1)^n}{2^{2n}n!(1+v)(2+v)\cdots(n+v)}c_0.
$$

One Frobenius solution of Bessel's equation of order v is therefore

$$
y_1(x) = \sum_{n=0}^{\infty}\frac{(-1)^n}{2^{2n}n!(1+v)(2+v)\cdots(n+v)}x^{2n+v},
$$

which is called a Bessel function of the first kind of order v. These functions occur in many applications, including the analysis of radiation from cylindrical containers and wave motion of a circular fixed-frame membrane. When $v - (-v) = 2v$ is not a positive integer, then by Case 2 of Theorem 4.1, we obtain a second Frobenius solution by using $-v$ instead of v in this analysis. ◆

EXAMPLE 4.6   Case 3 of Theorem 4.1, and Bessel Functions of the Second Kind

Bessel's equation of order $v = 0$ is

$$
x^2 y'' + xy' + x^2 y = 0.
$$

From Example 4.5 with $v = 0$, the indicial equation has only one root $r = 0$ and one Frobenius solution (with $c_0 = 1$):

$$
y_1(x) = \sum_{n=0}^{\infty}(-1)^n \frac{1}{2^{2n}(n!)^2}x^{2n}.
$$

From Case 3 of Theorem 4.1, attempt a second, linearly independent solution of the form

$$
y_2(x) = y_1(x)\ln(x) + \sum_{n=1}^{\infty}c_n^* x^n.
$$

Substitute this into the differential equation to get

$$
xy_1''\ln(x) + 2y_1' - \frac{1}{x}y_1 + \sum_{n=2}^{\infty}n(n-1)c_n^* x^{n-1}
$$

$$
+ y_1' + \frac{1}{x}y_1 + \sum_{n=1}^{\infty}kc_n^* x^{n-1} + xy_1\ln(x) + \sum_{n=1}^{\infty}c_n^* x^{n+1} = 0.
$$

Terms involving $\ln(x)$ and $y_1(x)$ cancel, and we are left with

$$2y_1' + \sum_{n=2}^{\infty} n(n-1)c_n^* x^{n-1} + \sum_{n=1}^{\infty} k c_n^* x^{n-1} + \sum_{n=1}^{\infty} c_n^* x^{n+1} = 0.$$

Since $n(n-1) = n^2 - n$, part of the first summation cancels all terms except the $n=1$ term in the second summation, and we have

$$2y_1' + \sum_{n=2}^{\infty} n^2 c_n^* x^{n-1} + c_1^* + \sum_{n=1}^{\infty} c_n^* x^{n+1} = 0.$$

Substitute the series for y_1' into this equation to get

$$2\sum_{n=1}^{\infty} \frac{(-1)^n}{2^{2n-1}n!(n-1)!} x^{2n-1} + \sum_{n=2}^{\infty} n^2 c_n^* x^{n-1} + c_1^* + \sum_{n=1}^{\infty} c_n^* x^{n+1} = 0.$$

Shift indices in the last series to write this equation as

$$\sum_{n=1}^{\infty} \frac{(-1)^n}{2^{2n-2}n!(n-1)!} x^{n-1} + c_1^* + 4c_2^* x + \sum_{n=3}^{\infty} (n^2 c_n^* + c_{n-2}^*) x^{n-1} = 0. \qquad (4.7)$$

The only constant term on the left side of this equation is c_1^*, which must therefore equal zero. The only even powers of x in this equation are on the right-most series when n is odd. The coefficients of these powers of x must also be zero, so

$$n^2 c_n^* + c_{n-2}^* = 0 \text{ for } n = 3, 5, 7, \cdots .$$

But then all odd-indexed coefficients are multiples of c_1^* and must also be zero:

$$c_{2n+1}^* = 0 \text{ for } n = 0, 1, 2, \cdots .$$

To determine the even-indexed coefficients, replace n by $2j$ in the second summation in equation (4.7) and n with j in the first summation to obtain

$$\sum_{j=1}^{\infty} \frac{(-1)^j}{2^{2j-1}j!(j-1)!} x^{2j-1} + 4c_2^* x + \sum_{j=1}^{\infty} (4j^2 c_{2j}^* + c_{2j-2}^*) x^{2j-1} = 0.$$

Combine terms and write this equation as

$$(4c_2^* - 1)x + \sum_{j=2}^{\infty} \left[\frac{(-1)^j}{2^{2j-2}j!(j-1)!} + 4j^2 c_{2j}^* + c_{2j-2}^* \right] x^{2j-1} = 0.$$

Equate the coefficient of each power of x to 0. First,

$$c_1^* = \frac{1}{4}$$

and the recurrence relation is

$$c_{2j}^* = \frac{(-1)^{j+1}}{2^{2j}(j!)^2 j} - \frac{1}{4j^2} c_{2j-2}^* \text{ for } j = 2, 3, \cdots .$$

If we look at some of these coefficients, a pattern emerges:

$$c_4^* = \frac{-1}{2^2 4^2} \left[1 + \frac{1}{2} \right]$$

$$c_6^* = \frac{1}{2^2 4^2 6^2} \left[1 + \frac{1}{2} + \frac{1}{3} \right],$$

and, in general,

$$c_{2j}^* = \frac{(-1)^{j+1}}{2^2 4^2 \cdots (2j)^2}\left[1 + \frac{1}{2} + \cdots + \frac{1}{j}\right] = \frac{(-1)^{j+1}}{2^{2j}(j!)^2}\phi(j),$$

where

$$\phi(j) = 1 + \frac{1}{2} + \cdots + \frac{1}{j}.$$

We therefore have a second solution of Bessel's equation of order zero:

$$y_2(x) = y_1(x)\ln(x) + \sum_{n=1}^{\infty} \frac{(-1)^{n+1}}{2^{2n}(n!)^2}\phi(n)x^{2n}$$

for $x > 0$. This solution is linearly independent from $y_1(x)$ and is called a Bessel function of the second kind of order zero. $\blacklozenge$

EXAMPLE 4.7   Case 4 of Theorem 4.1

The equation

$$xy'' - y = 0$$

has a regular singular point at 0. Substitute $y = \sum_{n=0}^{\infty} c_n x^{n+r}$ into this equation to obtain

$$\sum_{n=0}^{\infty}(n+r)(n+r-1)c_n x^{n+r-2} - \sum_{n=0}^{\infty} c_n x^{n+r} = 0.$$

Shift indices in the second summation to write this as

$$(r^2 - r)c_0 x^{r-1} + \sum_{n=1}^{\infty}[(n+r)(n+r-1)c_n - c_{n-1}]x^{n+r-1} = 0.$$

The indicial equation is $r^2 - r = 0$ with roots $r_1 = 1, r_2 = 0$. Then $r_1 - r_2 = 1$, a positive integer, so Case 4 of Theorem 4.1 applies. From the summation in the last equation, we have

$$(n+r)(n+r-1)c_n - c_{n-1} = 0.$$

Let $r = 1$ to obtain the recurrence relation

$$c_n = \frac{1}{n(n+1)}c_{n-1} \text{ for } n = 1, 2, 3, \cdots.$$

Some of the coefficients are

$$c_1 = \frac{1}{(1)(2)}c_0,$$

$$c_2 = \frac{1}{2(3)}c_1 = \frac{1}{2(2)(3)}c_0,$$

and

$$c_3 = \frac{1}{3(4)}c_2 = \frac{1}{2(3)(2)(3)(4)}c_0.$$

In general, we observe that

$$c_n = \frac{1}{n!(n+1)!}c_0 \text{ for } n = 1, 2, \cdots.$$

This gives us a Frobenius solution

$$y_1(x) = c_0 \sum_{n=0}^{\infty} \frac{1}{n!(n+1)!} x^{n+1}$$

$$= c_0 \left[x + \frac{1}{2}x^2 + \frac{1}{12}x^3 + \frac{1}{144}x^4 + \cdots \right].$$

Case 4 of Theorem 4.1 tells us that there is a second, linearly independent solution of the form

$$y_2(x) = k y_1(x) \ln(x) + \sum_{n=0}^{\infty} c_n^* x^{n+r_2}.$$

Of course, in this example, $r_2 = 0$. Substitute this form for y_2 into the differential equation to obtain

$$x \left[k y_1'' \ln(x) + 2k y_1' \frac{1}{x} - k y_1 \frac{1}{x^2} + \sum_{n=2}^{\infty} n(n-1)c_n^* x^{n-2} \right]$$

$$- k y_1 \ln(x) - \sum_{n=0}^{\infty} c_n^* x^n = 0. \tag{4.8}$$

Now

$$k \ln(x)[x y_1'' - y_1] = 0$$

because y_1 is a solution. For the remaining terms in equation (4.8), insert the series for $y_1(x)$ with $c_0 = 1$ for convenience, to obtain

$$2k \sum_{n=0}^{\infty} \frac{1}{(n!)^2} x^n - k \sum_{n=0}^{\infty} \frac{1}{n!(n+1)!} x^n + \sum_{n=2}^{\infty} c_n^* n(n-1) x^{n-1} - \sum_{n=0}^{\infty} c_n^* x^n = 0.$$

Shift indices in the third summation to write this equation as

$$2k \sum_{n=0}^{\infty} \frac{1}{(n!)^2} x^n - k \sum_{n=0}^{\infty} \frac{1}{n!(n+1)!} x^n + \sum_{n=1}^{\infty} c_{n+1}^* n(n+1) x^n - \sum_{n=0}^{\infty} c_n^* x^n = 0.$$

This enables us to combine terms:

$$(2k - 1 - c_0^*)x^0 + \sum_{n=1}^{\infty} \left[\frac{2k}{(n!)^2} - \frac{k}{n!(n+1)!} + n(n+1)c_{n+1}^* - c_n^* \right] x^n = 0.$$

Then

$$k - c_0^* = 0$$

and, for $n = 1, 2, \cdots,$

$$\frac{2k}{(n!)^2} - \frac{k}{n!(n+1)!} + n(n+1)c_{n+1}^* - c_n^* = 0.$$

These give us $k = c_0^*$ and the recurrence relation

$$c_{n+1}^* = \frac{1}{n(n+1)} \left[c_n^* - \frac{(2n+1)k}{n!(n+1)!} \right]$$

for $n = 1, 2, \cdots$. Since c_0^* can be any nonzero real number, we can obtain a particular second solution by choosing $k = c_0^* = 1$:

$$y_2(x) = y_1(x) \ln(x) + 1 - \frac{3}{4}x^2 - \frac{7}{36}x^3 - \frac{35}{1728}x^4 - \cdots. \quad \blacklozenge$$

In each of Problems 1 through 10, find the first five terms of each of two linearly independent solutions.

1. $xy'' + (1 - x)y' + y = 0$

2. $xy'' - 2xy' + 2y = 0$

3. $x(x - 1)y'' + 3y' - 2y = 0$

4. $4x^2 y'' + 4xy' + (4x^2 - 9)y = 0$

5. $4xy'' + 2y' + 2y = 0$

6. $4x^2 y'' + 4xy' - y = 0$

7. $x^2 y'' - 2xy' - (x^2 - 2)y = 0$

8. $xy'' - y' + 2y = 0$

9. $x(2 - x)y'' - 2(x - 1)y' + 2y = 0$

10. $x^2 y'' + x(x^3 + 1)y' - y = 0$

PART 2

Vectors, Linear Algebra, and Systems of Linear Differential Equations

CHAPTER 5

Algebra and Geometry of Vectors

5.1   **Vectors in the Plane and 3-Space**

Some quantities, such as temperature and mass, are completely specified by a number. Such quantities are called *scalars*. By contrast, a *vector* has both a magnitude and a sense of direction. If we push against an object, the effect is determined not only by the strength of the push, but its direction. Velocity and acceleration are also vectors.

We can include both magnitude and direction in one package by representing a vector as an arrow from the origin to a point (x, y, z) in 3-space, as in Figure 5.1. The choice of the point gives the direction of the vector (when viewed from the origin), and the length is its magnitude. To distinguish when we are thinking of a point as a vector (arrow from the origin to the point), we will denote this vector $<x, y, z>$. We call x the *first component of* $<x, y, z>$, y the *second component* and z the *third component*. These components are scalars.

Two vectors are equal exactly when their respective components are equal. That is,

$$<x_1, y_1, z_1> = <x_2, y_2, z_2>$$

exactly when $x_1 = x_2$, $y_1 = y_2$, and $z_1 = z_2$.

Since only direction and magnitude are important in specifying a vector, any arrow of the same length and orientation denotes the same vector. The arrows in Figure 5.2 represent the same vector.

The vector $<-x, -y, -z>$ is opposite in direction to $<x, y, z>$, as suggested in Figure 5.3.

It is convenient to denote vectors by bold-face letters such as $\mathbf{F}$, $\mathbf{G}$, and $\mathbf{H}$, and scalars (real numbers) in ordinary type.

The *length* (also called the *magnitude* or *norm*) of a vector $\mathbf{F} = <x, y, z>$ is the scalar

$$\|\mathbf{F}\| = \sqrt{x^2 + y^2 + z^2}.$$

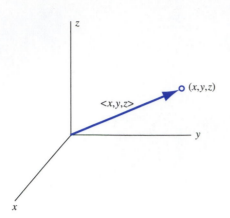

FIGURE 5.1   *Vector $<x, y, z>$ from the origin to the point (x, y, z).*

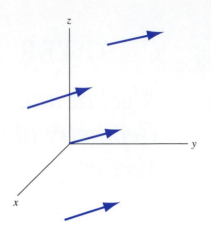

FIGURE 5.2   *Arrow representations of the same vector.*

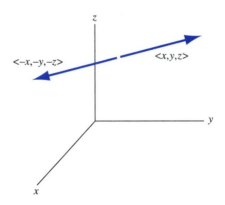

FIGURE 5.3 $< -x, -y, -z >$ *is opposite* $< x, y, z >$.

This is the distance from the origin to the point (x, y, z) and also the length of any arrow representing the vector $<x, y, z>$. For example, the norm of $\mathbf{G} = <-1, 4, 2>$ is $\| \mathbf{G} \| = \sqrt{21}$, which is the distance from the origin to the point $(-1, 4, 2)$.

> Multiply a vector $\mathbf{F} = < a, b, c >$ by a scalar α by multiplying each component of $\mathbf{F}$ by α. This produces a new vector denoted $\alpha \mathbf{F}$:
>
> $$\alpha \mathbf{F} = < \alpha a, \alpha b, \alpha c > .$$

We have

$$\| \alpha \mathbf{F} \| = |\alpha| \, \| \mathbf{F} \|$$

because

$$\| \alpha \mathbf{F} \| = \sqrt{(\alpha a)^2 + (\alpha b)^2 + (\alpha c)^2}$$

$$= \sqrt{(\alpha^2)(a^2 + b^2 + c^2)} = |\alpha|\sqrt{a^2 + b^2 + c^2} = |\alpha| \| \mathbf{F} \|.$$

This means that the length of $\alpha \mathbf{F}$ is $|\alpha|$ times the length of $\mathbf{F}$. We may therefore think of multiplication of a vector by a scalar as a scaling (stretching or shrinking) operation.

For example, $\frac{1}{2}\mathbf{F}$ is a vector having the direction of $\mathbf{F}$ and half the length of $\mathbf{F}$, while $2\mathbf{F}$ has the direction of $\mathbf{F}$ and length twice that of $\mathbf{F}$, and $-\frac{1}{2}\mathbf{F}$ has direction opposite that of $\mathbf{F}$ and half the length.

If $\alpha = 0$, then $\alpha \mathbf{F} = \,<0, 0, 0>$, which we call the zero vector and denote $\mathbf{O}$. This is the only vector with zero length and no specific direction, since it cannot be represented by an arrow.

Consistent with these interpretations of $\alpha \mathbf{F}$, we define two vectors $\mathbf{F}$ and $\mathbf{G}$ to be *parallel* if each is a nonzero scalar multiple of the other. Parallel vectors may differ in length and even be in opposite directions, but the straight lines through arrows representing them are parallel lines in 3-space.

Add two vectors by adding their respective components. If $\mathbf{F} = \,< a_1, a_2, a_3 >$ and $\mathbf{G} = \,< b_1, b_2, b_3 >$, then

$$\mathbf{F} + \mathbf{G} = \,< a_1 + a_2, b_1 + b_2, c_1 + c_2 >.$$

Vector addition and multiplication by scalars have the following properties:

1. $\mathbf{F} + \mathbf{G} = \mathbf{G} + \mathbf{F}$. (commutativity)
2. $\mathbf{F} + (\mathbf{G} + \mathbf{H}) = (\mathbf{F} + \mathbf{G}) + \mathbf{H}$.
3. $\mathbf{F} + \mathbf{O} = \mathbf{F}$.
4. $\alpha(\mathbf{F} + \mathbf{G}) = \alpha \mathbf{F} + \alpha \mathbf{G}$.
5. $(\alpha\beta)\mathbf{F} = \alpha(\beta\mathbf{F})$.
6. $(\alpha + \beta)\mathbf{F} = \alpha \mathbf{F} + \beta \mathbf{F}$.

It is sometimes useful to represent vector addition by the *parallelogram law*. If $\mathbf{F}$ and $\mathbf{G}$ are drawn as arrows from the same point, they form two sides of a parallelogram. The arrow along the diagonal of this parallelogram represents the sum $\mathbf{F} + \mathbf{G}$ (Figure 5.4). Since any arrows having the same lengths and direction represent the same vector, we can also draw the arrows in the sum as in Figure 5.5. It is sometimes convenient to draw one vector from the end of the other in this way when using the parallelogram law to represent a vector sum.

A vector of length 1 is called a *unit vector*. The unit vectors along the positive axes are shown in Figure 5.6 and are labeled

$$\mathbf{i} = \,< 1, 0, 0 >, \ \mathbf{j} = \,< 0, 1, 0 >, \ \text{and} \ \mathbf{k} = \,< 0, 0, 1 >.$$

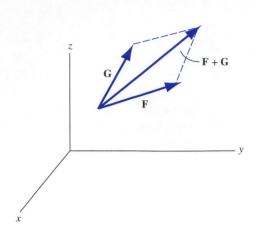

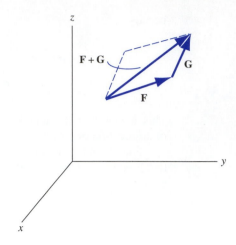

FIGURE 5.4 *Parallelogram law for vector addition.*

FIGURE 5.5 *Alternative view of the parallelogram law.*

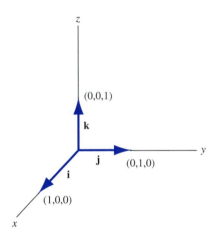

FIGURE 5.6 *Unit vectors* **i**, **j**, *and* **k**.

We can write any vector $\mathbf{F} = <a, b, c>$ as

$$\mathbf{F} = <a, b, c> = a<1, 0, 0> + b<0, 1, 0> + c<0, 0, 1>$$
$$= a\mathbf{i} + b\mathbf{j} + c\mathbf{k}.$$

We call $a\mathbf{i} + b\mathbf{j} + c\mathbf{k}$ the *standard representation* of $\mathbf{F}$. When a component of a vector is zero, we usually just omit this term in the standard representation. For example, we would usually write $\mathbf{F} = <-8, 0, 3>$ as $-8\mathbf{i} + 3\mathbf{k}$ instead of $-8\mathbf{i} + 0\mathbf{j} + 3\mathbf{k}$.

If a vector is represented by an arrow in the x, y-plane, we usually omit the third coordinate and use $\mathbf{i} = <1, 0>$ and $\mathbf{j} = <0, 1>$. For example, the vector $\mathbf{V}$ from the origin to the point $<2, -6, 0>$ can be represented as an arrow from the origin to the point $(2, -6)$ in the x, y-plane and can be written in standard form as

$$\mathbf{V} = 2\mathbf{i} - 6\mathbf{j}$$

where $\mathbf{i} = <1, 0>$ and $\mathbf{j} = <0, 1>$.

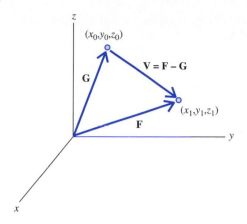

FIGURE 5.7 *Vector from* (x_0, y_0, z_0) *to* (x_1, y_1, z_1).

It is often useful use to know the components of the vector **V** represented by the arrow from one point to another, say from $P_0 = (x_0, y_0, z_0)$ to $P_1 = (x_1, y_1, z_1)$. For example, denote

$$\mathbf{G} = x_0\mathbf{i} + y_0\mathbf{j} + z_0\mathbf{k} \quad \text{and} \quad \mathbf{F} = x_1\mathbf{i} + y_1\mathbf{j} + z_1\mathbf{k}.$$

By the parallelogram law (Figure 5.7),

$$\mathbf{G} + \mathbf{V} = \mathbf{F}.$$

Therefore,

$$\mathbf{V} = \mathbf{F} - \mathbf{G} = (x_1 - x_0)\mathbf{i} + (y_1 - y_0)\mathbf{j} + (z_1 - z_0)\mathbf{k}.$$

The vector represented by the arrow from $(-1, 6, 3)$ to $(9, -1, -7)$ is $10\mathbf{i} - 7\mathbf{j} - 10\mathbf{k}$.

Using this idea, we can find a vector of any length in any given direction. For example, suppose we want a vector of length 7 in the direction from $(-1, 6, 5)$ to $(-8, 4, 9)$. First write a vector $\mathbf{V} = -7\mathbf{i} - 2\mathbf{j} + 4\mathbf{k}$ in the specified direction. A unit vector in this direction is

$$\mathbf{F} = \frac{1}{\|\mathbf{V}\|}\mathbf{V} = \frac{1}{\sqrt{69}}\mathbf{V}.$$

Then

$$7\mathbf{F} = \frac{7}{\sqrt{69}}(-7\mathbf{i} - 2\mathbf{j} + 4\mathbf{k})$$

has length 7 and is in the direction from $(-1, 6, 5)$ to $(-8, 4, 9)$.

5.1.1   Equation of a Line in 3-Space

We will show how to find equations of a line L in 3-space containing two given points. This is more subtle than the corresponding problem in the plane because there is no slope to exploit. To illustrate the idea, suppose the points are $(-2, -4, 7)$ and $(9, 1, -7)$. Form a vector between these two points (in either direction). The arrow from the first to the second point represents the vector

$$\mathbf{V} = 11\mathbf{i} + 5\mathbf{j} - 14\mathbf{k}.$$

Because P_0 and P_1 are on L, **V** is parallel to L and hence to any other vector aligned with L. Now suppose (x, y, z) is any point on L. Then the vector $(x + 2)\mathbf{i} + (y + 4)\mathbf{j} + (z - 7)\mathbf{k}$ from

$(-2, -4, 7)$ to (x, y, z) is also parallel to L and hence to $\mathbf{V}$. This vector must therefore be a scalar multiple of $\mathbf{V}$:

$$(x+2)\mathbf{i} + (y+4)\mathbf{j} + (z-7)\mathbf{k} = t\mathbf{V}$$
$$= 11t\mathbf{i} + 5t\mathbf{j} - 14t\mathbf{k}$$

for some scalar t. Since two vectors are equal only when their respective components are equal,

$$x + 2 = 11t, \quad y + 4 = 5t, \quad \text{and } z - 7 = -14t.$$

Usually we write these equations as

$$x = -2 + 11t, \quad y = -4 + 5t, \quad \text{and } z = 7 - 14t.$$

These are *parametric equations* of L. As t varies over the real numbers, the point $(-2 + 11t, -4 + 5t, 7 - 14t)$ varies over L. We obtain $(-2, -4, 7)$ when $t = 0$ and $(9, 1, -7)$ when $t = 1$.

This argument can be carried out in general, as indicated in Figure 5.8. Suppose (x_0, y_0, z_0) and (x_1, y_1, z_1) are distinct points. The vector from the first to the second is

$$(x_1 - x_0)\mathbf{i} + (y_1 - y_0)\mathbf{j} + (z_1 - z_0)\mathbf{k}.$$

The vector from (x_0, y_0, z_0) to an arbitrary point (x, y, z) on this line is

$$(x - x_0)\mathbf{i} + (y - y_0)\mathbf{j} + (z - z_0)\mathbf{k}.$$

These two vectors are parallel (being along the same line), so each is a scalar multiple of the other:

$$(x - x_0)\mathbf{i} + (y - y_0)\mathbf{j} + (z - z_0)\mathbf{k} = t[(x_1 - x_0)\mathbf{i} + (y_1 - y_0)\mathbf{j} + (z_1 - z_0)\mathbf{k}].$$

Equating respective components gives us

$$x = x_0 + (x_1 - x_0)t, \ y = y_0 + (y_1 - y_0)t, \ z = z_0 + (z_1 - z_0)t,$$

where t takes on all real values. These are the parametric equations of this line.

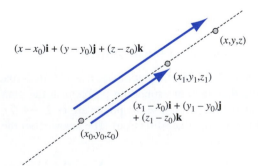

FIGURE 5.8   *Equations of the line between two given points.*

SECTION 5.1   **PROBLEMS**

In each of Problems 1 through 5, compute $\mathbf{F}+\mathbf{G}$, $\mathbf{F}-\mathbf{G}$, $2\mathbf{F}$, $3\mathbf{G}$, and $\|\mathbf{F}\|$.

1. $\mathbf{F}=2\mathbf{i}-3\mathbf{j}+5\mathbf{k}, \mathbf{G}=\sqrt{2}\mathbf{i}+6\mathbf{j}-5\mathbf{k}$
2. $\mathbf{F}=\mathbf{i}-3\mathbf{k}, \mathbf{G}=4\mathbf{j}$
3. $\mathbf{F}=2\mathbf{i}-5\mathbf{j}, \mathbf{G}=\mathbf{i}+5\mathbf{j}-\mathbf{k}$
4. $\mathbf{F}=\sqrt{2}\mathbf{i}-\mathbf{j}-6\mathbf{k}, \mathbf{G}=8\mathbf{i}+2\mathbf{k}$
5. $\mathbf{F}=\mathbf{i}+\mathbf{j}+\mathbf{k}, \mathbf{G}=2\mathbf{i}-2\mathbf{j}+2\mathbf{k}$

In each of Problems 6 through 9, find a vector having the given length and in the direction from the first point to the second.

6. $5, (0, 1, 4), (-5, 2, 2)$
7. $9, (1, 2, 1), (-4, -2, 3)$
8. $12, (-4, 5, 1), (6, 2, -3)$
9. $4, (0, 0, 1), (-4, 7, 5)$

In each of Problems 10 through 15, find parametric equations of the line containing the given points.

10. $(1, 0, 4), (2, 1, 1)$
11. $(3, 0, 0), (-3, 1, 0)$
12. $(2, 1, 1), (2, 1, -2)$
13. $(0, 1, 3), (0, 0, 1)$
14. $(1, 0, -4), (-2, -2, 5)$
15. $(2, -3, 6), (-1, 6, 4)$

5.2   The Dot Product

The *dot product* $\mathbf{F}\cdot\mathbf{G}$ of $\mathbf{F}$ and $\mathbf{G}$ is the real number formed by multiplying the two first components, then the two second components, then the two third components, and adding these three numbers. If $\mathbf{F}=a_1\mathbf{i}+b_1\mathbf{j}+c_1\mathbf{k}$ and $\mathbf{G}=a_2\mathbf{i}+b_2\mathbf{j}+c_2\mathbf{k}$, then

$$\mathbf{F}\cdot\mathbf{G}=a_1a_2+b_1b_2+c_1c_2.$$

Again, note that this dot product is a number, not a vector. For example,

$$(\sqrt{3}\mathbf{i}+4\mathbf{j}-\pi\mathbf{k})\cdot(-2\mathbf{i}+6\mathbf{j}+3\mathbf{k})=-2\sqrt{3}+24-3\pi.$$

The dot product has the following properties.

1. $\mathbf{F}\cdot\mathbf{G}=\mathbf{G}\cdot\mathbf{F}$.
2. $(\mathbf{F}+\mathbf{G})\cdot\mathbf{H}=\mathbf{F}\cdot\mathbf{H}+\mathbf{G}\cdot\mathbf{H}$.
3. $\alpha(\mathbf{F}\cdot\mathbf{G})=(\alpha\mathbf{F})\cdot\mathbf{G}=\mathbf{F}\cdot(\alpha\mathbf{G})$.
4. $\mathbf{F}\cdot\mathbf{F}=\|\mathbf{F}\|^2$.
5. $\mathbf{F}\cdot\mathbf{F}=0$ if and only if $\mathbf{F}=\mathbf{O}$.
6. $\|\alpha\mathbf{F}+\beta\mathbf{G}\|^2=\alpha^2\|\mathbf{F}\|^2+2\alpha\beta\mathbf{F}\cdot\mathbf{G}+\beta^2\|\mathbf{G}\|^2$.

From property 4 the dot product of a vector with itself is the square of the length of the vector, a fact that is used often in computations. Property 5 follows easily from this, since the

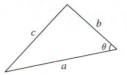

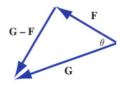

FIGURE 5.9 *The law of cosines and the angle between vectors.*

only vector having zero length is the zero vector. For Property 6, use properties 1 through 4 to write

$$\| \alpha\mathbf{F} + \beta\mathbf{G} \|^2 = (\alpha\mathbf{F} + \beta\mathbf{G}) \cdot (\alpha\mathbf{F} + \beta\mathbf{G})$$
$$= \alpha^2\mathbf{F} \cdot \mathbf{F} + \alpha\beta\mathbf{F} \cdot \mathbf{G} + \alpha\beta\mathbf{G} \cdot \mathbf{F} + \beta^2\mathbf{G} \cdot \mathbf{G}$$
$$= \alpha^2 \| \mathbf{F} \|^2 + 2\alpha\beta\mathbf{F} \cdot \mathbf{G} + \beta^2 \| \mathbf{G} \|^2 .$$

The dot product can be used to find an angle between two vectors. First recall the law of cosines. In the triangle of Figure 5.9 (top) with θ the angle opposite the side of length c,

$$a^2 + b^2 - 2ab\cos(\theta) = c^2.$$

Apply this to the vector triangle of Figure 5.9 (bottom), which has sides of length $a = \| \mathbf{G} \|$, $b = \| \mathbf{F} \|$, and $c = \| \mathbf{G} - \mathbf{F} \|$. Using property 6 of the dot product, we obtain

$$\| \mathbf{G} \|^2 + \| \mathbf{F} \|^2 - 2 \| \mathbf{F} \| \| \mathbf{G} \| \cos(\theta) = \| \mathbf{G} - \mathbf{F} \|^2$$
$$= \| \mathbf{G} \|^2 + \| \mathbf{F} \|^2 - 2\mathbf{G} \cdot \mathbf{F}.$$

Assuming that neither $\mathbf{F}$ nor $\mathbf{G}$ is the zero vector, this gives us

$$\cos(\theta) = \frac{\mathbf{F} \cdot \mathbf{G}}{\| \mathbf{F} \| \| \mathbf{G} \|}. \tag{5.1}$$

Since $|\cos(\theta)| \leq 1$ for all θ, equation (5.1) implies the *Cauchy-Schwarz inequality*:

$$|\mathbf{F} \cdot \mathbf{G}| \leq \| \mathbf{F} \| \| \mathbf{G} \| .$$

EXAMPLE 5.1

The angle θ between $\mathbf{F} = -\mathbf{i} + 3\mathbf{j} + \mathbf{k}$ and $\mathbf{G} = 2\mathbf{j} - 4\mathbf{k}$ is given by

$$\cos(\theta) = \frac{(-\mathbf{i} + 3\mathbf{j} + \mathbf{k}) \cdot (2\mathbf{j} - 4\mathbf{k})}{\| -\mathbf{i} + 3\mathbf{j} + \mathbf{k} \| \| 2\mathbf{j} - 4\mathbf{k} \|}$$
$$= \frac{(-1)(0) + (3)(2) + (1)(-4)}{\sqrt{1^2 + 3^2 + 1^2}\sqrt{2^2 + 4^2}} = \frac{2}{\sqrt{220}},$$

about 1.436 radians. ◆

EXAMPLE 5.2

Lines L_1 and L_2 have parametric equations

$$L_1 : x = 1 + 6t, \, y = 2 - 4t, \, z = -1 + 3t$$

and

$$L_2 : x = 4 - 3p, \, y = 2p, \, z = -5 + 4p.$$

The parameters t and p can take on any real values. We want an angle θ between these lines.

The strategy is to take a vector $\mathbf{V}_1$ along L_1 and a vector $\mathbf{V}_2$ along L_2 and find the angle between these vectors. For $\mathbf{V}_1$, find two points on L_1, say $(1, 2, -1)$ when $t = 0$ and $(7, -2, 2)$ when $t = 1$ and form

$$\mathbf{V}_1 = (7 - 1)\mathbf{i} + (-2 - 2)\mathbf{j} + (2 - (-1))\mathbf{k} = 6\mathbf{i} - 4\mathbf{j} + 3\mathbf{k}.$$

On L_2 take $(4, 0, -5)$ with $p = 0$ and $(1, 2, -1)$ with $p = 1$, forming

$$\mathbf{V}_2 = 3\mathbf{i} - 2\mathbf{j} - 4\mathbf{k}.$$

Now compute

$$\cos(\theta) = \frac{6(3) - 4(-2) + 3(-4)}{\sqrt{36 + 16 + 9}\sqrt{9 + 4 + 16}} = \frac{14}{\sqrt{1769}}.$$

An angle between L_1 and L_2 is $\arccos(14/\sqrt{1769})$, approximately 1.23 radians. ◆

Two nonzero vectors $\mathbf{F}$ and $\mathbf{G}$ are *orthogonal* (perpendicular) when the angle θ between them is $\pi/2$ radians. This occurs exactly when

$$\cos(\theta) = 0 = \frac{\mathbf{F} \cdot \mathbf{G}}{\|\mathbf{F}\| \|\mathbf{G}\|}$$

which occurs when $\mathbf{F} \cdot \mathbf{G} = 0$. It is convenient to also agree that $\mathbf{O}$ is orthogonal to every vector. With this convention, two vectors are orthogonal if and only if their dot product is zero.

EXAMPLE 5.3

Let $\mathbf{F} = -4\mathbf{i} + \mathbf{j} + 2\mathbf{k}$, $\mathbf{G} = 2\mathbf{i} + 4\mathbf{k}$ and $\mathbf{H} = 6\mathbf{i} - \mathbf{j} - 2\mathbf{k}$. Then $\mathbf{F} \cdot \mathbf{G} = 0$, so $\mathbf{F}$ and $\mathbf{G}$ are orthogonal. But $\mathbf{F} \cdot \mathbf{H}$ and $\mathbf{G} \cdot \mathbf{H}$ are not zero, so $\mathbf{F}$ and $\mathbf{H}$ are not orthogonal, and $\mathbf{G}$ and $\mathbf{H}$ are not orthogonal. ◆

EXAMPLE 5.4

Two lines are given parametrically by

$$L_1 : x = 2 - 4t, \, y = 6 + t, \, z = 3t$$

and

$$L_2 : x = -2 + p, \, y = 7 + 2p, \, z = 3 - 4p.$$

We want to know if these lines are orthogonal (whether or not they intersect).

The idea is to form a vector along each line and test these vectors for orthogonality. For a vector along L_1, take two points on this line, say $(2, 6, 0)$ when $t = 0$ and $(-2, 7, 3)$ when $t = 1$. Then $\mathbf{V}_1 = -4\mathbf{i} + \mathbf{j} + 3\mathbf{k}$ is parallel to L_1. Similarly, $(-2, 7, 3)$ is on L_2 when $p = 0$, and $(-1, 9, -1)$ is on L_2 when $p = 1$, so $\mathbf{V}_2 = \mathbf{i} + 2\mathbf{j} - 4\mathbf{k}$ is parallel to L_2. Compute $\mathbf{V}_1 \cdot \mathbf{V}_2 = -14 \neq 0$. Therefore, L_1 and L_2 are not orthogonal. ◆

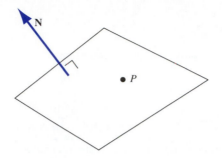

FIGURE 5.10   *A point P and a normal vector N determine a plane.*

Orthogonality is also useful for determining the equation of a plane in 3-space. Any plane has an equation of the form

$$ax + by + cz = d.$$

As suggested by Figure 5.10, if we specify a point on the plane and a vector orthogonal to the plane, then the plane is completely determined. Example 5.5 shows how to find the equation of the plane given this information.

EXAMPLE 5.5

We will find the equation of the plane Π containing the point $(-6, 1, 1)$ and orthogonal to the vector $\mathbf{N} = -2\mathbf{i} + 4\mathbf{j} + \mathbf{k}$. Such a vector $\mathbf{N}$ is said to be *normal* to Π and is called a *normal vector* to Π.

Here is a strategy. Because $(-6, 1, 1)$ is on Π, a point (x, y, z) is on Π exactly when the vector between $(-6, 1, 1)$ and (x, y, z) lies in Π. But then $(x + 6)\mathbf{i} + (y - 1)\mathbf{j} + (z - 1)\mathbf{k}$ must be orthogonal to $\mathbf{N}$, so

$$\mathbf{N} \cdot ((x + 6)\mathbf{i} + (y - 1)\mathbf{j} + (z - 1)\mathbf{k}) = 0.$$

Then

$$-2(x + 6) + 4(y - 1) + (z - 1) = 0,$$

or

$$-2x + 4y + z = 17.$$

This is the equation of Π. ◆

Following this reasoning in general, the equation of a plane containing a point $P_0 : (x_0, y_0, z_0)$ and having a normal vector $\mathbf{N} = a\mathbf{i} + b\mathbf{j} + c\mathbf{k}$ is

$$\mathbf{N} \cdot [(x - x_0)\mathbf{i} + (y - y_0)\mathbf{j} + (z - z_0)\mathbf{k}] = 0$$

or

$$a(x - x_0) + b(y - y_0) + c(z - z_0) = 0. \tag{5.2}$$

It is also sometimes convenient to notice that the vector $a\mathbf{i} + b\mathbf{j} + c\mathbf{k}$ is always a normal vector to a plane $ax + by + cz = d$ for any d.

In each of Problems 1 through 6, compute the dot product of the vectors and the cosine of the angle between them. Also determine if the vectors are orthogonal.

1. $\mathbf{i}, 2\mathbf{i} - 3\mathbf{j} + \mathbf{k}$

2. $2\mathbf{i} - 6\mathbf{j} + \mathbf{k}, \mathbf{i} - \mathbf{j}$

3. $-4\mathbf{i} - 2\mathbf{i} + 3\mathbf{k}, 6\mathbf{i} - 2\mathbf{j} - \mathbf{k}$

4. $8\mathbf{i} - 3\mathbf{j} + 2\mathbf{k}, -8\mathbf{i} - 3\mathbf{j} + \mathbf{k}$

5. $\mathbf{i} - 3\mathbf{k}, 2\mathbf{j} + 6\mathbf{k}$

6. $\mathbf{i} + \mathbf{j} + 2\mathbf{k}, \mathbf{i} - \mathbf{j} + 2\mathbf{k}$

In each of Problems 7 through 12, find the equation of the plane containing the given point and orthogonal to the given vector.

7. $(-1, 1, 2), 3\mathbf{i} - \mathbf{j} + 4\mathbf{k}$

8. $(-1, 0, 0), \mathbf{i} - 2\mathbf{j}$

9. $(2, -3, 4), 8\mathbf{i} - 6\mathbf{j} + 4\mathbf{k}$

10. $(-1, -1, -5), -3\mathbf{i} + 2\mathbf{j}$

11. $(0, -1, 4), 7\mathbf{i} + 6\mathbf{j} - 5\mathbf{k}$

12. $(-2, 1, -1), 4\mathbf{i} + 3\mathbf{j} + \mathbf{k}$

5.3 The Cross Product

The dot product produces a scalar from two vectors. The cross product produces a vector from two vectors.

> Let $\mathbf{F} = a_1\mathbf{i} + b_1\mathbf{j} + c_1\mathbf{k}$ and $\mathbf{G} = a_2\mathbf{i} + b_2\mathbf{j} + c_2\mathbf{k}$.
> The *cross product* of $\mathbf{F}$ with $\mathbf{G}$ is the vector $\mathbf{F} \times \mathbf{G}$ defined by
> $$\mathbf{F} \times \mathbf{G} = (b_1c_2 - b_2c_1)\mathbf{i} + (a_2c_1 - a_1c_2)\mathbf{j} + (a_1b_2 - a_2b_1)\mathbf{k}.$$

There is a simple device for remembering and computing these components. Form the determinant

$$\begin{vmatrix} \mathbf{i} & \mathbf{j} & \mathbf{k} \\ a_1 & b_1 & c_1 \\ a_2 & b_2 & c_2 \end{vmatrix}$$

having the standard unit vectors in the first row, the components of $\mathbf{F}$ in the second row, and the components of $\mathbf{G}$ in the third row. If this determinant is expanded by the first row, we get exactly $\mathbf{F} \times \mathbf{G}$:

$$\begin{vmatrix} \mathbf{i} & \mathbf{j} & \mathbf{k} \\ a_1 & b_1 & c_1 \\ a_2 & b_2 & c_2 \end{vmatrix}$$

$$= \begin{vmatrix} b_1 & c_1 \\ b_2 & c_2 \end{vmatrix}\mathbf{i} - \begin{vmatrix} a_1 & c_1 \\ a_2 & c_2 \end{vmatrix}\mathbf{j} + \begin{vmatrix} a_1 & a_2 \\ b_1 & b_2 \end{vmatrix}\mathbf{k}$$

$$= (b_1c_2 - b_2c_1)\mathbf{i} + (a_2c_1 - a_1c_2)\mathbf{j} + (a_1b_2 - a_2b_1)\mathbf{k}$$
$$= \mathbf{F} \times \mathbf{G}.$$

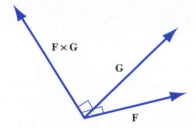

FIGURE 5.11 $\mathbf{F} \times \mathbf{G}$ *is orthogonal*
to $\mathbf{F}$ *and to* $\mathbf{G}$.

Interchanging two rows of a determinant changes its sign. This means that the cross product is *anticommutative*:

$$\mathbf{F} \times \mathbf{G} = -\mathbf{G} \times \mathbf{F}.$$

It is a routine calculation to verify that

$$\mathbf{F} \cdot (\mathbf{F} \times \mathbf{G}) = 0 \text{ and } \mathbf{F} \cdot (\mathbf{F} \times \mathbf{G}) = 0.$$

This means that $\mathbf{F} \times \mathbf{G}$ is orthogonal to both $\mathbf{F}$ and to $\mathbf{G}$ (Figure 5.11). One of the primary uses of the cross product is to produce a vector orthogonal to two given vectors. This can be used to find the equation of a plane containing three given points.

EXAMPLE 5.6

Find the equation of a plane containing the points $P : (-1, 4, 2)$, $Q : (6, -2, 8)$, and $R : (5, -1, -1)$.

Since we know a point (in fact, three points) in the plane, it is enough to find a normal vector $\mathbf{N}$ to the plane and then use equation (5.2). Use the three given points to form two vectors in the plane:

$$\mathbf{PQ} = 7\mathbf{i} - 6\mathbf{j} + 6\mathbf{k} \text{ and } \mathbf{PR} = 6\mathbf{i} - 5\mathbf{j} - 3\mathbf{k}.$$

The cross product of these vectors is orthogonal to the plane of these vectors, so

$$\mathbf{N} = \mathbf{PQ} \times \mathbf{PR} = 48\mathbf{i} + 57\mathbf{j} + \mathbf{k}$$

is a normal vector to this plane. By equation (5.2), the equation of the plane is

$$48(x + 1) + 57(y - 4) + (z - 2) = 0$$

or

$$48x + 57y + z = 182. \quad \blacklozenge$$

We will summarize some properties of the cross product.

1. $\mathbf{F} \times \mathbf{G} = -\mathbf{G} \times \mathbf{F}$.
2. $\mathbf{F} \times \mathbf{G}$ is orthogonal to both $\mathbf{F}$ and $\mathbf{G}$.
3. $\| \mathbf{F} \times \mathbf{G} \| = \| \mathbf{F} \| \| \mathbf{G} \| \sin(\theta)$, in which θ is the angle between $\mathbf{F}$ and $\mathbf{G}$.

4. If **F** and **G** are nonzero vectors, then $\mathbf{F} \times \mathbf{G} = \mathbf{O}$ if and only if **F** and **G** are parallel.

5. $\mathbf{F} \times (\mathbf{G} + \mathbf{H}) = \mathbf{F} \times \mathbf{G} + \mathbf{F} \times \mathbf{H}$.

6. $\alpha(\mathbf{F} \times \mathbf{G}) = (\alpha\mathbf{F}) \times \mathbf{G} = \mathbf{F} \times (\alpha\mathbf{G})$.

Property 4 follows from property 3, since two nonzero vectors are parallel exactly when the angle θ between them is zero, and in this, case $\sin(\theta) = 0$.

Property 4 provides a test for three points to be *collinear*, that is, to lie on a single line. Let P, Q, and R be the points. These points will be collinear exactly when the vector **F** from P to Q is parallel to the vector **G** from P to R. By property 4 this occurs when $\mathbf{F} \times \mathbf{G} = \mathbf{O}$.

SECTION 5.3   PROBLEMS

In each of Problems 1 through 4, compute $\mathbf{F} \times \mathbf{G}$ and $\mathbf{G} \times \mathbf{F}$ and verify the anticommutativity of the cross product.

1. $\mathbf{F} = -3\mathbf{i} + 6\mathbf{j} + \mathbf{k}, \mathbf{G} = -\mathbf{i} - 2\mathbf{j} + \mathbf{k}$

2. $\mathbf{F} = 6\mathbf{i} - \mathbf{k}, \mathbf{G} = \mathbf{j} + 2\mathbf{k}$

3. $\mathbf{F} = 2\mathbf{i} - 3\mathbf{j} + 4\mathbf{k}, \mathbf{G} = -3\mathbf{i} + 2\mathbf{j}$

4. $\mathbf{F} = 8\mathbf{i} + 6\mathbf{j}, \mathbf{G} = 14\mathbf{j}$

In each of Problems 5 through 9, determine whether the points are collinear. If they are not, determine an equation for the plane containing these points.

5. $(-1, 1, 6), (2, 0, 1), (3, 0, 0)$

6. $(4, 1, 1), (-2, -2, 3), (6, 0, 1)$

7. $(1, 0, -2), (0, 0, 0), (5, 1, 1)$

8. $(0, 0, 2), (-4, 1, 0), (2, -1, -1)$

9. $(-4, 2, -6), (1, 1, 3), (-2, 4, 5)$

In each of Problems 10, 11, and 12, find a vector normal to the given plane. There are infinitely many such vectors.

10. $8x - y + z = 12$

11. $x - y + 2z = 0$

12. $x - 3y + 2z = 9$

5.4   The Vector Space R^n

For systems involving n variables we may consider n-vectors

$$< x_1, x_2, \ldots, x_n >$$

with n components. Each x_j is a real number. The totality of such n-vectors is denoted R^n and is called "n-space". R^2 is the familiar plane and R^3 is 3-space. R^n has an algebraic structure which will prove useful when we consider matrices, systems of linear algebraic equations, and systems of linear differential equations.

Add *n*-vectors in the obvious ways:

$$<x_1, x_2, \ldots, x_n> + <y_1, y_2, \ldots, y_n> = <x_1+y_1, x_2+y_2, \ldots, x_n+y_n>$$

and multiply them by scalars:

$$\alpha <x_1, x_2, \ldots, x_n> = <\alpha x_1, \alpha x_2, \ldots, \alpha x_n>.$$

The dot product of two *n*-vectors is defined by

$$<x_1, x_2, \ldots, x_n> \cdot <y_1, y_2, \ldots, y_n> = x_1 y_1 + x_2 y_2 + \ldots + x_n y_n.$$

There is no cross product for *n*-vectors if $n > 3$.

As in the plane and 3-space, two *n*-vectors are equal exactly when their respective components are equal: $x_1 = y_1, x_2 = y_2, \ldots, x_n = y_n$.

The norm (length) of an *n*-vector $\mathbf{F} = <x_1, x_2, \ldots, x_n>$ is

$$\| \mathbf{F} \| = \sqrt{x_1^2 + \cdots + x_n^2}.$$

We can define *standard unit vectors* along the axes in R^n by

$$\mathbf{e}_1 = <1, 0, 0, \ldots, 0>,$$
$$\mathbf{e}_2 = <0, 1, 0, \ldots, 0>, \ldots,$$
$$\mathbf{e}_n = <0, 0, \ldots, 0, 1>.$$

This enables us to write any *n*-vector in *standard form*

$$<x_1, x_2, \ldots, x_n> = x_1 \mathbf{e}_1 + x_2 \mathbf{e}_2 + \cdots + x_n \mathbf{e}_n.$$

By analogy with 3-vectors, define two *n*-vectors to be *orthogonal* exactly when their dot product is zero. For example, the vectors $\mathbf{e}_1, \mathbf{e}_2, \ldots, \mathbf{e}_n$ are mutually orthogonal (each is orthogonal to all of the others).

A *linear combination* of *k* vectors $\mathbf{F}_1, \ldots, \mathbf{F}_k$ in R^n is a sum

$$\alpha_1 \mathbf{F}_1 + \alpha_2 \mathbf{F}_2 + \cdots + \alpha_k \mathbf{F}_k.$$

in which each α_j is a real number. Such a linear combination is again an *n*-vector.

A (finite) set of vectors in R^n is called *linearly dependent* if one of the vectors in the set is a linear combination of the others. Otherwise, the vectors are said to be *linearly independent*.

EXAMPLE 5.7

The vectors

$$\mathbf{F} = <3, -1, 0, 4>, \quad \mathbf{G} = <3, -8, -1, 10>, \quad \text{and} \quad \mathbf{H} = <6, -1, 1, 2>$$

are linearly dependent in R^4 because $\mathbf{G} = 3\mathbf{F} - \mathbf{H}$. ◆

In this example, we could also write

$$\mathbf{G} - 3\mathbf{F} + \mathbf{H} = <0, 0, 0, 0>.$$

This is a linear combination of $\mathbf{F}$, $\mathbf{G}$, and $\mathbf{H}$ with at least one nonzero coefficient and adding up to the zero vector. It is not difficult to show that this is just another formulation of linear dependence: *k* vectors in R^n are linearly dependent if and only if there is a linear combination of these *k* vectors with at least one nonzero coefficient adding up to the zero vector in R^n.

If the only linear combination of $\mathbf{F}_1, \ldots, \mathbf{F}_k$ that sums to the zero vector is the linear combination with all zero coefficients, then $\mathbf{F}_1, \ldots, \mathbf{F}_k$ are linearly independent. This means that no one of the vectors is a linear combination of the others.

EXAMPLE 5.8

We will determine whether or not the vectors $<1, 0, 3, 1>$, $<0, 1, -6, -1>$, and $<0, 2, 1, 0>$ are linearly independent in R^4. Suppose some linear combination of these vectors is equal to the zero vector in R^4:

$$\alpha <1, 0, 3, 1> + \beta <0, 1, -6, -1> + \gamma <0, 2, 1, 0> = <0, 0, 0, 0>.$$

Then

$$<\alpha, \beta + 2\gamma, 3\alpha - 6\beta + \gamma, \alpha - \beta> = <0, 0, 0, 0>.$$

Equating respective components, we obtain the system of algebraic equations

$$\alpha = 0$$
$$\beta + 2\gamma = 0$$
$$3\alpha - 6\beta + \gamma = 0$$
$$\alpha - \beta = 0.$$

From the first equation, $\alpha = 0$. From the fourth equation, $\beta = 0$. Then from the second equation, $\gamma = 0$. The only linear combination of these vectors that equals the zero vector has all of its coefficients zero. Therefore the vectors are linearly independent. ◆

Any set of mutually orthogonal nonzero vectors in R^n is linearly independent. To see why this is true, suppose that $\mathbf{F}_1, \ldots, \mathbf{F}_k$ are nonzero and mutually orthogonal, and suppose that a linear combination of these vectors equals the zero vector,

$$\alpha_1 \mathbf{F}_1 + \alpha_2 \mathbf{F}_2 + \cdots + \alpha_k \mathbf{F}_k = <0, 0, \ldots, 0>.$$

Take the dot product of this sum with $\mathbf{F}_1$ to get

$$\alpha_1 \mathbf{F}_1 \cdot \mathbf{F}_1 + \alpha_2 \mathbf{F}_2 \cdot \mathbf{F}_1 + \cdots + \mathbf{F}_k \cdot \mathbf{F}_1 = \mathbf{F}_1 \cdot <0, 0, \ldots, 0> = 0.$$

Because of the orthogonality, all terms on the left are zero except the first, and we are left with

$$\alpha_1 \| \mathbf{F}_1 \|^2 = 0.$$

Since $\mathbf{F}_1$ is not the zero vector, $\| \mathbf{F}_1 \| \neq 0$, and therefore, $\alpha_1 = 0$.

By taking the dot product with the other $\mathbf{F}_j$'s, we conclude that each $\alpha_j = 0$, so the vectors are linearly independent.

Subspaces

A set S of vectors in R^n is a *subspace* of R^n if it satisfies the following conditions:

1. S is not empty. That is, S has at least one vector in it.
2. Any linear combination of vectors in S is again in S.

EXAMPLE 5.9

Let S consist of all vectors $<a, b, 0, 0>$ in R^4. Then S is a subspace of R^4.

By contrast, let T consist of all vectors $<a, b, 0, 1>$ in R^4. Then T is not a subspace, because a sum of two such vectors will have fourth component 2, and therefore will not be in T. ◆

If W is a subspace of R^n, then a *basis* for W is a set of linearly independent vectors $\mathbf{v}_1, \cdots, \mathbf{v}_k$ in W, having the property that every vector in W is a linear combination of $\mathbf{v}_1, \cdots, \mathbf{v}_k$. For example,

$$\mathbf{F}_1 = <1, 0, 0, 0> \text{ and } \mathbf{F}_2 = <0, 1, 0, 0>$$

form a basis for the subspace S of R^4 in Example 5.9. These vectors are linearly independent, and every vector in S is of the form

$$<a, b, 0, 0> = a<1, 0, 0, 0> + b<0, 1, 0, 0> = a\mathbf{F}_1 + b\mathbf{F}_2,$$

a linear combination of $\mathbf{F}_1$ and $\mathbf{F}_2$.

It can be shown that any two bases of a given subspace have the same number of vectors in them. The number of vectors in a basis for a subspace of R^n is called the *dimension* of that subspace. The subspace of R^4 in Example 5.9 has a dimension of 2. We may also think of R^n as a subspace of itself (any linear combination of n-vectors is again an n-vector). R^n has a dimension of n, because the standard unit vectors constitute a basis.

SECTION 5.4   *PROBLEMS*

In each of Problems 1 through 10, determine whether the vectors are linearly independent or dependent in the appropriate R^n.

1. $3\mathbf{i} + 2\mathbf{j}, \mathbf{i} - \mathbf{j}$ in R^3

2. $2\mathbf{i}, 3\mathbf{j}, 5\mathbf{i} - 12\mathbf{k}, \mathbf{i} + \mathbf{j} + \mathbf{k}$ in R^3

3. $<8, 0, 2, 0, 0, 0, 0>, <0, 0, 0, 0, 1, -1, 0>$ in R^7

4. $<1, 0, 0, 0>, <0, 1, 1, 0>, <-4, 6, 6, 0>$ in R^4

5. $<1, 2, -3, 1>, <4, 0, 0, 2>, <6, 4, -6, 4>$ in R^4

6. $<0, 1, 1, 1>, <-3, 2, 4, 4>, <-2, 2, 34, 2>, <1, 1, -6, -2>$ in R^4

7. $<1, -2>, <4, 1>, <6, 6>$ in R^2

8. $<-1, 1, 0, 0, 0>, <0, -1, 1, 0, 0>, <0, 1, 1, 1, 0>$ in R^5

9. $<-2, 0, 0, 1, 1>, <1, 0, 0, 0, 0>, <0, 0, 0, 0, 2>, <1, -1, 3, 3, 1>$ in R^5

10. $<3, 0, 0, 4>, <2, 0, 0, 8>$ in R^4

In each of Problems 11 through 15, show that the set S is a subspace of the appropriate R^n and find a basis for this subspace and the dimension of the subspace.

11. S consists of all vectors $<x, y, -y, -x>$ in R^4.

12. S consists of all vectors $<x, y, 2x, 3y>$ in R^4.

13. S consists of all vectors in R^4 with zero second component.

14. S consists of all vectors in R^6 of the form $<x, x, y, y, 0, z>$.

15. S consists of all vectors $<0, x, 0, 2x, 0, 3x, 0>$ in R^7.

CHAPTER 6

Matrices and Systems of Linear Equations

6.1   Matrices

An $n \times m$ *matrix* is a rectangular array of objects arranged in n rows and m columns. We will denote matrices in boldface. For example,

$$\mathbf{A} = \begin{pmatrix} 2 & 1 & \pi & -4 \\ 6 & 1 & 8 & \sqrt{2} \end{pmatrix}$$

is a 2×4 matrix (two rows, four columns).

The object located in the row i and column j place of a matrix is called its i, j *element*. Often we write $\mathbf{A} = [a_{ij}]$, meaning that the i, j element of $\mathbf{A}$ is a_{ij}. In the above matrix $\mathbf{A}$, $a_{11} = 2$, $a_{13} = \pi$, $a_{23} = 8$, and $a_{24} = \sqrt{2}$.

The columns of an $n \times m$ matrix of real numbers are made up of n-vectors, and the rows are m-vectors. In the above matrix , the four columns are vectors in R^2, and the two rows are vectors in R^4.

In general, two matrices $\mathbf{A} = [a_{ij}]$ and $\mathbf{B} = [b_{ij}]$ are *equal* if they have the same number of rows, the same number of columns, and for each i and j, $a_{ij} = b_{ij}$. That is, the matrices have the same dimensions, and objects located in the same positions of the respective matrices must be equal.

There are three operations we will define for matrices: addition, multiplication by a real or complex number, and multiplication. These are defined as follows.

Addition of Matrices

If $\mathbf{A} = [a_{ij}]$ and $\mathbf{B} = [b_{ij}]$ are both $n \times m$ matrices, then their sum is defined to be the $n \times m$ matrix

$$\mathbf{A} + \mathbf{B} = [a_{ij} + b_{ij}].$$

We add two matrices of the same dimensions by adding objects in the same locations in the matrices. For example,

$$\begin{pmatrix} 1 & 2 & -3 \\ 4 & 0 & 2 \end{pmatrix} + \begin{pmatrix} -1 & 6 & 3 \\ 8 & 12 & 14 \end{pmatrix} = \begin{pmatrix} 0 & 8 & 0 \\ 12 & 12 & 16 \end{pmatrix}.$$

We may also think of this as adding respective row vectors or as adding respective column vectors.

Multiplication by a Number

Multiply a matrix by a number by multiplying each matrix element by the scalar. If $\mathbf{A} = [a_{ij}]$ then $c\mathbf{A} = [ca_{ij}]$.

For example,

$$\sqrt{2} \begin{pmatrix} -3 \\ 4 \\ 2t \\ \sin(2t) \end{pmatrix} = \begin{pmatrix} -3\sqrt{2} \\ 4\sqrt{2} \\ 2t\sqrt{2} \\ \sqrt{2}\sin(2t) \end{pmatrix}.$$

Multiplication of Matrices

Let $\mathbf{A} = [a_{ij}]$ be $n \times k$ and $\mathbf{B} = [b_{ij}]$ be $k \times m$. Then the product $\mathbf{AB}$ is the $n \times m$ matrix whose (i, j) element is

$$a_{i1}b_{1j} + a_{i2}b_{2j} + \ldots + a_{ik}b_{kj}.$$

This (i, j) element can be written

$$\sum_{s=1}^{k} a_{is}b_{sj}.$$

We can think of the elements of $\mathbf{AB}$ as dot products of row vectors of $\mathbf{A}$ with column vectors of $\mathbf{B}$:

$$(i, j) \text{ element of } \mathbf{AB} = (\text{row } i \text{ of } \mathbf{A}) \cdot (\text{column } j \text{ of } \mathbf{B}).$$

This explains why the number of columns of $\mathbf{A}$ must equal the number of rows of $\mathbf{B}$ for the product $\mathbf{AB}$ to be defined. We can only take the dot product of two vectors of the same dimension.

EXAMPLE 6.1

Let

$$A = \begin{pmatrix} 1 & 3 \\ 2 & 5 \end{pmatrix} \text{ and } B = \begin{pmatrix} 1 & 1 & 3 \\ 2 & 1 & 4 \end{pmatrix}.$$

Then A is 2×2 and B is 2×3, so we can compute AB, which will be a 2×3 matrix (number of rows of A, number of columns of B). In terms of dot products of rows with columns,

$$AB = \begin{pmatrix} 1 & 3 \\ 2 & 5 \end{pmatrix} \begin{pmatrix} 1 & 1 & 3 \\ 2 & 1 & 4 \end{pmatrix}$$

$$= \begin{pmatrix} <1,3> \cdot <1,2> & <1,3> \cdot <1,1> & <1,3> \cdot <3,4> \\ <2,5> \cdot <1,2> & <2,5> \cdot <1,1> & <2,5> \cdot <3,4> \end{pmatrix} = \begin{pmatrix} 7 & 4 & 15 \\ 12 & 7 & 26 \end{pmatrix}.$$

In this example, BA is not defined because the number of columns of B does not equal the number of rows of A. ◆

EXAMPLE 6.2

Let

$$A = \begin{pmatrix} 1 & 1 & 2 & 1 \\ 4 & 1 & 6 & 2 \end{pmatrix} \text{ and } B = \begin{pmatrix} -1 & 8 \\ 2 & 1 \\ 1 & 1 \\ 12 & 6 \end{pmatrix}.$$

Because A is 2×4 and B is 4×2, then AB is defined and is 2×2:

$$AB = \begin{pmatrix} <1,1,2,1> \cdot <-1,2,1,12> & <1,1,2,1> \cdot <8,1,1,6> \\ <4,1,6,2> \cdot <-1,2,1,12> & <4,1,6,2> \cdot <8,1,1,6> \end{pmatrix} = \begin{pmatrix} 15 & 17 \\ 28 & 51 \end{pmatrix}.$$

In this example, BA is also defined and is a 4×4 matrix:

$$BA = \begin{pmatrix} -1 & 8 \\ 2 & 1 \\ 1 & 1 \\ 12 & 6 \end{pmatrix} \begin{pmatrix} 1 & 1 & 2 & 1 \\ 4 & 1 & 6 & 2 \end{pmatrix} = \begin{pmatrix} 31 & 7 & 46 & 15 \\ 6 & 3 & 10 & 4 \\ 5 & 2 & 8 & 3 \\ 36 & 18 & 60 & 24 \end{pmatrix}.$$

This also shows that even when both AB and BA are defined, these matrices may not be equal and may not even have the same dimensions. ◆

We will list some properties of these matrix operations, assuming that the indicated operations are defined:

1. $A + B = B + A$.
2. $A(B + C) = AB + AC$.
3. $(A + B)C = AC + AC$.
4. $(AB)C = A(BC)$.
5. $cAB = (cA)B = A(cB)$ for any number c.

Zero Matrix

The *zero matrix* $\mathbf{O}_{nm}$ is the $n \times m$ matrix having every element equal to zero. This matrix has the property that

$$\mathbf{A} + \mathbf{O}_{nm} = \mathbf{O}_{nm} + \mathbf{A} = \mathbf{A}$$

for every $n \times m$ matrix $\mathbf{A}$.

With real numbers, if $ab = 0$, we know that at least one of a or b is zero. This is not true with matrices.

EXAMPLE 6.3

$$\begin{pmatrix} 1 & 2 \\ 0 & 0 \end{pmatrix} \begin{pmatrix} 6 & 4 \\ -3 & -2 \end{pmatrix} = \begin{pmatrix} 0 & 0 \\ 0 & 0 \end{pmatrix} = \mathbf{O}_{22}.$$

This product of matrices is a zero matrix, even though neither factor is a zero matrix. ◆

Square and Identity Matrices

A *square matrix* is one having the same number of rows and columns. If $\mathbf{A} = [a_{ij}]$ is $n \times n$, the *main diagonal* of $\mathbf{A}$ consists of the matrix elements $a_{11}, a_{22}, \ldots, a_{nn}$. These are the matrix elements from the upper-left corner to the lower-right corner.

The *identity matrix* $\mathbf{I}_n$ is the $n \times n$ matrix having each (i, j) element equal to 0 if $i \neq j$ and each main diagonal element equal to 1. For example,

$$\mathbf{I}_2 = \begin{pmatrix} 1 & 0 \\ 0 & 1 \end{pmatrix} \text{ and } \mathbf{I}_3 = \begin{pmatrix} 1 & 0 & 0 \\ 0 & 1 & 0 \\ 0 & 0 & 1 \end{pmatrix}.$$

If $\mathbf{A}$ is $n \times n$, then

$$\mathbf{A}\mathbf{I}_n = \mathbf{I}_n\mathbf{A} = \mathbf{A}.$$

If $\mathbf{A} = [a_{ij}]$ is $n \times m$, then the *transpose* of $\mathbf{A}$ is the $m \times n$ matrix $\mathbf{A}^t = [a_{ji}]$. We obtain $\mathbf{A}^t$ by interchanging the rows and columns of $\mathbf{A}$. For example, the 2×3 matrix

$$\mathbf{C} = \begin{pmatrix} -1 & 6 & 1 + x^2 \\ x & 7 & 1 \end{pmatrix}$$

has the 3×2 transpose

$$\mathbf{C}^t = \begin{pmatrix} -1 & x \\ 6 & 7 \\ 1 + x^2 & 1 \end{pmatrix}.$$

If we take the transpose of a transpose, then we interchange the rows and columns, then interchange them again, leaving every element in its original position. This means that

$$(\mathbf{A}^t)^t = \mathbf{A}.$$

It is less obvious that, if we take the transpose of a product, then we obtain the product of the transposes, in the reverse order.

$$(\mathbf{AB})^t = \mathbf{B}^t \mathbf{A}^t.$$

This is consistent with the definition of the matrix product. If $\mathbf{A}$ is $n \times m$ and $\mathbf{B}$ is $m \times k$, then $\mathbf{AB}$ is $n \times k$, so $(\mathbf{AB})^t$ is $k \times n$. However, $\mathbf{A}^t$ is $m \times n$ and $\mathbf{B}^t$ is $k \times m$, so $\mathbf{A}^t \mathbf{B}^t$ is defined only if $n = k$, while $\mathbf{B}^t \mathbf{A}^t$ is always defined and is $k \times n$.

In some contexts, it is useful to observe that the dot product of two n-vectors can be written as a matrix product. Write the n-vectors

$$\mathbf{X} = <x_1, x_2, \ldots, x_n> \text{ and } \mathbf{Y} = <y_1, y_2, \ldots, y_n>.$$

as $n \times 1$ column matrices

$$\mathbf{X} = \begin{pmatrix} x_1 \\ x_2 \\ \vdots \\ x_n \end{pmatrix} \text{ and } \mathbf{Y} = \begin{pmatrix} y_1 \\ y_2 \\ \vdots \\ y_n \end{pmatrix}.$$

Then $\mathbf{X}^t$ is a $1 \times n$ matrix, and $\mathbf{X}^t \mathbf{Y}$ is a 1×1 matrix, which we think of as just a scalar:

$$\mathbf{X}^t \mathbf{Y} = \begin{pmatrix} x_1 & x_2 & \ldots & x_n \end{pmatrix} \begin{pmatrix} y_1 \\ y_2 \\ \vdots \\ y_n \end{pmatrix}$$

$$= (x_1 y_1 + x_2 y_2 + \ldots + x_n y_n) = \mathbf{X} \cdot \mathbf{Y}.$$

SECTION 6.1   PROBLEMS

In each of Problems 1 through 6, perform the requested computation.

1. $\mathbf{A} = \begin{pmatrix} 1 & -1 & 3 \\ 2 & -4 & 6 \\ -1 & 1 & 2 \end{pmatrix}$, $\mathbf{B} = \begin{pmatrix} -4 & 0 & 0 \\ -2 & -1 & 6 \\ 8 & 15 & 4 \end{pmatrix}$; $2\mathbf{A} - 3\mathbf{B}$

2. $\mathbf{A} = \begin{pmatrix} -2 & 2 \\ 0 & 1 \\ 14 & 2 \\ 6 & 8 \end{pmatrix}$, $\mathbf{B} = \begin{pmatrix} 4 & 4 \\ 2 & 1 \\ 14 & 16 \\ 1 & 25 \end{pmatrix}$, $-5\mathbf{A} + 3\mathbf{B}$

3. $\mathbf{A} = \begin{pmatrix} x & 1-x \\ 2 & e^x \end{pmatrix}$, $\mathbf{B} = \begin{pmatrix} 1 & -6 \\ x & \cos(x) \end{pmatrix}$, $\mathbf{A}^2 + 2\mathbf{AB}$

4. $\mathbf{A} = (14), \mathbf{B} = (-12), -3\mathbf{A} - 5\mathbf{B}$

5. $\mathbf{A} = \begin{pmatrix} 1 & -2 & 1 & 7 & -9 \\ 8 & 2 & -5 & 0 & 0 \end{pmatrix}$,
   $\mathbf{B} = \begin{pmatrix} -5 & 1 & 8 & 21 & 7 \\ 12 & -6 & -2 & -1 & 9 \end{pmatrix}$, $4\mathbf{A} + 8\mathbf{B}$

6. $\mathbf{A} = \begin{pmatrix} -2 & 3 \\ 1 & 1 \end{pmatrix}$, $\mathbf{B} = \begin{pmatrix} 0 & 8 \\ -5 & 1 \end{pmatrix}$, $\mathbf{A}^3 - \mathbf{B}^2$

In each of Problems 7 through 12, determine which of $\mathbf{AB}$ and $\mathbf{BA}$ are defined. Carry out all such products.

7. $\mathbf{A} = \begin{pmatrix} -4 & 6 & 2 \\ -2 & -2 & 3 \\ 1 & 1 & 8 \end{pmatrix}$, $\mathbf{B} = \begin{pmatrix} -2 & 4 & 6 & 12 & 5 \\ -3 & -3 & 1 & 1 & 4 \\ 0 & 0 & 1 & 6 & -9 \end{pmatrix}$

8. $\mathbf{A} = \begin{pmatrix} -2 & -4 \\ 3 & -1 \end{pmatrix}$, $\mathbf{B} = \begin{pmatrix} 6 & 8 \\ 1 & -4 \end{pmatrix}$

9. $\mathbf{A} = (-1 \quad 6 \quad 2 \quad 14 \quad -22)$, $\mathbf{B} = \begin{pmatrix} -3 \\ 2 \\ 6 \\ 0 \\ -4 \end{pmatrix}$

10. $\mathbf{A} = \begin{pmatrix} -3 & 1 \\ 6 & 2 \\ 18 & -22 \\ 1 & 6 \end{pmatrix}$, $\mathbf{B} = \begin{pmatrix} -16 & 0 & 0 & 28 \\ 0 & 1 & 1 & 26 \end{pmatrix}$

11. $\mathbf{A} = \begin{pmatrix} -21 & 4 & 8 & -3 \\ 12 & 1 & 0 & 14 \\ 1 & 16 & 0 & -8 \\ 13 & 4 & 8 & 0 \end{pmatrix}$, $\mathbf{B} = \begin{pmatrix} -9 & 16 & 3 & 2 \\ 5 & 9 & 14 & 0 \end{pmatrix}$

12. $\mathbf{A} = \begin{pmatrix} -2 & 4 \\ 3 & 9 \end{pmatrix}$, $\mathbf{B} = \begin{pmatrix} 1 & -3 & 7 & 2 \\ 5 & 9 & 1 & 0 \end{pmatrix}$

13. Find nonzero 2×2 matrices $\mathbf{A}$, $\mathbf{B}$, and $\mathbf{C}$ such that $\mathbf{BA} = \mathbf{CA}$ but $\mathbf{B} \neq \mathbf{C}$.

6.2   Linear Homogeneous Systems

We will apply matrices to the solution of the *linear homogeneous system* of n equations in m unknowns:

$$a_{11}x_1 + a_{12}x_2 + \ldots + a_{1m}x_m = 0$$

$$a_{21}x_1 + a_{22}x_2 + \ldots + a_{2m}x_m = 0$$

$$\cdots\cdots\cdots$$

$$a_{n1}x_1 + a_{n2}x_2 + \ldots + a_{nm}x_m = 0.$$

The numbers a_{ij} are the *coefficients* of the system, and $\mathbf{A} = [a_{ij}]$ is the *matrix of coefficients* of the system. Row i contains the coefficients of equation i and column j contains the coefficients of x_j. Define

$$\mathbf{X} = \begin{pmatrix} x_1 \\ x_2 \\ \vdots \\ x_m \end{pmatrix}$$

and write the $n \times 1$ zero matrix $\mathbf{O}_{n1}$ as $\mathbf{O}$. Then the system can be written as the matrix equation

$$\mathbf{AX} = \mathbf{O}.$$

Our objective is to define operations on the matrix of coefficients $\mathbf{A}$ to produce a new system that has the same solutions as the original system, but is easier to solve. This will provide a method for solving the system by matrix manipulations.

There are three operations we can carry out so that the new system obtained has the same solutions as the original system. These are

1. Interchange two equations of the system.
2. Multiply an equation of the system by a nonzero number.
3. Add any constant multiple of one equation to another equation.

Performing these operations on the equations is equivalent to performing them on the matrix of coefficients. We therefore define the equivalent matrix operations:

1. Type 1 operation: interchange any two rows of the matrix of coefficients.
2. Type 2 operation: multiply any row of the matrix by a nonzero constant.
3. Type 3 operation: add any constant multiple of one row of the matrix to another row.

These are called *elementary row operations*. If a matrix $\mathbf{C}$ is obtained from a matrix $\mathbf{B}$ by a sequence of successively performed elementary row operations, then we say that $\mathbf{C}$ and $\mathbf{B}$ are *row equivalent*.

Reduced Matrix

Now define a matrix to be *reduced*, or in *reduced form* if it satisfies the following conditions:

1. If a row has a nonzero element, its first nonzero element (looking from left to right) is 1. The first nonzero element of any row is called the *leading entry* of that row.

2. If a column contains the leading entry of some row, all the other elements of that column are 0.

3. If there is a row with all zero elements, then this row must be below any row having a nonzero element.

4. The leading entries of the rows move from left to right as we move down the matrix. That is, if the leading entry of row r_1 is in column c_1, and the leading entry of row r_2 is in column c_2, and $r_1 < r_2$, then $c_1 < c_2$.

EXAMPLE 6.4

The following matrices are all reduced:

$$\begin{pmatrix} 1 & -4 & 1 & 0 \\ 0 & 0 & 0 & 1 \end{pmatrix}, \begin{pmatrix} 0 & 1 & 3 & 0 \\ 0 & 0 & 0 & 1 \\ 0 & 0 & 0 & 9 \end{pmatrix}, \begin{pmatrix} 0 & 1 & 2 & 0 & 0 \\ 0 & 0 & 0 & 1 & 0 \\ 0 & 0 & 0 & 0 & 0 \\ 0 & 0 & 0 & 0 & 0 \end{pmatrix}$$

and

$$\begin{pmatrix} 1 & 0 & 0 & 3 & 1 \\ 0 & 1 & 0 & -2 & 4 \\ 0 & 0 & 1 & 0 & 1 \\ 0 & 0 & 0 & 0 & 0 \end{pmatrix}.$$

The matrix

$$\begin{pmatrix} 0 & 1 & 5 & 0 & 0 \\ 0 & 0 & 1 & 0 & 0 \\ 0 & 0 & 0 & 1 & 0 \\ 0 & 0 & 0 & 0 & 1 \end{pmatrix}$$

is not reduced. The leading entry (first nonzero entry) of row two is 1 in the $2, 3$ place. But condition 2 of the definition of reduced matrix would require that all the other elements of that column be zero, and there is a 5 in the $1, 3$ place. ♦

Here is the fundamental result on reduced matrices.

THEOREM 6.1

If $\mathbf{A}$ is an $n \times m$ matrix of numbers, then there is a reduced matrix $\mathbf{A}_R$ that can be obtained from $\mathbf{A}$ by a sequence of elementary row operations, hence is row equivalent to $\mathbf{A}$. Further, if a different sequence of row operations is used to obtain a reduced matrix from $\mathbf{A}$, then this reduced matrix equals $\mathbf{A}_R$. ♦

This says that we can transform any matrix to reduced form by elementary row operations. Further, given $\mathbf{A}$, we always arrive at the same reduced matrix $\mathbf{A}_R$, no matter what row operations are used to reduce $\mathbf{A}$.

EXAMPLE 6.5

The last matrix of Example 6.4 can be reduced by adding -5 times row two to row one. Here is another example. Let

$$\mathbf{A} = \begin{pmatrix} -2 & 1 & 3 \\ 0 & 1 & 1 \\ 2 & 0 & 1 \end{pmatrix}.$$

We will reduce $\mathbf{A}$ as follows.

1. Multiply row one by $-1/2$ to get

$$\mathbf{A}_1 = \begin{pmatrix} 1 & -1/2 & -3/2 \\ 0 & 1 & 1 \\ 2 & 0 & 1 \end{pmatrix}.$$

2. Add (-2) row one to row three:

$$\mathbf{A}_2 = \begin{pmatrix} 1 & -1/2 & -3/2 \\ 0 & 1 & 1 \\ 0 & 1 & 4 \end{pmatrix}.$$

In $\mathbf{A}_2$, column two has a nonzero element below row one with the highest being in the $2, 2$ position. Since we want a 1 here, we do not have to multiply this row by anything. However, we want zeros above and below this 1 in the $1, 2$ and $3, 2$ positions. Therefore,

3. Add $1/2$ times row two to row one and -1 times row two to row three to get

$$\mathbf{A}_3 = \begin{pmatrix} 1 & 0 & -1 \\ 0 & 1 & 1 \\ 0 & 0 & 3 \end{pmatrix}.$$

In this matrix, column three has a nonzero element below row two in the $3, 3$ location.

4. Multiply row three by $1/3$:

$$\mathbf{A}_4 = \begin{pmatrix} 1 & 0 & -1 \\ 0 & 1 & 1 \\ 0 & 0 & 1 \end{pmatrix}.$$

Finally, we want zeros above the 1 in the $3, 3$ position of $\mathbf{A}_4$:

5. Add row three to row one and -1 times row three to row two:

$$\mathbf{A}_5 = \begin{pmatrix} 1 & 0 & 0 \\ 0 & 1 & 0 \\ 0 & 0 & 1 \end{pmatrix}.$$

This is a reduced matrix, so $\mathbf{A}_R = \mathbf{A}_5$.

Observe how we can use $\mathbf{A}_R$ to solve the 3×3 homogeneous system $\mathbf{AX} = \mathbf{O}$. This system is

$$-2x_1 + x_2 + 3x_3 = 0,$$
$$x_2 + x_3 = 0,$$
$$2x_1 + x_3 = 0.$$

Because elementary row operations produce new systems with the same solutions as the original, this system has the same solutions as the reduced system

$$\mathbf{A}_R \mathbf{X} = \mathbf{O}.$$

This reduced system is

$$\begin{pmatrix} 1 & 0 & 0 \\ 0 & 1 & 0 \\ 0 & 0 & 1 \end{pmatrix} \begin{pmatrix} x_1 \\ x_2 \\ x_3 \end{pmatrix} = \begin{pmatrix} 0 \\ 0 \\ 0 \end{pmatrix},$$

or

$$x_1 = 0$$
$$x_2 = 0$$
$$x_3 = 0.$$

with only the trivial solution $x_1 = x_2 = x_3 = 0$. Therefore, the original system $\mathbf{AX} = \mathbf{O}$ also has only the trivial solution. ◆

As the matrix of this example suggests, if $\mathbf{A}$ is $n \times n$, then $\mathbf{A}_R$ has n nonzero rows exactly when $\mathbf{A}_R = \mathbf{I}_n$, an observation that will be useful later.

EXAMPLE 6.6

We will solve the system

$$x_1 - 3x_2 + x_3 - 7x_4 + 4x_5 = 0$$
$$x_1 + 2x_2 - 3x_3 = 0$$
$$x_2 - 4x_3 + x_5 = 0.$$

The matrix of coefficients is

$$\mathbf{A} = \begin{pmatrix} 1 & -3 & 1 & -7 & 4 \\ 1 & 2 & -3 & 0 & 0 \\ 0 & 1 & -4 & 0 & 1 \end{pmatrix}.$$

We will find $\mathbf{A}_R$ to deal with the simpler system $\mathbf{A}_R\mathbf{X} = \mathbf{O}$. First subtract row one from row two:

$$\mathbf{A}_1 = \begin{pmatrix} 1 & -3 & 1 & -7 & 4 \\ 0 & 5 & -4 & 7 & -4 \\ 0 & 1 & -4 & 0 & 1 \end{pmatrix}.$$

Add 3 times row three to row one and -5 times row three to row two:

$$\mathbf{A}_2 = \begin{pmatrix} 1 & 0 & -11 & -7 & 7 \\ 0 & 0 & 16 & 7 & -9 \\ 0 & 1 & -4 & 0 & 1 \end{pmatrix}.$$

Interchange rows two and three:

$$\mathbf{A}_3 = \begin{pmatrix} 1 & 0 & -11 & -7 & 7 \\ 0 & 1 & -4 & 0 & 1 \\ 0 & 0 & 16 & 7 & -9 \end{pmatrix}.$$

Multiply row three by $1/16$:

$$\mathbf{A}_4 = \begin{pmatrix} 1 & 0 & -11 & -7 & 7 \\ 0 & 1 & -4 & 0 & 1 \\ 0 & 0 & 1 & 7/16 & -9/16 \end{pmatrix}.$$

Add 4 times row three to row two and 11 times row three to row one:

$$\mathbf{A}_5 = \begin{pmatrix} 1 & 0 & 0 & -35/16 & 13/16 \\ 0 & 1 & 0 & 28/16 & -20/16 \\ 0 & 0 & 1 & 7/16 & -9/16 \end{pmatrix}.$$

This is $\mathbf{A}_R$, and the reduced system $\mathbf{A}_R \mathbf{X} = \mathbf{O}$ is

$$x_1 - \tfrac{35}{16} x_4 + \tfrac{13}{16} x_5 = 0,$$

$$x_2 + \tfrac{28}{16} x_4 - \tfrac{20}{16} x_5 = 0,$$

$$x_3 + \tfrac{7}{16} x_4 - \tfrac{9}{16} x_5 = 0.$$

Now the point: this system is easy to solve:

$$x_1 = \tfrac{35}{16} x_4 - \tfrac{13}{16} x_5,$$

$$x_2 = -\tfrac{28}{16} x_4 + \tfrac{20}{16} x_5,$$

and

$$x_3 = -\tfrac{7}{16} x_4 + \tfrac{9}{16} x_5,$$

in which x_4 and x_5 can be given any values and these determine x_1, x_2, and x_3.

We can write this solution more neatly by setting $x_4 = 16\alpha$ and $x_5 = 16\beta$, with α and β as arbitrary numbers. Now

$$x_1 = 35\alpha - 13\beta,$$

$$x_2 = -28\alpha + 20\beta,$$

$$x_3 = -7\alpha + 9\beta,$$

$$x_4 = 16\alpha,$$

and

$$x_5 = 16\beta,$$

in which α and β are any numbers. We call this the *general solution* of the reduced system, and this is also the general solution of the original system. This formulation of the solution also makes it clear that the solution is two-dimensional in the sense that it depends on two arbitrary constants. By varying α and β, we obtain all solutions of the system. We can also write this general solution as a column matrix

$$\mathbf{X} = \begin{pmatrix} 35\alpha - 13\beta \\ -28\alpha + 20\beta \\ -7\alpha + 9\beta \\ 16\alpha \\ 16\beta \end{pmatrix} = \alpha \begin{pmatrix} 35 \\ -28 \\ -7 \\ 16 \\ 0 \end{pmatrix} + \beta \begin{pmatrix} -13 \\ 20 \\ 9 \\ 0 \\ 16 \end{pmatrix}. \; \blacklozenge$$

These examples suggest a strategy for solving the homogeneous system $\mathbf{AX} = \mathbf{O}$.

1. Find $\mathbf{A}_R$.

2. Solve the simpler system $\mathbf{A}_R \mathbf{X} = \mathbf{O}$.

3. The general solution of this system is the same as the general solution of the original system. The system may have only the trivial solution.

EXAMPLE 6.7

Solve the system

$$-x_1 + x_3 + x_4 + 2x_5 = 0,$$
$$x_2 + 3x_3 + 4x_5 = 0,$$
$$x_1 + 2x_2 + x_3 + x_4 + x_5 = 0,$$

and

$$-3x_1 + x_2 + 4x_5 = 0.$$

The matrix of coefficients is

$$\mathbf{A} = \begin{pmatrix} -1 & 0 & 1 & 1 & 2 \\ 0 & 1 & 3 & 0 & 4 \\ 1 & 2 & 1 & 1 & 1 \\ -3 & 1 & 0 & 0 & 4 \end{pmatrix}.$$

Routine manipulations yield the reduced form

$$\mathbf{A}_R = \begin{pmatrix} 1 & 0 & 0 & 0 & -9/8 \\ 0 & 1 & 0 & 0 & 5/8 \\ 0 & 0 & 1 & 0 & 9/8 \\ 0 & 0 & 0 & 1 & -1/4 \end{pmatrix}.$$

This is the coefficient matrix of the reduced system

$$x_1 - \tfrac{9}{8}x_5 = 0,$$
$$x_2 + \tfrac{5}{8}x_5 = 0,$$
$$x_3 + \tfrac{9}{8}x_5 = 0,$$

and

$$x_4 - \tfrac{1}{4}x_5 = 0.$$

Notice that x_1 through x_4 depend on x_5, which can be chosen arbitrarily. Set $x_5 = \alpha$ to write the general solution

$$x_1 = \frac{9}{8}\alpha, \ x_2 = -\frac{5}{8}\alpha, \ x_3 = -\frac{9}{8}\alpha, \ x_4 = \frac{1}{4}\alpha, \text{ and } x_5 = \alpha.$$

More neatly, if we let $\beta = \alpha/8$ (β is still any number), then

$$x_1 = 9\beta, \ x_2 = -5\beta, \ x_3 = -9\beta, \ x_4 = 2\beta, \text{ and } x_5 = 8\beta.$$

As a column matrix, this solution is

$$\mathbf{X} = \beta \begin{pmatrix} 9 \\ -5 \\ -9 \\ 2 \\ 8 \end{pmatrix}.$$

We may think of this solution as one-dimensional, depending on one arbitrary constant. ◆

We may think of the solutions of an $n \times m$ linear system as vectors in R^m. Since a linear combination of solutions of a homogeneous system is again a solution, these solutions form a subspace of R^m called the *solution space* of the system.

It is possible to read the dimension of this solution space directly from $\mathbf{A}_R$, hence, to know how many arbitrary constants to expect in the general solution. Define the *rank* of an $n \times m$ matrix $\mathbf{A}$ to be the number of nonzero rows in $\mathbf{A}_R$. This number is denoted rank($\mathbf{A}$). In Example 6.7, rank($\mathbf{A}$) $= 4$, and in Examples 6.5 and 6.6, rank($\mathbf{A}$) $= 3$.

In terms of rank, the following result gives the dimension of the solution space of $\mathbf{AX} = \mathbf{O}$.

THEOREM 6.2   *Dimension of the Solution Space*

The solution space of the system $\mathbf{AX} = \mathbf{O}$ has dimension $m - $ rank($\mathbf{A}$). ◆

In Example 6.7, $m = 5$ and rank($\mathbf{A}$) $= 4$, so the solution space has a dimension of $5 - 4 = 1$, as we saw with the general solution depending on one arbitrary constant. In Example 6.6, $m = 5$ and rank($\mathbf{A}$) $= 3$, so the solution space has a dimension of $5 - 3 = 2$, and the general solution depends on two arbitrary constants.

Theorem 6.2 provides a simple condition for a homogeneous system to have a nontrivial solution.

COROLLARY 6.1

Let $\mathbf{A}$ be $n \times m$. Then the homogeneous system $\mathbf{AX} = \mathbf{O}$ has a nontrivial solution if and only if $m - $ rank($\mathbf{A}$) > 0. ◆

The rationale for this is that the system can have a nontrivial solution only when the dimension of the solution space is positive, having something in it other than the zero vector. Since this dimension is $m - $ rank($\mathbf{A}$), there will be a nontrivial solution exactly when this number is positive.

Corollary 6.1 implies that $\mathbf{AX} = \mathbf{O}$ has only the trivial solution exactly when $m - $ rank($\mathbf{A}$) $= 0$. In particular, when $\mathbf{A}$ is square, then $m = n$, and this occurs exactly when the $n \times n$ matrix $\mathbf{A}_R$ has n nonzero rows, which in turn happens exactly when $\mathbf{A}_R = \mathbf{I}_n$.

COROLLARY 6.2

If $\mathbf{A}$ is $n \times n$, then $\mathbf{AX} = \mathbf{O}$ has only the trivial solution if and only if $\mathbf{A}_R = \mathbf{I}_n$. ◆

SECTION 6.2   *PROBLEMS*

In each of Problems 1 through 10, find the general solution of the system by reducing the coefficient matrix. Write the general solution in terms of one or more column matrices. Also determine the dimension of the solution space.

1. $x_1 + 2x_2 - x_3 + x_4 = 0$
   $x_2 - x_3 + x_4 = 0$

2. $-3x_1 + x_2 - x_3 + x_4 + x_5 = 0$
   $x_2 + x_3 + 4x_5 = 0$
   $-3x_3 + 2x_4 + x_5 = 0$

3. $-2x_1 + x_2 + 2x_3 = 0$
   $x_1 - x_2 = 0$
   $x_1 + x_2 = 0$

4. $4x_1 + x_2 - 3x_3 + x_4 = 0$
   $2x_1 - x_3 = 0$

5. $x_1 - x_2 + 3x_3 - x_4 + 4x_5 = 0$
   $2x_1 - 2x_2 + x_3 + x_4 = 0$
   $x_1 - 2x_3 + x_5 = 0$
   $x_3 + x_4 - x_5 = 0$

6. $6x_1 - x_2 + x_3 = 0$
    $x_1 - x_4 + 2x_5 = 0$
    $x_1 - 2x_5 = 0$

7. $-10x_1 - x_2 + 4x_3 - x_4 + x_5 - x_6 = 0$
    $x_2 - x_3 + 3x_4 = 0$
    $2x_1 - x_2 + x_5 = 0$
    $x_2 - x_4 + x_6 = 0$

8. $8x_1 - 2x_3 + x_6 = 0$
    $2x_1 - x_2 + 3x_4 - x_6 = 0$
    $x_2 + x_3 - 2x_5 - x_6 = 0$
    $x_4 - 3x_5 + 2x_6 = 0$

9. $x_1 - 2x_2 + x_5 - x_6 + x_7 = 0$
    $x_3 - x_4 + x_5 - 2x_6 + 3x_7 = 0$
    $x_1 - x_5 + 2x_6 = 0$
    $2x_1 - 3x_4 + x_5 = 0$

10. $4x_1 - 3x_2 + x_4 - 3x_6 = 0$
    $2x_2 + 4x_4 - x_5 - 6x_6 = 0$
    $3x_1 - 2x_2 + 4x_5 - x_6 = 0$
    $2x_1 + x_2 - 3x_3 + 4x_4 = 0$

11. Can a system $\mathbf{AX} = \mathbf{O}$ having at least as many equations as unknowns, have a nontrivial solution?

6.3   Nonhomogeneous Systems of Linear Equations

We will now consider the nonhomogeneous linear system of n equations in m unknowns:

$$a_{11}x_1 + a_{12}x_2 + \ldots + a_{1m}x_m = b_1$$
$$a_{21}x_1 + a_{22}x_2 + \ldots + a_{2m}x_m = b_2$$
$$\vdots$$
$$a_{n1}x_1 + a_{n2}x_2 + \ldots + a_{nm}x_m = b_n.$$

In matrix form,

$$\mathbf{AX} = \mathbf{B} \qquad (6.1)$$

where $\mathbf{A}$ is the coefficient matrix and

$$\mathbf{X} = \begin{pmatrix} x_1 \\ x_2 \\ \vdots \\ x_m \end{pmatrix} \text{ and } \mathbf{B} = \begin{pmatrix} b_1 \\ b_2 \\ \vdots \\ b_n \end{pmatrix}.$$

The system is *nonhomogeneous* if at least one $b_j \neq 0$. Nonhomogeneous systems differ from linear systems in two significant ways.

1. A nonhomogeneous system may have no solution. For example, the system

$$2x_1 - 3x_2 = 6$$
$$4x_1 - 6x_2 = 8$$

can have no solution. If $2x_1 - 3x_2 = 6$, then $4x_1 - 6x_2$ must equal 12, not 8.

We call $\mathbf{AX} = \mathbf{B}$ *consistent* if there is a solution. If there is no solution, the system is *inconsistent*.

2. A linear combination of solutions of a nonhomogeneous system $\mathbf{AX} = \mathbf{B}$ need not be a solution. Therefore, the solutions do not have the vector space structure seen for solutions of $\mathbf{AX} = \mathbf{O}$, and there is no solution space.

Nevertheless, solutions of $\mathbf{AX} = \mathbf{B}$ do have a structure that parallels that for linear second order differential equations. We will call $\mathbf{AX} = \mathbf{O}$ the *associated homogeneous system* of the nonhomogeneous system $\mathbf{AX} = \mathbf{B}$. As seen in Examples 6.6 and 6.7, we can always write the general solution of $\mathbf{AX} = \mathbf{O}$ in the form

$$\mathbf{H} = \alpha_1 \mathbf{X}_1 + \ldots + \alpha_k \mathbf{X}_k$$

where $k = m - rank(\mathbf{A})$ is the dimension of the solution space of the associated homogeneous system and $\alpha_1, \ldots, \alpha_k$ are arbitrary constants.

THEOREM 6.3

Let $\mathbf{H}$ be the general solution of the associated homogeneous system. Let $\mathbf{U}_p$ be any particular solution of $\mathbf{AX} = \mathbf{B}$. Then the expression $\mathbf{H} + \mathbf{U}_p$ contains every solution of the nonhomogeneous system $\mathbf{AX} = \mathbf{B}$.

Proof   To see why this is true, suppose $\mathbf{U}$ is any solution of $\mathbf{AX} = \mathbf{B}$. Then

$$\mathbf{A}(\mathbf{U} - \mathbf{U}_p) = \mathbf{AU} - \mathbf{AU}_p = \mathbf{B} - \mathbf{B} = \mathbf{O}.$$

Therefore $\mathbf{U} - \mathbf{U}_p$ is a solution of $\mathbf{AX} = \mathbf{O}$, and so has the form

$$\mathbf{U} - \mathbf{U}_p = c_1 \mathbf{X}_1 + \ldots + c_k \mathbf{X}_k.$$

for some constants $c_1, \ldots, c_k$. But then

$$\mathbf{U} = c_1 \mathbf{X}_1 + \ldots + c_k \mathbf{X}_k + \mathbf{U}_p,$$

as is claimed in the theorem.   ♦

Theorem 6.3 suggests a strategy for finding all solutions of $\mathbf{AX} = \mathbf{B}$ when the system is consistent.

1. Find the general solution $\mathbf{H}$ of $\mathbf{AX} = \mathbf{O}$.
2. Find any (particular) solution of $\mathbf{AX} = \mathbf{B}$.
3. The general solution of $\mathbf{AX} = \mathbf{B}$ is $\mathbf{H} + \mathbf{U}_p$.

Theorem 6.3 has a simple consequence in terms of uniqueness of the solution of a consistent honhomogeneous system.

COROLLARY 6.3

In the notation of Theorem 6.3, a consistent nonhomogeneous system $\mathbf{AX} = \mathbf{B}$ has a unique solution if and only if $\mathbf{H} = \mathbf{O}$.

EXAMPLE 6.8

For the system

$$-x_1 + x_2 + 3x_3 = -2$$
$$x_2 + 2x_3 = 4,$$

$$\mathbf{A} = \begin{pmatrix} -1 & 1 & 3 \\ 0 & 1 & 2 \end{pmatrix} \text{ and } \mathbf{B} = \begin{pmatrix} -2 \\ 4 \end{pmatrix}.$$

By the method of Section 6.2, the general solution of the associated homogeneous equation $\mathbf{AX} = \mathbf{O}$ is

$$\mathbf{H} = \alpha \begin{pmatrix} 1 \\ -2 \\ 1 \end{pmatrix}.$$

By a method we will describe shortly, a particular solution of $\mathbf{AX} = \mathbf{B}$ is

$$\mathbf{U}_p = \begin{pmatrix} 6 \\ 4 \\ 0 \end{pmatrix}.$$

Therefore, every solution of $\mathbf{AX} = \mathbf{B}$ is contained in the expression

$$\mathbf{X} = \alpha \begin{pmatrix} 1 \\ -2 \\ 1 \end{pmatrix} + \begin{pmatrix} 6 \\ 4 \\ 0 \end{pmatrix}$$

in which α is any real number. This is the general solution of $\mathbf{AX} = \mathbf{B}$. ◆

We will now outline a procedure for attempting to find a particular solution $\mathbf{U}_p$ of $\mathbf{AX} = \mathbf{B}$. This procedure will also tell us when there is no solution.

Step 1. Define the $n \times m + 1$ *augmented matrix* $[\mathbf{A}\vdots\mathbf{B}]$ by adjoining the column matrix $\mathbf{B}$ as an additional column to $\mathbf{A}$. The augmented matrix contains the coefficients of the unknowns of the system (in the first m columns), as well as the numbers on the right side of the equations (elements of $\mathbf{B}$).

Step 2. Reduce $[\mathbf{A}\vdots\mathbf{B}]$. Since we reduce a matrix to obtain leading entries of 1 from upper left toward the lower right, this results eventually in a reduced matrix

$$[\mathbf{A}\vdots\mathbf{B}]_R = [\mathbf{A}_R\vdots\mathbf{C}],$$

in which the first m columns are the reduced form of $\mathbf{A}$, and the last column is whatever results from $\mathbf{B}$ after the row operations used to reduce $\mathbf{A}$ have been applied to $[\mathbf{A}\vdots\mathbf{B}]$.

Solutions of the reduced system $\mathbf{A}_R\mathbf{X} = \mathbf{C}$ are the same as solutions of the original system $\mathbf{AX} = \mathbf{B}$, because the operations performed on the coefficients of the unknowns are also performed on the b_j's.

Step 3. From $[\mathbf{A}_R\vdots\mathbf{C}]$, either read a particular solution $\mathbf{U}_p$ of $[\mathbf{A}_R\vdots\mathbf{C}]$ or that this system has no solution (more about this shortly). In the latter case, we are done. If, however, there is a solution, proceed to step 4.

Step 4. Find the general solution $\mathbf{H}$ of $\mathbf{AX} = \mathbf{O}$, using $\mathbf{A}_R$ found as the first m columns of $[\mathbf{A}_R\vdots\mathbf{C}]$.

Step 5. The general solution of $\mathbf{AX} = \mathbf{B}$ is $\mathbf{X} = \mathbf{H} + \mathbf{U}_p$.

EXAMPLE 6.9

We will solve the system

$$\begin{pmatrix} -3 & 2 & 2 \\ 1 & 4 & -6 \\ 0 & -2 & 2 \end{pmatrix} \mathbf{X} = \begin{pmatrix} 8 \\ 1 \\ -2 \end{pmatrix}.$$

The first step is to reduce the augmented matrix

$$[\mathbf{A}\vdots\mathbf{B}] = \begin{pmatrix} -3 & 2 & 2 & \vdots & 8 \\ 1 & 4 & -6 & \vdots & 1 \\ 0 & -2 & 2 & \vdots & -2 \end{pmatrix}.$$

Interchange rows one and two:

$$\begin{pmatrix} 1 & 4 & -6 & \vdots & 1 \\ -3 & 2 & 2 & \vdots & 8 \\ 0 & -2 & 2 & \vdots & -2 \end{pmatrix}.$$

Add 3 times row one to row two:

$$\begin{pmatrix} 1 & 4 & -6 & \vdots & 1 \\ 0 & 14 & -16 & \vdots & 11 \\ 0 & -2 & 2 & \vdots & -2 \end{pmatrix}.$$

Multiply row two by $1/14$:

$$\begin{pmatrix} 1 & 4 & -6 & \vdots & 1 \\ 0 & 1 & -8/7 & \vdots & 11/14 \\ 0 & -2 & 2 & \vdots & -2 \end{pmatrix}.$$

Add 3 times row two to row one and 2 times row two to row one:

$$\begin{pmatrix} 1 & 0 & -10/7 & \vdots & -15/7 \\ 0 & 1 & -8/7 & \vdots & 11/14 \\ 0 & 0 & -2/7 & \vdots & -3/7 \end{pmatrix}.$$

Multiply row three by $-7/2$:

$$\begin{pmatrix} 1 & 0 & -10/7 & \vdots & -15/7 \\ 0 & 1 & -8/7 & \vdots & 11/14 \\ 0 & 0 & 1 & \vdots & 3/2 \end{pmatrix}.$$

Add $10/7$ times row three to row one and $8/7$ times row three to row two:

$$\begin{pmatrix} 1 & 0 & 0 & \vdots & 0 \\ 0 & 1 & 0 & \vdots & 5/2 \\ 0 & 0 & 1 & \vdots & 3/2 \end{pmatrix}.$$

This is $[\mathbf{A}\vdots\mathbf{B}]_R$. Notice that the first three rows and columns are actually $\mathbf{A}_R$, and the last column is whatever resulted from the row operations applied to reduce the first three rows and columns. The reduced augmented matrix represents the reduced system $\mathbf{A}_R\mathbf{X} = \mathbf{C}$, which is the system

$$\begin{pmatrix} 1 & 0 & 0 \\ 0 & 1 & 0 \\ 0 & 0 & 1 \end{pmatrix}\mathbf{X} = \begin{pmatrix} 0 \\ 5/2 \\ 3/2 \end{pmatrix}.$$

We can solve this reduced system by inspection, obtaining $x_1 = 0$, $x_2 = 5/2$, and $x_3 = 3/2$. Thus,

$$\mathbf{U}_p = \begin{pmatrix} 0 \\ 5/2 \\ 3/2 \end{pmatrix}$$

is a particular solution of the original system as well. From the reduced homogeneous system $\mathbf{A}_R\mathbf{X} = \mathbf{O}$, we read the general solution as $\mathbf{H} = \mathbf{O}$, the trivial solution (rank$(\mathbf{A}) = 3$).

The unique solution of $\mathbf{AX} = \mathbf{B}$ is therefore

$$\mathbf{X} = \mathbf{H} + \mathbf{U}_p = \mathbf{U}_p.$$

This solution is unique because the associated homogeneous system has only the trivial solution. ◆

EXAMPLE 6.10

We have seen that the system

$$2x_1 - 3x_2 = 6$$
$$4x_1 - 6x_2 = 8$$

has no solution. We will see how this conclusion reveals itself when we work with the augmented matrix, which is

$$[\mathbf{A} \vdots \mathbf{B}] = \begin{pmatrix} 2 & 3 & \vdots & 6 \\ 4 & -6 & \vdots & 8 \end{pmatrix}.$$

Reduce this matrix to obtain

$$[\mathbf{A} \vdots \mathbf{B}]_R = [\mathbf{A}_R \vdots \mathbf{C}] = \begin{pmatrix} 1 & -3/2 & \vdots & 2 \\ 0 & 0 & \vdots & -4 \end{pmatrix}.$$

The second equation of the reduced system is

$$0x_1 + 0x_2 = -4$$

which can have no solution. ◆

In Example 6.10, the augmented matrix has rank 2, while the matrix of the homogeneous system has rank 1. In general, whenever the rank of $\mathbf{A}$ is less than the rank of $[\mathbf{A} \vdots \mathbf{B}]$, then $\mathbf{A}_R$ will have at least one row of zeros, while the corresponding row in the reduced augmented matrix $[\mathbf{A}_R \vdots \mathbf{C}]$ has a nonzero element in this row in the $\mathbf{C}$ column. This corresponds to an equation of the form

$$0x_1 + 0x_2 + \ldots + 0x_m = c_j \neq 0$$

and this has no solution for the x_i's. We will record this observation.

THEOREM 6.4

The nonhomogeneous system $\mathbf{AX} = \mathbf{B}$ has a solution if and only if $\mathbf{A}$ and $[\mathbf{A} \vdots \mathbf{B}]$ have the same rank. ◆

EXAMPLE 6.11

We will solve the system

$$x_1 - x_2 + 2x_3 = 3,$$
$$-4x_1 + x_2 + 7x_3 = -5,$$

and

$$-2x_1 - x_2 + 11x_3 = 14.$$

The augmented matrix is

$$[\mathbf{A} \vdots \mathbf{B}] = \begin{pmatrix} 1 & -1 & 2 & \vdots & 3 \\ -4 & 1 & 7 & \vdots & -5 \\ -2 & -1 & 11 & \vdots & 14 \end{pmatrix}.$$

Reduce this augmented matrix to obtain

$$[\mathbf{A} \vdots \mathbf{B}]_R = [\mathbf{A}_R \vdots \mathbf{C}] = \begin{pmatrix} 1 & 0 & -3 & \vdots & 0 \\ 0 & 1 & -5 & \vdots & 0 \\ 0 & 0 & 0 & \vdots & 1 \end{pmatrix}.$$

A has rank 2, because its reduced matrix has two nonzero rows. But $[\mathbf{A} \vdots \mathbf{B}]$ has rank 3, because its reduced form has three nonzero rows. Therefore, this system is inconsistent. We can also observe from the reduced system that the last equation is

$$0x_1 + 0x_2 + 0x_3 = 1$$

with no solution. ◆

EXAMPLE 6.12

Solve the system

$$x_1 - x_2 + 2x_4 + x_5 + 6x_6 = -3,$$
$$x_2 + x_3 + 3x_4 + 2x_5 + 4x_6 = 1,$$

and

$$x_1 - 4x_2 + 3x_3 + x_4 + 2x_6 = 0.$$

The augmented matrix is

$$[\mathbf{A} \vdots \mathbf{B}] = \begin{pmatrix} 1 & 0 & -1 & 2 & 1 & 6 & \vdots & -3 \\ 0 & 1 & 1 & 3 & 2 & 4 & \vdots & 1 \\ 1 & -4 & 3 & 1 & 0 & 2 & \vdots & 0 \end{pmatrix}.$$

Reduce this to obtain

$$[\mathbf{A}\vdots\mathbf{B}]_R = [\mathbf{A}_R\vdots\mathbf{C}] = \begin{pmatrix} 1 & 0 & 0 & 27/8 & 15/8 & 60/8 & \vdots & -17/8 \\ 0 & 1 & 0 & 13/8 & 9/8 & 20/8 & \vdots & 1/8 \\ 0 & 0 & 1 & 11/8 & 7/8 & 12/8 & \vdots & 7/8 \end{pmatrix}.$$

The first six columns are $\mathbf{A}_R$, and we read from $[\mathbf{A}\vdots\mathbf{B}]_R$ that both $\mathbf{A}$ and $[\mathbf{A}\vdots\mathbf{B}]$ have rank 3, so the system is consistent. From the reduced augmented matrix, we read that

$$x_1 + \tfrac{27}{8}x_4 + \tfrac{15}{8}x_5 + \tfrac{60}{8}x_6 = -\tfrac{17}{8},$$
$$x_2 + \tfrac{13}{8}x_4 + \tfrac{9}{8}x_5 + \tfrac{20}{8}x_6 = \tfrac{1}{8},$$

and

$$x_3 + \tfrac{11}{8}x_4 + \tfrac{7}{8}x_5 + \tfrac{12}{8}x_6 = \tfrac{7}{8}.$$

From these, we read that

$$x_1 = -\tfrac{27}{8}x_4 - \tfrac{15}{8}x_5 - \tfrac{60}{8}x_6 - \tfrac{17}{8},$$
$$x_2 = -\tfrac{13}{8}x_4 - \tfrac{9}{8}x_5 - \tfrac{20}{8}x_6 + \tfrac{1}{8},$$

and

$$x_3 = -\tfrac{11}{8}x_4 - \tfrac{7}{8}x_5 - \tfrac{12}{8}x_6 + \tfrac{7}{8}.$$

We could have gone directly to these equations without the intermediate step. These actually give the general solution, with x_1, x_2, and x_3 in terms of x_4, x_5, and x_6, which can be given any real values. The solution is

$$\mathbf{X} = \begin{pmatrix} -\tfrac{27}{8}x_4 - \tfrac{15}{8}x_5 - \tfrac{60}{8}x_6 - \tfrac{17}{8} \\ -\tfrac{13}{8}x_4 - \tfrac{9}{8}x_5 - \tfrac{20}{8}x_6 + \tfrac{1}{8} \\ -\tfrac{11}{8}x_4 - \tfrac{7}{8}x_5 - \tfrac{12}{8}x_6 + \tfrac{7}{8} \\ x_4 \\ x_5 \\ x_6 \end{pmatrix}.$$

To write this in a more revealing way, let $x_4 = 8\alpha$, $x_5 = 8\beta$, and $x_6 = 8\gamma$ to write

$$\mathbf{X} = \alpha\begin{pmatrix} -27 \\ -13 \\ -11 \\ 8 \\ 0 \\ 0 \end{pmatrix} + \beta\begin{pmatrix} -15 \\ -9 \\ -7 \\ 0 \\ 8 \\ 0 \end{pmatrix} + \gamma\begin{pmatrix} -60 \\ -20 \\ -12 \\ 0 \\ 0 \\ 8 \end{pmatrix} + \begin{pmatrix} -17/8 \\ 1/8 \\ 7/8 \\ 0 \\ 0 \\ 0 \end{pmatrix} = \mathbf{H} + \mathbf{U}_p$$

with $\mathbf{H}$ as the general solution of $\mathbf{AX} = \mathbf{O}$ and $\mathbf{U}_p$ as a particular solution of $\mathbf{AX} = \mathbf{C}$. ◆

In each of Problems 1 through 10, find the general solution of the system or show that the system is inconsistent. Write the solution in matrix form.

1.
$$3x_1 - 2x_2 + x_3 = 6$$
$$x_1 + 10x_2 - x_3 = 2$$
$$-3x_1 - 2x_2 + x_3 = 0$$

2.
$$4x_1 - 2x_2 + 3x_3 + 10x_4 = 1$$
$$x_1 - 3x_4 = 8$$
$$2x_1 - 3x_2 + x_4 = 16$$

3.
$$2x_1 - 3x_2 + x_4 - x_6 = 0$$
$$3x_1 - 2x_3 + x_5 = 1$$
$$x_2 - x_4 + 6x_6 = 3$$

4.
$$2x_1 - 3x_2 = 1$$
$$-x_1 + 3x_2 = 0$$
$$x_1 - 4x_2 = 3$$

5.
$$3x_2 - 4x_4 = 10$$
$$x_1 - 3x_2 + 4x_3 - x_6 = 8$$
$$x_2 + x_3 - 6x_4 + x_6 = -9$$
$$x_1 - x_2 + x_6 = 0$$

6.
$$2x_1 - 3x_2 + x_4 = 1$$
$$3x_2 + x_3 - x_4 = 0$$
$$2x_1 - 3x_2 + 10x_3 = 0$$

7.
$$8x_2 - 4x_3 + 10x_6 = 1$$
$$x_3 + x_5 - x_6 = 2$$
$$x_4 - 3x_5 + 2x_6 = 0$$

8.
$$2x_1 - 3x_3 = 1$$
$$x_1 - x_2 + x_3 = 1$$
$$2x_1 - 4x_2 + x_3 = 2$$

9.
$$14x_3 - 3x_5 + x_7 = 2$$
$$x_1 + x_2 + x_3 - x_4 + x_6 = -4$$

10.
$$3x_1 - 2x_2 = -1$$
$$4x_1 + 3x_2 = 4$$

6.4   Matrix Inverses

Let $\mathbf{A}$ be an $n \times n$ matrix. An $n \times n$ matrix $\mathbf{B}$ is an inverse of $\mathbf{A}$ if

$$\mathbf{AB} = \mathbf{BA} = \mathbf{I}_n.$$

It is easy to find matrices that have no inverse. For example, let

$$\mathbf{A} = \begin{pmatrix} 1 & 0 \\ 2 & 0 \end{pmatrix}.$$

Suppose $\mathbf{B}$ is an inverse of $\mathbf{A}$. Then

$$\mathbf{B} = \begin{pmatrix} a & b \\ c & d \end{pmatrix}.$$

But then

$$\mathbf{AB} = \begin{pmatrix} 1 & 0 \\ 2 & 0 \end{pmatrix} \begin{pmatrix} a & b \\ c & d \end{pmatrix} = \begin{pmatrix} a & b \\ 2a & 2b \end{pmatrix} = \begin{pmatrix} 1 & 0 \\ 0 & 1 \end{pmatrix},$$

implying that

$$a = 1, b = 0, 2a = 0 \text{ and } b = 1$$

and this is impossible. On the other hand, some matrices do have inverses. For example,

$$\begin{pmatrix} 2 & 1 \\ 1 & 4 \end{pmatrix} \begin{pmatrix} 4/7 & -1/7 \\ -1/7 & 2/7 \end{pmatrix} = \begin{pmatrix} 4/7 & -1/7 \\ -1/7 & 2/7 \end{pmatrix} \begin{pmatrix} 2 & 1 \\ 1 & 4 \end{pmatrix} = \begin{pmatrix} 1 & 0 \\ 0 & 1 \end{pmatrix}.$$

Singular and Nonsingular Matrices

A matrix that has an inverse is called *nonsingular*. A matrix with no inverse is *singular*.

A matrix can have only one inverse. Suppose that **B** and **C** are inverses of **A**. Then

$$\mathbf{B} = \mathbf{BI}_n = \mathbf{B(AC)} = \mathbf{(BA)C} = \mathbf{I}_n\mathbf{C} = \mathbf{C}.$$

In view of this, we will denote the inverse of **A** as $\mathbf{A}^{-1}$. Here are additional facts about matrix inverses.

1. $\mathbf{I}_n$ is nonsingular and is its own inverse.

2. If **A** and **B** are nonsingular $n \times n$ matrices, then so is **AB**. Further,

$$(\mathbf{AB})^{-1} = \mathbf{B}^{-1}\mathbf{A}^{-1}.$$

   That is, the inverse of a product is the product of the inverses in the reverse order. This is true because

$$(\mathbf{B}^{-1}\mathbf{A}^{-1})(\mathbf{AB}) = \mathbf{B}^{-1}(\mathbf{A}^{-1}\mathbf{A})\mathbf{B} = \mathbf{B}^{-1}\mathbf{B} = \mathbf{I}_n.$$

3. If **A** is nonsingular, so is $\mathbf{A}^{-1}$, and

$$(\mathbf{A}^{-1})^{-1} = \mathbf{A}.$$

4. If **A** is nonsingular, so is its transpose $\mathbf{A}^t$, and

$$(\mathbf{A}^t)^{-1} = (\mathbf{A}^{-1})^t.$$

   This says that the inverse of a transpose is the transpose of the inverse.

5. If **A** and **B** are $n \times n$ matrices and either one is singular, then their products **AB** and **BA** in either order are singular.

6. An $n \times n$ matrix **A** is nonsingular if and only if the reduced form of **A** is $\mathbf{I}_n$. Alternatively, **A** is nonsingular if and only if its rank is n.

Matrix inverses relate to systems of linear equations in the following way.

THEOREM 6.5

Let **A** be $n \times n$. Then,

1. A homogeneous system $\mathbf{AX} = \mathbf{O}$ has a nontrivial solution if and only if **A** is singular.

2. A consistent nonhomogeneous system $\mathbf{AX} = \mathbf{B}$ has a unique solution if and only if **A** is nonsingular. In this case, the solution is

$$\mathbf{X} = \mathbf{A}^{-1}\mathbf{B}. \; \blacklozenge$$

For condition 1, if **A** is nonsingular, we can multiply $\mathbf{AX} = \mathbf{O}$ on the left by $\mathbf{A}^{-1}$ to get

$$\mathbf{A}^{-1}(\mathbf{AX}) = \mathbf{X} = \mathbf{A}^{-1}\mathbf{O} = \mathbf{O}.$$

Therefore, the homogeneous system can have only the trivial solution when **A** is nonsingular.

For condition 2, notice that if **A** is nonsingular we can multiply the equation on the left by $\mathbf{A}^{-1}$ to get the unique solution

$$\mathbf{X} = \mathbf{A}^{-1}\mathbf{B}.$$

There are several ways to determine if a square matrix is nonsingular and, when it is, to find its inverse. We will outline one method here and another when we develop determinants. Let $\mathbf{A}$ be an $n \times n$ real matrix. Form the $n \times 2n$ matrix $[\mathbf{A} \vdots \mathbf{I}_n]$ whose first n columns are $\mathbf{A}$ and whose second n columns are $\mathbf{I}_n$. For example, if

$$\mathbf{A} = \begin{pmatrix} 2 & 3 \\ -1 & 9 \end{pmatrix}$$

then

$$[\mathbf{A} \vdots \mathbf{I}_2] = \begin{pmatrix} 2 & 3 & \vdots & 1 & 0 \\ -1 & 9 & \vdots & 0 & 1 \end{pmatrix}.$$

Now reduce $\mathbf{A}$, carrying out the row operations across the entire matrix $[\mathbf{A} \vdots \mathbf{I}_n]$. $\mathbf{A}$ is nonsingular exactly when $\mathbf{A}_R = \mathbf{I}_n$ turns up in the first n columns. In this event, the second n columns form $\mathbf{A}^{-1}$.

EXAMPLE 6.13

Let

$$\mathbf{A} = \begin{pmatrix} 5 & -1 \\ 6 & 8 \end{pmatrix}.$$

Form

$$[\mathbf{A} \vdots \mathbf{I}_2] = \begin{pmatrix} 5 & -1 & \vdots & 1 & 0 \\ 6 & 8 & \vdots & 0 & 1 \end{pmatrix}.$$

Reduce $\mathbf{A}$, carrying out each row operation on the entire row of the augmented matrix. First multiply row one by $1/5$:

$$\begin{pmatrix} 1 & -1/5 & \vdots & 1/5 & 0 \\ 6 & 8 & \vdots & 0 & 1 \end{pmatrix}.$$

Add -6 times row one to row two:

$$\begin{pmatrix} 1 & -1/5 & \vdots & 1/5 & 0 \\ 0 & 46/5 & \vdots & -6/5 & 1 \end{pmatrix}.$$

Multiply row two by $5/46$:

$$\begin{pmatrix} 1 & -1/5 & \vdots & 1/5 & 0 \\ 6 & 1 & \vdots & -6/46 & 5/46 \end{pmatrix}.$$

Add $1/5$ times row two to row one:

$$\begin{pmatrix} 1 & 0 & \vdots & 8/46 & 1/46 \\ 0 & 1 & \vdots & -6/46 & 5/46 \end{pmatrix}.$$

This is in reduced form. The first two columns make up $\mathbf{A}_R$. Since these columns also form $\mathbf{I}_2$, then $\mathbf{A}$ is nonsingular. Further, we can read $\mathbf{A}^{-1}$ from the last two columns:

$$\mathbf{A}^{-1} = \begin{pmatrix} 8/46 & 1/46 \\ -6/46 & 5/46 \end{pmatrix}. \quad \blacklozenge$$

EXAMPLE 6.14

Let

$$A = \begin{pmatrix} -3 & 21 \\ 4 & -28 \end{pmatrix}.$$

Form

$$[A \vdots I_2] = \begin{pmatrix} -3 & 21 & \vdots & 1 & 0 \\ 4 & -28 & \vdots & 0 & 1 \end{pmatrix}.$$

Reduce this by multiplying row one by $-1/3$ and then adding -4 times row one to row two to get

$$\begin{pmatrix} 1 & -7 & \vdots & -1/3 & 0 \\ 0 & 0 & \vdots & 4/3 & 1 \end{pmatrix}.$$

The left two columns, which form A_R, do not equal I_2, so A is singular and has no inverse. Alternatively, from the left two columns, rank$(A) = 1 < 2$, so A is singular. ♦

We will illustrate the use of a matrix inverse to solve a nonhomogeneous system.

EXAMPLE 6.15

We will solve the system

$$2x_1 - x_2 + 3x_3 = 4$$
$$x_1 + 9x_2 - 2x_3 = -8$$
$$4x_1 - 8x_2 + 11x_3 = 15.$$

The matrix of coefficients is

$$A = \begin{pmatrix} 2 & -1 & 3 \\ 1 & 9 & -2 \\ 4 & -8 & 11 \end{pmatrix}.$$

A routine reduction yields

$$[A \vdots I_3]_R = \begin{pmatrix} 1 & 0 & 0 & \vdots & 83/53 & -13/53 & -25/53 \\ 0 & 1 & 0 & \vdots & -19/53 & 10/53 & 7/53 \\ 0 & 0 & 1 & \vdots & -44/53 & 12/53 & 19/53 \end{pmatrix}.$$

The first three columns are I_3, hence A is nonsingular and the last three columns of the reduced augmented matrix give us the inverse:

$$A^{-1} = \frac{1}{53} \begin{pmatrix} 83 & -13 & -25 \\ -19 & 10 & 7 \\ -44 & 12 & 19 \end{pmatrix}.$$

The unique solution of the system is obtained by multiplying it on the left by A^{-1}:

$$X = A^{-1}B = \frac{1}{53} \begin{pmatrix} 83 & -13 & -25 \\ -19 & 10 & 7 \\ -44 & 12 & 19 \end{pmatrix} \begin{pmatrix} 4 \\ -8 \\ 15 \end{pmatrix} = \begin{pmatrix} 61/53 \\ -51/53 \\ 13/53 \end{pmatrix}. \quad ♦$$

SECTION 6.4   PROBLEMS

In each of Problems 1 through 8, find the inverse of the matrix or show that the matrix is singular.

1. $\begin{pmatrix} -1 & 2 \\ 2 & 1 \end{pmatrix}$

2. $\begin{pmatrix} 12 & 3 \\ 4 & 1 \end{pmatrix}$

3. $\begin{pmatrix} -5 & 2 \\ 1 & 2 \end{pmatrix}$

4. $\begin{pmatrix} -1 & 0 \\ 4 & 4 \end{pmatrix}$

5. $\begin{pmatrix} 6 & 2 \\ 3 & 3 \end{pmatrix}$

6. $\begin{pmatrix} 1 & 1 & -3 \\ 2 & 16 & 1 \\ 0 & 0 & 4 \end{pmatrix}$

7. $\begin{pmatrix} -3 & 4 & 1 \\ 1 & 2 & 0 \\ 1 & 1 & 3 \end{pmatrix}$

8. $\begin{pmatrix} -2 & 1 & -5 \\ 1 & 1 & 4 \\ 0 & 3 & 3 \end{pmatrix}$

In each of Problems 9 through 12, use a matrix inverse to find the unique solution of the system.

9. $x_1 - x_2 + 3x_3 - x_4 = 1$
$x_2 - 3x_3 + 5x_4 = 2$
$x_1 - x_3 + x_4 = 0$
$x_1 + 2x_3 - x_4 = -5$

10. $8x_1 - x_2 - x_3 = 4$
$x_1 + 2x_2 - 3x_3 = 0$
$2x_1 - x_2 + 4x_3 = 5$

11. $2x_1 - 6x_2 + 3x_3 = -4$
$-x_1 + x_2 + x_3 = 5$
$2x_1 + 6x_2 - 5x_3 = 8$

12. $12x_1 + x_2 - 3x_3 = 4$
$x_1 - x_2 + 3x_3 = -5$
$-2x_1 + x_2 + x_3 = 0$

CHAPTER 7

Determinants

7.1   Definition of the Determinant

The determinant of a square matrix is formed as a sum of products of elements from its rows and columns, according to a scheme we will now define.

Permutation

A *permutation* of the integers $1, 2, \ldots, n$ is a rearrangement of these integers. For example, suppose p is the permutation that rearranges

$$1, 2, 3, 4, 5, 6 \rightarrow 3, 1, 4, 5, 2, 6.$$

Then $p(1) = 3$, $p(2) = 1$, $p(3) = 4$, $p(4) = 5$, $p(5) = 2$, and $p(6) = 6$.

A permutation is characterized as even or odd according to a rule we will illustrate. Consider the permutation

$$p : 1, 2, 3, 4, 5 \rightarrow 2, 5, 1, 4, 3.$$

For each k in the permuted list on the right, count the number of integers to the right of k that are smaller than k. There is one number to the right of 2 smaller than 2, three numbers to the right of 5 smaller than 5, no numbers to the right of 1 smaller than 1, one number to the right of 4 smaller than 4, and no numbers to the right of 3 smaller than 3. Since $1 + 3 + 0 + 1 + 0 = 5$ is odd, p is an *odd permutation*. If this sum had been even, p would be an *even permutation*.

If p is a permutation on $1, 2, \ldots, n$, define

$$\alpha(p) = \begin{cases} 1 & \text{if } p \text{ is an even permutation} \\ -1 & \text{if } p \text{ is an odd permutation.} \end{cases}$$

145

Determinant

The *determinant* of an $n \times n$ matrix $\mathbf{A}$ is now defined by

$$\det \mathbf{A} = \sum_p \alpha(p) a_{1p(1)} a_{2p(2)} \ldots a_{np(n)} \tag{7.1}$$

with this sum extending over all permutations p of $1, 2, \ldots, n$. Thus $\det \mathbf{A}$ is a sum, each term of which is plus or minus a product formed from one element of each row and column of $\mathbf{A}$.

We also denote $\det \mathbf{A}$ as $|\mathbf{A}|$.

EXAMPLE 7.1

We will evaluate the determinant of

$$\mathbf{A} = \begin{pmatrix} a_{11} & a_{21} \\ a_{21} & a_{22} \end{pmatrix}.$$

There are only two permutations on $1, 2$, namely

$$p : 1, 2 \to 1, 2 \text{ and } q : 1, 2 \to 2, 1.$$

It is easy to check that p is even and q is odd. Therefore,

$$|\mathbf{A}| = \alpha(p) a_{1p(1)} a_{2p(1)} + \alpha(q) a_{1q(1)} a_{2q(2)}$$
$$= a_{11} a_{22} - a_{12} a_{21}. \quad \blacklozenge$$

While the definition serves as a starting point, we will never use it to evaluate a determinant. Instead, we will develop properties of determinants that can be used to evaluate them with reasonable efficiency.

1. Let $\mathbf{A}$ be an $n \times n$ matrix.

$$|\mathbf{A}^t| = |\mathbf{A}|.$$

This is because each term in the sum of equation (7.1) is a product of matrix elements, one from each row and column, and we obtain the same terms from both $\mathbf{A}$ and $\mathbf{A}^t$.

2. If $\mathbf{A}$ has a zero row or column, then $|\mathbf{A}| = 0$.

This is because a zero row or column puts a zero factor in each term of the sum in equation (7.1).

3. Let $\mathbf{B}$ be formed from $\mathbf{A}$ by interchanging two rows or columns. Then

$$|\mathbf{B}| = -|\mathbf{A}|.$$

Interchanging two rows or columns changes the sign of the determinant.

4. If two rows of $\mathbf{A}$ are the same, or if two columns of $\mathbf{A}$ are the same, then

$$|\mathbf{A}| = 0.$$

This follows immediately from property 3, because in this case, interchanging the two identical rows (or columns) leaves the matrix unchanged but reverses the sign of its determinant, forcing this determinant to be zero.

5. If **B** is formed from **A** by multiplying a row or column by a nonzero number α, then

$$|\mathbf{B}| = \alpha|\mathbf{A}|.$$

This is because multiplying a row or column of **A** by α puts a factor of α in every term of the sum in equation (7.1).

6. If, for some number α, one row (or column) of **A** is α times another row (or column), then $|\mathbf{A}| = 0$.

   This conclusion follows from property 2 if $\alpha = 0$, so suppose that $\alpha \neq 0$. Now the conclusion follows from properties 4 and 5. For suppose that row k of **A** is α times row i. Form **B** from **A** by multiplying row k by $1/\alpha$. Then **B** has two identical rows, hence zero determinant by property 4. But then by property 5, $|\mathbf{B}| = (1/\alpha)|\mathbf{A}| = 0$, so $|\mathbf{A}| = 0$.

7. Suppose each element of row k of **A** is written as a sum

$$a_{kj} = b_{kj} + c_{kj}.$$

Define a matrix **B** from **A** by replacing each a_{kj} of **A** by b_{kj}. Define a matrix **C** from **A** by replacing each a_{kj} by c_{kj}. Then

$$|\mathbf{A}| = |\mathbf{B}| + |\mathbf{C}|.$$

In determinant notation,

$$|\mathbf{A}| = \begin{vmatrix} a_{11} & \cdots & a_{1j} & \cdots & a_{1n} \\ \vdots & \vdots & \vdots & \vdots & \vdots \\ b_{k1} + c_{k1} & \cdots & b_{kj} + c_{kj} & \cdots & b_{kn} + c_{kn} \\ \vdots & \vdots & \vdots & \vdots & \vdots \\ a_{n1} & \cdots & a_{nj} & \cdots & a_{nn} \end{vmatrix} \tag{7.2}$$

$$= \begin{vmatrix} a_{11} & \cdots & a_{1j} & \cdots & a_{1n} \\ \vdots & \vdots & \vdots & \vdots & \vdots \\ b_{k1} & \cdots & b_{kj} & \cdots & b_{kn} \\ \vdots & \vdots & \vdots & \vdots & \vdots \\ a_{n1} & \cdots & a_{nj} & \cdots & a_{nn} \end{vmatrix} + \begin{vmatrix} a_{11} & \cdots & a_{1j} & \cdots & a_{1n} \\ \vdots & \vdots & \vdots & \vdots & \vdots \\ c_{k1} & \cdots & c_{kj} & \cdots & c_{kn} \\ \vdots & \vdots & \vdots & \vdots & \vdots \\ a_{n1} & \cdots & a_{nj} & \cdots & a_{nn} \end{vmatrix}.$$

The conclusion follows by replacing each a_{kj} in the defining sum (7.1) with $b_{kj} + c_{kj}$.

   This conclusion also holds if each element of a column is written as a sum of two terms.

8. If **D** is formed from **A** by adding α times one row (or column) to another row (or column), then

$$|\mathbf{D}| = |\mathbf{A}|.$$

This result follows from property 7. To see this, we will deal with rows to be specific. Suppose α times row i is added to row k of **A** to form **D**. On the right side of equation (7.2), replace each b_{kj} with αa_{ij} and each c_{kj} with a_{kj}, resulting in

$$|\mathbf{D}| = \begin{vmatrix} a_{11} & \cdots & a_{1j} & \cdots & a_{1n} \\ \vdots & \vdots & \vdots & \vdots & \vdots \\ \alpha a_{i1} & \cdots & \alpha a_{ij} & \cdots & \alpha a_{in} \\ \vdots & \vdots & \vdots & \vdots & \vdots \\ a_{n1} & \cdots & a_{nj} & \cdots & a_{nn} \end{vmatrix} + \begin{vmatrix} a_{11} & \cdots & a_{1j} & \cdots & a_{1n} \\ \vdots & \vdots & \vdots & \vdots & \vdots \\ a_{k1} & \cdots & a_{kj} & \cdots & a_{kn} \\ \vdots & \vdots & \vdots & \vdots & \vdots \\ a_{n1} & \cdots & a_{nj} & \cdots & a_{nn} \end{vmatrix}.$$

$$= |\mathbf{B}| + |\mathbf{A}|.$$

Now look at the two determinants on the right side of the last equation. If we multiply row k of the first determinant on the right by $1/\alpha$, then rows i and k of the resulting determinant are identical, so this determinant is zero. Therefore

$$|\mathbf{D}| = |\mathbf{A}|.$$

9. If $\mathbf{A}$ and $\mathbf{B}$ are $n \times n$ matrices, then

$$|\mathbf{AB}| = |\mathbf{A}||\mathbf{B}|.$$

The determinant of a product is the product of the matrices.

Properties 3, 5, and 8 tell us the effects of elementary row operations on the determinant of a matrix. However, now these operations can be applied to columns as well. The reason we only operated on rows when dealing with systems of linear equations is that each row represents one equation of the system, and multiplying an entire equation (row) by a nonzero number does not change the solutions of the system. However, in that context, each column represents only one unknown, and if we multiply just one unknown of a system by a nonzero number, we obtain a new system with possibly different solutions.

SECTION 7.1   PROBLEMS

1. Let $\mathbf{A} = [a_{ij}]$ be an $n \times n$ matrix and let α be a number. Form $\mathbf{B} = [\alpha a_{ij}]$ by multiplying each element of $\mathbf{A}$ by α. How are $|\mathbf{A}|$ and $|\mathbf{B}|$ related?

2. Let $\mathbf{A} = [a_{ij}]$ be an $n \times n$ matrix. Let α be a nonzero number. Form the matrix $\mathbf{B} = [\alpha^{i-j} a_{ij}]$. How are $|\mathbf{A}|$ and $|\mathbf{B}|$ related? *Hint:* It is useful to look at the 2×2 and 3×3 cases to get some idea of what $\mathbf{B}$ looks like.

3. An $n \times n$ matrix is *skew-symmetric* if $\mathbf{A} = -\mathbf{A}^t$. Explain why the determinant of a skew-symmetric matrix having an odd number of rows and columns must be zero.

4. List all six permutations on $1, 2$, and 3 and determine which are even and which are odd (you should get half of each). Using this, evaluate the determinant of the 3×3 matrix $\mathbf{A} = [a_{ij}]$.

7.2   Evaluation of Determinants I

In this section we will show how to exploit elementary row and column operations to evaluate a determinant. The key is the following observation: if a row k or column r of an $n \times n$ matrix $\mathbf{A}$ has all zero elements, except possibly for a_{kr} in row k and column r, then

$$|\mathbf{A}| = (-1)^{k+r} a_{kr} |\mathbf{A}_{kr}| \tag{7.3}$$

where $\mathbf{A}_{kr}$ is the $n-1 \times n-1$ matrix obtained by deleting row k and column r of $\mathbf{A}$. This follows from equation (7.1), and it suggests a strategy for evaluating $|\mathbf{A}|$. Use row and/or column operations to obtain from $\mathbf{A}$ a new matrix $\mathbf{B}$ having at most one nonzero element in some row or column. From properties 3, 5 and 8 of determinants in Section 7.1, $|\mathbf{A}|$ is a multiple of $|\mathbf{B}|$, which in turn is a multiple of the $n-1 \times n-1$ determinant formed by deleting the row and column of $\mathbf{B}$ containing the nonzero element. Repeat this strategy on this $n-1 \times n-1$ determinant to reduce the problem to one or more $n-2 \times n-2$ determinants, and so on until we have small enough determinants that their evaluation is relatively easy.

EXAMPLE 7.2

Let

$$
\mathbf{A} = \begin{pmatrix} -6 & 0 & 1 & 3 & 2 \\ -1 & 5 & 0 & 1 & 7 \\ 8 & 3 & 2 & 1 & 7 \\ 0 & 1 & 5 & -3 & 2 \\ 1 & 15 & -3 & 9 & 4 \end{pmatrix}.
$$

One (of many) ways to evaluate $|\mathbf{A}|$ is to exploit the 1 in the $(1, 3)$ position to get zeros in the other locations in column three. Add -2 times row one to row three, -5 times row one to row four, and 3 times row one to row five to get

$$
\mathbf{B} = \begin{pmatrix} -6 & 0 & 1 & 3 & 2 \\ -1 & 5 & 0 & 1 & 7 \\ 20 & 3 & 0 & -5 & 3 \\ 30 & 1 & 0 & -18 & -8 \\ -17 & 15 & 0 & 18 & 10 \end{pmatrix}.
$$

Adding a multiple of one row to another does not change the value of the determinant, so

$$
|\mathbf{A}| = |\mathbf{B}|.
$$

Further, by equation (7.3),

$$
|\mathbf{B}| = (-1)^{1+3}(1)|\mathbf{C}| = |\mathbf{C}|,
$$

where $\mathbf{C}$ is the 4×4 matrix formed by deleting row one and column three of $\mathbf{B}$:

$$
\mathbf{C} = \begin{pmatrix} -1 & 5 & 1 & 7 \\ 20 & 3 & -5 & 3 \\ 30 & 1 & -18 & -8 \\ -17 & 15 & 18 & 10 \end{pmatrix}.
$$

Now work on $\mathbf{C}$. Again, there are many ways to proceed. We will use the -1 in the $(1, 1)$ position to get zeros in row one. Add 5 times column one to column two, add column one to column three and add 7 times column one to column four of $\mathbf{C}$ to get

$$
\mathbf{D} = \begin{pmatrix} -1 & 0 & 0 & 0 \\ 20 & 103 & 15 & 143 \\ 30 & 151 & 12 & 202 \\ -17 & 70 & 1 & -109 \end{pmatrix}.
$$

Because we added a multiple of one column to another,

$$
|\mathbf{C}| = |\mathbf{D}|.
$$

And, using equation (7.3) again,

$$|\mathbf{D}| = (-1)^{1+1}(-1)|\mathbf{E}| = -|\mathbf{E}|,$$

where $\mathbf{E}$ is the 3×3 matrix formed from $\mathbf{D}$ by deleting row one and column one:

$$\mathbf{E} = \begin{pmatrix} 103 & 15 & 143 \\ 151 & 12 & 202 \\ -70 & 1 & -109 \end{pmatrix}.$$

To evaluate $\mathbf{E}$ we will use the 1 in the $(3, 2)$ place. Add -1 times row three to row one and -12 times row three to row two to get

$$\mathbf{F} = \begin{pmatrix} 1153 & 0 & 1778 \\ 991 & 0 & 1510 \\ -70 & 1 & -109 \end{pmatrix}.$$

Then

$$|\mathbf{E}| = |\mathbf{F}|.$$

Further,

$$|\mathbf{F}| = (-1)^{3+2}(1)|\mathbf{G}| = -|\mathbf{G}|$$

where $\mathbf{G}$ is the 2×2 matrix obtained by deleting row three and column two of $\mathbf{F}$:

$$\mathbf{G} = \begin{pmatrix} 1153 & 1778 \\ 991 & 1510 \end{pmatrix}.$$

This is 2×2 which we evaluate easily:

$$|\mathbf{G}| = (1153)(1510) - (1778)(991) = -20,968.$$

Working back through the chain of determinants, we have

$$|\mathbf{A}| = |\mathbf{B}| = |\mathbf{C}||\mathbf{D}|$$
$$= -|\mathbf{E}| = -|\mathbf{F}| = |\mathbf{G}| = -20,968. \quad \blacklozenge$$

EXAMPLE 7.3

Evaluate $|\mathbf{A}|$ with

$$\mathbf{A} = \begin{pmatrix} -2 & 5 & -4 & 7 \\ 3 & 10 & 2 & -5 \\ 2 & 2 & -6 & 3 \\ -4 & -7 & 9 & 14 \end{pmatrix}.$$

We can multiply row one by $-1/2$ to get a 1 in the $(1, 1)$ position of the new matrix $\mathbf{B}$:

$$\mathbf{B} = \begin{pmatrix} 1 & -5/2 & 2 & -7/2 \\ 3 & 10 & 2 & -5 \\ 2 & 2 & -6 & 3 \\ -4 & -7 & 9 & 14 \end{pmatrix}.$$

Then

$$|\mathbf{B}| = -\frac{1}{2}|\mathbf{A}|.$$

Add -3 times row one to row two, -2 times row one to row three, and 4 times row one to row three of **B** to get

$$\mathbf{C} = \begin{pmatrix} 1 & -5/2 & 2 & -7/2 \\ 0 & 35/2 & -4 & 11/2 \\ 0 & 7 & -10 & 10 \\ 0 & -17 & -17 & 0 \end{pmatrix}.$$

Then

$$|\mathbf{C}| = |\mathbf{B}|.$$

Further,

$$|\mathbf{C}| = (-1)^{1+1}(1)|\mathbf{D}| = |\mathbf{D}|$$

where **D** is the 3×3 matrix formed from **C** by deleting row one and column one:

$$\mathbf{D} = \begin{pmatrix} 35/2 & -4 & 11/2 \\ 7 & -10 & 10 \\ -17 & 17 & 0 \end{pmatrix}.$$

Then

$$|\mathbf{D}| = |\mathbf{C}|.$$

Next add column two to column one of **D** to get

$$\mathbf{E} = \begin{pmatrix} 27/2 & -4 & 11/2 \\ -3 & -10 & 10 \\ 0 & 17 & 0 \end{pmatrix}.$$

Then

$$|\mathbf{E}| = |\mathbf{D}|$$

and

$$|\mathbf{E}| = (-1)^{2+3}(17)|\mathbf{F}|,$$

where **F** is the 2×2 matrix we get from **E** by deleting row three and column 2:

$$\mathbf{F} = \begin{pmatrix} 27/2 & 11/2 \\ -3 & 10 \end{pmatrix}.$$

Now

$$|\mathbf{F}| = (27/2)(10) - (11/2)(-3) = 303/2.$$

Working through these steps we get

$$|\mathbf{A}| = -2|\mathbf{B}| = -2|\mathbf{C}| = -2|\mathbf{E}|$$

$$= -2(-17)|\mathbf{F}| = 34(303/2) = 5,151. \quad \blacklozenge$$

In each of Problems 1 through 8, use the method of this section to evaluate the determinant. In each problem, there are many different sequences of operations that can be used to make the evaluation.

1. $\begin{vmatrix} -2 & 4 & 1 \\ 1 & 6 & 3 \\ 7 & 0 & 4 \end{vmatrix}$

2. $\begin{vmatrix} 2 & -3 & 7 \\ 14 & 1 & 1 \\ -13 & -1 & 5 \end{vmatrix}$

3. $\begin{vmatrix} -4 & 5 & 6 \\ -2 & 3 & 5 \\ 2 & -2 & 6 \end{vmatrix}$

4. $\begin{vmatrix} 2 & -5 & 8 \\ 4 & 3 & 8 \\ 13 & 0 & -4 \end{vmatrix}$

5. $\begin{vmatrix} 17 & -2 & 5 \\ 1 & 12 & 0 \\ 14 & 7 & -7 \end{vmatrix}$

6. $\begin{vmatrix} -3 & 3 & 9 & 6 \\ 1 & -2 & 15 & 6 \\ 7 & 1 & 1 & 5 \\ 2 & 1 & -1 & 3 \end{vmatrix}$

7. $\begin{vmatrix} 0 & 1 & 1 & -4 \\ 6 & -3 & 2 & 2 \\ 1 & -5 & 1 & -2 \\ 4 & 8 & 2 & 2 \end{vmatrix}$

8. $\begin{vmatrix} 2 & 7 & -1 & 0 \\ 3 & 1 & 1 & 8 \\ -2 & 0 & 3 & 1 \\ 4 & 8 & -1 & 0 \end{vmatrix}$

7.3   Evaluation of Determinants II

In the preceding section, we evaluated determinants by using row and column operations to produce rows and/or columns with all but one entry zero. In this section, we exploit this idea from a different perspective to write the determinant as a sum of numbers times smaller determinants. This method, called expansion by cofactors, can be used recursively until we have determinants of small enough size to be easily evaluated.

Choose a row k of $\mathbf{A} = [a_{ij}]$. An extension of property 7 of determinants from Section 7.1 enables us to write

$$|\mathbf{A}| = |[a_{ij}]| = \begin{vmatrix} a_{11} & a_{12} & \cdots & \cdots & a_{1n} \\ \vdots & \vdots & \vdots & \vdots & \vdots \\ a_{k1} & a_{k2} & \cdots & \cdots & a_{kn} \\ \vdots & \vdots & \vdots & \vdots & \vdots \\ a_{n1} & a_{n2} & \cdots & \cdots & a_{nn} \end{vmatrix} = \begin{vmatrix} a_{11} & a_{12} & \cdots & \cdots & a_{1n} \\ \vdots & \vdots & \vdots & \vdots & \vdots \\ a_{k1} & 0 & \cdots & \cdots & 0 \\ \vdots & \vdots & \vdots & \vdots & \vdots \\ a_{n1} & a_{n2} & \cdots & \cdots & a_{nn} \end{vmatrix}$$

$$+ \begin{vmatrix} a_{11} & a_{12} & \cdots & \cdots & a_{1n} \\ \vdots & \vdots & \vdots & \vdots & \vdots \\ 0 & a_{k2} & \cdots & \cdots & 0 \\ \vdots & \vdots & \vdots & \vdots & \vdots \\ a_{n1} & a_{n2} & \cdots & \cdots & a_{nn} \end{vmatrix} + \cdots + \begin{vmatrix} a_{11} & a_{12} & \cdots & \cdots & a_{1n} \\ \vdots & \vdots & \vdots & \vdots & \vdots \\ 0 & 0 & \cdots & \cdots & a_{kn} \\ \vdots & \vdots & \vdots & \vdots & \vdots \\ a_{n1} & a_{n2} & \cdots & \cdots & a_{nn} \end{vmatrix}.$$

Each of the n determinants on the right has a row with exactly one possibly nonzero element and can be expanded by that element, as in Section 7.2. To write this expansion, define the *minor* of a_{ij} to be the determinant of the $n - 1 \times n - 1$ matrix formed by deleting row i and column j of $\mathbf{A}$. This minor is denoted M_{ij}. The *cofactor* of a_{ij} is the number $(-1)^{i+j} M_{ij}$. Now this sum of determinants gives us the following.

THEOREM 7.1 *Cofactor Expansion by a Row*

For any k with $1 \le k \le n$,

$$|\mathbf{A}| = \sum_{j=1}^{n} (-1)^{k+j} a_{kj} M_{kj}. \qquad (7.4)$$

Equation (7.4) states that the determinant of $\mathbf{A}$ is the sum, along any row k, of the matrix elements of that row, each multiplied by its cofactor. This holds for any row of the matrix, although of course, this sum is easier to evaluate if we choose a row with as many zero elements as possible. Equation (7.4) is called *expansion by cofactors along row k*. If we write out a few terms for fixed k we get

$$|\mathbf{A}| = (-1)^{k+1} a_{k1} M_{k1} + (-1)^{k+2} a_{k2} M_{k2} + \cdots + (-1)^{k+n} a_{kn} M_{kn}.$$

EXAMPLE 7.4

Let

$$\mathbf{A} = \begin{pmatrix} -6 & 3 & 7 \\ 12 & -5 & -6 \\ 2 & 4 & -6 \end{pmatrix}.$$

If we expand by cofactors along row one, we get

$$|\mathbf{A}| = \sum_{j=1}^{3} (-1)^{1+j} a_{1j} M_{1j}$$

$$= (-1)^{1+1}(-6) \begin{vmatrix} -5 & -9 \\ 4 & -6 \end{vmatrix} + (-1)^{1+2}(3) \begin{vmatrix} 12 & -9 \\ 2 & -6 \end{vmatrix}$$

$$+ (-1)^{1+3}(7) \begin{vmatrix} 12 & -5 \\ 2 & 4 \end{vmatrix}$$

$$= (-6)(30 + 36) - 3(-72 + 18) + 7(-48 + 10) = 172.$$

If we expand by row three, we get

$$|\mathbf{A}| = \sum_{j=1}^{3} (-1)^{3+j} a_{3j} M_{3j}$$

$$= (-1)^{3+1}(2) \begin{vmatrix} 3 & 7 \\ -5 & -9 \end{vmatrix} + (-1)^{3+2}(4) \begin{vmatrix} -6 & 7 \\ 12 & -9 \end{vmatrix}$$

$$+ (-1)^{3+3}(-6) \begin{vmatrix} -6 & 3 \\ 12 & -5 \end{vmatrix}$$

$$= (2)(-27 + 35) - 4(54 - 84) - 6(30 - 36) = 172.$$

We can also do a cofactor expansion along a column. Now fix j and sum the elements of column j times their cofactors.

THEOREM 7.2   *Cofactor Expansion by a Column*

For any j with $1 \leq j \leq n$,

$$|\mathbf{A}| = \sum_{i=1}^{n} (-1)^{i+j} a_{ij} M_{ij}. \tag{7.5}$$

◆

EXAMPLE 7.5

We will expand the determinant of the matrix of Example 7.4 using column one:

$$|\mathbf{A}| = \sum_{i=1}^{3} (-1)^{i+1} a_{i1} M_{i1}$$

$$= (-1)^{1+1}(-6) \begin{vmatrix} -5 & -9 \\ 4 & -6 \end{vmatrix} + (-1)^{2+1}(12) \begin{vmatrix} 3 & 7 \\ 4 & -6 \end{vmatrix}$$

$$+ (-1)^{3+1}(2) \begin{vmatrix} 3 & 7 \\ -5 & -9 \end{vmatrix}$$

$$= (-6)(30 + 36) - 12(-18 - 28) + 2(-27 + 35) = 172.$$

If we expand by column two we get

$$|\mathbf{A}| = \sum_{i=1}^{3} (-1)^{i+2} a_{i2} M_{i2}$$

$$= (-1)^{1+2}(3) \begin{vmatrix} 12 & -9 \\ 2 & -6 \end{vmatrix} + (-1)^{2+2}(-5) \begin{vmatrix} -6 & 7 \\ 2 & -6 \end{vmatrix}$$

$$+ (-1)^{3+2}(4) \begin{vmatrix} -6 & 7 \\ 12 & -9 \end{vmatrix}$$

$$= (-3)(-72 + 18) - 5(36 - 14) - 4(54 - 84) = 172. \quad ◆$$

Sometimes we use row and column operations to produce a row or column with some zero elements, then write a cofactor expansion by that row or column. Each zero element eliminates one term from the cofactor expansion.

SECTION 7.3   *PROBLEMS*

In each of Problems 1 through 8, evaluate the determinant using a cofactor expansion by a row and again by a column. Elementary row and/or column operations may be performed first to simplify the cofactor expansion.

1. $\begin{vmatrix} -4 & 2 & -8 \\ 1 & 1 & 0 \\ 1 & -3 & 0 \end{vmatrix}$

2. $\begin{vmatrix} 1 & 1 & 6 \\ 2 & -2 & 1 \\ 3 & -1 & 4 \end{vmatrix}$

3. $\begin{vmatrix} 7 & -3 & 1 \\ 1 & -2 & 4 \\ -3 & 1 & 0 \end{vmatrix}$

4. $\begin{vmatrix} 5 & -4 & 3 \\ -1 & 1 & 6 \\ -2 & -2 & 4 \end{vmatrix}$

5. $\begin{vmatrix} -5 & 0 & 1 & 6 \\ 2 & -1 & 3 & 7 \\ 4 & 4 & -5 & -8 \\ 1 & -1 & 6 & 2 \end{vmatrix}$

6. $\begin{vmatrix} 4 & 3 & -5 & 6 \\ 1 & -5 & 15 & 2 \\ 0 & -5 & 1 & 7 \\ 8 & 9 & 0 & 15 \end{vmatrix}$

7. $\begin{vmatrix} -3 & 1 & 14 \\ 0 & 1 & 16 \\ 2 & -3 & 4 \end{vmatrix}$

8. $\begin{vmatrix} 14 & 13 & -2 & 5 \\ 7 & 1 & 1 & 7 \\ 0 & 2 & 12 & 3 \\ 1 & -6 & 5 & 23 \end{vmatrix}$

9. Show that
$$\begin{vmatrix} 1 & a & a^2 \\ 1 & b & b^2 \\ 1 & c & c^2 \end{vmatrix} = (a-b)(c-a)(b-c).$$

This is called *Vandermonde's determinant*.

10. Show that
$$\begin{vmatrix} a & b & c & d \\ b & c & d & a \\ c & d & a & b \\ d & a & b & c \end{vmatrix}$$
$$= (a+b+c+d)(b-a+d-c)\begin{vmatrix} 0 & 1 & -1 & 1 \\ 1 & c & d & a \\ 1 & d & a & b \\ 1 & a & b & c \end{vmatrix}.$$

11. A square matrix $\mathbf{A} = [a_{ij}]$ is *upper triangular* if all elements above the main diagonal are zero. This occurs when $a_{ij} = 0$ for $i > j$. Show that the determinant of an upper triangular matrix is the product of its main diagonal elements.

7.4 A Determinant Formula for $\mathbf{A}^{-1}$

Existence of an inverse for a matrix is intimately connected the vanishing or nonvanishing of the determinant, which is in turn tied to the rank of the matrix.

We know from Section 6.4 that an $n \times n$ matrix $\mathbf{A}$ is nonsingular when rank$(\mathbf{A}) = n$, because in this case $\mathbf{A}_R = \mathbf{I}_n$. Since $|\mathbf{I}_n| \neq 0$, and the reduced form of a matrix is obtained by elementary row operations, then in this case, we also have $|\mathbf{A}| \neq 0$.

We can summarize this discussion as follows.

THEOREM 7.3

Let $\mathbf{A}$ be an $n \times n$ matrix of numbers. Then the following are equivalent.

1. $\mathbf{A}$ is nonsingular (has an inverse).
2. $|\mathbf{A}| \neq 0$.
3. rank$(\mathbf{A}) = n$. ◆

Thus, nonvanishing of $|\mathbf{A}|$ is necessary and sufficient for $\mathbf{A}$ to have an inverse. In this case, we can write an explicit formula for the elements of $\mathbf{A}^{-1}$.

THEOREM 7.4 *Elements of a Matrix Inverse*

Let $\mathbf{A}$ be a nonsingular $n \times n$ matrix and define an $n \times n$ matrix $\mathbf{B} = [b_{ij}]$ by

$$b_{ij} = \frac{1}{|\mathbf{A}|}(-1)^{i+j} M_{ji}.$$

Then $\mathbf{B} = \mathbf{A}^{-1}$. ◆

Note that the i, j element of **B** is defined in terms of $(-1)^{i+j} M_{ji}$, which is the cofactor of a_{ji} (not a_{ij}).

We can see why this construction yields $\mathbf{A}^{-1}$ by multiplying the two matrices. The i, j element of **AB** is

$$(\mathbf{AB})_{ij} = \sum_{k=1}^{n} a_{ik} b_{kj} = \frac{1}{|\mathbf{A}|} \sum_{k=1}^{n} (-1)^{j+k} a_{ik} M_{jk}. \tag{7.6}$$

If $i = j$, the sum in equation (7.6) is exactly the cofactor expansion of $|\mathbf{A}|$ by row i. The main diagonal elements of **AB** are therefore 1.

If $i \neq j$, the sum in equation (7.6) is the cofactor expansion by row j of the determinant of the matrix formed from **A** by replacing row j by row i. But this matrix has two identical rows, so its determinant is zero and the off-diagonal elements of **AB** are all zero. This means that

$$\mathbf{AB} = \mathbf{I}_n.$$

Similarly, $\mathbf{BA} = \mathbf{I}_n$.

EXAMPLE 7.6

Let

$$\mathbf{A} = \begin{pmatrix} -2 & 4 & 1 \\ 6 & 3 & -3 \\ 2 & 9 & -5 \end{pmatrix}.$$

It is routine to compute $|\mathbf{A}| = 120$ so **A** is nonsingular. We will determine $\mathbf{A}^{-1}$ by computing the elements of the matrix **B** of Theorem 7.4:

$$b_{11} = \frac{1}{120} M_{11} = \frac{1}{120} \begin{vmatrix} 3 & -3 \\ 9 & -5 \end{vmatrix} = \frac{12}{120} = \frac{1}{10},$$

$$b_{12} = \frac{1}{120}(-1) M_{21} = -\frac{1}{120} \begin{vmatrix} 4 & 1 \\ 9 & -5 \end{vmatrix} = \frac{29}{120},$$

$$b_{13} = \frac{1}{120} M_{31} = \frac{1}{120} \begin{vmatrix} 4 & 1 \\ 3 & -3 \end{vmatrix} = -\frac{1}{8},$$

$$b_{21} = -\frac{1}{120} M_{12} = -\frac{1}{120} \begin{vmatrix} 6 & -3 \\ 2 & -5 \end{vmatrix} = \frac{1}{5},$$

$$b_{22} = \frac{1}{120} M_{22} = \frac{1}{120} \begin{vmatrix} -2 & 1 \\ 2 & -5 \end{vmatrix} = \frac{1}{15},$$

$$b_{23} = -\frac{1}{120} M_{32} = -\frac{1}{120} \begin{vmatrix} -2 & 1 \\ 6 & -3 \end{vmatrix} = 0,$$

$$b_{31} = \frac{1}{120} M_{13} = \frac{1}{120} \begin{vmatrix} 6 & 3 \\ 2 & 9 \end{vmatrix} = \frac{2}{5},$$

$$b_{32} = -\frac{1}{120} M_{23} = -\frac{1}{120} \begin{vmatrix} -2 & 4 \\ 2 & 9 \end{vmatrix} = \frac{13}{60},$$

and

$$b_{33} = \frac{1}{120} M_{33} = \frac{1}{120} \begin{vmatrix} -2 & 4 \\ 6 & 3 \end{vmatrix} = -\frac{1}{4}.$$

Then

$$\mathbf{B} = \mathbf{A}^{-1} = \begin{pmatrix} 1/10 & 29/120 & -1/8 \\ 1/5 & 1/15 & 0 \\ 2/5 & 13/60 & -1/4 \end{pmatrix}. \quad \blacklozenge$$

SECTION 7.4   PROBLEMS

In each of Problems 1 through 8, test the matrix for singularity by evaluating its determinant. If the matrix is nonsingular, use Theorem 7.4 to compute the inverse.

1. $\begin{pmatrix} 2 & -1 \\ 1 & 6 \end{pmatrix}$

2. $\begin{pmatrix} 3 & 0 \\ 1 & 4 \end{pmatrix}$

3. $\begin{pmatrix} -1 & 1 \\ 1 & 4 \end{pmatrix}$

4. $\begin{pmatrix} 2 & 5 \\ -7 & -3 \end{pmatrix}$

5. $\begin{pmatrix} 6 & -1 & 3 \\ 0 & 1 & -4 \\ 2 & 2 & -3 \end{pmatrix}$

6. $\begin{pmatrix} -14 & 1 & -3 \\ 2 & -1 & 3 \\ 1 & 1 & 7 \end{pmatrix}$

7. $\begin{pmatrix} 0 & -4 & 3 \\ 2 & -1 & 6 \\ 1 & -1 & 7 \end{pmatrix}$

8. $\begin{pmatrix} 11 & 0 & -5 \\ 0 & 1 & 0 \\ 4 & -7 & 9 \end{pmatrix}$

7.5   Cramer's Rule

Cramer's rule is a determinant formula for the unique solution of a nonhomogeneous system $\mathbf{AX} = \mathbf{B}$ when $\mathbf{A}$ is nonsingular. Of course, this solution can be found as $\mathbf{X} = \mathbf{A}^{-1}\mathbf{B}$, but the following method is sometimes convenient.

THEOREM 7.5   Cramer's Rule

Let $\mathbf{A}$ be a nonsingular $n \times n$ matrix of numbers and $\mathbf{B}$ an $n \times 1$ matrix of numbers. Then the unique solution of $\mathbf{AX} = \mathbf{B}$ is determined by

$$x_k = \frac{1}{|\mathbf{A}|} |\mathbf{A}(k; \mathbf{B})| \tag{7.7}$$

for $k = 1, 2, \ldots, n$, where $\mathbf{A}(k; \mathbf{B})$ is the matrix obtained from $\mathbf{A}$ by replacing column k of $\mathbf{A}$ with $\mathbf{B}$. $\blacklozenge$

It is easy to see why this works. Let

$$\mathbf{B} = \begin{pmatrix} b_1 \\ b_2 \\ \vdots \\ b_n \end{pmatrix}.$$

Multiply column k of $\mathbf{A}$ by x_k. This multiplies the determinant of $\mathbf{A}$ by x_k:

$$x_k|\mathbf{A}| = \begin{vmatrix} a_{11} & a_{12} & \cdots & a_{1k}x_k & \cdots & a_{1n} \\ a_{21} & a_{22} & \cdots & a_{2k}x_k & \cdots & a_{2n} \\ \vdots & \vdots & \vdots & \vdots & \vdots & \vdots \\ a_{n1} & a_{n2} & \cdots & a_{nk}x_k & \cdots & a_{nn} \end{vmatrix}.$$

For each $j \neq k$, add x_j times column j to column k in the last determinant. Since this operation does not change the value of a determinant, then

$$x_k|\mathbf{A}| = \begin{vmatrix} a_{11} & a_{12} & \cdots & a_{11}x_1 + \cdots + a_{1n}x_n & \cdots & a_{1n} \\ a_{21} & a_{22} & \cdots & a_{21}x_1 + \cdots + a_{2n}x_n & \cdots & a_{2n} \\ \vdots & \vdots & \vdots & \vdots & & \vdots \\ a_{n1} & a_{n2} & \cdots & a_{n1}x_1 + \cdots + a_{nn}x_n & \cdots & a_{nn} \end{vmatrix}$$

$$= \begin{vmatrix} a_{11} & a_{12} & \cdots & b_1 & \cdots & a_{1n} \\ a_{21} & a_{22} & \cdots & b_2 & \cdots & a_{2n} \\ \vdots & \vdots & \vdots & \vdots & \vdots & \vdots \\ a_{n1} & a_{n2} & \cdots & b_n & \cdots & a_{nn} \end{vmatrix} = |\mathbf{A}(k; \mathbf{B})|.$$

and this gives us equation (7.7).

EXAMPLE 7.7

Solve the system

$$x_1 - 3x_2 - 4x_3 = 1,$$
$$-x_1 + x_2 - 3x_3 = 14,$$

and

$$x_2 - 3x_3 = 5.$$

The matrix of coefficients is

$$\mathbf{A} = \begin{pmatrix} 1 & -3 & -4 \\ -1 & 1 & -3 \\ 0 & 1 & -3 \end{pmatrix}.$$

We find that $|\mathbf{A}| = 13$. By Cramer's rule,

$$x_1 = \frac{1}{13}\begin{vmatrix} 1 & -3 & -4 \\ 14 & 1 & -3 \\ 5 & 1 & -3 \end{vmatrix} = -\frac{117}{13} = -9,$$

$$x_2 = \frac{1}{13}\begin{vmatrix} 1 & 1 & -4 \\ -1 & 14 & -3 \\ 0 & 5 & -3 \end{vmatrix} = -\frac{10}{13},$$

and

$$x_3 = \frac{1}{13}\begin{vmatrix} 1 & -3 & 1 \\ -1 & 1 & 14 \\ 0 & 1 & 5 \end{vmatrix} = -\frac{25}{13}.\ \blacklozenge$$

SECTION 7.5   *PROBLEMS*

In each of Problems 1 through 8, solve the system using Cramer's rule or show that the rule does not apply because the matrix of coefficients is singular.

1. $15x_1 - 4x_2 = 5$
   $8x_1 + x_2 = -4$

2. $x_1 + 4x_2 = 3$
   $x_1 + x_2 = 0$

3. $8x_1 - 4x_2 + 3x_3 = 0$
   $x_1 + 5x_2 - x_3 = -5$
   $-2x_1 + 6x_2 + x_3 = -4$

4. $5x_1 - 6x_2 + x_3 = 4$
   $-x_1 + 3x_2 - 4x_3 = 5$
   $2x_1 + 3x_2 + x_3 = -8$

5. $x_1 + x_2 - 3x_3 = 0$
   $x_2 - 4x_3 = 0$
   $x_1 - x_2 - x_3 = 5$

6. $6x_1 + 4x_2 - x_3 + 3x_4 - x_5 = 7$
   $x_1 - 4x_2 + x_5 = -5$
   $x_1 - 3x_2 + x_3 - 4x_5 = 0$
   $-2x_1 + x_3 - 2x_5 = 4$
   $x_3 - x_4 - x_5 = 8$

7. $2x_1 - 4x_2 + x_3 - x_4 = 6$
   $x_2 - 3x_3 = 10$
   $x_1 - 4x_3 = 0$
   $x_2 - x_3 + 2x_4 = 4$

8. $2x_1 - 3x_2 + x_4 = 2$
   $x_2 - x_3 + x_4 = 2$
   $x_3 - 2x_4 = 5$
   $x_1 - 3x_2 + 4x_3 = 0$

CHAPTER 8

Eigenvalues and Diagonalization

8.1   Eigenvalues and Eigenvectors

In this chapter, the term "number" refers to a real or complex number. Let $\mathbf{A}$ be an $n \times n$ matrix of numbers.

> A number λ is an *eigenvalue* of $\mathbf{A}$ if there is a nonzero $n \times 1$ matrix $\mathbf{E}$ such that
>
> $$\mathbf{AE} = \lambda \mathbf{E}. \tag{8.1}$$
>
> We call $\mathbf{E}$ an *eigenvector* associated with the eigenvalue λ.

We may think of an $n \times 1$ matrix of numbers as an n-vector with real and/or complex components. If we think of $\mathbf{A}$ as mapping an n-vector $\mathbf{X}$ to an n-vector $\mathbf{AX}$, then equation (8.1) holds when $\mathbf{A}$ moves $\mathbf{E}$ to a parallel vector $\lambda \mathbf{E}$. This is the geometric significance of an eigenvector.

If c is a nonzero number and $\mathbf{AE} = \lambda \mathbf{E}$, then

$$\mathbf{A}(c\mathbf{E}) = c\mathbf{AE} = c\lambda \mathbf{E} = \lambda(c\mathbf{E}).$$

This means that $c\mathbf{E}$ is also an eigenvector with eigenvalue λ. Nonzero constant multiples of eigenvectors are eigenvectors.

EXAMPLE 8.1

Let

$$\mathbf{A} = \begin{pmatrix} 1 & 0 \\ 0 & 0 \end{pmatrix}.$$

Because

$$\begin{pmatrix} 1 & 0 \\ 0 & 0 \end{pmatrix} \begin{pmatrix} 0 \\ 4 \end{pmatrix} = \begin{pmatrix} 0 \\ 0 \end{pmatrix} = 0 \begin{pmatrix} 0 \\ 4 \end{pmatrix},$$

then 0 is an eigenvalue of **A** with eigenvector

$$\mathbf{E} = \begin{pmatrix} 0 \\ 4 \end{pmatrix}.$$

For any nonzero number α,

$$\begin{pmatrix} 0 \\ 4\alpha \end{pmatrix}$$

is also an eigenvector. Zero can be an eigenvalue, but an eigenvector must be a nonzero vector (at least one nonzero element). ◆

EXAMPLE 8.2

Let

$$\mathbf{A} = \begin{pmatrix} 1 & -1 & 0 \\ 0 & 1 & 1 \\ 0 & 0 & -1 \end{pmatrix}.$$

Then

$$\mathbf{A} \begin{pmatrix} 6 \\ 0 \\ 0 \end{pmatrix} = \begin{pmatrix} 6 \\ 0 \\ 0 \end{pmatrix}.$$

Therefore, 1 is an eigenvalue with eigenvector

$$\begin{pmatrix} 6 \\ 0 \\ 0 \end{pmatrix}$$

or any nonzero constant times this matrix. Similarly,

$$\mathbf{A} \begin{pmatrix} 1 \\ 2 \\ -4 \end{pmatrix} = \begin{pmatrix} -1 \\ -2 \\ 4 \end{pmatrix} = (-1) \begin{pmatrix} 1 \\ 2 \\ -4 \end{pmatrix}.$$

Therefore, -1 is an eigenvalue with eigenvector

$$\begin{pmatrix} 1 \\ 2 \\ -4 \end{pmatrix}.$$

Again, any nonzero multiple of this eigenvector is also an eigenvector with eigenvalue -1. ◆

We would like to be able to find all of the eigenvalues of a given matrix. The key lies in the equation

$$|\lambda \mathbf{I}_n - \mathbf{A}| = 0.$$

If expanded, this determinant is a polynomial of degree n in the unknown λ. This polynomial is called the *characteristic polynomial* of $\mathbf{A}$ and is denoted $p_{\mathbf{A}}(\lambda)$. The equation $p_{\mathbf{A}}(\lambda) = 0$ is the *characteristic equation* of $\mathbf{A}$. It has n roots (perhaps some repeated, some or all complex). We claim that these n numbers are all of the eigenvalues of $\mathbf{A}$. To see why this is true, suppose λ_1 is a root of the characteristic polynomial. The vanishing of the determinant $|\lambda_1 \mathbf{I}_n - \mathbf{A}|$ means that $\lambda_1 \mathbf{I}_n - \mathbf{A}$ has rank less than n (is nonsingular), hence that the homogeneous system

$$(\lambda_1 \mathbf{I}_n - \mathbf{A})\mathbf{X} = \mathbf{O}$$

of n linear equations in n unknowns has a nontrivial solution $\mathbf{E}_1$. But this means that

$$(\lambda_1 \mathbf{I}_n - \mathbf{A})\mathbf{E}_1 = \mathbf{O}.$$

Write this equation as

$$\lambda_1 \mathbf{I}_n \mathbf{E}_1 - \mathbf{A}\mathbf{E}_1 = \mathbf{O}.$$

Now, $\lambda_1 \mathbf{I}_n \mathbf{E}_1$ is just $\lambda_1 \mathbf{E}_1$, so the last equation reads

$$\mathbf{A}\mathbf{E}_1 = \lambda_1 \mathbf{E}_1,$$

and this makes λ_1 an eigenvalue of $\mathbf{A}$ with corresponding eigenvector $\mathbf{E}_1$. We will summarize this conclusion.

THEOREM 8.1   *Eigenvalues of* $\mathbf{A}$

Let $\mathbf{A}$ be an $n \times n$ matrix of numbers. Then

1. λ is an eigenvalue of $\mathbf{A}$ if and only if λ is a root of the characteristic polynomial of $\mathbf{A}$. This occurs exactly when

$$|\lambda \mathbf{I}_n - \mathbf{A}| = 0.$$

   Therefore $\mathbf{A}$ has n eigenvalues, counting each eigenvalues as many times as it appears as a root of $p_{\mathbf{A}}(\lambda)$.

2. If λ is an eigenvalue of $\mathbf{A}$, then any nontrivial solution of $(\lambda \mathbf{I}_n - \mathbf{A})\mathbf{X} = \mathbf{O}$ is an eigenvector of $\mathbf{A}$ associated with λ. ♦

EXAMPLE 8.3

Let

$$\mathbf{A} = \begin{pmatrix} 1 & -1 & 0 \\ 0 & 1 & 1 \\ 0 & 0 & -1 \end{pmatrix},$$

as in Example 8.2. The characteristic polynomial is

$$p_{\mathbf{A}}(\lambda) = |\lambda \mathbf{I}_3 - \mathbf{A}| = \begin{vmatrix} \lambda - 1 & 1 & 0 \\ 0 & \lambda - 1 & -1 \\ 0 & 0 & \lambda + 1 \end{vmatrix} = (\lambda - 1)^2 (\lambda + 1).$$

This polynomial has roots $1, 1$, and -1, and these are the eigenvalues of $\mathbf{A}$. The root 1 has multiplicity 2 and must be listed twice as an eigenvalue of $\mathbf{A}$. $\mathbf{A}$ has three eigenvalues.

To find an eigenvector associated with the eigenvalue 1, solve

$$((1)\mathbf{I}_3 - \mathbf{A})\mathbf{X} = \begin{pmatrix} 0 & 1 & 0 \\ 0 & 0 & -1 \\ 0 & 0 & 2 \end{pmatrix} \begin{pmatrix} x_1 \\ x_2 \\ x_3 \end{pmatrix} = \begin{pmatrix} 0 \\ 0 \\ 0 \end{pmatrix}.$$

This system of three equations in three unknowns has the general solution

$$\begin{pmatrix} \alpha \\ 0 \\ 0 \end{pmatrix}$$

and this is an eigenvector associated with 1 for any $\alpha \neq 0$.

For eigenvectors associated with -1, solve

$$((-1)\mathbf{I}_3 - \mathbf{A})\mathbf{X} = \begin{pmatrix} -2 & 1 & 0 \\ 0 & -2 & -1 \\ 0 & 0 & 0 \end{pmatrix} \mathbf{X} = \mathbf{O}.$$

This system has a general solution of

$$\begin{pmatrix} \beta \\ 2\beta \\ -4\beta \end{pmatrix},$$

and this is an eigenvector associated with -1 for any $\beta \neq 0$. ◆

EXAMPLE 8.4

Let

$$\mathbf{A} = \begin{pmatrix} 1 & -2 \\ 2 & 0 \end{pmatrix}.$$

The characteristic polynomial is

$$p_{\mathbf{A}}(\lambda) = |\lambda\mathbf{I}_2 - \mathbf{A}| = \left| \begin{pmatrix} \lambda & 0 \\ 0 & \lambda \end{pmatrix} - \begin{pmatrix} 1 & -2 \\ 2 & 0 \end{pmatrix} \right| = \begin{vmatrix} \lambda - 1 & 2 \\ -2 & \lambda \end{vmatrix} = \lambda^2 - \lambda + 4,$$

with roots

$$\frac{1 + \sqrt{15}i}{2} \quad \text{and} \quad \frac{1 - \sqrt{15}i}{2},$$

and these are the eigenvalues of $\mathbf{A}$.

For an eigenvector corresponding to $(1 + \sqrt{15}i)/2$ solve

$$((1 + \sqrt{15}i)/2)\mathbf{I}_2 - \mathbf{A})\mathbf{X} = \mathbf{O},$$

which is

$$\left[\frac{1 + \sqrt{15}i}{2} \begin{pmatrix} 1 & 0 \\ 0 & 1 \end{pmatrix} - \begin{pmatrix} 1 & -2 \\ 2 & 0 \end{pmatrix} \right] \mathbf{X} = \mathbf{O}.$$

This is the system

$$\begin{pmatrix} \frac{-1 + \sqrt{15}i}{2} & 2 \\ -2 & \frac{1 + \sqrt{15}i}{2} \end{pmatrix} \begin{pmatrix} x_1 \\ x_2 \end{pmatrix} = \begin{pmatrix} 0 \\ 0 \end{pmatrix}$$

with a general solution of

$$\alpha \begin{pmatrix} 1 \\ (1 - \sqrt{15}i)/4 \end{pmatrix}.$$

This is an eigenvector associated with $(1 + \sqrt{15}i)/2$ for any $\alpha \neq 0$.

Similarly, solve the 2×2 system

$$(((1 - \sqrt{15}i)/2)\mathbf{I}_2 - \mathbf{A})\mathbf{X} = \mathbf{O}$$

to obtain

$$\mathbf{X} = \beta \begin{pmatrix} 1 \\ (1 + \sqrt{15}i)/4 \end{pmatrix}.$$

This is an eigenvector associated with $(1 - \sqrt{15}i)/2$ for any $\beta \neq 0$. ◆

If $\mathbf{A}$ has real numbers as elements and $\lambda = \alpha + i\beta$ is an eigenvalue, then the conjugate $\overline{\lambda} = \alpha - i\beta$ is also an eigenvalue. This is because the characteristic polynomial of $\mathbf{A}$ has real coefficients in this case, so complex roots occur in conjugate pairs. Further, if $\mathbf{E}$ is an eigenvector corresponding to λ, then $\overline{\mathbf{E}}$ is an eigenvector corresponding to $\overline{\lambda}$, where we take the conjugate of a matrix by taking the conjugate of each of its elements. This can be seen by taking the conjugate of $\mathbf{A}\mathbf{E} = \lambda\mathbf{E}$ and using the fact that, with real elements, $\overline{\mathbf{A}} = \mathbf{A}$. This observation can be seen in Example 8.4.

There is a general expression for the eigenvalues of a matrix that is sometimes useful.

LEMMA 8.1

Let $\mathbf{A}$ be $n \times n$. Let λ be an eigenvalue of $\mathbf{A}$ with eigenvector $\mathbf{E}$. Then

$$\lambda = \frac{\overline{\mathbf{E}}^t \mathbf{A}\mathbf{E}}{\overline{\mathbf{E}}^t \mathbf{E}}. \tag{8.2}$$

◆

Before giving the one-line proof of this expression, let us examine what the right side means. Let

$$\mathbf{E} = \begin{pmatrix} e_1 \\ e_2 \\ \vdots \\ e_n \end{pmatrix}.$$

Then

$$\overline{\mathbf{E}}^t \mathbf{A}\mathbf{E} = \begin{pmatrix} \overline{e_1} & \overline{e_2} & \cdots & \overline{e_n} \end{pmatrix} \begin{pmatrix} a_{11} & a_{12} & \cdots & a_{1n} \\ a_{21} & a_{22} & \cdots & a_{2n} \\ \vdots & \vdots & \vdots & \vdots \\ a_{n1} & a_{n2} & \cdots & a_{nn} \end{pmatrix} \begin{pmatrix} e_1 \\ e_2 \\ \vdots \\ e_n \end{pmatrix}.$$

This is a product of a $1 \times n$ matrix with an $n \times n$ matrix and an $n \times 1$ matrix, and hence it is a 1×1 matrix, which we think of as a number. If we carry out this matrix product, we obtain the number

$$\overline{\mathbf{E}}^t \mathbf{A}\mathbf{E} = \sum_{i=1}^{n} \sum_{j=1}^{n} a_{ij} \overline{e_i} e_j.$$

For the denominator of equation (8.2), we have a $1 \times n$ matrix multiplied by an $n \times 1$ matrix, which is also a 1×1 matrix or number. Specifically,

$$\overline{\mathbf{E}}^t \mathbf{E} = \begin{pmatrix} \overline{e_1} & \overline{e_2} \cdots & \overline{e_n} \end{pmatrix} \begin{pmatrix} e_1 \\ e_2 \\ \vdots \\ e_n \end{pmatrix} = \sum_{j=1}^n \overline{e}_j e_j = \sum_{j=1}^n |e_j|^2.$$

Therefore, the conclusion of Lemma 8.1 can be written

$$\lambda = \frac{\sum_{i=1}^n \sum_{j=1}^n a_{ij} \overline{e}_i e_j}{\sum_{j=1}^n |e_j|^2}. \tag{8.3}$$

Proof of Lemma 8.1   Since $\mathbf{AE} = \lambda \mathbf{E}$, then

$$\overline{\mathbf{E}}^t \mathbf{AE} = \lambda \overline{\mathbf{E}}^t \mathbf{E},$$

and this proves the lemma. ♦

It is sometimes useful to know whether the eigenvectors of a matrix are linearly dependent or independent. The following theorem partly answers this question.

THEOREM 8.2

Suppose the $n \times n$ matrix $\mathbf{A}$ has n distinct eigenvalues. Then $\mathbf{A}$ has n linearly independent eigenvectors.

Proof   We will show by induction that any k distinct eigenvalues have associated with them k linearly independent eigenvectors. For $k=1$, there is nothing to prove. Now suppose any $k-1$ distinct eigevalues have associated with them $k-1$ linearly independent eigenvectors. Suppose $\lambda_1, \ldots, \lambda_k$ are k distinct eigenvalues with, respectively, eigenvectors $\mathbf{E}_1, \ldots, \mathbf{E}_k$. We want to show that these eigenvectors are linearly independent.

If these eigenvectors were linearly dependent, there would be numbers $c_1, \ldots, c_k$ that are not all zero, such that

$$c_1 \mathbf{E}_1 + c_2 \mathbf{E}_2 + \cdots + c_k \mathbf{E}_k = \mathbf{O}.$$

By relabeling if necessary, we will assume for convenience that $c_1 \neq 0$. Now

$$(\lambda_1 \mathbf{I}_n - \mathbf{A})(c_1 \mathbf{E}_1 + c_2 \mathbf{E}_2 + \cdots + c_k \mathbf{E}_k) = \mathbf{O}$$
$$= c_1(\lambda_1 \mathbf{I}_n - \mathbf{A})\mathbf{E}_1 + c_2(\lambda_1 \mathbf{I}_n - \mathbf{A})\mathbf{E}_2 + \cdots + c_k(\lambda_1 \mathbf{I}_n - \mathbf{A})\mathbf{E}_k$$
$$= c_1(\lambda_1 \mathbf{E}_1 - \mathbf{AE}_1) + c_2(\lambda_1 \mathbf{E}_2 - \mathbf{AE}_2) + \cdots + c_k(\lambda_1 \mathbf{E}_k - \mathbf{AE}_k)$$
$$= c_1(\lambda_1 \mathbf{E}_1 - \lambda_1 \mathbf{E}_1) + c_2(\lambda_1 \mathbf{E}_2 - \lambda_2 \mathbf{E}_2) + \cdots + c_k(\lambda_1 \mathbf{E}_k - \lambda_k \mathbf{E}_k)$$
$$= c_2(\lambda_1 - \lambda_2)\mathbf{E}_2 + \cdots + c_k(\lambda_1 - \lambda_k)\mathbf{E}_k.$$

But $\mathbf{E}_2, \ldots, \mathbf{E}_k$ are linearly independent by the induction hypothesis. Therefore,

$$c_2 = \cdots = c_k = 0.$$

But then $c_1 \mathbf{E}_1 = \mathbf{O}$. Since an eigenvector cannot be the zero vector, then $c_1 = 0$ is also a contradiction. This completes the proof by induction. ♦

To illustrate, in Example 8.4, $\mathbf{A}$ was a 2×2 matrix and had two distinct eigenvalues. The eigenvectors produced for each eigenvalue were linearly independent.

The converse of Theorem 8.2 is not true. If **A** does not have n distinct eigenvalues, **A** may or may not have n linearly independent eigenvectors.

EXAMPLE 8.5

Let

$$\mathbf{A} = \begin{pmatrix} 5 & -4 & 4 \\ 12 & -11 & 12 \\ 4 & -4 & 5 \end{pmatrix}.$$

The eigenvalues of **A** are $-3, 1$, and 1, with 1 a repeated root of the characteristic polynomial. Corresponding to -3, we find an eigenvector

$$\begin{pmatrix} 1 \\ 3 \\ 1 \end{pmatrix}.$$

Now look for an eigenvector corresponding to 1. We must solve the system

$$((1)\mathbf{I}_2 - \mathbf{A})\mathbf{X} = \begin{pmatrix} -4 & 4 & -4 \\ -12 & 12 & -12 \\ -4 & 4 & -4 \end{pmatrix} \begin{pmatrix} x_1 \\ x_2 \\ x_3 \end{pmatrix} = \begin{pmatrix} 0 \\ 0 \\ 0 \end{pmatrix}.$$

This system has the general solution

$$\alpha \begin{pmatrix} 1 \\ 0 \\ -1 \end{pmatrix} + \beta \begin{pmatrix} 0 \\ 1 \\ 1 \end{pmatrix},$$

in which α and β are any numbers. With $\alpha = 1$ and $\beta = 0$ and then with $\alpha = 0$ and $\beta = 1$, we obtain two linearly independent eigenvectors associated with eigenvalue 1:

$$\begin{pmatrix} 1 \\ 0 \\ -1 \end{pmatrix} \text{ and } \begin{pmatrix} 0 \\ 1 \\ 1 \end{pmatrix}.$$

For this matrix **A**, we can produce three linearly independent eigenvectors, even though the eigenvalues are not distinct. This is not always possible, as we will see shortly. ◆

Symmetric Matrices

Eigenvalues and eigenvectors of special classes of matrices may exhibit special properties. Symmetric matrices form one such special class. $\mathbf{A} = [a_{ij}]$ is *symmetric* if $a_{ij} = a_{ji}$ whenever $i \neq j$. This means that $\mathbf{A} = \mathbf{A}^t$.

If we think of the elements a_{ii} as forming the main diagonal of **A**, symmetry means that each off-diagonal element is equal to its reflection across this main diagonal. For example,

$$\begin{pmatrix} -7 & -2-i & 1 & 14 \\ -2-i & 2 & -9 & 47i \\ 1 & -9 & -4 & \pi \\ 14 & 47i & \pi & 22 \end{pmatrix}$$

is symmetric.

It is a significant property of symmetric matrices that those with real elements must have all real eigenvalues.

THEOREM 8.3   *Eigenvalues of Real Symmetric Matrices*

The eigenvalues of a real symmetric matrix are real.

Proof   Use equation (8.2). Clearly the denominator is a positive number. All we need to show is that the numerator is real. We will use the facts that, since $\mathbf{A}$ is real and symmetric, then $\overline{\mathbf{A}} = \mathbf{A} = \mathbf{A}^t$. Now

$$\overline{(\overline{\mathbf{E}}^t \mathbf{A} \mathbf{E})} = \overline{\overline{\mathbf{E}}^t} \, \overline{\mathbf{A} \mathbf{E}} = \mathbf{E}^t \mathbf{A} \overline{\mathbf{E}}.$$

But $\mathbf{E}^t \mathbf{A} \overline{\mathbf{E}}$ is a 1×1 matrix, hence it equals its own transpose. Then

$$\mathbf{E}^t \mathbf{A} \overline{\mathbf{E}} = (\mathbf{E}^t \mathbf{A} \overline{\mathbf{E}})^t = \overline{\mathbf{E}}^t \mathbf{A} (\mathbf{E}^t)^t = \overline{\mathbf{E}}^t \mathbf{A} \mathbf{E}.$$

From the last two equations, $\overline{\mathbf{E}}^t \mathbf{A} \mathbf{E}$ equals its own conjugate, hence it is a real number (1×1 matrix). Therefore, the numerator and denominator in equation (8.2) for λ are both real, so λ is real. ◆

One ramification of real eigenvalues and real matrix elements is that associated eigenvectors also will be real. We claim that if $\mathbf{A}$ is also symmetric, then eigenvectors associated with distinct eigenvalues must be orthogonal when thought of as vectors in R^n.

THEOREM 8.4   *Orthogonality of Eigenvectors*

Let $\mathbf{A}$ be a real symmetric matrix. Then eigenvectors associated with distinct eigenvalues are orthogonal.

Proof   We can derive this result by a useful interplay between matrix and vector notation. Let λ and μ be distinct eigenvalues of $\mathbf{A}$ with eigenvectors, respectively,

$$\mathbf{E} = \begin{pmatrix} e_1 \\ e_2 \\ \vdots \\ e_n \end{pmatrix} \text{ and } \mathbf{G} = \begin{pmatrix} g_1 \\ g_2 \\ \vdots \\ g_n \end{pmatrix}.$$

The dot product of these eigenvectors can be thought of as the matrix product:

$$\mathbf{E} \cdot \mathbf{G} = e_1 g_1 + e_2 g_2 + \cdots + e_n g_n = \mathbf{E}^t \mathbf{G}.$$

Now use the facts that $\mathbf{AE} = \lambda \mathbf{E}$, $\mathbf{AG} = \mu \mathbf{G}$, and $\mathbf{A} = \mathbf{A}^t$ to write

$$\lambda \mathbf{E}^t \mathbf{G} = (\mathbf{AE})^t \mathbf{G} = (\mathbf{E}^t \mathbf{A}^t) \mathbf{G}$$

$$= (\mathbf{E}^t \mathbf{A}) \mathbf{G} = \mathbf{E}^t (\mathbf{AG}) = \mathbf{E}^t (\mu \mathbf{G}).$$

But then

$$(\lambda - \mu) \mathbf{E}^t \mathbf{G} = 0.$$

Since $\lambda \neq \mu$, then $\mathbf{E}^t \mathbf{G} = \mathbf{E} \cdot \mathbf{G} = 0$, so $\mathbf{E}$ and $\mathbf{G}$ are orthogonal. ◆

EXAMPLE 8.6

$$A = \begin{pmatrix} 3 & 0 & -2 \\ 0 & 2 & 0 \\ -2 & 0 & 0 \end{pmatrix}$$

is a 3×3 symmetric matrix. The eigenvalues are $2, -1$, and 4 with associated eigenvectors

$$\begin{pmatrix} 0 \\ 1 \\ 0 \end{pmatrix}, \begin{pmatrix} 1 \\ 0 \\ 2 \end{pmatrix}, \text{ and } \begin{pmatrix} 2 \\ 0 \\ -1 \end{pmatrix}.$$

These eigenvectors are mutually orthogonal. ◆

Finding eigenvalues of a matrix may be difficult because finding the roots of a polynomial can be difficult. There is a method due to Gershgorin that enables us to place the eigenvalues inside disks in the plane. This is sometimes useful to get some idea of how the eigenvalues of a matrix are distributed.

THEOREM 8.5 *Gershgorin*

Let $\mathbf{A}$ be an $n \times n$ matrix of numbers. For $k = 1, 2, \ldots, n$ let

$$r_k = \sum_{j=1, j \neq k}^{n} |a_{kj}|.$$

Let C_k be the circle of radius r_k centered at (α_k, β_k), where $a_{kk} = \alpha_k + \beta_k i$. Then each eigenvalue of $\mathbf{A}$, when plotted as a point in the complex plane, lies on or within one of the circles $C_1, \ldots, C_n$. ◆

C_k is the circle centered at the kth diagonal element a_{kk} of $\mathbf{A}$ having a radius equal to the sum of the magnitudes of the elements across row k, excluding the diagonal element of that row.

EXAMPLE 8.7

Let

$$A = \begin{pmatrix} 12i & 1 & 3 \\ 2 & -6 & 2+i \\ 3 & 1 & 5 \end{pmatrix}.$$

The characteristic polynomial of $\mathbf{A}$ is

$$p_{\mathbf{A}}(\lambda) = \lambda^3 + (1 - 12i)\lambda^2$$
$$- (43 + 13i)\lambda - 68 + 381i.$$

The Gershgorin circles have centers and radii:

$$C_1 : (0, 12), \quad r_1 = 1 + 3 = 4,$$
$$C_2 : (-6, 0), \quad r_2 = 2 + \sqrt{5}$$
$$C_3 : (5, 0), \quad r_3 = 3 + 1 = 4.$$

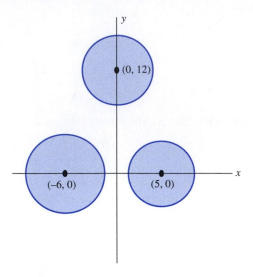

FIGURE 8.1 *Gerschgorin circles in Example 8.7.*

Figure 8.1 shows these Gershgorin circles. The eigenvalues are in the disks determined by these circles. ◆

Gershgorin's theorem is not a way of approximating eigenvalues, since some of the disks may have large radii. However, sometimes information that is revealed by these disks can be useful. For example, in studies of the stability of fluid flow it is important to know whether eigenvalues occur in the right half-plane.

SECTION 8.1   *PROBLEMS*

In each of Problems 1 through 12, find the eigenvalues of the matrix. For each eigenvalue, find an eigenvector. Sketch the Gershgorin circles for the matrix and locate the eigenvalues as points in the plane.

1. $\begin{pmatrix} 1 & 3 \\ 2 & 1 \end{pmatrix}$

2. $\begin{pmatrix} -2 & 0 \\ 1 & 4 \end{pmatrix}$

3. $\begin{pmatrix} -5 & 0 \\ 1 & 2 \end{pmatrix}$

4. $\begin{pmatrix} 6 & -2 \\ -3 & 4 \end{pmatrix}$

5. $\begin{pmatrix} 1 & -6 \\ 2 & 2 \end{pmatrix}$

6. $\begin{pmatrix} 0 & 1 \\ 0 & 0 \end{pmatrix}$

7. $\begin{pmatrix} 2 & 0 & 0 \\ 1 & 0 & 2 \\ 0 & 0 & 3 \end{pmatrix}$

8. $\begin{pmatrix} -2 & 1 & 0 \\ 1 & 3 & 0 \\ 0 & 0 & -1 \end{pmatrix}$

9. $\begin{pmatrix} -3 & 1 & 1 \\ 0 & 0 & 0 \\ 0 & 1 & 0 \end{pmatrix}$

10. $\begin{pmatrix} 0 & 0 & -1 \\ 0 & 0 & 1 \\ 2 & 0 & 0 \end{pmatrix}$

11. $\begin{pmatrix} -4 & 1 & 0 & 1 \\ 0 & 1 & 0 & 0 \\ 0 & 0 & 2 & 0 \\ 1 & 0 & 0 & 3 \end{pmatrix}$

12. $\begin{pmatrix} 5 & 1 & 0 & 9 \\ 0 & 1 & 0 & 9 \\ 0 & 0 & 0 & 9 \\ 0 & 0 & 0 & 0 \end{pmatrix}$

In each of Problems 13 through 18, find the eigenvalues and associated eigenvectors of the symmetric matrix. Verify that eigenvectors associated with distinct eigenvalues are orthogonal.

13. $\begin{pmatrix} 4 & -2 \\ -2 & 1 \end{pmatrix}$

14. $\begin{pmatrix} -3 & 5 \\ 5 & 4 \end{pmatrix}$

15. $\begin{pmatrix} 6 & 1 \\ 1 & 4 \end{pmatrix}$

16. $\begin{pmatrix} -13 & 1 \\ 1 & 4 \end{pmatrix}$

17. $\begin{pmatrix} 0 & 1 & 0 \\ 1 & -2 & 0 \\ 0 & 0 & 3 \end{pmatrix}$

18. $\begin{pmatrix} 0 & 1 & 1 \\ 1 & 2 & 0 \\ 1 & 0 & 2 \end{pmatrix}$

19. Suppose λ is an eigenvalue of $\mathbf{A}$ with eigenvector $\mathbf{E}$. Let k be a positive integer. Show that λ^k is an eigenvalue of $\mathbf{A}^k$ with eigenvector $\mathbf{E}$.

20. Let $\mathbf{A}$ be an $n \times n$ matrix of numbers. Show that the constant term in the characteristic polynomial of $\mathbf{A}$ is $(-1)^n|\mathbf{A}|$. Use this to show that any singular matrix must have zero as an eigenvalue.

8.2   Diagonalization

Recall that the elements a_{ii} of a matrix make up its *main diagonal*. All other matrix elements are called *off-diagonal elements*.

A square matrix is called a *diagonal matrix* if all off-diagonal elements are zero.

A diagonal matrix has the appearance

$$\mathbf{D} = \begin{pmatrix} d_1 & 0 & 0 & \cdots & 0 & 0 \\ 0 & d_2 & 0 & \cdots & 0 & 0 \\ 0 & 0 & d_3 & \cdots & 0 & 0 \\ \vdots & \vdots & \vdots & \vdots & \vdots & \vdots \\ 0 & 0 & 0 & \cdots & 0 & d_n \end{pmatrix}$$

in which we have written $a_{ii} = d_i$.

Diagonal matrices have many pleasant properties. Let $\mathbf{A}$ and $\mathbf{B}$ be $n \times n$ diagonal matrices with diagonal elements, respectively, a_{ii} and b_{ii}.

1. A sum of diagonal matrices is diagonal with diagonal elements $a_{ii} + b_{ii}$.

2. A product of diagonal matrices is diagonal with diagonal elements $a_{ii}b_{ii}$.

3. The determinant of a diagonal matrix is the product of its diagonal elements,

$$|\mathbf{A}| = a_{11}\,a_{22}\,\ldots\,a_{nn}.$$

4. From property 3, a diagonal matrix is nonsingular exactly when each diagonal element is nonzero (so the matrix has nonzero determinant). In this event, the inverse of $\mathbf{A}$ is the diagonal matrix having diagonal elements $1/a_{ii}$.

5. The eigenvalues of a diagonal matrix are its diagonal elements.

6. If **A** is a diagonal matrix, then

$$\begin{pmatrix} 0 \\ 0 \\ \vdots \\ 1 \\ 0 \\ \vdots \\ 0 \end{pmatrix}$$

with all zero elements except for 1 in the $(i, 1)$ place is an eigenvector corresponding to the eigenvalue a_{ii}.

"Most" matrices are not diagonal. However, sometimes it is possible to transform a matrix to a diagonal one. This will enable us to transform certain types of problems to simpler ones. Define an $n \times n$ matrix **A** of numbers to be *diagonalizable* if there is an $n \times n$ matrix **P** such that $\mathbf{P}^{-1}\mathbf{AP}$ is a diagonal matrix. In this case, we say that **P** *diagonalizes* **A**.

We will see that not every matrix is diagonalizable. The following result not only tells us exactly when **A** is diagonalizable, but also how to choose **P** to diagonalize **A**, and what $\mathbf{P}^{-1}\mathbf{AP}$ must look like.

THEOREM 8.6   *Diagonalization of a Matrix*

Let **A** be $n \times n$. Then **A** is diagonalizable if and only if **A** has n linearly independent eigenvectors. Furthermore, if **P** is the $n \times n$ matrix having these eigenvectors as columns, then $\mathbf{P}^{-1}\mathbf{AP}$ is the $n \times n$ diagonal matrix having the eigenvalues of **A** down its main diagonal in the order in which the eigenvectors were chosen as columns of **P**.

In addition, if **Q** is any matrix that diagonalizes **A**, then necessarily the diagonal matrix $\mathbf{Q}^{-1}\mathbf{AQ}$ has the eigenvalues of **A** along its main diagonal, and the columns of **Q** must be eigenvectors of **A** in the order in which the eigenvalues appear on the main diagonal of **A**.

Proof   Let the eigenvalues of **A** be $\lambda_1, \ldots,$ and λ_n (not necessarily distinct), and let corresponding eigenvectors be $\mathbf{V}_1, \ldots,$ and $\mathbf{V}_n$. These are assumed to be linearly independent. The dimension of the column space of **P** is n, so **A** has rank n and is nonsingular. Now compute $\mathbf{P}^{-1}\mathbf{AP}$ column by column as follows. First,

$$\text{column } j \text{ of } \mathbf{AP} = \mathbf{A}(\text{column } j \text{ of } \mathbf{P}) = \mathbf{AV}_j = \lambda_j \mathbf{V}_j.$$

Therefore, **AP** has the form

$$\mathbf{AP} = \begin{pmatrix} | & | & \cdots & | \\ \lambda_1 \mathbf{V}_1 & \lambda_2 \mathbf{V}_2 & \cdots & \lambda_n \mathbf{V}_n \\ | & | & \cdots & | \end{pmatrix}.$$

Then

$$\text{column } j \text{ of } \mathbf{P}^{-1}\mathbf{AP} = \mathbf{P}^{-1}(\text{column } j \text{ of } \mathbf{AP})$$
$$= \mathbf{P}^{-1}[\lambda_j \mathbf{V}_j] = \lambda_j \mathbf{P}^{-1}\mathbf{V}_j.$$

But $\mathbf{V}_j$ is column j of $\mathbf{P}$, so

$$\mathbf{P}^{-1}\mathbf{V}_j = \text{column } j \text{ of } \mathbf{P}^{-1}\mathbf{AP} = \begin{pmatrix} 0 \\ 0 \\ \vdots \\ 1 \\ \vdots \\ 0 \end{pmatrix}.$$

The matrix on the right is all zeros except for 1 in row j. Combining the last two equations, we have

$$\text{column } j \text{ of } \mathbf{P}^{-1}\mathbf{AP} = \lambda_j \mathbf{P}^{-1}\mathbf{V}_j = \lambda_j \begin{pmatrix} 0 \\ 0 \\ \vdots \\ 1 \\ \vdots \\ 0 \end{pmatrix} = \begin{pmatrix} 0 \\ 0 \\ \vdots \\ \lambda_j \\ \vdots \\ 0 \end{pmatrix}.$$

We now know the columns of $\mathbf{P}^{-1}\mathbf{AP}$, and putting these together yields the diagonal matrix

$$\mathbf{P}^{-1}\mathbf{AP} = \begin{pmatrix} \lambda_1 & 0 & 0 & \cdots & 0 \\ 0 & \lambda_2 & 0 & \cdots & 0 \\ 0 & 0 & \lambda_3 & \cdots & 0 \\ \vdots & \vdots & \vdots & \vdots & \vdots \\ 0 & 0 & 0 & \cdots & \lambda_n \end{pmatrix}.$$

The second part of the theorem is proved by a similar strategy, looking at the columns of $\mathbf{Q}^{-1}\mathbf{AQ}$. ◆

EXAMPLE 8.8

Let

$$\mathbf{A} = \begin{pmatrix} -1 & 4 \\ 0 & 3 \end{pmatrix}.$$

A has eigenvalues $-1, 3$ and corresponding linearly independent eigenvectors

$$\begin{pmatrix} 1 \\ 0 \end{pmatrix} \text{ and } \begin{pmatrix} 1 \\ 1 \end{pmatrix}.$$

Form

$$\mathbf{P} = \begin{pmatrix} 1 & 1 \\ 0 & 1 \end{pmatrix}.$$

Determine

$$\mathbf{P}^{-1} = \begin{pmatrix} 1 & -1 \\ 0 & 1 \end{pmatrix}.$$

A simple computation shows that

$$\mathbf{P}^{-1}\mathbf{AP} = \begin{pmatrix} -1 & 0 \\ 0 & 3 \end{pmatrix},$$

which is a diagonal matrix with the eigenvalues of **A** on the main diagonal in the order in which the eigenvectors were used to form the columns.

If we reverse the order of these eigenvectors as columns and define

$$\mathbf{Q} = \begin{pmatrix} 1 & 1 \\ 1 & 0 \end{pmatrix},$$

then

$$\mathbf{Q}^{-1}\mathbf{A}\mathbf{Q} = \begin{pmatrix} 3 & 0 \\ 0 & -1 \end{pmatrix}$$

with the eigenvalues along the main diagonal, but now in the order reflecting the order of the eigenvectors used in forming the columns of **Q**. ◆

EXAMPLE 8.9

Let

$$\mathbf{A} = \begin{pmatrix} -1 & 1 & 3 \\ 2 & 1 & 4 \\ 1 & 0 & -2 \end{pmatrix}.$$

The eigenvalues are -1, $(-1+\sqrt{29})/2$, and $(-1-\sqrt{29})/2$, and corresponding eigenvectors are

$$\begin{pmatrix} 1 \\ -3 \\ 1 \end{pmatrix}, \begin{pmatrix} 3+\sqrt{29} \\ 10+2\sqrt{29} \\ 2 \end{pmatrix}, \text{ and } \begin{pmatrix} 3-\sqrt{29} \\ 10-2\sqrt{29} \\ 2 \end{pmatrix}.$$

These are linearly independent, because the eigenvalues are distinct. Use these eigenvectors as columns of **P** to form

$$\mathbf{P} = \begin{pmatrix} 1 & 3+\sqrt{29} & 3-\sqrt{29} \\ -3 & 10+2\sqrt{29} & 10-2\sqrt{29} \\ 1 & 2 & 2 \end{pmatrix}.$$

We find that

$$\mathbf{P}^{-1} = \frac{\sqrt{29}}{812} \begin{pmatrix} 232/\sqrt{29} & -116/\sqrt{29} & 232/\sqrt{29} \\ -16-2\sqrt{29} & -1+\sqrt{29} & -19+5\sqrt{29} \\ -16-2\sqrt{29} & 1+\sqrt{29} & 19+5\sqrt{29} \end{pmatrix}.$$

Then

$$\mathbf{P}^{-1}\mathbf{A}\mathbf{P} = \begin{pmatrix} -1 & 0 & 0 \\ 0 & (-1+\sqrt{29})/2 & 0 \\ 0 & 0 & (-1-\sqrt{29})/2 \end{pmatrix},$$

with the eigenvalues down the main diagonal in the order of the eigenvalues listed for columns of **P**. ◆

Although *n* distinct eigenvalues guarantee that **A** is diagonalizable, an $n \times n$ matrix with fewer than *n* distinct eigenvalues *may* still be diagonalizable. This will occur if we are able to find *n* linearly independent eigenvectors.

EXAMPLE 8.10

Let

$$A = \begin{pmatrix} 5 & -4 & 4 \\ 12 & -11 & 12 \\ 4 & -4 & 5 \end{pmatrix}$$

as in Example 8.5. We found the eigenvalues $-3, 1$, and 1 with a repeated eigenvalue. Nevertheless, we were able to find three linearly independent eigenvectors. Use these as columns to form

$$P = \begin{pmatrix} 1 & 1 & 0 \\ 3 & 0 & 1 \\ 1 & -1 & 1 \end{pmatrix}.$$

Then P diagonalizes A:

$$P^{-1}AP = \begin{pmatrix} -3 & 0 & 0 \\ 0 & 1 & 0 \\ 0 & 0 & 1 \end{pmatrix}.$$

We know this from Theorem 8.6, without computing the product $P^{-1}AP$, which requires the determination of P^{-1}. ◆

If A has fewer than n linearly independent eigenvectors, then A is not diagonalizable.

EXAMPLE 8.11

Let

$$B = \begin{pmatrix} 1 & -1 \\ 0 & 1 \end{pmatrix}.$$

B has eigenvalues $1, 1$, and all eigenvectors are constant multiples of

$$\begin{pmatrix} 1 \\ 0 \end{pmatrix}.$$

If P diagonalizes A, then P must have eigenvectors as columns. But then P must have the form

$$P = \begin{pmatrix} \alpha & \beta \\ 0 & 0 \end{pmatrix}$$

for some nonzero α and β. But this matrix is singular with no inverse, because $|P| = 0$. Therefore B is not diagonalizable. ◆

One of the main features of Theorem 8.6 is that it tells us what $P^{-1}AP$ must look like without writing P or computing P^{-1}. This is important, because computing a matrix inverse may take some effort.

EXAMPLE 8.12

Let

$$A = \begin{pmatrix} -2 & 0 & 0 & 5 \\ 1 & 3 & 0 & 0 \\ 0 & 4 & 4 & 0 \\ 2 & 0 & 0 & -3 \end{pmatrix}.$$

A has eigenvalues 3, 4, $(-5+\sqrt{41})/2$, and $(-5-\sqrt{41})/2$. Because these are distinct, **A** has four linearly independent eigenvectors and therefore is diagonalizable. There is a matrix **P** such that

$$\mathbf{P}^{-1}\mathbf{AP} = \begin{pmatrix} 3 & 0 & 0 & 0 \\ 0 & 4 & 0 & 0 \\ 0 & 0 & (-5+\sqrt{41})/2 & 0 \\ 0 & 0 & 0 & (-5-\sqrt{41})/2 \end{pmatrix}.$$

We do not have to actually write down **P** (this would require finding eigenvectors) or compute **P**$^{-1}$ to draw this conclusion. ◆

SECTION 8.2   PROBLEMS

In each of Problems 1 through 8, produce a matrix **P** that diagonalizes the given matrix or show that the matrix is not diagonalizable. Determine **P**$^{-1}$**AP**. *Hint*: Keep in mind that it is not necessary to compute **P** to know this product matrix.

1. $\begin{pmatrix} 0 & -1 \\ 4 & 3 \end{pmatrix}$

2. $\begin{pmatrix} 5 & 3 \\ 1 & 3 \end{pmatrix}$

3. $\begin{pmatrix} 1 & 0 \\ -4 & 1 \end{pmatrix}$

4. $\begin{pmatrix} -5 & 3 \\ 0 & 9 \end{pmatrix}$

5. $\begin{pmatrix} 5 & 0 & 0 \\ 1 & 0 & 3 \\ 0 & 0 & -2 \end{pmatrix}$

6. $\begin{pmatrix} 0 & 0 & 0 \\ 1 & 0 & 2 \\ 0 & 1 & 3 \end{pmatrix}$

7. $\begin{pmatrix} 1 & 0 & 0 & 0 \\ 0 & 4 & 1 & 0 \\ 0 & 0 & -3 & 1 \\ 0 & 0 & 1 & -2 \end{pmatrix}$

8. $\begin{pmatrix} -2 & 0 & 0 & 0 \\ -4 & -2 & 0 & 0 \\ 0 & 0 & -2 & 0 \\ 0 & 0 & 0 & -2 \end{pmatrix}$

9. Let **A** have eigenvalues $\lambda_1, \cdots, \lambda_n$ and suppose that **P** diagonalizes **A**. Show that, for any positive integer k,

$$\mathbf{A}^k = \mathbf{P} \begin{pmatrix} \lambda_1^k & 0 & \cdots & 0 \\ 0 & \lambda_2^k & \cdots & 0 \\ \vdots & \vdots & \vdots & \vdots \\ 0 & 0 & \cdots & \lambda_n^{\ k} \end{pmatrix} \mathbf{P}^{-1}.$$

In each of Problems 10, 11, and 12, use the idea of Problem 9 to compute the indicated power of the matrix.

10. $\mathbf{A} = \begin{pmatrix} -3 & -3 \\ -2 & 4 \end{pmatrix}; \mathbf{A}^{16}$

11. $\mathbf{A} = \begin{pmatrix} -1 & 0 \\ 1 & -5 \end{pmatrix}; \mathbf{A}^{18}$

12. $\mathbf{A} = \begin{pmatrix} -2 & 3 \\ 3 & -4 \end{pmatrix}; \mathbf{A}^{31}$

8.3   Some Special Matrices

Symmetric matrices have special properties not shared by matrices in general. We will briefly examine other types of matrices having special properties. We will use the fact that the operations of forming an inverse and taking a complex conjugate can be performed in either order. That is, if **A** is an $n \times n$ matrix of numbers, then

$$(\overline{\mathbf{A}})^{-1} = \overline{(\mathbf{A}^{-1})}.$$

This follows from the observation that

$$\mathbf{I}_n = \overline{\mathbf{I}_n} = \overline{\mathbf{A}\mathbf{A}^{-1}} = \overline{\mathbf{A}}\,\overline{(\mathbf{A}^{-1})}.$$

Orthogonal Matrices

An $n \times n$ matrix is *orthogonal* if its transpose equals its inverse:

$$\mathbf{A}^{-1} = \mathbf{A}^t$$

or, equivalently,

$$\mathbf{A}\mathbf{A}^t = \mathbf{A}^t\mathbf{A} = \mathbf{I}_n.$$

For example, it is routine to check that

$$\mathbf{A} = \begin{pmatrix} 0 & 1/\sqrt{5} & 2/\sqrt{5} \\ 1 & 0 & 0 \\ 0 & 2/\sqrt{5} & -1/\sqrt{5} \end{pmatrix}$$

is orthogonal by verifying that $\mathbf{A}\mathbf{A}^t = \mathbf{I}_3$.

Because the transpose of a transpose is the original matrix, a matrix is orthogonal exactly when its transpose is orthogonal.

An orthogonal matrix must have determinant equal to 1 or -1.

THEOREM 8.7

If $\mathbf{A}$ is orthogonal, then $|\mathbf{A}| = \pm 1$.

Proof This is immediate. If $\mathbf{A}$ is orthogonal, then

$$|\mathbf{I}_n| = 1 = |\mathbf{A}\mathbf{A}^{-1}| = |\mathbf{A}\mathbf{A}^t| = |\mathbf{A}||\mathbf{A}^t| = |\mathbf{A}|^2. \quad \blacklozenge$$

In the 3×3 example just given of an orthogonal matrix, notice that the row vectors are mutually orthogonal unit vectors in R^3, as are the column vectors. This property completely characterizes real orthogonal matrices.

THEOREM 8.8 *Orthogonal Matrices*

Let $\mathbf{A}$ be an $n \times n$ real matrix. Then, $\mathbf{A}$ is orthogonal if and only if the row (column) vectors are mutually orthogonal unit vectors in R^n.

Proof The (i, j) element of a product $\mathbf{CD}$ is the dot product of row i of $\mathbf{C}$ with column j of $\mathbf{D}$. In particular,

$$(i, j) \text{ element of } \mathbf{A}\mathbf{A}^t = (\text{row } i \text{ of } \mathbf{A}) \cdot (\text{column } j \text{ of } \mathbf{A}^t)$$

$$= (\text{row } i \text{ of } \mathbf{A})(\text{row } j \text{ of } \mathbf{A}).$$

Now suppose that $\mathbf{A}$ is orthogonal. Then $\mathbf{A}\mathbf{A}^t = \mathbf{I}_n$, so if $i \neq j$, then the dot product of rows i and j is the (i, j) element of $\mathbf{I}_n$, which is zero. Therefore, the rows are orthogonal in R^n. Furthermore, the square of the magnitude of row i is the dot product of row i with itself, which is the (i, i) element of $\mathbf{I}_n$ or 1. Therefore, the row vectors are unit vectors.

Conversely, if the rows of $\mathbf{A}$ are mutually orthogonal unit vectors in R^n, then the i, j element of $\mathbf{AA}^t$ is 0 if $i \neq j$ and 1 if $i = j$, so $\mathbf{AA}^t = \mathbf{I}_n$, and $\mathbf{A}$ is an orthogonal matrix.

The same argument applies to the column vectors. ◆

We now know a lot about orthogonal matrices. To illustrate, we will determine all 2×2 orthogonal matrices. Suppose

$$\mathbf{Q} = \begin{pmatrix} a & b \\ c & d \end{pmatrix}$$

is orthogonal. What does this tell us about $a, b, c,$ and d? Because the row (column) vectors are mutually orthogonal unit vectors,

$$ac + bd = 0,$$

$$ab + cd = 0,$$

$$a^2 + b^2 = 1,$$

$$\text{and } c^2 + d^2 = 1.$$

Furthermore, $|\mathbf{Q}| = \pm 1$, so

$$ad - bc = 1 \text{ or } ad - bc = -1.$$

By analyzing all cases in these equations, we find that, for some θ in $[0, 2\pi)$, $a = \cos(\theta)$, $b = \sin(\theta)$, and $\mathbf{Q}$ must be one of

$$\begin{pmatrix} \cos(\theta) & \sin(\theta) \\ -\sin(\theta) & \cos(\theta) \end{pmatrix} \text{ or } \begin{pmatrix} \cos(\theta) & \sin(\theta) \\ \sin(\theta) & -\cos(\theta) \end{pmatrix}.$$

For example, with $\theta = \pi/6$, we obtain the two orthogonal matrices

$$\begin{pmatrix} \sqrt{3}/2 & 1/2 \\ -1/2 & \sqrt{3}/2 \end{pmatrix} \text{ and } \begin{pmatrix} \sqrt{3}/2 & 1/2 \\ 1/2 & \sqrt{3}/2 \end{pmatrix}.$$

Unitary Matrices

An $n \times n$ real or complex matrix $\mathbf{U}$ is *unitary* if its transpose equals the inverse of its conjugate (or conjugate of its inverse):

$$(\overline{\mathbf{U}})^{-1} = \mathbf{U}^t.$$

This is equivalent to

$$\overline{\mathbf{U}}\mathbf{U}^t = \mathbf{I}_n.$$

For example, it is easy to verify that

$$\mathbf{U} = \begin{pmatrix} i/\sqrt{2} & 1/\sqrt{2} \\ -i/\sqrt{2} & 1/\sqrt{2} \end{pmatrix}$$

is unitary.

If $\mathbf{U}$ is a unitary real matrix, then $\mathbf{U} = \overline{\mathbf{U}}$, so

$$\overline{\mathbf{U}}\mathbf{U}^t = \mathbf{U}\mathbf{U}^t = \mathbf{I}_n,$$

and $\mathbf{U}$ is a real orthogonal matrix. In this sense, unitary matrices are complex extensions of real orthogonal matrices. This analogy extends to a complex analogue of Theorem 8.7. Suppose $\mathbf{U}$ is a complex $n \times n$ matrix. Then the row (column) vectors are n-vectors having complex components. We say that the rows (or columns) are in C^n, which is the vector space of complex n-vectors. If

$$\mathbf{Z} = \begin{pmatrix} z_1 \\ z_2 \\ \vdots \\ z_n \end{pmatrix} \text{ and } \mathbf{W} = \begin{pmatrix} w_1 \\ w_2 \\ \vdots \\ w_n \end{pmatrix}$$

are two complex n-vectors, define their (complex) dot product to be

$$\mathbf{Z} \cdot \mathbf{W} = \mathbf{Z}^t \mathbf{W} = \overline{z_1} w_1 + \overline{z_2} w_2 + \cdots + \overline{z_n} w_n.$$

With this definition,

$$\mathbf{Z} \cdot \mathbf{Z} = \sum_{j-1}^{n} \overline{z_j} z_j = \sum_{j=1}^{n} |z_j|^2,$$

which is a real number interpreted as the length of $\mathbf{Z}$. As in R^n, we define two vectors in C^n to be orthogonal if their complex dot product is zero.

We can now state the unitary analogue of Theorem 8.6.

THEOREM 8.9   *Unitary Matrices*

Let $\mathbf{U}$ be a complex $n \times n$ matrix. Then $\mathbf{U}$ is unitary if and only if the rows (columns) are mutually orthogonal unit vectors in C^n.

Proof   The proof is very like that of Theorem 8.8, using the complex inner product of complex n-vectors. ◆

We claim that the eigenvalues of a unitary matrix must have a magnitude of 1.

THEOREM 8.10

Let λ be an eigenvalue of the unitary matrix $\mathbf{U}$. Then $|\lambda| = 1$.

Proof   Let $\mathbf{E}$ be an eigenvector for the eigenvalue λ. Then $\mathbf{U}\mathbf{E} = \lambda\mathbf{E}$, so $\overline{\mathbf{U}\mathbf{E}} = (\overline{\lambda})\overline{\mathbf{E}}$. Then

$$(\overline{\mathbf{U}\mathbf{E}})^t = (\overline{\lambda})\overline{\mathbf{E}}^t.$$

But then

$$(\overline{\mathbf{E}})^t \overline{\mathbf{U}}^t = (\overline{\lambda})\overline{\mathbf{E}}^t.$$

Since $\mathbf{U}$ is unitary, then $\overline{\mathbf{U}}^t = \mathbf{U}^{-1}$, so the last equation becomes

$$(\overline{\mathbf{E}})^t \mathbf{U}^{-1} = (\overline{\lambda})\overline{\mathbf{E}}^t.$$

Multiply both sides of this equation on the right by $\mathbf{U}\mathbf{E}$ to obtain

$$\overline{\mathbf{E}}^t \mathbf{E} = \overline{\lambda}\overline{\mathbf{E}}^t \mathbf{U}\mathbf{E} = \overline{\lambda}\overline{\mathbf{E}}^t \lambda\mathbf{E} = \overline{\lambda}\lambda\overline{\mathbf{E}}^t \mathbf{E}.$$

But $\mathbf{E}^t\mathbf{E}$ is the dot product of the eigenvector with itself, and so it is a positive number. This equation therefore implies that $\overline{\lambda}\lambda = |\lambda|^2 = 1$, as was claimed. ◆

Hermitian and Skew-Hermitian Matrices

> An $n \times n$ complex matrix $\mathbf{H}$ is *hermitian* if $\overline{\mathbf{H}} = \mathbf{H}^t$.

If, for example, $\mathbf{H}$ is real (hence also complex) and symmetric, then $\overline{\mathbf{H}} = \mathbf{H} = \mathbf{H}^t$, so $\mathbf{H}$ is automatically hermitian.

As an example, it is routine to verify that

$$\mathbf{H} = \begin{pmatrix} 15 & 8i & 6-2i \\ -8i & 0 & -4+i \\ 6+2i & -4-i & -3 \end{pmatrix}$$

is hermitian.

If $\mathbf{H}$ is hermitian and $\mathbf{Z}$ is any $n \times 1$ complex matrix, we claim that $\overline{\mathbf{Z}}^t\mathbf{H}\mathbf{Z}$ must be a real number. To verify this, use the fact that $\overline{\mathbf{H}}^t = \mathbf{H}$ to write

$$\overline{\overline{\mathbf{Z}}^t\mathbf{H}\mathbf{Z}} = \overline{(\overline{\mathbf{Z}}^t)}\overline{\mathbf{H}\mathbf{Z}} = \mathbf{Z}^t\overline{\mathbf{H}\mathbf{Z}}.$$

But $\overline{\mathbf{Z}}^t\mathbf{H}\mathbf{Z}$ is 1×1, and so equals its own transpose. Use this to continue from the last equation:

$$\mathbf{Z}^t\overline{\mathbf{H}\mathbf{Z}} = (\mathbf{Z}^t\overline{\mathbf{H}\mathbf{Z}})^t = \overline{\mathbf{Z}}^t\mathbf{H}\mathbf{Z}.$$

From the last two equations,

$$\overline{\overline{\mathbf{Z}}^t\mathbf{H}\mathbf{Z}} = \overline{\mathbf{Z}}^t\mathbf{H}\mathbf{Z}.$$

Therefore, $\overline{\mathbf{Z}}^t\mathbf{H}\mathbf{Z}$, being equal to its own complex conjugate, must be real. One consequence of this is that a hermitian matrix has only real eigenvalues.

THEOREM 8.11

All eigenvalues of a hermitian matrix are real.

Proof   From Lemma 8.1, if λ is an eigenvalue of hermitian $\mathbf{H}$ with eigenvector $\mathbf{E}$, then

$$\lambda = \frac{\overline{\mathbf{E}}^t\mathbf{H}\mathbf{E}}{\overline{\mathbf{E}}^t\mathbf{E}}.$$

This is real because the denominator is real, and we have just proved that the numerator is real ◆

> An $n \times n$ complex matrix S is *skew-hermitian* if $\overline{\mathbf{S}} = -\mathbf{S}^t$.

As an example,

$$S = \begin{pmatrix} 0 & 8i & 2i \\ 8i & 0 & 4i \\ 2i & 4i & 0 \end{pmatrix}$$

is skew-hermitian.

We claim that, if S is skew-hermitian and Z is an $n \times 1$ complex matrix, then $\overline{Z}^t S Z$ must be zero or pure imaginary. To verify this, use the fact that $\overline{S}^t = -S$ with the argument just used for hermitian matrices to show that

$$\overline{(\overline{Z}^t S Z)} = -\overline{Z}^t S Z.$$

If we write $\overline{Z}^t S Z = a + ib$, this means that

$$a - ib = -a - ib$$

which means that $a = 0$, hence $\overline{Z}^t S Z = ib$ is zero or pure imaginary. This leads to the following.

THEOREM 8.12

Each eigenvalue of a skew-hermitian matrix is zero or pure imaginary.

Proof If λ is an eigenvalue of S with eigenvector E, then (Lemma 8.1)

$$\lambda = \frac{\overline{E}^t H E}{\overline{E}^t E}.$$

The denominator of this quotient is a positive number, and the numerator is zero or pure imaginary, proving the theorem. ◆

In the complex plane, eigenvalues of unitary matrices lie on the unit circle, eigenvalues of hermitian matrices lie on the horizontal (real) axis, and eigenvalues of skew-hermitian matrices are on the vertical (imaginary) axis. This is indicated in Figure 8.2.

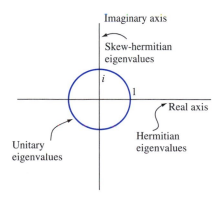

FIGURE 8.2 *Eigenvalues of unitary, hermitian, and skew-hermitian matrices.*

SECTION 8.3   PROBLEMS

In each of Problems 1 through 9, determine whether the matrix is unitary, hermitian, skew-hermitian, or none of these. Find the eigenvalues and, with each eigenvalue, an associated eigenvector. If the matrix is diagonalizable, write a matrix that diagonalizes it. In Problems 5 and 7, the eigenvalues must be approximated, so only "approximate" eigenvectors can be found. It is instructive to try to diagonalize the given matrix using these approximate eigenvectors as columns of a diagonalizing matrix.

1. $\begin{pmatrix} 0 & 2i \\ 2i & 4 \end{pmatrix}$

2. $\begin{pmatrix} 3 & 4i \\ 4i & -5 \end{pmatrix}$

3. $\begin{pmatrix} 0 & 1 & 0 \\ -1 & 0 & 1-i \\ 0 & -1-i & 0 \end{pmatrix}$

4. $\begin{pmatrix} 1/\sqrt{2} & i/\sqrt{2} & 0 \\ -1/\sqrt{2} & i/\sqrt{2} & 0 \\ 0 & 0 & 1 \end{pmatrix}$

5. $\begin{pmatrix} 3 & 2 & 0 \\ 2 & 0 & i \\ 0 & -i & 0 \end{pmatrix}$

6. $\begin{pmatrix} -1 & 0 & 3-i \\ 0 & 1 & 0 \\ 3+i & 0 & 0 \end{pmatrix}$

7. $\begin{pmatrix} i & 1 & 0 \\ -1 & 0 & 2i \\ 0 & 2i & 0 \end{pmatrix}$

8. $\begin{pmatrix} 3i & 0 & 0 \\ -1 & 0 & i \\ 0 & -i & 0 \end{pmatrix}$

9. $\begin{pmatrix} 8 & -1 & i \\ -1 & 0 & 0 \\ -i & 0 & 0 \end{pmatrix}$

10. Suppose **A** is unitary, hermitian, or skew-hermitian. Show that

$$\overline{\mathbf{A}\mathbf{A}^t} = \overline{\mathbf{A}}\mathbf{A}.$$

11. Prove that the main diagonal elements of a skew-hermitian matrix must be zero or pure imaginary.

12. Prove that the main diagonal elements of a hermitian matrix must be real.

13. Prove that the product of two unitary matrices is unitary.

14. Let **A** be a real symmetric $n \times n$ matrix with n distinct eigenvalues. Prove that **A** can be diagonalized by an orthogonal matrix.

SYSTEMS OF LINEAR DIFFERENTIAL
EQUATIONS SOLUTION OF $\mathbf{X}' = \mathbf{A}\mathbf{X}$ WHEN A
IS CONSTANT SOLUTION OF $\mathbf{X}' = \mathbf{A}\mathbf{X} + \mathbf{G}$

CHAPTER 9

Systems of Linear Differential Equations

9.1   **Systems of Linear Differential Equations**

We will apply matrices to the solution of a system of n linear differential equations in n unknown functions,

$$x_1'(t) = a_{11}(t)x_1(t) + a_{12}(t)x_2(t) + \ldots + a_{1n}(t)x_n(t) + g_1(t)$$

$$x_2'(t) = a_{21}(t)x_1(t) + a_{22}(t)x_2(t) + \ldots + a_{2n}(t)x_n(t) + g_2(t)$$

$$\vdots$$

$$x_n'(t) = a_{n1}(t)x_1(t) + a_{n2}(t)x_2(t) + \ldots + a_{nn}(t)x_n(t) + g_n(t).$$

The functions $a_{ij}(t)$ are continuous and $g_j(t)$ are piecewise continuous on some interval (perhaps the whole real line). Define matrices

$$\mathbf{A}(t) = [a_{ij}(t)], \ \mathbf{X}(t) = \begin{pmatrix} x_1(t) \\ x_2(t) \\ \vdots \\ x_n(t) \end{pmatrix}, \text{ and } \mathbf{G}(t) = \begin{pmatrix} g_1(t) \\ g_2(t) \\ \vdots \\ g_n(t) \end{pmatrix}.$$

Differentiate a matrix by differentiating each element. Matrix differentiation follows the usual rules of calculus. The derivative of a sum is the sum of the derivatives, and the product rule has the same form whenever the product is defined:

$$(\mathbf{W}\mathbf{N})' = \mathbf{W}'\mathbf{N} + \mathbf{W}\mathbf{N}'.$$

With this notation, the system of linear differential equations is

$$\mathbf{X}'(t) = \mathbf{A}(t)\mathbf{X}(t) + \mathbf{G}(t),$$

or

$$\mathbf{X}' = \mathbf{A}\mathbf{X} + \mathbf{G}. \tag{9.1}$$

183

We will refer to this as a *linear system*. This system is *homogeneous* if $\mathbf{G}(t)$ is the $n \times 1$ zero matrix, which occurs when each $g_j(t)$ is identically zero. Otherwise the system is *nonhomogeneous*.

We will outline a procedure for finding all solutions of the system in equation (9.1). This will be analogous to the solution of second-order linear differential equations (Chapter 2). We will then show how to carry out this procedure in the constant coefficient case where $\mathbf{A}$ is a constant matrix.

9.1.1   The Homogeneous System $\mathbf{X}' = \mathbf{AX}$.

If $\mathbf{\Phi}_1$ and $\mathbf{\Phi}_2$ are solutions of $\mathbf{X}' = \mathbf{AX}$, then so is any linear combination

$$c_1 \mathbf{\Phi}_1 + c_2 \mathbf{\Phi}_2.$$

This extends to any finite sum of solutions.

A set of k solutions is *linearly dependent* if one is a linear combination of the others. If this is not the case, then the solutions are *linearly independent*. The following fundamental result is the system analog of Theorem 2.2.

THEOREM 9.1   *General Solution of* $\mathbf{X}' = \mathbf{AX}$

Suppose each $a_{ij}(t)$ is continuous on an open interval I. Then there exist n linearly independent solutions $\mathbf{\Phi}_1, \mathbf{\Phi}_2, \ldots, \mathbf{\Phi}_n$ of $\mathbf{AX} = \mathbf{O}$ on I.

Further, given any n linearly independent solutions $\mathbf{\Phi}_1, \mathbf{\Phi}_2, \ldots, \mathbf{\Phi}_n$, any solution of $\mathbf{X}' = \mathbf{AX}$ on this interval can be written as a linear combination

$$c_1 \mathbf{\Phi}_1 + c_2 \mathbf{\Phi}_2 + \ldots + c_n \mathbf{\Phi}_n$$

for some numbers $c_1, c_2, \ldots, c_n$.   ◆

For this reason, we call $c_1 \mathbf{\Phi}_1 + c_2 \mathbf{\Phi}_2 + \ldots + c_n \mathbf{\Phi}_n$ the *general solution* of $\mathbf{X}' = \mathbf{AX}$ on I. Every solution is contained in this expression by varying the choices of the constants.

EXAMPLE 9.1

The 2×2 system

$$\mathbf{X}' = \begin{pmatrix} 1 & -4 \\ 1 & 5 \end{pmatrix} \mathbf{X}.$$

is the matrix formulation of the system

$$x_1' = x_1 - 4x_2$$
$$x_2' = x_1 + 5x_2.$$

One can verify by substitution that

$$\mathbf{\Phi}_1(t) = \begin{pmatrix} -2e^{3t} \\ e^{3t} \end{pmatrix} \text{ and } \mathbf{\Phi}_2(t) = \begin{pmatrix} (1-2t)e^{3t} \\ te^{3t} \end{pmatrix}$$

are solutions for all real t. Every solution is a linear combination of $\mathbf{\Phi}_1$ and $\mathbf{\Phi}_1$, and the general solution of the system is

$$\mathbf{\Phi}(t) = c_1 \mathbf{\Phi}_1(t) + c_2 \mathbf{\Phi}_2(t),$$

with c_1 and c_2 arbitrary constants.   ◆

There is a determinant test to tell whether n solutions of $\mathbf{X}' = \mathbf{A}\mathbf{X}$ are linearly independent.

THEOREM 9.2 *Test for Linear Independence*

Let $\mathbf{\Phi}_1, \mathbf{\Phi}_2, \ldots, \mathbf{\Phi}_n$ be n solutions of $\mathbf{X}' = \mathbf{A}\mathbf{X}$ on an open interval I with $\mathbf{A}$ as an $n \times n$ matrix. Then

1. $\mathbf{\Phi}_1, \mathbf{\Phi}_2, \ldots, \mathbf{\Phi}_n$ are linearly independent if and only if, for any t_0 in I, the n vectors $\mathbf{\Phi}_1(t_0), \mathbf{\Phi}_2(t_0), \ldots, \mathbf{\Phi}_n(t_0)$ are linearly independent in R^n.

2. Write

$$
\mathbf{\Phi}_j(t) = \begin{pmatrix} \varphi_{1j}(t) \\ \varphi_{2j}(t) \\ \vdots \\ \varphi_{nj}(t) \end{pmatrix}
$$

for $j = 1, 2, \ldots, n$. Then $\mathbf{\Phi}_1, \mathbf{\Phi}_2, \ldots, \mathbf{\Phi}_n$ are linearly independent on I if and only if, for any t_0 in this interval,

$$
\begin{vmatrix} \varphi_{11}(t_0) & \varphi_{12}(t_0) & \ldots & \varphi_{1n}(t_0) \\ \varphi_{21}(t_0) & \varphi_{22}(t_0) & \ldots & \varphi_{2n}(t_0) \\ \vdots & \vdots & \vdots & \vdots \\ \varphi_{n1}(t_0) & \varphi_{n2}(t_0) & \ldots & \varphi_{nn}(t_0) \end{vmatrix} \neq 0. \quad \blacklozenge
$$

Conclusion 2 follows from 1. Each $\mathbf{\Phi}_j(t_0)$ is a vector in R^n, and these columns form an $n \times n$ matrix of real numbers. The determinant of this matrix is nonzero exactly when these columns are linearly independent (so the matrix has rank n and nonzero determinant).

EXAMPLE 9.2

In Example 9.1, both solutions are defined for all t so we can consider I as the entire real line. Choose any t_0, for example, $t_0 = 0$. Now

$$
\mathbf{\Phi}(0) = \begin{pmatrix} -2 \\ 1 \end{pmatrix} \text{ and } \mathbf{\Phi}_2(0) = \begin{pmatrix} 1 \\ 0 \end{pmatrix}.
$$

These are linearly independent vectors in R^2. Therefore, the solutions $\mathbf{\Phi}_1$ and $\mathbf{\Phi}_2$ are linearly independent, as we observed in Example 9.1.

To illustrate conclusion 2 of the theorem, evaluate the 2×2 determinant with $\mathbf{\Phi}_1(0)$ and $\mathbf{\Phi}_2(0)$ as columns:

$$
\begin{vmatrix} -2 & 1 \\ 1 & 0 \end{vmatrix} = -1 \neq 0. \quad \blacklozenge
$$

We know the general solution of $\mathbf{X}' = \mathbf{A}\mathbf{X}$ if we have n linearly independent solutions. These solutions are $n \times 1$ matrices. We can form an $n \times n$ matrix $\mathbf{\Omega}$ using these n solutions as columns. Such a matrix is called a *fundamental matrix* for the system.

In terms of this fundamental matrix, we can write the general solution in the compact form

$$c_1 \Phi_1 + c_2 \Phi_2 + \ldots + c_n \Phi_n = \Omega C. \quad \blacklozenge$$

EXAMPLE 9.3

In Example 9.1, we had two linearly independent solutions of a 2×2 system. Form a fundamental matrix with these as columns:

$$\Omega(t) = \begin{pmatrix} -2e^{3t} & (1-2t)e^{3t} \\ e^{3t} & te^{3t} \end{pmatrix}.$$

The general solution $c_1 \Phi_1 + c_2 \Phi_2$ can be written as ΩC:

$$\Omega C = \begin{pmatrix} -2e^{3t} + (1-2t)e^{3t} \\ e^{3t} + te^{3t} \end{pmatrix} \begin{pmatrix} c_1 \\ c_2 \end{pmatrix}$$

$$= \begin{pmatrix} c_1(-2e^{3t}) & c_2(1-2t)e^{3t} \\ c_1 e^{3t} & c_2 te^{3t} \end{pmatrix} = c_1 \begin{pmatrix} -2e^{3t} \\ e^{3t} \end{pmatrix} + c_2 \begin{pmatrix} (1-2t)e^{3t} \\ te^{3t} \end{pmatrix}$$

$$= c_1 \Phi_1(t) + c_2 \Phi_2(t). \quad \blacklozenge$$

In an initial value problem, $x_1(t_0), \ldots, x_n(t_0)$ are given. This information specifies the $n \times 1$ matrix $\mathbf{X}(t_0)$. We usually solve an initial value problem by finding the general solution of the system and then solving for the constants to find the particular solution satisfying the initial conditions. It is often convenient to use a fundamental matrix to carry out this plan.

EXAMPLE 9.4

Solve the initial value problem

$$\mathbf{X}' = \begin{pmatrix} 1 & -4 \\ 1 & 5 \end{pmatrix} \mathbf{X}; \quad \mathbf{X}(0) = \begin{pmatrix} -2 \\ 3 \end{pmatrix}.$$

The general solution is ΩC with Ω as the fundamental matrix of Example 9.3. To solve the initial value problem, we must choose $\mathbf{C}$ so that

$$\mathbf{X}(0) = \Omega(0)\mathbf{C} = \begin{pmatrix} -2 \\ 3 \end{pmatrix}.$$

This is the algebraic system

$$\begin{pmatrix} -2 & 1 \\ 1 & 0 \end{pmatrix} \mathbf{C} = \begin{pmatrix} -2 \\ 3 \end{pmatrix}.$$

The solution for $\mathbf{C}$ is

$$\mathbf{C} = \begin{pmatrix} -2 & 1 \\ 1 & 0 \end{pmatrix}^{-1} \begin{pmatrix} -2 \\ 3 \end{pmatrix} = \begin{pmatrix} 0 & 1 \\ 1 & 2 \end{pmatrix} \begin{pmatrix} -2 \\ 3 \end{pmatrix} = \begin{pmatrix} 3 \\ 4 \end{pmatrix}.$$

The unique solution of the initial value problem is

$$\mathbf{X}(t) = \Omega(t) \begin{pmatrix} 3 \\ 4 \end{pmatrix} = \begin{pmatrix} -2e^{3t} - 8te^{3t} \\ 3e^{3t} + 4te^{3t} \end{pmatrix}. \quad \blacklozenge$$

9.1.2 The Nonhomogeneous Linear System

The following is the analog of Theorem 2.3 for nonhomogeneous linear systems.

THEOREM 9.3

Let $\mathbf{\Omega}$ be a fundamental matrix for the homogeneous system $\mathbf{X}' = \mathbf{AX}$. Let $\mathbf{\Psi}_p$ be any particular solution of the nonhomogeneous system $\mathbf{X}' = \mathbf{AX} + \mathbf{G}$. Then the general solution of the nonhomogeneous system is

$$\mathbf{X} = \mathbf{\Omega C} + \mathbf{\Psi}_p. \quad \blacklozenge$$

SECTION 9.1   PROBLEMS

In each of Problems 1 through 5, (a) verify that the given functions satisfy the system, (b) write the system in matrix form $\mathbf{X}' = \mathbf{AX}$ for an appropriate $\mathbf{A}$, (c) write n linearly independent $n \times 1$ matrix solutions $\mathbf{\Phi}_1, \ldots, \mathbf{\Phi}_n$, for appropriate n, (d) use the determinant test of Theorem 9.2(2) to verify that these solutions are linearly independent, (e) form a fundamental matrix for the system, and (f) use the fundamental matrix to solve the initial value problem.

1. $x_1' = 5x_1 + 3x_2, x_2' = x_1 + 3x_2,$
    $x_1(t) = -c_1 e^{2t} + 3c_2 e^{6t}, x_2(t) = c_1 e^{2t} + c_2 e^{6t},$
    $x_1(0) = 0, x_2(0) = 4$

2. $x_1' = 2x_1 + x_2, x_2' = -3x_1 + 6x_2,$
    $x_1(t) = c_1 e^{4t} \cos(t) + c_2 e^{4t} \sin(t)$
    $x_2(t) = 2c_1 e^{4t}[\cos(t) - \sin(t)] + 2c_2 e^{4t}[\cos(t) + \sin(t)],$
    $x_1(0) = -2, x_2(0) = 1$

3. $x_1' = 3x_1 + 8x_2, x_2' = x_1 - x_2,$
    $x_1(t) = 4c_1 e^{(1+2\sqrt{3})t} + 4c_2 e^{(1-2\sqrt{3}t)},$
    $x_2(t) = (-1 + \sqrt{3})c_1 e^{(1+\sqrt{3})t} + (-1 - \sqrt{3})c_2 e^{(1-2\sqrt{3})6t},$
    $x_1(0) = 2, x_2(0) = 2$

4. $x_1' = x_1 - x_2, x_2' = 4x_1 + 2x_2,$
    $x_1(t) = 2e^{3t/2}\left[c_1 \cos(\sqrt{15}t/2) + c_2 \sin(\sqrt{15}t/2)\right],$
    $x_2(t) = c_1 e^{3t/2}\left[-\cos(\sqrt{15}t/2) + \sqrt{15}\sin(\sqrt{15}t/2)\right]$
    $\qquad -c_2 e^{3t/2}\left[\sin(\sqrt{15}t/2) + \sqrt{15}\cos(\sqrt{15}t/2)\right],$
    $x_1(0) = -2, x_2(0) = 7$

5. $x_1' = 5x_1 - 4x_2 + 4x_3, x_2' = 12x_1 - 11x_2 + 12x_3,$
    $x_3'(t) = 4x_1 - 4x_2 + 5x_3$
    $x_1(t) = -c_1 e^t + c_3 e^{-3t}, x_2(t) = c_2 e^{2t} + c_3 e^{-3t},$
    $x_3(t) = (c_3 - c_1)e^t + c_3 e^{-3t},$
    $x_1(0) = 1, x_2(0) = -3, x_3(0) = 5$

9.2   Solution of X' = AX when A Is Constant

From Section 9.1, we know what to look for to solve a linear system. Now we will focus on the special case that $\mathbf{A}$ is a real, constant matrix. Taking a cue from the constant coefficient, second-order differential equation, we attempt solutions of the form $\mathbf{X} = \mathbf{E}e^{\lambda t}$ with $\mathbf{E}$ an $n \times 1$ matrix of numbers and λ a number. For this to be a solution, we need

$$(\mathbf{E}e^{\lambda t})' = \mathbf{E}\lambda e^{\lambda t} = \mathbf{AE}e^{\lambda t}.$$

This will be true if

$$\mathbf{AE} = \lambda \mathbf{E},$$

which holds if λ is an eigenvalue of $\mathbf{A}$ with associated eigenvector $\mathbf{E}$.

THEOREM 9.4

Let $\mathbf{A}$ be an $n \times n$ matrix of real numbers. If λ is an eigenvalue with associated eigenvector $\mathbf{E}$, then $\mathbf{E}e^{\lambda t}$ is a solution of $\mathbf{X}' = \mathbf{A}\mathbf{X}$. ◆

Often we write $e^{\lambda t}\mathbf{E}$ as $\mathbf{E}e^{\lambda t}$. Both expressions mean the elements of the eigenvector multiplied by the scalar $e^{\lambda t}$.

We need n linearly independent solutions to write the general solution of $\mathbf{X}' = \mathbf{A}\mathbf{X}$. The next theorem addresses this.

THEOREM 9.5

Let $\mathbf{A}$ be an $n \times n$ matrix of real numbers. Suppose $\mathbf{A}$ has eigenvalues $\lambda_1, \ldots, \lambda_n$ and there are n corresponding eigenvectors $\mathbf{E}_1, \ldots, \mathbf{E}_n$ that are linearly independent. Then $\mathbf{E}_1 e^{\lambda_1 t}, \ldots, \mathbf{E}_n e^{\lambda_n t}$ are linearly independent solutions. ◆

When the eigenvalues are distinct, we can always find n linearly independent eigenvectors. But even when the eigenvalues are not distinct, it still may be possible to find n linearly independent eigenvectors, and in this case, we have n linearly independent solutions, hence the general solution. We can also use these solutions as columns of a fundamental matrix.

EXAMPLE 9.5

We will solve the system

$$\mathbf{X}' = \begin{pmatrix} 4 & 2 \\ 3 & 3 \end{pmatrix} \mathbf{X}.$$

$\mathbf{A}$ has eigenvalues of 1 and 6 with corresponding eigenvectors

$$\begin{pmatrix} 1 \\ -3/2 \end{pmatrix} \text{ and } \begin{pmatrix} 1 \\ 1 \end{pmatrix}.$$

These are linearly independent (the eigenvalues are distinct), so we have two linearly independent solutions

$$\begin{pmatrix} 1 \\ -3/2 \end{pmatrix} e^t \text{ and } \begin{pmatrix} 1 \\ 1 \end{pmatrix} e^{6t}.$$

The general solution is

$$\mathbf{X} = c_1 \begin{pmatrix} 1 \\ -3/2 \end{pmatrix} e^t + c_2 \begin{pmatrix} 1 \\ 1 \end{pmatrix} e^{6t}.$$

We can also write the fundamental matrix

$$\mathbf{\Omega}(t) = \begin{pmatrix} e^t & e^{6t} \\ -3e^t/2 & e^{6t} \end{pmatrix}$$

in terms of which the general solution is $\mathbf{X}(t) = \mathbf{\Omega}(t)\mathbf{C}$.

If we write out the components individually, the general solution is

$$x_1(t) = c_1 e^t + c_2 e^{6t}$$

and

$$x_2(t) = -\frac{3}{2} c_1 e^t + c_2 e^{6t}. \quad \blacklozenge$$

EXAMPLE 9.6

Consider the system

$$\mathbf{X}' = \begin{pmatrix} 5 & 14 & 4 \\ 12 & -11 & 12 \\ 4 & -4 & 5 \end{pmatrix} \mathbf{X}.$$

The eigenvalues of $\mathbf{A}$ are -3, 1, and 1. Even though there is a repeated eigenvalue in this example, $\mathbf{A}$ has three linearly independent eigenvectors. They are

$$\begin{pmatrix} 1 \\ 3 \\ 1 \end{pmatrix} \text{ associated with eigenvalue } -3$$

and

$$\begin{pmatrix} 1 \\ 1 \\ 0 \end{pmatrix} \text{ and } \begin{pmatrix} -1 \\ 0 \\ 1 \end{pmatrix} \text{ associated with eigenvalue } 1.$$

The general solution is

$$\mathbf{X}(t) = c_1 \begin{pmatrix} 1 \\ 3 \\ 1 \end{pmatrix} e^{3t} + c_2 \begin{pmatrix} 1 \\ 1 \\ 0 \end{pmatrix} e^t + c_3 \begin{pmatrix} -1 \\ 0 \\ 1 \end{pmatrix} e^t.$$

We can also write the general solution $\mathbf{X}(t) = \mathbf{\Omega}(t)\mathbf{C}$, where

$$\mathbf{\Omega}(t) = \begin{pmatrix} e^{-3t} & e^t & -e^t \\ 3e^{-3t} & e^t & 0 \\ e^{-3t} & 0 & e^t \end{pmatrix}. \quad \blacklozenge$$

EXAMPLE 9.7   A Mixing Problem

Two tanks are connected by pipes as in Figure 9.1. Tank 1 initially contains 20 liters of water in which 150 grams of chlorine are dissolved. Tank 2 initially contains 50 grams of chlorine dissolved in 10 liters of water. Beginning at time $t = 0$, pure water is pumped into tank 1 at a rate of 3 liters per minute, while chlorine/water solutions are exchanged between the tanks and also flow out of both tanks at the rates shown. We want to determine the amount of chlorine in each tank at time t.

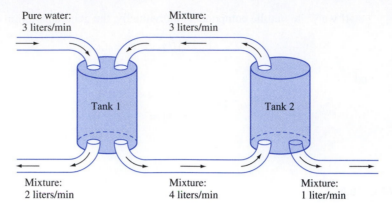

FIGURE 9.1   *Exchange of mixtures in tanks in Example 9.7.*

Let $x_j(t)$ be the number of grams of chlorine in tank j at time t. Reading from Figure 9.1,

rate of change of $x_1(t) = x_1'(t) =$ rate in minus rate out

$$= 3 \left(\frac{\text{liter}}{\text{min}} \right) \cdot 0 \left(\frac{\text{gram}}{\text{liter}} \right) + 3 \left(\frac{\text{liter}}{\text{min}} \right) \cdot \frac{x_2}{10} \left(\frac{\text{gram}}{\text{liter}} \right)$$

$$- 2 \left(\frac{\text{liter}}{\text{min}} \right) \cdot \frac{x_1}{20} \left(\frac{\text{gram}}{\text{liter}} \right) - 4 \left(\frac{\text{liter}}{\text{min}} \right) \cdot \frac{x_1}{20} \left(\frac{\text{gram}}{\text{liter}} \right)$$

$$= -\frac{6}{20} x_1 + \frac{3}{10} x_2.$$

Similarly, with the dimensions excluded,

$$x_2'(t) = 4 \frac{x_1}{20} - 3 \frac{x_2}{10} - \frac{x_2}{10} = \frac{4}{20} x_1 - \frac{4}{10} x_2.$$

The system we must solve is $\mathbf{X}' = \mathbf{AX}$ with

$$\mathbf{A} = \begin{pmatrix} -3/10 & 3/10 \\ 1/5 & -2/5 \end{pmatrix}.$$

The initial conditions are

$$x_1(0) = 150 \text{ and } x_2(0) = 50$$

or

$$\mathbf{X}(0) = \begin{pmatrix} 150 \\ 50 \end{pmatrix}.$$

The eigenvalues of $\mathbf{A}$ are $-1/10$ and $-1/5$ with corresponding eigenvalues, respectively,

$$\begin{pmatrix} 3/2 \\ 1 \end{pmatrix} \text{ and } \begin{pmatrix} -1 \\ 1 \end{pmatrix}.$$

The fundamental matrix is

$$\boldsymbol{\Omega}(t) = \begin{pmatrix} (3/2)e^{-t/10} & -e^{-3t/5} \\ e^{-t/10} & e^{-3t/5} \end{pmatrix}.$$

The general solution is $\mathbf{X}(t) = \boldsymbol{\Omega}(t)\mathbf{C}$. To solve the initial value problem, we need $\mathbf{C}$ so that

$$\mathbf{X}(0) = \begin{pmatrix} 150 \\ 50 \end{pmatrix} = \boldsymbol{\Omega}\mathbf{C} = \begin{pmatrix} 3/2 & -1 \\ 1 & 1 \end{pmatrix} \mathbf{C}.$$

Then

$$\mathbf{C} = \begin{pmatrix} 3/2 & -1 \\ 1 & 1 \end{pmatrix}^{-1} \begin{pmatrix} 150 \\ 50 \end{pmatrix}$$

$$= \begin{pmatrix} 2/5 & 2/5 \\ -2/5 & 3/5 \end{pmatrix} \begin{pmatrix} 150 \\ 50 \end{pmatrix} = \begin{pmatrix} 80 \\ -30 \end{pmatrix}.$$

The solution of the initial value problem is

$$\mathbf{X}(t) = \begin{pmatrix} (3/2)e^{-t/10} & -e^{-3t/5} \\ e^{-t/10} & e^{-3t/5} \end{pmatrix} \begin{pmatrix} 80 \\ -30 \end{pmatrix}$$

$$= \begin{pmatrix} 120e^{-t/10} + 30e^{-3t/5} \\ 80e^{-t/10} - 30e^{-3t/5} \end{pmatrix}.$$

As $t \to \infty$, $x_1(t) \to 0$, and $x_2(t) \to 0$, as we might expect. ◆

9.2.1 The Complex Eigenvalue Case

A real matrix $\mathbf{A}$ may have some complex eigenvalues and eigenvectors. In such a case, it is possible to write a solution strictly in terms of real quantities, just as we did for the constant coefficient, linear second-order equation when the characteristic equation has complex roots.

THEOREM 9.6 *Real Solutions When Complex Eigenvalues Occur*

Let $\mathbf{A}$ be an $n \times n$ matrix of real numbers. Let $\alpha + i\beta$ be a complex eigenvalue with corresponding eigenvector $\mathbf{U} + i\mathbf{V}$, in which $\mathbf{U}$ and $\mathbf{V}$ are real $n \times 1$ matrices. Then

$$e^{\alpha t}[\cos(\beta t)\mathbf{U} - \sin(\beta t)\mathbf{V}]$$

and

$$e^{\alpha t}[\sin(\beta t)\mathbf{U} + \cos(\beta t)\mathbf{V}]$$

are linearly independent solutions of $\mathbf{X}' = \mathbf{A}\mathbf{X}$.

Proof We know that $\alpha - i\beta$ is also an eigenvalue of $\mathbf{A}$ with eigenvector $\mathbf{U} - i\mathbf{V}$. Use Theorem 9.4 and Euler's formula to define two solutions:

$$\boldsymbol{\Phi}_1(t) = e^{\alpha t}(\cos(\beta t) + i\sin(\beta t))(\mathbf{U} + i\mathbf{V})$$

$$= e^{\alpha t}(\cos(\beta t) + i\sin(\beta t))(\mathbf{U} + i\mathbf{V})$$

$$= e^{\alpha t}(\cos(\beta t)\mathbf{U} - \sin(\beta t)\mathbf{V} + ie^{\alpha t}(\sin(\beta t)\mathbf{U} + \cos(\beta t)\mathbf{V}.$$

and

$$\boldsymbol{\Phi}_2(t) = e^{\alpha t}(\cos(\beta t) - i\sin(\beta t))(\mathbf{U} - i\mathbf{V})$$

$$= e^{\alpha t}(\cos(\beta t) - i\sin(\beta t))(\mathbf{U} - i\mathbf{V})$$

$$= e^{\alpha t}(\cos(\beta t)\mathbf{U} - \sin(\beta t))\mathbf{V} + ie^{\alpha t}(-\cos(\beta t)\mathbf{V} - \sin(\beta t)\mathbf{U}).$$

Since a linear combination of solutions is a solution, we can define two new solutions

$$\frac{1}{2}(\boldsymbol{\Phi}_1(t) + \boldsymbol{\Phi}_2(t))$$

and

$$\frac{1}{2i}(\boldsymbol{\Phi}_1(t) - \boldsymbol{\Phi}_2(t)),$$

and these are the solutions given in the theorem. ◆

Theorem 9.6 enables us to replace two complex terms

$$e^{(\alpha+i\beta)}(\mathbf{U}+i\mathbf{V}) \text{ and } e^{(\alpha-i\beta)}(\mathbf{U}-i\mathbf{V})$$

in the general solution with the two solutions given in the theorem, which involve only real quantities.

EXAMPLE 9.8

We will solve the system $\mathbf{X}' = \mathbf{AX}$ with

$$\mathbf{A} = \begin{pmatrix} 2 & 0 & 1 \\ 0 & -2 & -2 \\ 0 & 2 & 0 \end{pmatrix}.$$

The eigenvalues are 2, $-1+\sqrt{3}i$, and $-1-\sqrt{3}i$. Corresponding eigenvectors are, respectively,

$$\begin{pmatrix} 1 \\ 0 \\ 0 \end{pmatrix}, \begin{pmatrix} 1 \\ -2\sqrt{3}i \\ -3+\sqrt{3}i \end{pmatrix}, \text{ and } \begin{pmatrix} 1 \\ 2\sqrt{3}i \\ -3-\sqrt{3}i \end{pmatrix}.$$

One solution is

$$\begin{pmatrix} 1 \\ 0 \\ 0 \end{pmatrix} e^{2t}.$$

For the other two solutions, write

$$\begin{pmatrix} 1 \\ -2\sqrt{3}i \\ -3+\sqrt{3}i \end{pmatrix} = \begin{pmatrix} 1 \\ 0 \\ -3 \end{pmatrix} + i\begin{pmatrix} 0 \\ -2\sqrt{3} \\ \sqrt{3} \end{pmatrix} = \mathbf{U}+i\mathbf{V}.$$

By Theorem 9.6, with $\alpha=-1$ and $\beta=\sqrt{3}$, we have the two additional solutions

$$e^{-t}(\cos(\sqrt{3}t)\mathbf{U}-\sin(\sqrt{3}t)\mathbf{V}) = \begin{pmatrix} e^{-t}\cos(\sqrt{3}t) \\ 2\sqrt{3}e^{-t}\sin(\sqrt{3}t) \\ e^{-t}(-3\cos(\sqrt{3}t)-\sqrt{3}\sin(\sqrt{3}t)) \end{pmatrix}$$

and

$$e^{-t}(\sin(\sqrt{3}t)\mathbf{U}+\cos(\sqrt{3}t)\mathbf{V}) = \begin{pmatrix} e^{-t}\sin(\sqrt{3}t) \\ -2\sqrt{3}e^{-t}\cos(\sqrt{3}t) \\ e^{-t}(-3\sin(\sqrt{3}t)+\sqrt{3}\cos(\sqrt{3}t)) \end{pmatrix}.$$

Using these as columns, we obtain the fundamental matrix

$$\mathbf{\Omega}(t) = \begin{pmatrix} e^{2t} & e^{-t}\cos(\sqrt{3}t) & e^{-t}\sin(\sqrt{3}t) \\ 0 & 2\sqrt{3}e^{-t}\sin(\sqrt{3}t) & -2\sqrt{3}e^{-t}\cos(\sqrt{3}t) \\ 0 & e^{-t}(-3\cos(\sqrt{3}t)-\sqrt{3}t\sin(\sqrt{3}t)) & e^{-t}(\sqrt{3}\cos(\sqrt{3}t)-3\sin(\sqrt{3}t)) \end{pmatrix}.$$

The general solution is $\mathbf{X}(t) = \mathbf{\Omega}(t)\mathbf{C}.$ ◆

9.2.2   Fewer Than *n* Independent Eigenvectors

Two examples will give us a sense of how to proceed when **A** does not have *n* linearly independent eigenvectors.

EXAMPLE 9.9

We will solve $\mathbf{X}' = \mathbf{A}\mathbf{X}$ when

$$\mathbf{A} = \begin{pmatrix} 1 & 3 \\ -3 & 7 \end{pmatrix}.$$

A has eigenvalues of 4 and 4, and all eigenvectors are scalar multiples of

$$\begin{pmatrix} 1 \\ 1 \end{pmatrix}$$

One solution is

$$\mathbf{\Phi}_1(t) = \begin{pmatrix} 1 \\ 1 \end{pmatrix} e^{4t}.$$

We need a second, linearly independent solution. Set

$$\mathbf{E}_1 = \begin{pmatrix} 1 \\ 1 \end{pmatrix}$$

and attempt a second solution of the form

$$\mathbf{\Phi}_2(t) = \mathbf{E}_1 t e^{4t} + \mathbf{E}_2 e^{4t},$$

in which $\mathbf{E}_2$ is a 2×1 constant matrix to be determined. For $\mathbf{\Phi}_2(t)$ to be a solution, we must have $\mathbf{\Phi}_2'(t) = \mathbf{A}\mathbf{\Phi}_2(t)$. This is the equation

$$\mathbf{E}_1[e^{4t} + 4te^{4t}] + 4\mathbf{E}_2 e^{4t} = \mathbf{A}\mathbf{E}_1 t e^{4t} + \mathbf{A}\mathbf{E}_2 e^{4t}.$$

Divide this by e^{4t} to get

$$\mathbf{E}_1 + 4t\mathbf{E}_1 + 4\mathbf{E}_2 = \mathbf{A}\mathbf{E}_1 t + \mathbf{A}\mathbf{E}_2.$$

But $\mathbf{A}\mathbf{E}_1 = 4\mathbf{E}_1$, so the terms involving t cancel, leaving

$$\mathbf{A}\mathbf{E}_2 - 4\mathbf{E}_2 = \mathbf{E}_1$$

or

$$(\mathbf{A} - 4\mathbf{I}_2)\mathbf{E}_2 = \mathbf{E}_1.$$

If

$$\mathbf{E}_2 = \begin{pmatrix} a \\ b \end{pmatrix},$$

then we have the linear system of two equations in two unknowns:

$$(\mathbf{A} - 4\mathbf{I}_2)\begin{pmatrix} a \\ b \end{pmatrix} = \begin{pmatrix} 1 \\ 1 \end{pmatrix}.$$

This is the system

$$\begin{pmatrix} -3 & 3 \\ -3 & 3 \end{pmatrix}\begin{pmatrix} a \\ b \end{pmatrix} = \begin{pmatrix} 1 \\ 1 \end{pmatrix}$$

with a general solution of

$$\mathbf{E}_2 = \begin{pmatrix} \alpha \\ (1+3\alpha)/3 \end{pmatrix}.$$

Since we need only one $\mathbf{E}_2$, let $\alpha = 1$ to get

$$\mathbf{E}_2 = \begin{pmatrix} 1 \\ 4/3 \end{pmatrix}.$$

Therefore, a second solution is

$$\boldsymbol{\Phi}_2(t) = \mathbf{E}_1 t e^{4t} + \mathbf{E}_2 e^{4t} = \begin{pmatrix} 1 \\ 1 \end{pmatrix} t e^{4t} + \begin{pmatrix} 1 \\ 4/3 \end{pmatrix} e^{4t} = \begin{pmatrix} 1+t \\ 4/3+t \end{pmatrix} e^{4t}.$$

$\boldsymbol{\Phi}_1$ and $\boldsymbol{\Phi}_2$ are linearly independent solutions and can be used as columns of the fundamental matrix

$$\boldsymbol{\Omega}(t) = \begin{pmatrix} e^{4t} & (1+t)e^{4t} \\ e^{4t} & (4/3+t)e^{4t} \end{pmatrix}.$$

The general solution is $\mathbf{X}(t) = \boldsymbol{\Omega}(t)\mathbf{C}$. ◆

EXAMPLE 9.10

We will solve $\mathbf{X}' = \mathbf{A}\mathbf{X}$ when

$$\mathbf{A} = \begin{pmatrix} -2 & -1 & -5 \\ 25 & -7 & 0 \\ 0 & 1 & 3 \end{pmatrix}.$$

$\mathbf{A}$ has eigenvalues of -2, -2, and -2, and all eigenvectors are scalar multiples of

$$\mathbf{E}_1 = \begin{pmatrix} -1 \\ -5 \\ 1 \end{pmatrix}.$$

One solution is $\boldsymbol{\Phi}_1(t) = \mathbf{E}_1 e^{-2t}$. We will try a second solution, linearly independent from the first, of the form

$$\boldsymbol{\Phi}_2(t) = \mathbf{E}_1 t e^{-2t} + \mathbf{E}_2 e^{-2t},$$

in which $\mathbf{E}_2$ must be determined. Substitute this proposed solution into the differential equation to get

$$\mathbf{E}_1[e^{-2t} - 2te^{-2t}] + \mathbf{E}_2[-2e^{-2t}] = \mathbf{A}\mathbf{E}_1 t e^{-2t} + \mathbf{A}\mathbf{E}_2 e^{-2t}.$$

Divide by e^{-2t} and recall that $\mathbf{A}\mathbf{E}_1 = -2\mathbf{E}_1$ to cancel some terms in the last equation, leaving

$$\mathbf{A}\mathbf{E}_2 + 2\mathbf{E}_2 = \mathbf{E}_1.$$

This is the system

$$(\mathbf{A} + 2\mathbf{I}_3)\mathbf{E}_2 = \mathbf{E}_1.$$

If we write

$$\mathbf{E}_2 = \begin{pmatrix} a \\ b \\ c \end{pmatrix},$$

then we have the system of algebraic equations

$$\begin{pmatrix} 0 & -1 & -5 \\ 25 & -5 & 0 \\ 0 & 1 & 5 \end{pmatrix} \begin{pmatrix} a \\ b \\ c \end{pmatrix} = \begin{pmatrix} -1 \\ -5 \\ 1 \end{pmatrix}$$

with a general solution of

$$\begin{pmatrix} -\alpha \\ 1 - 5\alpha \\ \alpha \end{pmatrix}$$

for α any real number. Choose $\alpha = 1$ to get

$$\mathbf{E}_2 = \begin{pmatrix} -1 \\ -4 \\ 1 \end{pmatrix}.$$

This gives us a second solution of the differential equation

$$\boldsymbol{\Phi}_2(t) = \mathbf{E}_1 t e^{-2t} + \mathbf{E}_2 e^{-2t}$$

$$= \begin{pmatrix} -1 \\ -5 \\ 1 \end{pmatrix} t e^{2t} + \begin{pmatrix} -1 \\ -4 \\ 1 \end{pmatrix} e^{-2t} = \begin{pmatrix} -1 - t \\ -4 - 5t \\ 1 + t \end{pmatrix} e^{-2t}.$$

We need one more solution, linearly independent from the first two. Try for a third solution of the form

$$\boldsymbol{\Phi}_3(t) = \frac{1}{2} \mathbf{E}_1 t^2 e^{-2t} + \mathbf{E}_2 t e^{-2t} + \mathbf{E}_3 e^{-2t}.$$

Substitute this into $\mathbf{X}' = \mathbf{A}\mathbf{X}$ to get

$$\mathbf{E}_1 [t e^{-2t} - t^2 e^{-2t}] + \mathbf{E}_2 [e^{-2t} - 2t e^{-2t}] + \mathbf{E}_3 [-2 e^{-2t}]$$

$$= \frac{1}{2} \mathbf{A} \mathbf{E}_1 t^2 e^{-2t} + \mathbf{A} \mathbf{E}_2 t e^{-2t} + \mathbf{A} \mathbf{E}_3 e^{-2t}.$$

Divide e^{-2t} and use the fact that $\mathbf{A}\mathbf{E}_1 = -2\mathbf{E}_1$ and

$$\mathbf{A}\mathbf{E}_2 = \begin{pmatrix} 1 \\ 3 \\ -1 \end{pmatrix}$$

to get

$$\mathbf{E}_1 t - \mathbf{E}_1 t^2 + \mathbf{E}_2 - 2\mathbf{E}_2 t - 2\mathbf{E}_3 = -\mathbf{E}_1 t^2 + \begin{pmatrix} 1 \\ 3 \\ -1 \end{pmatrix} t + \mathbf{A}\mathbf{E}_3. \qquad (9.2)$$

Now

$$\mathbf{E}_1 t - 2\mathbf{E}_2 t = (\mathbf{E}_1 - 2\mathbf{E}_2)t = \begin{pmatrix} -1 - 2(-1) \\ -5 - 2(-4) \\ 1 - 2(1) \end{pmatrix} t = \begin{pmatrix} 1 \\ 3 \\ -1 \end{pmatrix} t.$$

Three terms cancel in equation (9.2), and it reduces to

$$\mathbf{E}_2 - 2\mathbf{E}_3 = \mathbf{A}\mathbf{E}_3.$$

Write this equation as

$$(\mathbf{A} + 2\mathbf{I}_3)\mathbf{E}_3 = \mathbf{E}_2.$$

This is the system

$$\begin{pmatrix} 0 & -1 & -5 \\ 25 & -5 & 0 \\ 0 & 1 & 5 \end{pmatrix} \mathbf{E}_3 = \begin{pmatrix} -11 \\ -4 \\ 1 \end{pmatrix}$$

with a general solution of

$$\mathbf{E}_3 = \begin{pmatrix} (1 - 25\alpha)/25 \\ 1 - 5\alpha \\ \alpha \end{pmatrix}.$$

Let $\alpha = 1$ to get

$$\mathbf{E}_3 = \begin{pmatrix} -24/25 \\ -4 \\ 1 \end{pmatrix}.$$

A third solution is

$$\mathbf{\Phi}_3(t) = \frac{1}{2}\begin{pmatrix} -1 \\ -5 \\ 1 \end{pmatrix} t^2 e^{-2t} + \begin{pmatrix} -1 \\ -4 \\ 1 \end{pmatrix} t e^{-2t} + \begin{pmatrix} -24/25 \\ -4 \\ 1 \end{pmatrix} e^{-2t}$$

$$= \begin{pmatrix} -24/25 - t - t^2/2 \\ -4 - 4t - 5t^2/2 \\ 1 + t + t^2/2 \end{pmatrix} e^{-2t}.$$

We now have three linearly independent solutions and can use these as columns of the fundamental matrix

$$\mathbf{\Omega}(t) = \begin{pmatrix} -e^{-2t} & (-1-t)e^{-2t} & (-24/25 - t - t^2/2)e^{-2t} \\ -5e^{-2t} & (-4-5t)e^{-2t} & (-4 - 4t - 5t^2/2)e^{-2t} \\ e^{-2t} & (1+t)e^{-2t} & (1 + t + t^2/2)e^{-2t} \end{pmatrix}.$$

The general solution is $\mathbf{X}(t) = \mathbf{\Omega}(t)\mathbf{C}$. ◆

These examples suggest a procedure to follow. Suppose we know the n eigenvalues of $\mathbf{A}$, counting multiplicities. If these are all distinct, then corresponding eigenvectors are linearly independent, and we can write the general solution.

Thus, suppose an eigenvalue λ has multiplicity $k > 1$. If there are k linearly independent eigenvectors associated with λ, then we can produce k linearly independent solutions corresponding to λ.

If there are associated with λ only r linearly independent eigenvectors and $r < k$, then we need from λ a total of $r - k$ more solutions of the system that are linearly independent from the others.

If $r - k = 1$, then we need one such solution, which can be obtained as in Example 9.9.

If $r - k = 2$, we can produce two more linearly independent solutions associated with λ by following Example 9.10.

If $r - k = 3$, produce two more linearly independent solutions as in Example 9.10, then a third of the form

$$\mathbf{E}_4 = \frac{1}{3!}\mathbf{E}_1 t^3 e^{\lambda t} + \frac{1}{2}\mathbf{E}_2 t^2 e^{\lambda t} + \mathbf{E}_3 t e^{\lambda t} + \mathbf{E}_4 e^{\lambda t}.$$

For larger $r - k$, continue this pattern to produce the needed $r - k$ additional solutions, obtaining r linearly independent solutions associated with λ. When this is done for each eigenvalue of $\mathbf{A}$, we have a total of n linearly independent solutions.

| SECTION 9.2 | *PROBLEMS* |

In each of Problems 1 through 10, find a fundamental matrix for the system and write the general solution as a matrix. If initial values are given, solve the initial value problem.

1. $x_1' = 3x_1, x_2' = 5x_1 - 4x_2$

2. $x_1' = 4x_1 + 2x_2, x_2' = 3x_1 + 3x_2$

3. $x_1' = x_1 + x_2, x_2' = x_1 + x_2$

4. $x_1' = 2x_1 + x_2 - 2x_3, x_2' = 3x_1 - 2x_2,$
   $x_3' = 3x_1 - x_2 - 3x_3$

5. $x_1' = x_1 + 2x_2 + x_3, x_2' = 6x_1 - x_2, x_3' = -x_1 - 2x_2 - x_3$

6. $x_1' = 3x_1 - 4x_2, x_2' = 2x_1 - 3x_2; x_1(0) = 7, x_2(0) = 5$

7. $x_1' = x_1 - 2x_2, x_2' = -6x_1; x_1(0) = 1, x_2(0) = -19$

8. $x_1' = 2x_1 - 10x_2, x_2' = -x_1 - x_2; x_1(0) = -3, x_2(0) = 6$

9. $x_1' = 3x_1 - x_2 + x_3, x_2' = x_1 + x_2 - x_3,$
   $x_3' = x_1 - x_2 + x_3; x_1(0) = 1, x_2(0) = 5, x_3(0) = 1$

10. $x_1' = 2x_1 + x_2 - x_3, x_2' = 3x_1 - 2x_2,$
    $x_3' = 3x_1 + x_2 - 3x_3; x_1(0) = 1, x_2(0) = 7, x_3(0) = 3$

In each of Problems 11 through 15, find a real-valued fundamental matrix for the system $\mathbf{X}' = \mathbf{AX}$ with the given coefficient matrix.

11. $\begin{pmatrix} 2 & -4 \\ 1 & 2 \end{pmatrix}$

12. $\begin{pmatrix} 0 & 5 \\ -1 & -2 \end{pmatrix}$

13. $\begin{pmatrix} 3 & -5 \\ 1 & -1 \end{pmatrix}$

14. $\begin{pmatrix} 1 & -1 & 1 \\ 1 & -1 & 0 \\ 1 & 0 & -1 \end{pmatrix}$

15. $\begin{pmatrix} -2 & 1 & 0 \\ -5 & 0 & 0 \\ 0 & 3 & -2 \end{pmatrix}$

In each of Problems 16 through 21, find a fundamental matrix for the system with the given coefficient matrix.

16. $\begin{pmatrix} 2 & 0 \\ 5 & 2 \end{pmatrix}$

17. $\begin{pmatrix} 3 & 2 \\ 0 & 3 \end{pmatrix}$

18. $\begin{pmatrix} 1 & 5 & 0 \\ 0 & 1 & 0 \\ 4 & 8 & 1 \end{pmatrix}$

19. $\begin{pmatrix} 2 & 5 & 6 \\ 0 & 8 & 9 \\ 0 & 1 & -2 \end{pmatrix}$

20. $\begin{pmatrix} 0 & 1 & 0 & 0 \\ 0 & 0 & 1 & 0 \\ 0 & 0 & 0 & 1 \\ -1 & -2 & 0 & 0 \end{pmatrix}$

21. $\begin{pmatrix} 1 & 5 & -2 & 6 \\ 0 & 3 & 0 & 4 \\ 0 & 3 & 0 & 4 \\ 0 & 0 & 0 & 1 \end{pmatrix}$

9.3   Solution of $\mathbf{X}' = \mathbf{AX} + \mathbf{G}$

We know from Theorem 9.3 that the general solution is the sum of the general solution of the homogeneous problem $\mathbf{X}' = \mathbf{AX}$ plus any particular solution of the nonhomogeneous system. We therefore need a method for finding a particular solution. We will develop two methods.

9.3.1   Variation of Parameters

Variation of parameters for systems follows the same line of reasoning as variation of parameters for second-order linear differential equations. If $\mathbf{\Omega}(t)$ is a fundamental matrix for the homogeneous system $\mathbf{X}' = \mathbf{AX}$, then the general solution of the homogeneous system is $\mathbf{\Omega C}$. Using this as a template, look for a particular solution of the nonhomogeneous system of the form $\mathbf{\Psi}_p(t) = \mathbf{\Omega}(t)\mathbf{U}(t)$, where $\mathbf{U}(t)$ is an $n \times 1$ matrix to be determined.

Substitute this proposed particular solution into the nonhomogeneous system to obtain

$$(\mathbf{\Omega U})' = \mathbf{\Omega}'\mathbf{U} + \mathbf{\Omega U}' = \mathbf{A}(\mathbf{\Omega U}) + \mathbf{G} = (\mathbf{A\Omega})\mathbf{U} + \mathbf{G}. \qquad (9.3)$$

Ω is a fundamental matrix for the homogeneous system, so $\Omega' = A\Omega$. Therefore, $\Omega'U = (A\Omega)U$, and equation (9.3) becomes

$$\Omega U' = G.$$

Since Ω is nonsingular,

$$U' = \Omega^{-1}G.$$

Then

$$U(t) = \int \Omega^{-1}(t)G(t)dt,$$

in which we integrate a matrix by integrating each element of the matrix. Once we have $U(t)$, we have the general solution

$$X(t) = \Omega(t)C + \Omega(t)U(t)$$

of the nonhomogeneous system.

EXAMPLE 9.11

We will solve the system

$$X' = \begin{pmatrix} 1 & -10 \\ -1 & 4 \end{pmatrix} X + \begin{pmatrix} t \\ 1 \end{pmatrix}.$$

The eigenvalues of A are -1 and 6 with corresponding eigenvectors

$$\begin{pmatrix} 5 \\ 1 \end{pmatrix} \text{ and } \begin{pmatrix} -2 \\ 1 \end{pmatrix}.$$

A fundamental matrix for $X' = AX$ is

$$\Omega(t) = \begin{pmatrix} 5e^{-t} & -2e^{6t} \\ e^{-t} & e^{6t} \end{pmatrix}.$$

Compute

$$\Omega^{-1}(t) = \frac{1}{7} \begin{pmatrix} e^t & 2e^t \\ -e^{-6t} & 5e^{-6t} \end{pmatrix}.$$

Then

$$U'(t) = \Omega^{-1}(t)G(t) = \frac{1}{7} \begin{pmatrix} e^t & 2e^t \\ -e^{-6t} & 5e^{-6t} \end{pmatrix} \begin{pmatrix} t \\ 1 \end{pmatrix}$$

$$= \frac{1}{7} \begin{pmatrix} 2e^t + te^t \\ 5e^{-6t} - te^{-6t} \end{pmatrix}.$$

Then

$$U(t) = \int \Omega^{-1}(t)G(t)dt$$

$$= \begin{pmatrix} (t+1)e^t/7 \\ (-29/252)e^{-6t} + (1/42)te^{-6t} \end{pmatrix}.$$

The general solution of the nonhomogeneous system is

$$\mathbf{X}(t) = \boldsymbol{\Omega}(t)\mathbf{C} + \boldsymbol{\Omega}(t)\mathbf{U}(t) = \begin{pmatrix} 5e^{-t} & -2e^{6t} \\ e^{-t} & e^{6t} \end{pmatrix} \mathbf{C}$$

$$+ \begin{pmatrix} 5e^{-t} & -2e^{6t} \\ e^{-t} & e^{6t} \end{pmatrix} \begin{pmatrix} (t+1)e^{t}/7 \\ (-29/252)e^{-6t} + (1/42)te^{-6t} \end{pmatrix}$$

$$= \begin{pmatrix} 5e^{-t} & -2e^{6t} \\ e^{-t} & e^{6t} \end{pmatrix} \mathbf{C} + \frac{1}{3} \begin{pmatrix} 17/6 + (49/7)t \\ 1/12 + t/2 \end{pmatrix}. \quad \blacklozenge$$

Although in this example the coefficient matrix **A** was constant, this is not required to apply the method of variation of parameters.

9.3.2   Solution by Diagonalizing A

If **A** is a diagonalizable matrix of real numbers, then we can solve the system $\mathbf{X}' = \mathbf{AX} + \mathbf{G}$ by the change of variables $\mathbf{X} = \mathbf{PZ}$, where **P** diagonalizes **A**.

EXAMPLE 9.12

We will solve the system

$$\mathbf{X}' = \begin{pmatrix} 3 & 3 \\ 1 & 5 \end{pmatrix} \mathbf{X} + \begin{pmatrix} 8 \\ 4e^{3t} \end{pmatrix}.$$

The eigenvalues of **A** are 2 and 6 with eigenvectors, respectively,

$$\begin{pmatrix} -3 \\ 1 \end{pmatrix} \text{ and } \begin{pmatrix} 1 \\ 1 \end{pmatrix}.$$

Form **P** using these eigenvectors as columns:

$$\mathbf{P} = \begin{pmatrix} -3 & 1 \\ 1 & 1 \end{pmatrix}.$$

Then

$$\mathbf{P}^{-1}\mathbf{AP} = \mathbf{D} = \begin{pmatrix} 2 & 0 \\ 0 & 6 \end{pmatrix}$$

with the eigenvalues down the main diagonal. Compute

$$\mathbf{P}^{-1} = \begin{pmatrix} -1/4 & 1/4 \\ 1/4 & 3/4 \end{pmatrix}.$$

Now make the change of variables $\mathbf{X} = \mathbf{PZ}$ into the differential equation:

$$\mathbf{X}' = (\mathbf{PZ})' = \mathbf{PZ}' = \mathbf{A}(\mathbf{PZ}) + \mathbf{G}.$$

Then

$$\mathbf{PZ}' = (\mathbf{AP})\mathbf{Z} + \mathbf{G}.$$

Multiply this equation on the left by $\mathbf{P}^{-1}$ to get

$$\mathbf{Z}' = (\mathbf{P}^{-1}\mathbf{AP})\mathbf{Z} + \mathbf{P}^{-1}\mathbf{G}$$

or

$$\mathbf{Z}' = \mathbf{DZ} + \mathbf{P}^{-1}\mathbf{G}.$$

This is

$$\begin{pmatrix} z_1' \\ z_2' \end{pmatrix} = \begin{pmatrix} 2 & 0 \\ 0 & 6 \end{pmatrix} \begin{pmatrix} z_1 \\ z_2 \end{pmatrix} + \begin{pmatrix} -1/4 & 1/4 \\ 1/4 & 3/4 \end{pmatrix} \begin{pmatrix} 8 \\ 4e^{3t} \end{pmatrix}$$

$$= \begin{pmatrix} 2z_1 - 2 + e^{3t} \\ 6z_2 + 2 + 3e^{3t} \end{pmatrix}.$$

This is an *uncoupled system* in the sense that we have a differential equation for just z_1 and a second differential equation for just z_2. Solve each of these first-order linear differential equations to obtain

$$z_1(t) = c_1 e^{2t} + e^{3t} + 1$$

and

$$z_2(t) = c_2 e^{6t} - e^{3t} - \frac{1}{3}.$$

Then

$$\mathbf{X}(t) = \mathbf{PZ}(t) = \begin{pmatrix} -3 & 1 \\ 1 & 1 \end{pmatrix} \begin{pmatrix} c_1 e^{2t} + e^{3t} + 1 \\ c_2 e^{6t} - e^{3t} - 1/3 \end{pmatrix}$$

$$= \begin{pmatrix} -3c_1 e^{2t} + c_2 e^{6t} - 4e^{3t} - 10/3 \\ c_1 e^{2t} + c_2 e^{6t} + 2/3 \end{pmatrix}$$

$$= \begin{pmatrix} -3e^{2t} & e^{6t} \\ e^{2t} & e^{6t} \end{pmatrix} \mathbf{C} + \begin{pmatrix} -4e^{3t} - 10/3 \\ 2/3 \end{pmatrix}.$$

This is the general solution in the form $\mathbf{\Omega}(t)\mathbf{C} + \mathbf{\Psi}_p$, which is a sum of the general solution of the associated homogeneous equation and a particular solution of the nonhomogeneous equation. ◆

SECTION 9.3   PROBLEMS

In each of Problems 1 through 9, find the general solution with **A** and **G** given. In Problems 6 through 9, also solve the initial value problem.

1. $\begin{pmatrix} 5 & 2 \\ -2 & 1 \end{pmatrix}, \begin{pmatrix} -3e^t \\ e^{3t} \end{pmatrix}$

2. $\begin{pmatrix} 2 & -4 \\ 1 & -2 \end{pmatrix}, \begin{pmatrix} 1 \\ 3t \end{pmatrix}$

3. $\begin{pmatrix} 7 & -1 \\ 1 & 5 \end{pmatrix}, \begin{pmatrix} 2e^{6t} \\ 6te^{6t} \end{pmatrix}$

4. $\begin{pmatrix} 2 & 0 & 0 \\ 0 & 6 & -4 \\ 0 & 4 & -2 \end{pmatrix}, \begin{pmatrix} e^{2t}\cos(3t) \\ -2 \\ -2 \end{pmatrix}$

5. $\begin{pmatrix} 1 & 0 & 0 & 0 \\ 4 & 3 & 0 & 0 \\ 0 & 0 & 3 & 0 \\ -1 & 2 & 9 & 1 \end{pmatrix}, \begin{pmatrix} 0 \\ -2e^t \\ 0 \\ e^t \end{pmatrix}$

6. $\begin{pmatrix} 2 & 0 \\ 5 & 2 \end{pmatrix}, \begin{pmatrix} 2 \\ 10t \end{pmatrix}; \begin{pmatrix} 0 \\ 3 \end{pmatrix}$

7. $\begin{pmatrix} 5 & -4 \\ 4 & -3 \end{pmatrix}, \begin{pmatrix} 2e^t \\ 2e^t \end{pmatrix}; \begin{pmatrix} -1 \\ 3 \end{pmatrix}$

8. $\begin{pmatrix} 2 & -3 & 1 \\ 0 & 2 & 4 \\ 0 & 0 & 1 \end{pmatrix}, \begin{pmatrix} 10e^{2t} \\ 6e^{2t} \\ -e^{2t} \end{pmatrix}; \begin{pmatrix} 5 \\ 11 \\ -2 \end{pmatrix}$

9. $\begin{pmatrix} 1 & -3 & 0 \\ 3 & -5 & 0 \\ 4 & 7 & -2 \end{pmatrix}, \begin{pmatrix} te^{-2t} \\ te^{-2t} \\ t^2 e^{-2t} \end{pmatrix}; \begin{pmatrix} 6 \\ 2 \\ 3 \end{pmatrix}$

In each of Problems 10 through 16, find the general solution of the system. In Problems 15 and 16, also solve the initial value problem.

10. $x_1' = -2x_1 + x_2, x_2' = -4x_1 + 3x_2 + 10\cos(t)$

11. $x_1' = 3x_1 + 3x_2 + 8, x_2' = x_1 + 5x_2 + 4e^{3t}$

12. $x_1' = x_1 + x_2 + 6e^{3t}, x_2' = x_1 + x_2 + 4$

13. $x_1' = 6x_1 + 5x_2 - 4\cos(3t), x_2' = x_1 + 2x_2 + 8$

14. $x_1' = 3x_1 - 2x_2 + 3e^{2t}, x_2' = 9x_1 - 3x_2 + e^{2t}$

15. $x_1' = x_1 + x_2 + 6e^{2t}, x_2' = x_1 + x_2 + 2e^{2t}$;
    $x_1(0) = 6, x_2(0) = 0$

16. $x_1' = x_1 - 2x_2 + 2t, x_2' = -x_1 + 2x_2 + 5$;
    $x_1(0) = 13, x_2(0) = 12$

$$f](s) = \int_0^\infty e^{-st} t^{-1/2}\, dt \quad = 2 \int_0^\infty e^{-sx^2}\, dx \ (\text{set } x = t^{1/2}) \quad = \frac{2}{\sqrt{s}} \int_0^\infty e^{-z^2}\, dz \ (\text{set } z = x$$

$$= \sqrt{\frac{\pi}{s}}$$

$$\mathcal{L}[f](s) = \int_0^\infty e$$

PART 3

Vector Analysis

$$= 2 \int_0^\infty e^{-s}\ dx$$

$$= \sqrt{\frac{\pi}{s}}$$

$$(s) = \int_0^\infty e^{-st} t^{-1/2}\, dt$$

$$= 2 \int_0^\infty e^{-sx^2}\, dx \ (\text{set } x = t^{1/2})$$

$$= \frac{2}{\sqrt{s}} \int_0^\infty e^{-z^2}\, dz \ (\text{set } z = x\sqrt{s})$$

$$[f](s) = \int_0^\infty e^{-st} t^{-1/2}$$

CHAPTER 10

Vector Differential Calculus

10.1   Vector Functions of One Variable

A vector function of one variable is a function of the form $\mathbf{F}(t) = x(t)\mathbf{i} + y(t)\mathbf{j} + z(t)\mathbf{k}$. This vector function is continuous at t_0 if each component function is continuous at t_0.

We may think of $\mathbf{F}(t)$ as the *position vector* of a curve in 3-space. For each t for which the vector is defined, draw $\mathbf{F}(t)$ as an arrow from the origin to the point $(x(t), y(t), z(t))$. This arrow sweeps out a curve C as t varies. When thought of in this way, the coordinate functions are parametric equations of this curve.

EXAMPLE 10.1

$\mathbf{H}(t) = t^2\mathbf{i} + \sin(t)\mathbf{j} - t^2\mathbf{k}$ is the position vector for the curve given parametrically by

$$x = t^2, \ y = \sin(t), \ \text{and} \ z = -t^2.$$

205

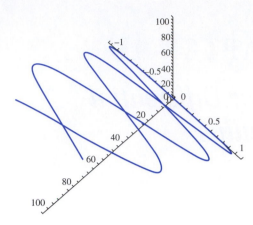

FIGURE 10.1 *Graph of the curve of Example 10.1.*

Figure 10.1 shows part of a graph of this curve. ◆

$\mathbf{F}(t) = x(t)\mathbf{i} + y(t)\mathbf{j} + z(t)\mathbf{k}$ is differentiable at t if each component function is differentiable at t, and in this case,

$$\mathbf{F}'(t) = x'(t)\mathbf{i} + y'(t)\mathbf{j} + z'(t)\mathbf{k}.$$

We differentiate a vector function by differentiating each component.

To give an interpretation to the vector $\mathbf{F}'(t_0)$, look at the limit of the difference quotient:

$$
\begin{aligned}
\mathbf{F}'(t_0) &= \lim_{h \to 0} \frac{\mathbf{F}(t_0 + h) - \mathbf{F}(t_0)}{h} \\
&= \left(\lim_{h \to 0} \frac{x(t_0 + h) - x(t_0)}{h} \right) \mathbf{i} + \left(\lim_{h \to 0} \frac{y(t_0 + h) - y(t_0)}{h} \right) \mathbf{j} \\
&\quad + \left(\lim_{h \to 0} \frac{z(t_0 + h) - z(t_0)}{h} \right) \mathbf{k} \\
&= x'(t_0)\mathbf{i} + y'(t_0)\mathbf{j} + z'(t_0)\mathbf{k}.
\end{aligned}
$$

Figure 10.2 shows the vectors $\mathbf{F}(t_0 + h)$, $\mathbf{F}(t_0)$, and $\mathbf{F}(t_0 + h) - \mathbf{F}(t_0)$ using the parallelogram law. As h is chosen smaller, the tip of the vector $\mathbf{F}(t_0 + h) - \mathbf{F}(t_0)$ slides along C toward $\mathbf{F}(t_0)$, and $(1/h)[\mathbf{F}(t_0 + h) - \mathbf{F}(t_0)]$ moves into the position of the tangent vector to C at the point $(f(t_0), g(t_0), h(t_0))$. In calculus, the derivative of a function gives the slope of the tangent to the graph at a point. In vector calculus, the derivative of the position vector of a curve gives the tangent vector to the curve at a point.

The length of a curve having differentiable position vector $\mathbf{F}(t)$ is

$$\text{length} = \int_a^b \sqrt{(x'(t))^2 + (y'(t))^2 + (z'(t))^2} \, dt = \int_a^b \| \mathbf{F}'(t) \| \, dt.$$

This is the integral of the length of the tangent vector.

Now imagine starting at $(x(a), y(a), z(a))$ at time $t = a$ and moving along the curve, reaching the point $(x(t), y(t), z(t))$ at time t. Let $s(t)$ be the distance along the curve from the starting point to this point (Figure 10.3). Then

$$s(t) = \int_a^t \| \mathbf{F}'(\xi) \| \, d\xi.$$

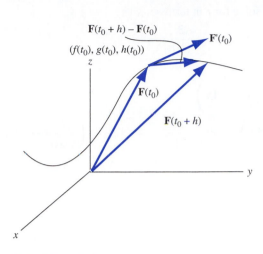

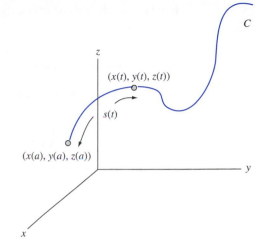

FIGURE 10.2 $\mathbf{F}'(t_0)$ *as a tangent vector.*

FIGURE 10.3 *Distance function along a curve.*

This function measures length along C and is strictly increasing, hence it has an inverse. At least in theory, we can solve for $t = t(s)$, writing the parameter t in terms of arc length along C. We can substitute this function into the position function to obtain

$$\mathbf{G}(s) = \mathbf{F}(t(s)).$$

$\mathbf{G}$ is also a position vector for C, except now the variable is s, which varies from 0 to L, the length of C. Therefore, $\mathbf{G}'(s)$ is also a tangent vector to C. We claim that this tangent vector in terms of arc length is always a unit vector. To see this, observe from the fundamental theorem of calculus that

$$s'(t) = \| \mathbf{F}'(t) \|.$$

Then

$$\mathbf{G}'(s) = \frac{d}{ds}\mathbf{F}(t(s)) = \frac{d}{dt}\mathbf{F}(t)\frac{dt}{ds}$$

$$= \frac{1}{ds/dt}\mathbf{F}'(t) = \frac{1}{\| \mathbf{F}'(t) \|}\mathbf{F}'(t),$$

and this vector has length 1.

EXAMPLE 10.2

Let C be defined by

$$x = \cos(t), \quad y = \sin(t), \quad \text{and } z = t/3$$

for $-4\pi \le t \le 4\pi$. C has a position vector

$$\mathbf{F}(t) = \cos(t)\mathbf{i} + \sin(t)\mathbf{j} + \frac{1}{3}t\mathbf{k}$$

and a tangent vector

$$\mathbf{F}'(t) = -\sin(t)\mathbf{i} + \cos(t)\mathbf{j} + \frac{1}{3}\mathbf{k}.$$

It is routine to compute $\| \mathbf{F}'(t) \| = \sqrt{10}/3$, so the distance function along C is

$$s(t) = \int_{-4\pi}^{t} \frac{1}{3}\sqrt{10}\,d\xi = \frac{1}{3}\sqrt{10}(t + 4\pi).$$

In this example, we can explicitly solve for t in terms of s as

$$t = t(s) = \frac{3}{\sqrt{10}}s - 4\pi.$$

Substitute this into $\mathbf{F}(t)$ to get

$$\mathbf{G}(s) = \mathbf{F}(t(s)) = \mathbf{F}\left(\frac{3}{\sqrt{10}}s - 4\pi\right)$$

$$= \cos\left(\frac{3}{\sqrt{10}}s - 4\pi\right)\mathbf{i} + \sin\left(\frac{3}{\sqrt{10}}s - 4\pi\right)\mathbf{j} + \frac{1}{3}\left(\frac{3}{\sqrt{10}}s - 4\pi\right)\mathbf{k}$$

$$= \cos\left(\frac{3}{\sqrt{10}}s\right)\mathbf{i} + \sin\left(\frac{3}{\sqrt{10}}s\right)\mathbf{j} + \left(\frac{3}{\sqrt{10}}s\right)\mathbf{k}.$$

Now compute

$$\mathbf{G}'(s) = -\frac{3}{\sqrt{10}}\cos\left(\frac{3}{\sqrt{10}}s\right)\mathbf{i} + \frac{3}{\sqrt{10}}\sin\left(\frac{3}{\sqrt{10}}s\right)\mathbf{j} + \frac{1}{\sqrt{10}}\mathbf{k},$$

and this is a unit tangent vector to C. ◆

Rules for differentiating various combinations of vectors are like those for functions of one variable. If the functions and vectors are differentiable and α is a number, then

1. $[\mathbf{F}(t) + \mathbf{G}(t)]' = \mathbf{F}'(t) + \mathbf{G}'(t)$.
2. $(\alpha\mathbf{F})'(t) = \alpha\mathbf{F}'(t)$.
3. $[f(t)\mathbf{F}(t)]' = f'(t)\mathbf{F}(t) + f(t)\mathbf{F}'(t)$.
4. $[\mathbf{F}(t) \cdot \mathbf{G}(t)]' = \mathbf{F}'(t) \cdot \mathbf{G}(t) + \mathbf{F}(t) \cdot \mathbf{G}'(t)$.
5. $[\mathbf{F}(t) \times \mathbf{G}(t)]' = \mathbf{F}'(t) \times \mathbf{G}(t) + \mathbf{F}(t) \times \mathbf{G}'(t)$.
6. $[\mathbf{F}(f(t))]' = f'(t)\mathbf{F}'(f(t))$.

Rules 3, 4, and 5 are product rules, which are reminiscent of the rule for differentiating a product of functions of one variable. In rule 4, the order of the factors is important, since the cross product is anti-commutative. Rule 6 is a chain rule for vector differentiation.

SECTION 10.1   PROBLEMS

In each of Problems 1 through 5, compute the requested derivative.

1. $\mathbf{F}(t) = \mathbf{i} + 3t^2\mathbf{j} + 2t\mathbf{k}$, $f(t) = 4\cos(3t)$; (d/dt) $[f(t)\mathbf{F}(t)]$

2. $\mathbf{F}(t) = t\mathbf{i} - 3t^2\mathbf{k}$, $\mathbf{G}(t) = \mathbf{i} + \cos(t)\mathbf{k}$; (d/dt) $[\mathbf{F}(t) \cdot \mathbf{G}(t)]$

3. $\mathbf{F}(t) = t\mathbf{i} + \mathbf{j} + 4\mathbf{k}$, $\mathbf{G}(t) = \mathbf{i} - \cos(t)\mathbf{j} + t\mathbf{k}$; $(d/dt)[\mathbf{F}(t) \times \mathbf{G}(t)]$

4. $\mathbf{F}(t) = \sinh(t)\mathbf{j} - t\mathbf{k}$, $\mathbf{G}(t) = t\mathbf{i} + t^2\mathbf{j} - t^2\mathbf{k}$; $(d/dt)[\mathbf{F}(t) \times \mathbf{G}(t)]$

5. $\mathbf{F}(t) = t\mathbf{i} - \cosh(t)\mathbf{j} + e^t\mathbf{k}$, $f(t) = 1 - 2t^3$; $(d/dt)[f(t)\mathbf{F}(t)]$

In each of Problems 6, 7, and 8, (a) write the position vector and tangent vector for the curve whose parametric equations are given, (b) find the length function $s(t)$ for the curve, (c) write the position vector as a function of s, and (d) verify by differentiation that this position vector in terms of s is a unit tangent to the curve.

6. $x = \sin(t)$, $y = \cos(t)$, $z = 45t$; $0 \leq t \leq 2\pi$

7. $x = 2t^2$, $y = 3t^2$, $z = 4t^2$; $1 \leq t \leq 3$

8. $x = y = z = t^3$; $-1 \leq t \leq 1$

9. Suppose $\mathbf{F}(t) = x(t)\mathbf{i} + y(t)\mathbf{j} + z(t)\mathbf{k}$ is the position vector for a particle moving along a curve in 3-space. Suppose that $\mathbf{F} \times \mathbf{F}' = \mathbf{O}$. Show that the particle always moves in the same direction.

10.2   Velocity and Curvature

Imagine a particle or object moving along a path C having a position vector $\mathbf{F}(t) = x(t)\mathbf{i} + y(t)\mathbf{j} + z(t)\mathbf{k}$, as t varies from a to b. We want to relate $\mathbf{F}$ to the dynamics of the particle. We will assume that the coordinate functions are twice differentiable.

Define the *velocity* $\mathbf{v}(t)$ of the particle at time t to be
$$\mathbf{v}(t) = \mathbf{F}'(t).$$
The *speed* $v(t)$ is the magnitude of the velocity:
$$v(t) = \|\,\mathbf{v}(t)\,\|.$$

Then
$$v(t) = \|\,\mathbf{F}'(t)\,\| = \frac{ds}{dt},$$
the rate of change of distance along the curve with respect to time.

The *acceleration* $\mathbf{a}(t)$ is the rate of change of the velocity with respect to time:
$$\mathbf{a}(t) = \mathbf{v}'(t) = \mathbf{F}''(t).$$

If $\mathbf{F}'(t) \neq \mathbf{O}$, then this vector is a tangent vector to C. We obtain a unit tangent vector $\mathbf{T}(t)$ by dividing $\mathbf{F}'(t)$ by its length. This leads to various expressions for the unit tangent vector to C, such as
$$\mathbf{T}(t) = \frac{1}{\|\,\mathbf{F}'(t)\,\|}\mathbf{F}'(t) = \frac{1}{ds/dt}\mathbf{F}'(t)$$
$$= \frac{1}{\|\,\mathbf{v}(t)\,\|}\mathbf{v}(t) = \frac{1}{v(t)}\mathbf{v}(t).$$

Thus, the unit tangent vector is also the velocity vector divided by the speed.

The *curvature* $\kappa(s)$ of C is defined as the rate of change of the unit tangent with respect to arc length along C:
$$\kappa(s) = \left\|\frac{d\mathbf{T}}{ds}\right\|.$$

This definition is motivated by Figure 10.4, which suggests that the more a curve bends at a point, the faster the unit tangent vector is changing direction there. This expression for the curvature, however, is difficult to work with, because we usually have the unit tangent vector as

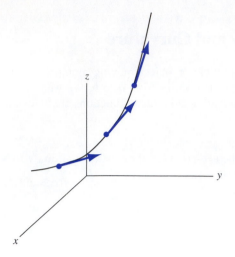

FIGURE 10.4   *Curvature as a rate of change of the unit tangent vector.*

a function of t, not of s. We therefore usually compute the curvature as a function of t by using the chain rule:

$$\kappa(t) = \left\| \frac{d\mathbf{T}}{dt} \frac{dt}{ds} \right\|$$

$$= \frac{1}{\| \mathbf{F}'(t) \|} \| \mathbf{T}'(t) \|.$$

EXAMPLE 10.3

Let C have a position vector

$$\mathbf{F}(t) = [\cos(t) + t \sin(t)]\mathbf{i} + [\sin(t) - t \cos(t)]\mathbf{j} + t^2\mathbf{k}.$$

for $t \geq 0$. Figure 10.5 is part of the graph of C. A tangent vector is given by

$$\mathbf{F}'(t) = t \cos(t)\mathbf{i} + t \sin(t)\mathbf{j} + 2t\mathbf{k}.$$

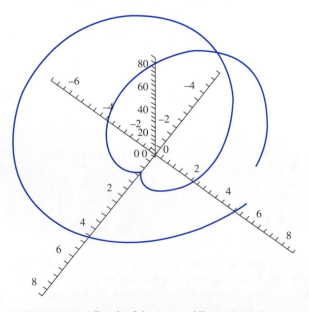

FIGURE 10.5   *Graph of the curve of Example 10.3.*

Since $\| \mathbf{F}'(t) \| = \sqrt{5}t$, the unit tangent vector is

$$\mathbf{T}(t) = \frac{1}{\| \mathbf{F}'(t) \|} \mathbf{F}'(t) = \frac{1}{\sqrt{5}}[\cos(t)\mathbf{i} + \sin(t)\mathbf{j} + 2\mathbf{k}].$$

Then

$$\mathbf{T}'(t) = \frac{1}{\sqrt{5}}[-\sin(t)\mathbf{i} + \cos(t)\mathbf{j}]$$

and the curvature of C is

$$\kappa(t) = \frac{1}{\| \mathbf{F}'(t) \|} \| \mathbf{T}'(t) \|$$

$$= \frac{1}{\sqrt{5}t} \sqrt{\frac{1}{5}[\sin^2(t) + \cos^2(t)]} = \frac{1}{5t}$$

for $t > 0$. ◆

SECTION 10.2   *PROBLEMS*

In each of Problems 1 through 8, a position vector is given. Determine the velocity, speed, acceleration, unit tangent vector, and curvature.

1. $\mathbf{F} = 3t\mathbf{i} - 2\mathbf{j} + t^2\mathbf{k}$

2. $\mathbf{F} = t\sin(t)\mathbf{i} + t\cos(t)\mathbf{j} + \mathbf{k}$

3. $\mathbf{F} = 2t\mathbf{i} - 2t\mathbf{j} + t\mathbf{k}$

4. $\mathbf{F} = e^t \sin(t)\mathbf{i} - \mathbf{j} + e^t \cos(t)\mathbf{k}$

5. $\mathbf{F} = 3e^{-t}(\mathbf{i} + \mathbf{j} - 2\mathbf{k})$

6. $\mathbf{F} = \alpha \cos(t)\mathbf{i} + \beta t\mathbf{j} + \alpha \sin(t)\mathbf{k}$

7. $\mathbf{F} = 2\sinh(t)\mathbf{j} - 2\cosh(t)\mathbf{k}$

8. $\mathbf{F} = \ln(t)(\mathbf{i} - \mathbf{j} + 2\mathbf{k})$

9. Show that any straight line has a curvature of zero. *Hint*: Any straight line can be parametrized as $x = a + bt$, $y = c + dt$, and $z = k + ht$.

10. Show that the curvature of a circle is constant. *Hint*: If the radius is r, then the curvature is $1/r$.

11. Define a vector $\mathbf{N}(s) = (1/\kappa(s))\mathbf{T}'(s)$, using the arc length along the curve C defined by $\mathbf{F}(t)$ as parameter. Prove that $\mathbf{N}$ is a unit vector that is orthogonal to the tangent to C. *Hint*: First show that $\mathbf{N}$ is a unit vector. For the orthogonality, differentiate the equation $\| \mathbf{T} \|^2 = 1$ to show that the tangent vector $\mathbf{T}$ is orthogonal to $\mathbf{T}'$.

10.3   Vector Fields and Streamlines

Vector functions $\mathbf{F}(x, y)$ in two variables and $\mathbf{G}(x, y, z)$ in three variables are called *vector fields*. At each point where the vector field is defined, we can draw an arrow representing the vector at that point. This suggests fields of arrows "growing" out of points in regions of the plane or 3-space.

Take partial derivatives of vector fields by differentiating each component. For example, if

$$\mathbf{F}(x, y, z) = \cos(x + y^2 + 2z)\mathbf{i} - xyz\mathbf{j} + xye^z\mathbf{k},$$

then

$$\frac{\partial \mathbf{F}}{\partial x} = -\sin(x + y^2 + 2z)\mathbf{i} - yz\mathbf{j} + ye^z\mathbf{k},$$

$$\frac{\partial \mathbf{F}}{\partial y} = -2y\sin(x + y^2 + 2z)\mathbf{i} - xz\mathbf{j} + xe^z\mathbf{k},$$

and

$$\frac{\partial \mathbf{F}}{\partial z} = -2\sin(x + y^2 + 2z)\mathbf{i} - xy\mathbf{j} + xye^z\mathbf{k}.$$

Given a vector field $\mathbf{F}$ in 3-space, a *streamline* of $\mathbf{F}$ is a curve with the property that, at each point (x, y, z) of the curve, $\mathbf{F}(x, y, z)$ is a tangent vector to the curve.

If $\mathbf{F}$ is the velocity field for a fluid flowing through some region, then the streamlines are called *flow lines* of the fluid. These describe trajectories of imaginary particles moving with the fluid. If $\mathbf{F}$ is a magnetic field, the streamlines are called *lines of force*. Iron filings put on a piece of cardboard held over a magnet will align themselves along these lines of force.

Given a vector field, we would like to find all of the streamlines. This is the problem of constructing a curve through each point of a region of space given the tangent to the curve at each point. To solve this problem, suppose that C is a streamline of $\mathbf{F} = f\mathbf{i} + g\mathbf{j} + h\mathbf{k}$. Let C have parametric equations $x = x(\xi)$, $y = y(\xi)$, and $z = z(\xi)$. The position vector for C is

$$\mathbf{R}(\xi) = x(\xi)\mathbf{i} + y(\xi)\mathbf{j} + z(\xi)\mathbf{k}.$$

Now

$$\mathbf{R}'(\xi) = x'(\xi)\mathbf{i} + y'(\xi)\mathbf{j} + z'(\xi)\mathbf{k}$$

is tangent to C at $(x(\xi), y(\xi), z(\xi))$ and is therefore parallel to the tangent vector $\mathbf{F}(x(\xi), y(\xi), z(\xi))$ at this point. These vectors must therefore be scalar multiples of each other, say

$$\mathbf{R}'(\xi) = t\mathbf{F}(x(\xi), y(\xi), z(\xi)).$$

Then

$$\frac{dx}{d\xi}\mathbf{i} + \frac{dy}{d\xi}\mathbf{j} + \frac{dz}{d\xi}\mathbf{k} = tf(x(\xi), y(\xi), z(\xi))\mathbf{i} + tg(x(\xi), y(\xi), z(\xi))\mathbf{j} + th(x(\xi), y(\xi), z(\xi))\mathbf{k}.$$

Equating respective components in this equation gives us

$$\frac{dx}{d\xi} = tf, \ \frac{dy}{d\xi} = tg \text{ and } \frac{dz}{d\xi} = th.$$

This is a system of differential equations for the parametric equations of the streamlines. If f, g, and h are nonzero, this system can be written as

$$\frac{dx}{f} = \frac{dy}{g} = \frac{dz}{h}.$$

EXAMPLE 10.4

We will find the streamlines of $\mathbf{F}(x, y, z) = x^2\mathbf{i} + 2y\mathbf{j} - \mathbf{k}$. If x and y are not zero, the streamlines satisfy

$$\frac{dx}{x^2} = \frac{dy}{2y} = \frac{dz}{-1}.$$

These differential equations can be solved in pairs. First integrate

$$\frac{dx}{x^2} = -dz.$$

to get

$$-\frac{1}{x} = -z + c$$

with c as an arbitrary constant. Next integrate

$$\frac{dy}{2y} = -dz$$

to get

$$\frac{1}{2}\ln|y| = -z + k.$$

It is convenient to express two of the variables in terms of the third. If we write x and y in terms of z, we have

$$x = \frac{1}{z - c} \text{ and } y = ae^{-2z},$$

in which a is constant. This gives us parametric equations of the streamlines with z as the parameter. If we want the streamline through a particular point, we must choose a and c accordingly. For example, suppose we want the streamline through $(-1, 6, 2)$. Then $z = 2$, and we need

$$-1 = \frac{1}{2 - c} \text{ and } 6 = ae^{-4}.$$

Then $c = 3$ and $a = 6e^4$, so the streamline through $(-1, 6, 2)$ has parametric equations

$$x = \frac{1}{z - 3}, \quad y = 6e^{4-2z}, \text{ and } z = z. \quad \blacklozenge$$

SECTION 10.3   PROBLEMS

In each of Problems 1 through 6, find the streamlines of the vector field and also the streamline through the given point.

1. $\mathbf{F} = \mathbf{i} - y^2\mathbf{j} + z\mathbf{k}$; $(2, 1, 1)$

2. $\mathbf{F} = \mathbf{i} - 2\mathbf{j} + \mathbf{k}$; $(0, 1, 1)$

3. $\mathbf{F} = (1/x)\mathbf{i} + e^x\mathbf{j} - \mathbf{k}$; $(2, 0, 4)$

4. $\mathbf{F} = \cos(y)\mathbf{i} + \sin(x)\mathbf{j}$; $(\pi/2, 0, -4)$

5. $\mathbf{F} = 2e^z\mathbf{i} - \cos(y)\mathbf{k}$; $(3, \pi/4, 0)$

6. $\mathbf{F} = 3x^2\mathbf{i} - y\mathbf{j} + z^3\mathbf{k}$; $(2, 1, 6)$

7. Construct a vector field whose streamlines are circles about the origin.

10.4   The Gradient Field

Let $\varphi(x, y, z)$ be a real-valued function of three variables. In the context of vector fields, φ is called a *scalar field*. The *gradient* of φ is the vector field

$$\nabla\varphi = \frac{\partial\varphi}{\partial x}\mathbf{i} + \frac{\partial\varphi}{\partial y}\mathbf{j} + \frac{\partial\varphi}{\partial z}\mathbf{k}.$$

The symbol $\nabla\varphi$ is read "del φ" and ∇ is called the *del operator*. If φ is a function of just (x, y), then $\nabla\varphi$ is a vector field in the plane.

For example, if $\varphi(x, y, z) = x^2 y \cos(yz)$, then

$$\nabla\varphi = 2xy \cos(yz)\mathbf{i} + [x^2 \cos(yz) - x^2 z \sin(yz)]\mathbf{j} - x^2 y^2 \sin(yz)\mathbf{k}.$$

If P is a point, then the gradient of φ evaluated at P is denoted $\nabla\varphi(P)$.

The gradient has the obvious properties

$$\nabla(\varphi + \psi) = \nabla(\varphi) + \nabla(\psi)$$

and, for any number c,

$$\nabla(c\varphi) = c\nabla(\varphi).$$

The gradient is related to the directional derivative. Let $P_0 : (x_0, y_0, z_0)$ be a point, and let $\mathbf{u} = a\mathbf{i} + b\mathbf{j} + c\mathbf{k}$ be a unit vector represented as an arrow from P_0. We want to measure the rate of change of $\varphi(x, y, z)$ as (x, y, z) varies from P_0 in the direction of $\mathbf{u}$. To do this, let $t > 0$. The point $P : (x_0 + at, y_0 + bt, z_0 + ct)$ is on the line through P_0 in the direction of $\mathbf{u}$, and P varies in this direction as t varies.

We measure the rate of change $D_{\mathbf{u}}\varphi(P_0)$ of $\varphi(x, y, z)$ in the direction of $\mathbf{u}$, at P_0, by setting

$$D_{\mathbf{u}}\varphi(P_0) = \frac{d}{dt}\,\varphi(x_0 + at, y_0 + bt, z_0 + ct)\Big]_{t=0}.$$

$D_{\mathbf{u}}\varphi(P_0)$ is the *directional derivative* of φ at P_0 in the direction of $\mathbf{u}$.

We can compute a directional derivative in terms of the gradient as follows. By the chain rule,

$$\begin{aligned}
D_{\mathbf{u}}\varphi(P_0) &= \frac{d}{dt}\varphi(x_0 + at, y_0 + bt, z_0 + ct)\bigg]_{t=0} \\
&= a\frac{\partial\varphi}{\partial x}(x_0, y_0, z_0) + b\frac{\partial\varphi}{\partial y}(x_0, y_0, z_0) + c\frac{\partial\varphi}{\partial z}(x_0, y_0, z_0) \\
&= a\frac{\partial\varphi}{\partial x}(P_0) + b\frac{\partial\varphi}{\partial y}(P_0) + c\frac{\partial\varphi}{\partial z}(P_0) \\
&= \nabla\varphi(P_0) \cdot (a\mathbf{i} + b\mathbf{j} + c\mathbf{k}) \\
&= \nabla\varphi(P_0) \cdot \mathbf{u}.
\end{aligned}$$

Therefore, $D_{\mathbf{u}}\varphi(P_0)$ is the dot product of the gradient of φ at the point with the unit vector specifying the direction.

EXAMPLE 10.5

Let $\varphi(x, y, z) = x^2 y - xe^z$ and $P_0 = (2, -1, \pi)$. We will compute the rate of change of $\varphi(x, y, z)$ at P_0 in the direction of $\mathbf{u} = (1/\sqrt{6})(\mathbf{i} - 2\mathbf{j} + \mathbf{k})$.

The gradient is

$$\nabla\varphi = (2xy - e^z)\mathbf{i} + x^2\mathbf{j} - xe^z\mathbf{k}.$$

Then

$$\nabla\varphi(2, -1, \pi) = (-4 - e^\pi)\mathbf{i} + 4\mathbf{j} - 2e^\pi\mathbf{k}.$$

The directional derivative of φ at P_0 in the direction of $\mathbf{u}$ is

$$D_{\mathbf{u}}\varphi(2, -1, \pi) = ((-4 - e^\pi)\mathbf{i} + 4\mathbf{j} - 2e^\pi\mathbf{k}) \cdot \frac{1}{\sqrt{6}}(\mathbf{i} - 2\mathbf{j} + \mathbf{k})$$

$$= \frac{1}{\sqrt{6}}(-4 - e^\pi - 8 - 2e^\pi)$$

$$= \frac{-3}{\sqrt{6}}(4 + e^\pi). \quad \blacklozenge$$

If a direction is specified by a vector that is not of length 1, divide it by its length before computing the directional derivative.

Now imagine standing at P_0 and observing $\varphi(x, y, z)$ as (x, y, z) moves away from P_0. In what direction will $\varphi(x, y, z)$ increase at the greatest rate? We claim that this is the direction of the gradient of φ at P_0. $\quad \blacklozenge$

THEOREM 10.1

Let φ and its first partial derivatives be continuous in some sphere about P_0, and suppose that $\nabla\varphi(P_0) \neq \mathbf{O}$. Then

1. At P_0, $\varphi(x, y, z)$ has its maximum rate of change in the direction of $\nabla\varphi(P_0)$. This maximum rate of change is $\| \nabla\varphi(P_0) \|$.

2. At P_0, $\varphi(x, y, z)$ has its minimum rate of change in the direction of $-\nabla\varphi(P_0)$. This minimum rate of change is $- \| \nabla\varphi(P_0) \|$.

Proof For position 1, let $\mathbf{u}$ be any unit vector from P_0 and consider

$$D_{\mathbf{u}}\varphi(P_0) = \nabla\varphi(P_0) \cdot \mathbf{u}$$

$$= \| \nabla\varphi(P_0) \| \| \mathbf{u} \| \cos(\theta)$$

$$= \| \nabla\varphi(P_0) \| \cos(\theta)$$

where θ is the angle between $\mathbf{u}$ and $\nabla\varphi(P_0)$. Clearly, $D_{\mathbf{u}}\varphi(P_0)$ has its maximum when $\cos(\theta) = 1$, which occurs when $\theta = 0$ and hence when $\mathbf{u}$ is in the same direction as $\nabla\varphi(P_0)$.

For position 2, $D_{\mathbf{u}}\varphi(P_0)$ has its minimum when $\cos(\theta) = -1$ and hence when $\theta = \pi$ and $\nabla\varphi(P_0)$ is opposite $\mathbf{u}$. ◆

EXAMPLE 10.6

Let $\varphi(x, y, z) = 2xz + z^2 e^y$ and $P_0 : (2, 1, 1)$. The gradient of φ is

$$\nabla\varphi(x, y, z) = 2z\mathbf{i} + z^2 e^y \mathbf{j} + (2x + 2ze^y)\mathbf{k},$$

so

$$\nabla\varphi(2, 1, 1) = 2\mathbf{i} + e\mathbf{j} + (4 + 2e)\mathbf{k}.$$

The maximum rate of increase of $\varphi(x, y, z)$ at $(2, 1, 1)$ is in the direction of $2\mathbf{i} + e\mathbf{j} + (4 + 2e)\mathbf{k}$, and this maximum rate of change is

$$\sqrt{4 + e^2 + (4 + 2e)^2} \text{ or } \sqrt{20 + 16e + 5e^2}. \quad ◆$$

10.4.1   Level Surfaces, Tangent Planes, and Normal Lines

Depending on φ and the number k, the locus of points (x, y, z) such that $\varphi(x, y, z) = k$ may be a surface in 3-space. Any such surface is called a *level surface* of φ. For example, if $\varphi(x, y, z) = x^2 + y^2 + z^2$, then the level surface of $\varphi(x, y, z) = k$ is a sphere of radius $\sqrt{k}$ if $k > 0$, a single point $(0, 0, 0)$ if $k = 0$, and vacuous if $k < 0$. Part of the level surface $\psi(x, y, z) = z - \sin(xy) = 0$ is shown in Figure 10.6.

Suppose $P_0 : (x_0, y_0, z_0)$ is on a level surface S given by $\varphi(x, y, z) = k$. Assume that there are smooth (having continuous tangents) curves on the surface passing through P_0, as typified by C in Figure 10.7. Each such curve has a tangent vector at P_0. The plane containing these tangent vectors is called the *tangent plane* to S at P_0. A vector orthogonal to this tangent plane at P_0 is called a *normal vector*, or *normal*, to S at P_0. We will determine this tangent plane and normal vector. The key lies in the following fact about the gradient vector.

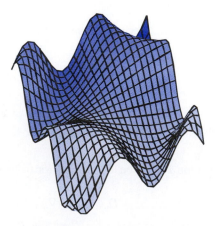

FIGURE 10.6   *Part of the graph of the level surface $z = \sin(xy)$.*

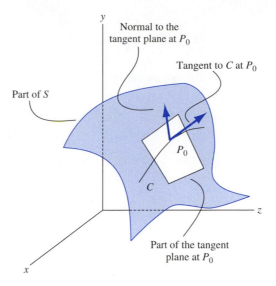

FIGURE 10.7 *Normal to a level surface.*

THEOREM 10.2 *Normal to a Level Surface*

Let φ and its first partial derivatives be continuous. Then $\nabla\varphi(P)$ is normal to the level surface $\varphi(x, y, z) = k$ at any point P on this surface such that $\nabla\varphi(P) \neq \mathbf{O}$. ◆

To understand this conclusion, let P_0 be on the level surface S and suppose a smooth curve C on the surface passes through P_0, as in Figure 10.7. Let C have parametric equations $x = x(t)$, $y = y(t)$, and $z = z(t)$ for $a \leq t \leq b$. Since P_0 is on C, for some t_0,

$$x(t_0) = x_0, \ \ y(t_0) = y_0, \ \text{and } z(t_0) = z_0.$$

Further, because C lies on the level surface,

$$\varphi(x(t), y(t), z(t)) = k$$

for $a \leq t \leq b$. Then

$$\frac{d}{dt}\varphi(x(t), y(t), z(t)) = 0 = \frac{\partial\varphi}{\partial x}x'(t) + \frac{\partial\varphi}{\partial y}y'(t) + \frac{\partial\varphi}{\partial z}z'(t)$$

$$= \nabla\varphi \cdot [x'(t)\mathbf{i} + y'(t)\mathbf{j} + z'(t)\mathbf{k}].$$

But $x'(t)\mathbf{i} + y'(t)\mathbf{j} + z'(t)\mathbf{k} = \mathbf{T}(t)$ is a tangent vector to C. Letting $t = t_0$, $\mathbf{T}(t_0)$ is tangent to C at P_0, and the last equation tells us that

$$\nabla\varphi(P_0) \cdot \mathbf{T}(t_0) = 0.$$

Therefore $\nabla\varphi(P_0)$ is normal to the tangent to C at P_0. But C is *any* smooth curve on S passing through P_0. Therefore $\nabla\varphi(P_0)$ is normal to every tangent vector at P_0 to any curve on S through P_0 and is therefore normal to the tangent plane to S at P_0.

Now we have a point P_0 on the tangent plane at P_0 and a vector $\nabla\varphi(P_0)$ orthogonal to this plane. The equation of the tangent plane is therefore

$$\nabla\varphi(P_0) \cdot [(x - x_0)\mathbf{i} + (y - y_0)\mathbf{j} + (z - z)_0\mathbf{k}] = 0$$

or

$$\frac{\partial \varphi}{\partial x}(P_0)(x - x_0) + \frac{\partial \varphi}{\partial y}(P_0)(y - y_0) + \frac{\partial \varphi}{\partial z}(P_0)(z - z_0) = 0. \tag{10.1}$$

A straight line through P_0 and parallel to the normal vector is called the *normal line* to S at P_0. Since the gradient of φ at P_0 is a normal vector, if (x, y, z) is on this normal line, then for some scalar t,

$$(x - x_0)\mathbf{i} + (y - y_0)\mathbf{j} + (z - z_0)\mathbf{k} = t\nabla\varphi(P_0).$$

The parametric equations of the normal line to S at P_0 are

$$x = x_0 + t\frac{\partial \varphi}{\partial x}(P_0), \quad y = y_0 + t\frac{\partial \varphi}{\partial y}(P_0), \quad \text{and } z = z_0 + t\frac{\partial \varphi}{\partial z}(P_0). \tag{10.2}$$

EXAMPLE 10.7

The level surface $\varphi(x, y, z) = z - \sqrt{x^2 + y^2}$ is a cone with its vertex at the origin (Figure 10.8). Compute

$$\nabla\varphi(1, 1, \sqrt{2}) = -\frac{1}{\sqrt{2}}\mathbf{i} - \frac{1}{\sqrt{2}}\mathbf{j} + \mathbf{k}.$$

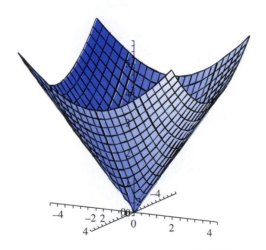

FIGURE 10.8   *Circular cone $z = \sqrt{x^2 + y^2}$.*

The tangent plane to the cone at $(1, 1, \sqrt{2})$ has the equation

$$-\frac{1}{\sqrt{2}}(x - 1) - \frac{1}{\sqrt{2}}(y - 1) + z - \sqrt{2} = 0,$$

or

$$x + y - \sqrt{2}z = 0.$$

The normal line to the cone at $(1, 1, \sqrt{2})$ has parametric equations

$$x = 1 - \frac{1}{\sqrt{2}}t, \quad y = 1 - \frac{1}{\sqrt{2}}t, \quad \text{and } z = \sqrt{2} + t. \quad \blacklozenge$$

In each of Problems 1 through 4, compute the gradient of the function and evaluate this gradient at the given point. Determine at this point the maximum and minimum rate of change of the function at this point.

1. $\varphi(x, y, z) = xyz$; $(1, 1, 1)$

2. $\varphi(x, y, z) = x^2 y - \sin(xz)$; $(1, -1, \pi/4)$

3. $\varphi(x, y, z) = 2xy + xe^z$; $(-2, 1, 6)$

4. $\varphi(x, y, z) = \cos(xyz)$; $(-1, 1, \pi/2)$

In each of Problems 5 through 8, compute the directional derivative of the function in the direction of the given vector.

5. $\varphi(x, y, z) = 8xy^2 - xz$; $(1/\sqrt{3})(\mathbf{i} + \mathbf{j} + \mathbf{k})$

6. $\varphi(x, y, z) = \cos(x - y) + e^z$; $\mathbf{i} - \mathbf{j} + 2\mathbf{k}$

7. $\varphi(x, y, z) = x^2 yz^3$; $2\mathbf{j} + \mathbf{k}$

8. $\varphi(x, y, z) = yz + xz + xy$; $\mathbf{i} - 4\mathbf{k}$

In each of Problems 9 through 14, find the equations of the tangent plane and normal line to the level surface at the point.

9. $x^2 + y^2 + z^2 = 4$; $(1, 1, \sqrt{2})$

10. $z = x^2 + y$; $(-1, 1, 2)$

11. $z^2 = x^2 - y^2$; $(1, 1, 0)$

12. $x^2 - y^2 + z^2 = 0$; $(1, 1, 0)$

13. $2x - \cos(xyz) = 3$; $(1, \pi, 1)$

14. $3x^4 + 3y^4 + 6z^4 = 12$; $(1, 1, 1)$

15. Suppose that $\nabla \varphi(x, y, z) = \mathbf{i} + \mathbf{k}$. What can be said about level surfaces of φ? Show that the streamlines of $\nabla \varphi$ are orthogonal to the level surfaces of φ.

10.5 Divergence and Curl

The gradient operator produces a vector field from a scalar function. We will discuss two other important vector operations. One produces a scalar field from a vector field, and the other produces a vector field from a vector field. Let

$$\mathbf{F}(x, y, z) = f(x, y, z)\mathbf{i} + g(x, y, z)\mathbf{j} + h(x, y, z)\mathbf{k}.$$

The *divergence* of $\mathbf{F}$ is the scalar field

$$\text{div } \mathbf{F} = \frac{\partial f}{\partial x} + \frac{\partial g}{\partial y} + \frac{\partial h}{\partial z}.$$

The *curl* of $\mathbf{F}$ is the vector field

$$\text{curl } \mathbf{F} = \left(\frac{\partial h}{\partial y} - \frac{\partial g}{\partial z} \right)\mathbf{i} + \left(\frac{\partial f}{\partial z} - \frac{\partial h}{\partial x} \right)\mathbf{j} + \left(\frac{\partial g}{\partial x} - \frac{\partial f}{\partial y} \right)\mathbf{k}.$$

Divergence, curl, and gradient can all be written as vector operations with the *del operator* ∇, which is a symbolic vector defined by

$$\nabla = \frac{\partial}{\partial x}\mathbf{i} + \frac{\partial}{\partial y}\mathbf{j} + \frac{\partial}{\partial z}\mathbf{k}.$$

The symbol ∇, which is called "del", is treated like a vector in carrying out calculations, and the "product" of $\partial/\partial x$, $\partial/\partial y$ and ∂z with a scalar function φ is interpreted to mean, respectively,

$\partial\varphi/\partial x$, $\partial\varphi/\partial y$, and $\partial\varphi/\partial z$. Now observe how gradient, divergence, and curl are obtained using this operator.

1. The product of the vector ∇ and the scalar function φ is the gradient of φ:

$$\nabla\varphi = \left(\frac{\partial}{\partial x}\mathbf{i} + \frac{\partial}{\partial y}\mathbf{j} + \frac{\partial}{\partial z}\mathbf{k}\right)\varphi$$

$$= \frac{\partial\varphi}{\partial x}\mathbf{i} + \frac{\partial\varphi}{\partial y}\mathbf{j} + \frac{\partial\varphi}{\partial z}\mathbf{k} = \text{gradient of } \varphi.$$

2. The dot product of ∇ and $\mathbf{F}$ is the divergence of $\mathbf{F}$:

$$\nabla\cdot\mathbf{F} = \left(\frac{\partial}{\partial x}\mathbf{i} + \frac{\partial}{\partial y}\mathbf{j} + \frac{\partial}{\partial z}\mathbf{k}\right)\cdot(f\mathbf{i} + g\mathbf{j} + h\mathbf{k})$$

$$= \frac{\partial f}{\partial x} + \frac{\partial g}{\partial y} + \frac{\partial h}{\partial z} = \text{divergence of } \mathbf{F}.$$

3. The cross product of ∇ with $\mathbf{F}$ is the curl of $\mathbf{F}$:

$$\nabla\times\mathbf{F} = \begin{vmatrix} \mathbf{i} & \mathbf{j} & \mathbf{k} \\ \partial/\partial x & \partial/\partial y & \partial/\partial z \\ f & g & h \end{vmatrix}$$

$$= \left(\frac{\partial h}{\partial y} - \frac{\partial g}{\partial z}\right)\mathbf{i} + \left(\frac{\partial f}{\partial z} - \frac{\partial h}{\partial x}\right)\mathbf{j} + \left(\frac{\partial g}{\partial x} - \frac{\partial f}{\partial y}\right)\mathbf{k} = \text{curl of } \mathbf{F}.$$

There are two relationships between gradient, divergence, and curl that are fundamental to vector analysis: the curl of a gradient is the zero vector, and the divergence of a curl is the number zero.

THEOREM 10.3

Let $\mathbf{F}$ be a continuous vector field whose components have continuous first and second partial derivatives, and let φ be a continuous scalar field with continuous first and second partial derivatives. Then

1. $\nabla\times(\nabla\varphi) = \mathbf{O}$.
2. $\nabla\cdot(\nabla\times\mathbf{F}) = 0$. ♦

Both of these identities can be verified by direct computation. For the first,

$$\nabla\times(\nabla\varphi) = \nabla\times\left(\frac{\partial\varphi}{\partial x}\mathbf{i} + \frac{\partial\varphi}{\partial y}\mathbf{j} + \frac{\partial\varphi}{\partial z}\mathbf{k}\right)$$

$$= \begin{vmatrix} \mathbf{i} & \mathbf{j} & \mathbf{k} \\ \partial/\partial x & \partial/\partial y & \partial/\partial z \\ \partial\varphi/\partial x & \partial\varphi/\partial y & \partial\varphi/\partial z \end{vmatrix}$$

$$= \left(\frac{\partial^2\varphi}{\partial y\partial z} - \frac{\partial^2\partial\varphi}{\partial z\partial y}\right)\mathbf{i} + \left(\frac{\partial^2\varphi}{\partial z\partial x} - \frac{\partial^2\partial\varphi}{\partial x\partial z}\right)\mathbf{j} + \left(\frac{\partial^2\varphi}{\partial x\partial y} - \frac{\partial^2\partial\varphi}{\partial y\partial x}\right)\mathbf{k}$$

$$= \mathbf{O}$$

because the mixed partials cancel in pairs in the components of $\nabla\times(\nabla\varphi)$.

Operator notation with ∇ can simplify such calculations. In this notation, $\nabla \times (\nabla \varphi) = \mathbf{O}$ is immediate, because $\nabla \times \nabla$ is the cross product of a "vector" with itself, which is always zero. Similarly, for 2, $\nabla \times \mathbf{F}$ is orthogonal to ∇, so its dot product with ∇ is zero.

10.5.1 A Physical Interpretation of Divergence

Suppose $\mathbf{F}(x, y, z, t)$ is the velocity of a fluid at point (x, y, z) and time t. Time plays no role in computing divergence, but is included here because a flow may depend on time. We will show that the divergence of $\mathbf{F}$ measures the outward flow of the fluid from any point.

Imagine a small rectangular box in the fluid, as in Figure 10.9. First look at the front and back faces II and I, respectively. The flux of the flow out of this box across II is the normal component of the velocity (dot product of $\mathbf{F}$ with $\mathbf{i}$) multiplied by the area of this face:

$$\text{flux outward across face II} = \mathbf{F}(x + \Delta x, y, z, t) \cdot \mathbf{i} \Delta y \Delta z$$
$$= f(x + \Delta x, y, z, t) \Delta y \Delta z.$$

On face I, the unit outer normal is $-\mathbf{i}$, so the flux outward across this face is $-f(x, y, z, t) \Delta y \Delta z$. The total outward flux across faces II and I is therefore

$$[f(x + \Delta x, y, z, t) - f(x, y, z, t)] \Delta y \Delta z.$$

A similar calculation holds for the pairs of other opposite sides. The total flux of fluid flowing out of the box across its faces is

$$\text{total flux} = [f(x + \Delta x, y, z, t) - f(x, y, z, t)] \Delta y \Delta z$$
$$+ [g(x, y + \Delta y, z, t) - g(x, y, z, t)] \Delta x \Delta z$$
$$+ [h(x, y, z + \Delta z, t) - h(x, y, z, t)] \Delta x \Delta y.$$

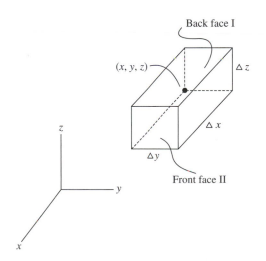

FIGURE 10.9 *Interpretation of divergence.*

The total flux per unit volume out of the box is obtained by dividing this quantity by $\Delta x \Delta y \Delta z$, obtaining

$$\text{flux per unit volume} = \frac{f(x + \Delta x, y, z, t) - f(x, y, z, t)}{\Delta x}$$

$$+ \frac{g(x, y + \Delta y, z, t) - g(x, y, z, t)}{\Delta y}$$

$$+ \frac{h(x, y, z + \Delta z, t) - h(x, y, z, t)}{\Delta z}.$$

In the limit as $(\Delta x, \Delta y, \Delta t) \to (0, 0, 0)$, this sum approaches div $\mathbf{F}(x, y, z, t)$.

10.5.2 A Physical Interpretation of Curl

The curl vector is interpreted as a measure of rotation (or swirl) about a point. In British literature, the curl is often called the rot (for rotation) of a vector.

To understand this interpretation, suppose an object rotates with uniform angular speed ω about a line L, as in Figure 10.10. The angular velocity vector $\mathbf{\Omega}$ has magnitude ω and is directed along L as a right-handed screw would progress if given the same sense of rotation as the object. Put L through the origin, and let $\mathbf{R} = x\mathbf{i} + y\mathbf{j} + z\mathbf{k}$ for any point (x, y, z) on the rotating object. Let $\mathbf{T}(x, y, z)$ be the tangential linear velocity and $R = \| \mathbf{R} \|$. Then

$$\| \mathbf{T} \| = \omega R \sin(\theta) = \| \mathbf{\Omega} \times \mathbf{R} \|$$

with θ the angle between $\mathbf{R}$ and $\mathbf{\Omega}$. Since $\mathbf{T}$ and $\mathbf{\Omega} \times \mathbf{R}$ have the same direction and magnitude, we conclude that $\mathbf{T} = \mathbf{\Omega} \times \mathbf{R}$. Now write $\mathbf{\Omega} = a\mathbf{i} + b\mathbf{j} + c\mathbf{k}$ to obtain

$$\mathbf{T} = \mathbf{\Omega} \times \mathbf{R} = (bz - cy)\mathbf{i} + (cx - az)\mathbf{j} + (ay - bx)\mathbf{k}.$$

Then

$$\nabla \times \mathbf{T} = \begin{vmatrix} \mathbf{i} & \mathbf{j} & \mathbf{k} \\ \partial/\partial z & \partial/\partial y & \partial/\partial z \\ bz - cy & cx - az & ay - bx \end{vmatrix}$$

$$= 2a\mathbf{i} + 2b\mathbf{j} + 2c\mathbf{k} = 2\mathbf{\Omega}.$$

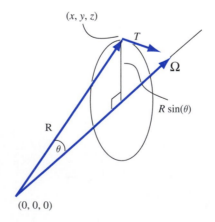

FIGURE 10.10 *Interpretation of curl.*

Therefore,

$$\mathbf{\Omega} = \frac{1}{2}\nabla \times \mathbf{T}.$$

The angular momentum of a uniformly rotating body is a constant times the curl of the linear velocity.

SECTION 10.5   *PROBLEMS*

In each of Problems 1 through 5, compute $\nabla \cdot \mathbf{F}$ and $\nabla \times \mathbf{F}$ and verify explicitly that $\nabla \cdot (\nabla \times \mathbf{F}) = 0$.

1. $\mathbf{F} = x\mathbf{i} + y\mathbf{j} + 2z\mathbf{k}$

2. $\mathbf{F} = \sinh(xyz)\mathbf{j}$

3. $\mathbf{F} = 2xy\mathbf{i} + xe^y\mathbf{j} + 2z\mathbf{k}$

4. $\mathbf{F} = x\mathbf{i} + y\mathbf{j} + 2z\mathbf{k}$

5. $\mathbf{F} = \sinh(x)\mathbf{i} + \cosh(xyz)\mathbf{j} - (x + y + z)\mathbf{k}$

In each of Problems 6 through 10, compute $\nabla\varphi$ and verify explicitly that $\nabla \times (\nabla\varphi) = \mathbf{O}$.

6. $\varphi(x, y, z) = 18xyz + e^x$

7. $\varphi(x, y, z) = x - y + 2z^2$

8. $\varphi(x, y, z) = \sin(xz)$

9. $\varphi(x, y, z) = -2x^3yz^2$

10. $\varphi(x, y, z) = x\cos(x + y + z)$

11. Let φ be a scalar field and $\mathbf{F}$ a vector field. Derive expressions for $\nabla \cdot (\varphi\mathbf{F})$ and $\nabla \times (\varphi\mathbf{F})$ in terms of operations applied to $\varphi(x, y, z)$ and to $\mathbf{F}(x, y, z)$.

CHAPTER 11

Vector Integral Calculus

11.1 Line Integrals

For line integrals, we need some preliminary observations about curves. Suppose a curve C has parametric equations

$$x = x(t), \ y = y(t), \ \text{and } z = z(t) \quad \text{for } a \leq t \leq b.$$

These are the *coordinate functions* of C. It is convenient to think of t as time and C as the trajectory of an object, which at time t is at $C(t) = (x(t), y(t), z(t))$. C has an orientation, since the object starts at the *initial point* $(x(a), y(a), z(a))$ at time $t = a$ and ends at the *terminal point* $(x(b), y(b), z(b))$ at time $t = b$. We often indicate this orientation by putting arrows along the graph (as in Figures 11.2 and 11.3).

We call C:

- *continuous* if each coordinate function is continuous;
- *differentiable* if each coordinate function is differentiable;
- *closed* if the initial and terminal points coincide:

$$(x(a), y(a), z(a)) = (x(b), y(b), z(b));$$

- *simple* if $a < t_1 < t_2 < b$ implies that

$$(x(t_1), y)t_1), z(t_1)) \neq (x(t_2), y(t_2), z(t_2));$$

and

- *smooth* if the coordinate functions have continuous derivatives which are never all zero for the same value of t.

If C is smooth and we let $\mathbf{R}(t) = x(t)\mathbf{i} + y(t)\mathbf{i} + z(t)\mathbf{k}$ be the position function for C, then $\mathbf{R}'(t)$ is a continuous tangent vector to C. Smoothness of C means that the curve has a continuous tangent vector as we move along it.

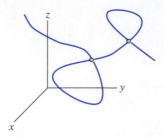

FIGURE 11.1   *A   nonsimple curve.*

A curve is *simple* if it does not intersect itself at different times. The curve whose graph is shown in Figure 11.1 is not simple. A closed curve has the same initial terminal points but is still called simple if it does not pass through any other point more than once.

We must be careful to distinguish between a curve and its graph, although informally we often use these terms interchangeably. The graph is a drawing, while the curve carries with it a sense of orientation from an initial to a terminal point. The graph of a curve does not carry all of this information.

EXAMPLE 11.1

Let C have coordinate functions

$$x = 4\cos(t), \quad y = 4\sin(t), \quad \text{and } z = 9 \quad \text{for } 0 \leq t \leq 2\pi.$$

The graph of C is a circle of radius 4 about the origin in the plane $z = 9$. C is simple, closed and smooth.

Let K be given by

$$x = 4\cos(t), \quad y = 4\sin(t), \quad \text{and } z = 9 \quad \text{for } 0 \leq t \leq 4\pi.$$

The graph of K is the same as the graph of C, except that a particle traversing K goes around this circle twice. This fact is not displayed in the graph alone.

Let L be the curve given by

$$x(t) = 4\cos(t), \quad y = 4\sin(t), \quad \text{and } z = 9 \quad \text{for } 0 \leq t \leq 3\pi.$$

The graph of L is again the circle of radius 4 about the origin in the plane $z = 9$. L is smooth and not simple, but L is also not closed, since the initial point is $(4, 0, 9)$ and the terminal point is $(-4, 0, 9)$. A particle moving along L traverses the complete circle from $(4, 0, 9)$ to $(4, 0, 9)$ and then continues on to $(-4, 0, 9)$, where it stops. Again, this behavior is not clear from the graph. ◆

Line Integral

We are now ready to define the line integral, which is an integral over a curve. Suppose C is a smooth curve with coordinate functions $x = x(t)$, $y = y(t)$, and $z = z(t)$ for $a \leq t \leq b$. Let f, g, and h be continuous at least at points on the graph of C. Then the *line integral* $\int_C f\,dx + g\,dy + h\,dz$ is defined by

$$\int_C f\,dx + g\,dy + h\,dz$$

$$= \int_a^b \left[f(x(t), y(t), z(t)) \frac{dx}{dt} + g(x(t), y(t), z(t)) \frac{dy}{dt} + h(x(t), y(t), z(t)) \frac{dx}{dt} \right] dt.$$

$\int_C f\,dx + g\,dy + h\,dz$ is a number obtained by replacing x, y, and z in $f(x, y, z)$, $g(x, y, z)$, and $h(x, y, z)$ with the coordinate functions $x(t)$, $y(t)$, and $z(t)$ of C, replacing

$$dx = x'(t)\,dt, \quad dy = y'(t)\,dt, \quad \text{and } dz = z'(t)\,dt$$

and integrating the resulting function of t from a to b.

EXAMPLE 11.2

We will evaluate $\int_C x\,dx - yz\,dy + e^z\,dz$ if C is the curve with coordinate functions

$$x = t^3, \quad y = -t, \text{ and } z = t^2 \quad \text{for } 1 \le t \le 2.$$

First,

$$dx = 3t^2\,dt, \quad dy = -dt, \text{ and } dz = 2t\,dt.$$

Put the coordinate functions of C into x, $-yz$, and e^z to obtain

$$\int_C x\,dx - yz\,dy + e^z\,dz$$

$$= \int_1^2 \left[t^3(3t^2) - (-t)(t^2)(-1) + e^{t^2}(2t) \right] dt$$

$$= \int_1^2 [3t^5 - t^3 + 2t e^{t^2}]\,dt$$

$$= \frac{111}{4} + e^4 - e. \quad \blacklozenge$$

EXAMPLE 11.3

Evaluate $\int_C xyz\,dx - \cos(yz)\,dy + xz\,dz$ along the straight line segment L from $(1, 1, 1)$ to $(-2, 1, 3)$.

Parametric equations of L are

$$x = 1 - 3t, \quad y = 1, \text{ and } z = 1 + 2t \quad \text{for } 0 \le t \le 1.$$

Then

$$dx = -3\,dt, \quad dy = 0, \text{ and } dz = 2\,dt.$$

The line integral is

$$\int_C xyz\,dx - \cos(yz)\,dy + xz\,dz$$

$$= \int_0^1 [(1-3t)(1+3t)(-3) - \cos(1+2t)(0) + (1-3t)(1+2t)(2)]\,dt$$

$$= \int_0^1 (-1+t+6t^2)\,dt = \frac{3}{2}. \quad \blacklozenge$$

We have a line integral in the plane if C is in the plane and the functions involve only x and y.

EXAMPLE 11.4

Evaluate $\int_K xy\,dx - y\sin(x)\,dy$ if K has coordinate functions $x = t^2$ and $y = t$ for $-1 \le t \le 2$. Here

$$dx = 2t\,dt \text{ and } dy = dt$$

so

$$\int_K xy\,dx - y\sin(x)\,dy = \int_{-1}^2 [t^2 t\,(2t) - t\sin(t^2)]\,dt$$

$$= \int_{-1}^2 [2t^4 - t\sin(t^2)]\,dt = \frac{66}{5} + \frac{1}{2}(\cos(4) - \cos(1)). \quad \blacklozenge$$

Line integrals have properties we normally expect of integrals.

1. The line integral of a sum is the sum of the line integrals:

$$\int_C (f + f^*)\,dx + (g + g^*)\,dy + (h + h^*)\,dz$$

$$= \int_C f\,dx + g\,dy + h\,dz + \int_C f^*\,dx + g^*\,dy + h^*\,dz.$$

2. Constants factor through a line integral:

$$\int_C (cf)\,dx + (cg)\,dy + (ch)\,dz = c\int_C f\,dx + g\,dy + h\,dz.$$

For definite integrals, $\int_a^b F(x)\,dx = -\int_b^a F(x)\,dx$. The analogue of this for line integrals is that reversing the direction on C changes the sign of the line integral. Suppose C is a smooth curve from P_0 to P_1. Let C have coordinate functions

$$x = x(t), y = y(t), z = z(t) \text{ for } a \le t \le b.$$

Define K as the curve with coordinate functions

$$\tilde{x} = x(a+b-t), \tilde{y} = y(a+b-t), \tilde{z} = z(a+b-t) \text{ for } a \le t \le b.$$

The graphs of C and K are the same, but the initial point of K is the terminal point of C, since

$$(\tilde{x}(a), \tilde{y}(a), \tilde{z}(a)) = (x(b), y(b), z(b)).$$

Similarly, the terminal point of K is the initial point of C.

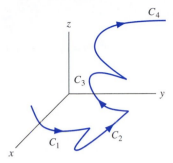

FIGURE 11.2 *A piecewise smooth curve.*

3. We denote a curve K formed from C in this way as $-C$. The effect of this reversal of orientation is to change the sign of a line integral.

$$\int_C f\,dx + g\,dy + h\,dz = -\int_{-C} f\,dx + g\,dy + h\,dz.$$

This can be proven by a simple change of variables in the integrals with respect to t defining these line integrals.

The next property of line integrals reflects the fact that

$$\int_a^b F(x)\,dx = \int_a^c F(x)\,dx + \int_c^b F(x)\,dx$$

for definite integrals. A curve C is *piecewise smooth* if it has a continuous tangent at all but finitely many points. Such a curve typically has the appearance of the graph in Figure 11.2, with a finite number of "corners" at which there may be no tangent.

We write

$$C = C_1 \oplus C_2 \oplus \cdots \oplus C_n$$

if, as in Figure 11.2, C begins with a smooth piece C_1. C_2 begins where C_1 ends, C_3 begins where C_2 ends, and so on. Each C_j is smooth, but where C_j joins with C_{j+1} there may be no tangent in the resulting curve. For C formed in this way,

$$\int_C f\,dx + g\,dy + h\,dz = \int_{C_1 \oplus C_2 \oplus \cdots \oplus C_n} f\,dx + g\,dy + h\,dz$$

$$= \sum_{j=1}^n \int_{C_j} f\,dx + g\,dy + h\,dz.$$

EXAMPLE 11.5

Let C be the curve consisting of the quarter circle $x^2 + y^2 = 1$ in the (x, y) plane from $(1, 0)$ to $(0, 1)$, followed by the horizontal line segment from $(0, 1)$ to $(2, 1)$. We will compute $\int_C dx + y^2 dy$.

Write $C = C_1 \oplus C_2$, where C_1 is the quarter circle part and C_2 the line segment part. Parametrize C_1 by $x = \cos(t)$ and $y = \sin(t)$ for $0 \le t \le \pi/2$. On C_1,

$$dx = -\sin(t)\,dt \quad \text{and} \quad dy = \cos(t)\,dt,$$

so

$$\int_{C_1} dx + y^2 \, dy = \int_0^{\pi/2} [-\sin(t) + \sin^2(t)\cos(t)] \, dt = -\frac{2}{3}.$$

Parametrize C_2 by $x = s$ and $y = 1$ for $0 \le s \le 2$. On C_2,

$$dx = ds \text{ and } dy = 0$$

so

$$\int_{C_2} dx + y^2 \, dy = \int_0^2 ds = 2.$$

Then

$$\int_C dx + y^2 \, dy = -\frac{2}{3} + 2 = \frac{4}{3}. \quad \blacklozenge$$

It is sometimes convenient to write a line integral in vector notation. Let $\mathbf{F} = f\mathbf{i} + g\mathbf{j} + h\mathbf{k}$ and form the position vector $\mathbf{R}(t) = x(t)\mathbf{i} + y(t)\mathbf{j} + z(t)\mathbf{k}$. Then

$$d\mathbf{R} = dx \, \mathbf{i} + dy \, \mathbf{j} + dz \, \mathbf{k}$$

and

$$\mathbf{F} \cdot d\mathbf{R} = f \, dx + g \, dy + h \, dz$$

suggesting the notation

$$\int_C f \, dx + g \, dy + h \, dz = \int_C \mathbf{F} \cdot d\mathbf{R}.$$

EXAMPLE 11.6

A force $\mathbf{F}(x, y, z) = x^2\mathbf{i} - zy\mathbf{j} + x\cos(z)\mathbf{k}$ moves an object along the path C given by $x = t^2$, $y = t$, and $z = \pi t$ for $0 \le t \le 3$. We want to calculate the work done by this force.

At any point on C, the particle will be moving in the direction of the tangent to C at that point. We may approximate the work done along a small segment of the curve starting at (x, y, z) by $\mathbf{F}(x, y, z) \cdot d\mathbf{R}$, with the dimensions of force times distance. The work done in moving the object along the entire path is approximated by the sum of these approximations along segments of the path. In the limit, as the lengths of these segments tend to zero, we obtain

$$\text{work} = \int_C \mathbf{F} \cdot d\mathbf{R} = \int_C x^2 \, dx - zy \, dy + x\cos(z) \, dz$$

$$= \int_0^3 [t^4(2t) - (\pi t)(t) + t^2\cos(\pi t)(\pi)] \, dt$$

$$= \int_0^3 [2t^5 - \pi t^2 + \pi t^2 \cos(\pi t)] \, dt$$

$$= 243 - 9\pi - \frac{6}{\pi}. \quad \blacklozenge$$

11.1.1 Line Integral With Respect to Arc Length

In some contexts, it is useful to have a line integral with respect to arc length along C. If $\varphi(x, y, z)$ is a scalar field and C is a smooth curve with coordinate functions $x = x(t)$, $y = y(t)$, and $z = z(t)$ for $a \leq t \leq b$, we define

$$\int_C \varphi(x, y, z)\, ds = \int_a^b \varphi(x(t), y(t), z(t))\sqrt{x'(t)^2 + y'(t)^2 + z'(t)^2}\, dt.$$

The rationale behind this definition is that

$$ds = \sqrt{x'(t)^2 + y'(t)^2 + z'(t)^2}\, dt$$

is the differential element of arc length along C.

To see how such a line integral arises, suppose C is a thin wire having density $\delta(x, y, z)$ at (x, y, z), and we want to compute the mass. Partition $[a, b]$ into n subintervals by inserting points

$$a = t_0 < t_1 < t_2 < \ldots < t_{n-1} < t_n = b$$

of length $\Delta t = (b - a)/n$, where n is a positive integer. We can make Δt as small as we want by choosing n large, so that values of $\delta(x, y, z)$ are approximated as closely as we want on $[t_{j-1}, t_j]$ by $\delta(P_j)$, where $P_j = (x(t_j), y(t_j), z(t_j))$. The length of wire between P_{j-1} and P_j is $\Delta s = s(P_j) - s(P_{j-1}) \approx ds_j$. The density of this piece of wire is approximately $\delta(P_j)ds_j$, and $\sum_{j=1}^n \delta(P_j)ds_j$ approximates the mass of the wire. In the limit, as $n \to \infty$, this gives

$$\text{mass of the wire} = \int_C \delta(x, y, z)\, ds.$$

A similar argument gives the coordinates $(\tilde{x}, \tilde{y}, \tilde{z})$ of the center of mass of the wire as

$$\tilde{x} = \frac{1}{m} \int_C x\delta(x, y, z)ds, \quad \tilde{y} = \frac{1}{m} \int_C y\delta(x, y, z)ds, \quad \tilde{z} = \frac{1}{m} \int_C z\delta(x, y, z)ds,$$

in which m is the mass.

EXAMPLE 11.7

A wire is bent into the shape of the quarter circle C given by $x = 2\cos(t)$, $y = 2\sin(t)$, and $z = 3$ for $0 \leq t \leq \pi/2$. The density function is $\delta(x, y, z) = xy^2$. We want the mass and center of mass of the wire.

The mass is

$$m = \int_C xy^2\, ds = \int_0^{\pi/2} 2\cos(t)[2\sin(t)]^2 \sqrt{4\sin^2(t) + 4\cos^2(t)}\, dt$$

$$= \int_0^{\pi/2} 16\cos(t)\sin^2(t)\, dt = \frac{16}{3}.$$

Now compute the coordinates of the center of mass. First,

$$\tilde{x} = \frac{1}{m}\int_C x\delta(x,y,z)\,ds$$

$$= \frac{3}{16}\int_0^{\pi/2}[2\cos(t)]^2[2\sin(t)]^2\sqrt{4\sin^2(t)+4\cos^2(t)}\,dt$$

$$= 6\int_0^{\pi/2}\cos^2(t)\sin^2(t)\,dt = \frac{3\pi}{8}.$$

Next,

$$\tilde{y} = \frac{1}{m}\int_C y\delta(x,y,z)\,ds$$

$$= \frac{3}{16}\int_0^{\pi/2}[2\cos(t)][2\sin(t)]^3\sqrt{4\sin^2(t)+4\cos^2(t)}\,dt$$

$$= 6\int_0^{\pi/2}\cos(t)\sin^3(t)\,dt = \frac{3}{2}$$

and

$$\tilde{z} = \frac{3}{16}\int_C zxy^2\,ds$$

$$= \frac{3}{16}\int_0^{16}3[2\cos(t)][2\sin(t)]^2\sqrt{4\sin^2(t)+4\cos^2(t)}\,dt$$

$$= 9\int_0^{\pi/2}\sin^2(t)\cos(t)\,dt = 3.$$

It should not be surprising that $\tilde{z}=3$, because the wire is in the $z=3$ plane. ◆

SECTION 11.1   PROBLEMS

In each of Problems 1 through 10, evaluate the line integral.

1. $\int_C x\,dx - dy + z\,dz$ with C given by $x=y=t, z=t^3$ for $0\le t\le 1$.

2. $\int_C -4x\,dx + y^2\,dy - yz\,dz$ with C given by $x = -t^2, y=0, z=-3t$ for $0\le t\le 1$.

3. $\int_C (x+y)\,ds$ with C given by $x=y=t, z=t^2$ for $0\le t\le 2$.

4. $\int_C x^2 z\,ds$ with C the line segment from $(0,1,1)$ to $(1,2,-1)$.

5. $\int_C \mathbf{F}\cdot d\mathbf{R}$ with $\mathbf{F}=\cos(x)\mathbf{i} - y\mathbf{j} + xz\mathbf{k}$ and $\mathbf{R}=t\mathbf{i} - t^2\mathbf{j}+\mathbf{k}$ for $0\le t\le 3$.

6. $\int_C 4xy\,ds$ with C given by $x=y=t, z=2t$ for $1\le t\le 2$.

7. $\int_C \mathbf{F}\cdot d\mathbf{R}$ with $\mathbf{F}=x\mathbf{i}+y\mathbf{j}-z\mathbf{k}$ and C the circle $x^2 + y^2=4, z=0$, going around once counterclockwise.

8. $\int_C yz\,ds$ with C the parabola $z=y^2, x=1$ for $0\le y\le 2$.

9. $\int_C -xyz\,dz$ with C given by $x=1, y=\sqrt{z}$ for $4\le z\le 9$.

10. $\int_C xz\,dy$ with C given by $x=y=t, z=-4t^2$ for $1\le t\le 3$.

11. Find the work done by $\mathbf{F}=x^2\mathbf{i} - 2yz\mathbf{j} + z\mathbf{k}$ in moving an object along the line segment from $(1,1,1)$ to $(4,4,4)$.

12. Find the mass and center of mass of a thin, straight wire extending from the origin to $(3,3,3)$ if $\delta(x,y,z)=x+y+z$ grams per centimeter.

13. Show that any Riemann integral $\int_a^b f(x)dx$ is a line integral $\int_C \mathbf{F}\cdot d\mathbf{R}$ for appropriate choices of $\mathbf{F}$ and $\mathbf{R}$.

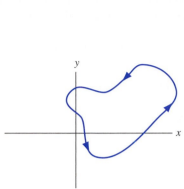

Exterior of C

FIGURE 11.3 *Orientation on a curve.*

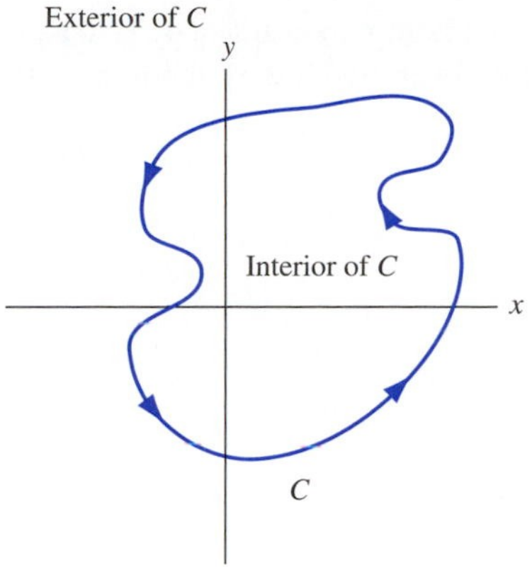

FIGURE 11.4 *Interior and exterior of a simple closed curve.*

11.2   Green's Theorem

Green's theorem is a relationship between double integrals and line integrals around closed curves in the plane. It was formulated independently by the British self-taught amateur natural philosopher George Green and the Ukrainian mathematician Michel Ostrogradsky and is used in potential theory and partial differential equations.

A closed curve C in the (x, y) plane is *positively oriented* if a point on the curve moves counterclockwise as the parameter describing C increases. If the point moves clockwise, then C is *negatively oriented*. We denote orientation by placing an arrow on the graph, as in Figure 11.3.

A simple closed curve C in the plane encloses a region, called the *interior* of C. The unbounded region that remains if the interior is cut out is the *exterior* of C (note Figure 11.4). If C is positively oriented then, as we walk around C in the positive direction, the interior is over our left shoulder.

We will use the term *path* for a piecewise smooth curve. We often denote a line integral over a closed path C as $\oint_C$ with a small oval on the integral sign. This is not obligatory and does not affect the meaning of the line integral.

THEOREM 11.1   *Green's Theorem*

Let C be a simple closed positively oriented path in the plane. Let D consist of all points on C and in its interior. Let f, g, $\partial f/\partial y$, and $\partial g/\partial x$ be continuous on D. Then

$$\oint_C f(x, y)\, dx + g(x, y)\, dy = \iint_D \left(\frac{\partial g}{\partial x} - \frac{\partial f}{\partial y} \right) dA. \quad \blacklozenge$$

EXAMPLE 11.8

Sometimes Green's theorem simplifies an integration. Suppose we want to compute the work done by $\mathbf{F}(x, y) = (y - x^2 e^x)\mathbf{i} + (\cos(2y^2) - x)\mathbf{j}$ in moving a particle counterclockwise about the rectangular path C having vertices $(0, 1)$, $(1, 1)$, $(1, 3)$, and $(0, 3)$.

If we attempt to evaluate $\int_C \mathbf{F} \cdot d\mathbf{R}$, we encounter integrals that cannot be done in elementary form. However, by Green's theorem with D the solid rectangle bounded by C is

$$\text{work} = \oint_C \mathbf{F} \cdot d\mathbf{R} = \iint_D \left(\frac{\partial}{\partial x}(\cos(2y^2) - x) - \frac{\partial}{\partial y}(y - x^2 e^x) \right) dA$$

$$= \iint_D -2 \, dA = (-2)[\text{area of } D] = -4. \quad \blacklozenge$$

EXAMPLE 11.9

Another use of Green's theorem is in deriving general results. Suppose we want to evaluate

$$\oint_C 2x \cos(2y) \, dx - 2x^2 \sin(2y) \, dy$$

for every positively oriented simple closed path C in the plane.

There are infinitely many such paths. However, $f(x, y)$ and $g(x, y)$ have the special property that

$$\frac{\partial}{\partial x}\left(-2x^2 \sin(2y)\right) - \frac{\partial}{\partial y}\left(2x \cos(2y)\right)$$

$$= -4x \sin(2y) + 4x \sin(2y) = 0.$$

By Green's theorem, for any such closed path C in the plane,

$$\oint_C 2x \cos(2y) \, dx - 2x^2 \sin(2y) \, dy = \iint_D 0 \, dA = 0. \quad \blacklozenge$$

SECTION 11.2   PROBLEMS

1. A particle moves once counterclockwise about the triangle with vertices $(0,0)$, $(4,0)$, and $(1,6)$, under the influence of the force $\mathbf{F} = xy\mathbf{i} + x\mathbf{j}$. Calculate the work done by this force.

2. A particle moves once counterclockwise around the circle of radius 6 about the origin and under the influence of the force $\mathbf{F} = (e^x - y + x \cosh(x))\mathbf{i} + (y^{3/2} + x)\mathbf{j}$. Calculate the work done.

3. A particle moves once counterclockwise about the rectangle with vertices $(1, 1)$, $(1, 7)$, $(3, 1)$, and $(3, 7)$ under the influence of the force $\mathbf{F} = (-\cosh(4x^4) + xy)\mathbf{i} + (e^{-y} + x)\mathbf{j}$. Calculate the work done.

In each of Problems 4 through 11, use Green's theorem to evaluate $\oint_C \mathbf{F} \cdot d\mathbf{R}$. All curves are oriented positively.

4. $\mathbf{F} = 2y\mathbf{i} - x\mathbf{j}$, and C is the circle of radius 4 about $(1, 3)$.

5. $\mathbf{F} = x^2\mathbf{i} - 2xy\mathbf{j}$, and C is the triangle with vertices $(1, 1)$, $(4, 1)$, and $(2, 6)$.

6. $\mathbf{F} = (x + y)\mathbf{i} + (x - y)\mathbf{j}$, and C is the ellipse $x^2 + 4y^2 = 1$.

7. $\mathbf{F} = 8xy^2\mathbf{j}$, and C is the circle of radius 4 about the origin.

8. $\mathbf{F} = (x^2 - y)\mathbf{i} + (\cos(2y) - e^{3y} + 4x)\mathbf{j}$, and C is any square with sides of length 5.

9. $\mathbf{F} = e^x \cos(y)\mathbf{i} - e^x \sin(y)\mathbf{j}$, and C is any simple closed path in the plane.

10. $\mathbf{F} = x^2 y\mathbf{i} - xy^2\mathbf{j}$, and C is the boundary of the region $x^2 + y^2 \leq 4$, $x \geq 0$, and $y \geq 0$.

11. $\mathbf{F} = xy\mathbf{i} + (xy^2 - e^{\cos(y)})\mathbf{j}$, and C is the triangle with vertices $(0, 0)$, $(3, 0)$, and $(0, 5)$.

12. Let D be the interior of a positively oriented simple closed path C.

    (a) Show that the area of D equals $\oint_C -y \, dx$.

    (b) Show that the area of D equals $\oint_C x \, dy$.

    (c) Show that the area of D equals

    $$\frac{1}{2}\oint_C -y \, dx + x \, dy.$$

11.3   An Extension of Green's Theorem

There is an extension of Green's theorem to include the case that there are finitely many points $P_1, \ldots, P_n$ enclosed by C at which f, g, $\partial f/\partial y$, and/or $\partial g/\partial x$ are not continuous, or perhaps not even defined. The idea is to excise these points by enclosing them in small disks which are thought of as cut out of D.

Enclose each P_j with a circle K_j of sufficiently small radius that no circle intersects either C or any of the other circles (Figure 11.5). Draw a channel consisting of two parallel line segments from C to K_1, then from K_1 to K_2, and so on, until the last channel is drawn from K_{n-1} to K_n. This is illustrated in Figure 11.6 for $n = 3$.

Now form the simple closed path C^* of Figure 11.7, consisting of "most of" C, "most of" each K_j, and the inserted channel lines. By "most of" C, we mean that a small arc of C and each circle between the channel cuts has been excised in forming C^*.

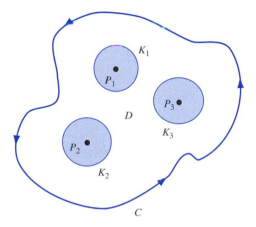

FIGURE 11.5 *Enclosing points with small circles interior to C.*

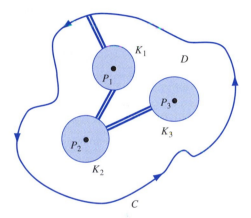

FIGURE 11.6 *Channels connecting C to K_1, K_1 to K_2, \ldots, K_{n-1} to K_n.*

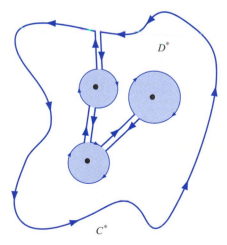

FIGURE 11.7 *The simple closed path C^* with each P_j exterior to C^*.*

Each P_j is *external* to C^*, while $f, g, \partial f/\partial y$, and $\partial g/\partial x$ are continuous on and in the interior of C^*. The orientation on C^* is also crucial. If we begin at a point of C just before the channel to K_1, we move counterclockwise on C until we reach the first channel cut, then go along this cut to K_1, then *clockwise* around part of K_1 until we reach a channel cut to K_2, and then clockwise around K_2 until we reach a cut to K_3. After going clockwise around part of K_3, we reach the other side of the cut from this circle to K_2, move clockwise around it to the cut to K_1, then clockwise around it to the cut back to C, and then continue counterclockwise around C. We are are taking the case $n = 3$ for clarity.

If D^* is the interior of C^*, then by Green's theorem,

$$\oint_{C^*} f\, dx + g\, dy = \iint_{D^*} \left(\frac{\partial g}{\partial x} - \frac{\partial f}{\partial y} \right) dA. \tag{11.1}$$

Now take a limit in Figure 11.7 as the channels are made narrower. The opposite sides of each channel merge to single line segments, which are integrated over in both directions in equation (11.1). The contributions to the sum in this equation from the channel cuts is therefore zero. Furthermore, as the channels narrow, the small arcs of C and each K_j cut out in making the channels are restored, and the line integrals in equation (11.1) are over all of C and the circles K_j. Recalling that in equation (11.1) the integrations over the $K_j's$ are clockwise, equation (11.1) can be written as

$$\oint_C f\, dx + g\, dy - \sum_{j=1}^{n} \oint_{K_j} f\, dx + g\, dy = \iint_{D^*} \left(\frac{\partial g}{\partial x} - \frac{\partial f}{\partial y} \right) dA. \tag{11.2}$$

in which all integrations (over C and each K_j) are now taken in the positive, counterclockwise sense. This accounts for the minus sign on each of the integrals $\oint_{K_j} f\, dx + g\, dy$ in equation (11.2). Finally, write equation (11.2) as

$$\oint_C f\, dx + g\, dy = \sum_{j=1}^{n} \oint_{K_j} f\, dx + g\, dy + \iint_{D^*} \left(\frac{\partial g}{\partial x} - \frac{\partial f}{\partial y} \right) dA. \tag{11.3}$$

This is the extended form of Green's theorem. When D contains points at which $f, g, \partial f/\partial y$ and/or $\partial g/\partial x$ are not continuous, then $\oint_C f\, dx + g\, dy$ is the sum of the line integrals $\oint_{K_j} f\, dx + g\, dy$ about small circles centered at the P_j's, together with

$$\iint_{D^*} \left(\frac{\partial g}{\partial x} - \frac{\partial f}{\partial y} \right) dA$$

over the region D^* formed by excising from D the disks bounded by the K_j's.

EXAMPLE 11.10

We will evaluate

$$\oint_C \frac{-y}{x^2 + y^2}\, dx + \frac{x}{x^2 + y^2}\, dy$$

in which C is any simple closed positively oriented path in the plane, but not passing through the origin.

With

$$f(x, y) = \frac{-y}{x^2 + y^2} \text{ and } g(x, y) = \frac{x}{x^2 + y^2}$$

we have

$$\frac{\partial g}{\partial x} = \frac{\partial f}{\partial y} = \frac{y^2 - x^2}{(x^2 + y^2)^2}.$$

This suggests that we consider two cases.

Case 1 If C does not enclose the origin, Green's theorem applies and

$$\oint_C \frac{-y}{x^2 + y^2}\,dx + \frac{x}{x^2 + y^2}\,dy = \iint_D \left(\frac{\partial g}{\partial x} - \frac{\partial f}{\partial y} \right) dA = 0.$$

Case 2 If C encloses the origin, then C encloses a point where f and g are not defined. Now use equation (11.3). Let K be a circle about the origin with a sufficiently small radius r so that K does not intersect C (Figure 11.8). Then

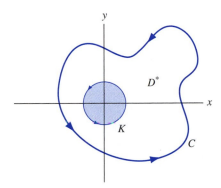

FIGURE 11.8 *Case 2 of Example 11.10.*

$$\oint_C f\,dx + g\,dy$$

$$= \oint_K f\,dx + g\,dy + \iint_{D^*} \left(\frac{\partial g}{\partial x} - \frac{\partial f}{\partial y} \right) dA$$

$$= \oint_K f\,dx + g\,dy$$

where D^* is the region between D and K, including both curves. Both of these line integrals are in the counterclockwise sense. The last line integral is over a circle and can be evaluated explicitly. Parametrize K by $x = r\cos(\theta)$ and $y = r\sin(\theta)$ for $0 \le \theta \le 2\pi$. Then

$$\oint_K f\,dx + g\,dy$$

$$= \int_0^{2\pi} \left(\frac{-r\sin(\theta)}{r^2}[-r\sin(\theta)] + \frac{r\cos(\theta)}{r^2}[r\cos(\theta)] \right) d\theta$$

$$= \int_0^{2\pi} d\theta = 2\pi.$$

We conclude that

$$\oint_C f\,dx + g\,dy = \begin{cases} 0 & \text{if } C \text{ does not enclose the origin} \\ 2\pi & \text{if } C \text{ encloses the origin.} \end{cases}$$

SECTION 11.3   PROBLEMS

In each of Problems 1 through 5, evaluate $\oint_C \mathbf{F} \cdot d\mathbf{R}$ over any simple closed path in the (x, y) plane that does not pass through the origin. This may require cases, as in Example 11.10.

1. $\mathbf{F} = \dfrac{x}{x^2 + y^2}\mathbf{i} + \dfrac{y}{x^2 + y^2}\mathbf{j}$

2. $\mathbf{F} = \left(\dfrac{1}{x^2 + y^2}\right)^{3/2}(x\mathbf{i} + y\mathbf{j})$

3. $\mathbf{F} = \left(\dfrac{-y}{x^2 + y^2} + x^2\right)\mathbf{i} + \left(\dfrac{x}{x^2 + y^2} - 2y\right)\mathbf{j}$

4. $\mathbf{F} = \left(\dfrac{-y}{x^2 + y^2} + 3x\right)\mathbf{i} + \left(\dfrac{x}{x^2 + y^2} - y\right)\mathbf{j}$

5. $\mathbf{F} = \left(\dfrac{x}{\sqrt{x^2 + y^2}} + 2x\right)\mathbf{i} + \left(\dfrac{y}{\sqrt{x^2 + y^2}} - 3y^2\right)\mathbf{j}$

11.4   Potential Theory

A vector field $\mathbf{F}$ is *conservative* if it is derivable from a potential function. This means that, for some scalar field φ,

$$\mathbf{F} = \nabla\varphi = \frac{\partial\varphi}{\partial x}\mathbf{i} + \frac{\partial\varphi}{\partial y}\mathbf{j} + \frac{\partial\varphi}{\partial z}\mathbf{k}.$$

We call φ a *potential function* or *potential* for $\mathbf{F}$. Of course, if φ is a potential, so is $\varphi + c$ for any constant c.

An important ramification of $\mathbf{F}$ being conservative is that the value of $\int_C \mathbf{F} \cdot d\mathbf{R}$ depends only on the endpoints of C. If C has differentiable coordinate functions $x = x(t)$, $y = y(t)$, and $z = z(t)$ for $a \leq t \leq b$, then

$$\int_C \mathbf{F} \cdot d\mathbf{R} = \int_C \frac{\partial\varphi}{\partial x}dx + \frac{\partial\varphi}{\partial y}dy + \frac{\partial\varphi}{\partial z}dy$$

$$= \int_a^b \left(\frac{\partial\varphi}{\partial x}\frac{dx}{dt} + \frac{\partial\varphi}{\partial y}\frac{dy}{dt} + \frac{\partial\varphi}{\partial z}\frac{dz}{dt}\right)$$

$$= \int_a^b \frac{d}{dt}\varphi(x(t), y(t), z(t))dt$$

$$= \varphi(x(b), y(b), z(b)) - \varphi(x(a), y(a), z(a)).$$

In summary, if $\mathbf{F} = \nabla\varphi$ and C is a path from P_0 to P_1, then

$$\int_C \mathbf{F} \cdot d\mathbf{R} = \varphi(P_1) - \varphi(P_0). \tag{11.4}$$

Equation (11.4) has two important ramifications for conservative $\mathbf{F}$.

1. The value of $\int_C \mathbf{F} \cdot d\mathbf{R}$ depends only on the endpoints of C and not on C itself. In particular, if C_1 and C_2 are paths having the same initial points and the same terminal points, then $\int_{C_1} \mathbf{F} \cdot d\mathbf{R} = \int_{C_2} \mathbf{F} \cdot d\mathbf{R}$. In these circumstances, we say that $\int_C \mathbf{F} \cdot d\mathbf{R}$ is *independent of path*.

2. If C is closed, $\oint_C \mathbf{F} \cdot d\mathbf{R} = 0$. This is because if C is closed then $P_0 = P_1$ in equation (11.4).

If a vector field is conservative, we can sometimes find a potential field by integration.

EXAMPLE 11.11

We will determine if the vector field

$$\mathbf{F}(x, y, z) = 3x^2yz^2\mathbf{i} + (x^3z^2 + e^z)\mathbf{j} + (2x^3yz + ye^z)\mathbf{k}$$

is conservative by attempting to find a potential function.

If $\mathbf{F} = \nabla \varphi$ for some φ, then

$$\frac{\partial \varphi}{\partial x} = 3x^2yz^2, \tag{11.5}$$

$$\frac{\partial \varphi}{\partial y} = x^3z^2 + e^z, \tag{11.6}$$

and

$$\frac{\partial \varphi}{\partial z} = 2x^3yz + ye^z. \tag{11.7}$$

Choose one of these equations, say (11.5). To reverse $\partial \varphi / \partial x$, integrate this equation with respect to x to get

$$\varphi(x, y, z) = \int 3x^2yz^2 \, dx = x^3yz^2 + \alpha(y, z).$$

The "constant of integration" may involve y and z because the integration reverses a partial differentiation in which y and z were held fixed. Now we know φ to within $\alpha(y, z)$. To determine $\alpha(y, z)$, choose one of the other equations, say (11.6), to get

$$\frac{\partial \varphi}{\partial y} = x^3z^2 + e^z = \frac{\partial}{\partial y}(x^3yz^2 + \alpha(y, z))$$

$$= x^3z^2 + \frac{\partial \alpha(y, z)}{\partial y}.$$

This requires that

$$\frac{\partial \alpha(y, z)}{\partial y} = e^z.$$

Integrate this with respect to y, holding z fixed to get

$$\int \frac{\partial \alpha(y, z)}{\partial y} dy = ye^z + \beta(z)$$

with $\beta(z)$ an as yet unknown function of z. We now have

$$\varphi(x, y, z) = x^3yz^2 + \alpha(y, z) = x^3yz^2 + ye^z + \beta(z),$$

and we have only to determine $\beta(z)$. For this, use the third equation (11.7) to write

$$\frac{\partial \varphi}{\partial z} = 2x^3yz + ye^z = 2x^3yz + ye^z + \beta'(z).$$

This forces $\beta'(z) = 0$, so $\beta(z) = k$, which is any number. With $\varphi(x, y, z) = x^3yz^2 + ye^z + k$ for any number k (which we can choose to be 0), we have $\mathbf{F} = \nabla \varphi$, and φ is a potential function for $\mathbf{F}$.

This enables us to easily evaluate $\int_C \mathbf{F} \cdot d\mathbf{R}$. If, for example, C is a path from $(0, 0, 0)$ to $(-1, 3, -2)$, then

$$\int_C \mathbf{F} \cdot d\mathbf{R} = \varphi(-1, 3, -2) - \varphi(0, 0, 0) = -12 + 3e^{-2}.$$

If C is a closed path, then $\oint_C \mathbf{F} \cdot d\mathbf{R} = 0$. ◆

There are nonconservative vector fields.

EXAMPLE 11.12

Let $\mathbf{F} = y\mathbf{i} + e^x\mathbf{j}$ be a vector field in the plane. If this is conservative, there would be a potential function $\varphi(x, y)$ such that

$$\frac{\partial \varphi}{\partial x} = y \text{ and } \frac{\partial \varphi}{\partial y} = e^x.$$

Integrate the first with respect to x, thinking of y as fixed, to get

$$\varphi(x, y) = \int y \, dx = xy + \alpha(y).$$

But then we would have to have

$$\frac{\partial \varphi}{\partial y} = e^x = \frac{\partial}{\partial y}(xy + \alpha(y)) = x + \alpha'(y).$$

This would make α depend on x. This is impossible, since $\alpha(y)$ was the "constant" of integration with respect to x. $\mathbf{F}$ has no potential and is therefore not conservative. ◆

Integration is an awkward way to determine whether or not a vector field is conservative. But there is another issue as well. Sometimes the region in which the vector field is defined makes a difference, as the following example shows.

EXAMPLE 11.13

Let

$$\mathbf{F}(x, y) = \frac{-y}{x^2 + y^2}\mathbf{i} + \frac{x}{x^2 + y^2}\mathbf{j}$$

for all (x, y) except the origin. This is a vector field in the plane. Routine integrations would appear to derive the potential function

$$\varphi(x, y) = -\arctan\left(\frac{x}{y}\right).$$

However, this potential is not defined for all (x, y).

If we restrict (x, y) to the right quarter plane $x > 0$ and $y > 0$, then φ is indeed a potential function and $\mathbf{F}$ is conservative in this region. However, suppose we attempt to consider $\mathbf{F}$ over the set D consisting of the entire plane with the origin removed. Then φ is not a potential function. Furthermore, $\mathbf{F}$ is not conservative over D, because $\int_C \mathbf{F} \cdot d\mathbf{R}$ is not independent of path in D. To see this, we will evaluate this integral over two paths from $(1, 0)$ to $(-1, 0)$, shown in Figure 11.9.

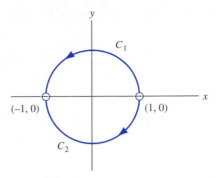

FIGURE 11.9 *Two paths of integration in Example 11.13.*

First, let C_1 be the half-circle given by $x = \cos(\theta)$ and $y = \sin(\theta)$ for $0 \le \theta \le \pi$. This is the upper half of the circle $x^2 + y^2 = 1$. Then

$$\int_{C_1} \mathbf{F} \cdot d\mathbf{R} = \int_0^\pi [(-\sin(\theta))(-\sin(\theta)) + \cos(\theta)(\cos(\theta))]d\theta$$

$$= \int_0^\pi d\theta = \pi.$$

Next, let C_2 be the half-circle from $(1, 0)$ to $(-1, 0)$ given by $x = \cos(\theta)$ and $y = -\sin(\theta)$ for $0 \le \theta \le \pi$. This is the lower half of the circle $x^2 + y^2 = 1$, and

$$\int_{C_2} \mathbf{F} \cdot d\mathbf{R} = \int_0^\pi [\sin(\theta)(-\sin(\theta)) + \cos(\theta)(-\cos(\theta))]d\theta$$

$$= -\int_0^\pi d\theta = -\pi.$$

In this example, $\int_C \mathbf{F} \cdot d\mathbf{R}$ depends not only on the vector field but also on the curve. ◆

We will state two definitive tests for a vector field to be conservative: one for vector fields in the plane and the other for 3-space. As Example 11.13 suggests, the region in which the vector field is defined makes a difference.

In the plane, a set D of points is called a *domain* if it satisfies two conditions:

1. If P is a point in D, there is a circle about P that encloses only points of D.
2. Between any two points of D, there is a path lying entirely in D.

For example, the interior of a solid rectangle is a domain.

A domain D is called *simply connected* if every simple closed path in D encloses only points of D. In Example 11.13, the plane with the origin removed was not simply connected, because a closed path about the origin encloses a point (the origin) not in the set.

Now here is a test for vector fields in the plane to be conservative.

THEOREM 11.2

Let $\mathbf{F} = f\mathbf{i} + g\mathbf{j}$ be a vector field defined over a simply connected domain D in the plane. Then $\mathbf{F}$ is conservative on D if and only if

$$\frac{\partial f}{\partial y} = \frac{\partial g}{\partial x}. \tag{11.8}$$

◆

In Example 11.13, the components of **F** satisfy equation (11.8), but the set (the plane with the origin removed) is not simply connected, so the theorem does not apply. In that example, we saw that there is no potential function over the entire punctured plane.

In 3-space, there is a similar test for a vector field to be conservative with adjustments to accommodate the extra dimension. A set S of points in R^3 is a *domain* if it satisfies the following two conditions:

1. If P is a point in S, there is a sphere about P that encloses only points of S.
2. Between any two points of S, there is a path lying entirely in S. For example, the interior of a cube is a domain.

Furthermore, S is *simply connected* if every simple closed path in S is the boundary of a surface in S. With this, we can state the following.

THEOREM 11.3

Let **F** be a vector field defined over a simply connected domain S in R^3. Then **F** is conservative on S if and only if

$$\nabla \times \mathbf{F} = \mathbf{O}. \quad \blacklozenge$$

Thus, the conservative vector fields in R^3 are those with zero curl. These are the irrotational vector fields.

With the right perspective, these tests in 2-space and 3-space can be combined into a single test. Given $\mathbf{F} = f\mathbf{i} + g\mathbf{j}$ in the plane, define

$$\mathbf{G}(x, y, z) = f(x, y)\mathbf{i} + g(x, y)\mathbf{j} + 0\mathbf{k}$$

to think of **F** as a vector field in 3-space. Now compute

$$\nabla \times \mathbf{G} = \begin{vmatrix} \mathbf{i} & \mathbf{j} & \mathbf{k} \\ \partial/\partial x & \partial/\partial y & \partial/\partial z \\ f(x, y) & g(x, y) & 0 \end{vmatrix} = \left(\frac{\partial g}{\partial x} - \frac{\partial f}{\partial y} \right)\mathbf{k}.$$

The 3-space condition $\nabla \times \mathbf{G} = \mathbf{O}$ therefore reduces to equation (11.8) if the vector field is in the plane.

SECTION 11.4   PROBLEMS

In each of Problems 1 through 10, determine whether **F** is conservative in the given region D. If D is not defined explicitly, it is understood to be the entire plane or 3-space. If the vector field is conservative, find a potential.

1. $\mathbf{F} = y^3\mathbf{i} + (3xy^2 - 4)\mathbf{j}$
2. $\mathbf{F} = (6y + e^{xy})\mathbf{i} + (6x + xe^{xy})\mathbf{j}$
3. $\mathbf{F} = 16x\mathbf{i} + (2 - y^2)\mathbf{j}$
4. $\mathbf{F} = 2xy\cos(x^2)\mathbf{i} + \sin(x^2)\mathbf{j}$
5. $\mathbf{F} = \left(\dfrac{2x}{x^2 + y^2} \right)\mathbf{i} + \left(\dfrac{2y}{x^2 + y^2} \right)\mathbf{j}$

   D is the plane with the origin removed.

6. $\mathbf{F} = 2x\mathbf{i} - 2y\mathbf{j} + 2z\mathbf{k}$
7. $\mathbf{F} = \mathbf{i} - 2\mathbf{j} + \mathbf{k}$
8. $\mathbf{F} = yz\cos(x)\mathbf{i} + (z\sin(x) + 1)\mathbf{j} + y\sin(x)\mathbf{k}$
9. $\mathbf{F} = (x^2 - 2)\mathbf{i} + xyz\mathbf{j} - yz^2\mathbf{k}$
10. $\mathbf{F} = e^{xyz}(1 + xyz)\mathbf{i} + x^2z\mathbf{j} + x^2y\mathbf{k}$

In each of Problems 11 through 20, determine a potential function to evaluate $\int_C \mathbf{F} \cdot d\mathbf{R}$ for C any path from the first point to the second.

11. $\mathbf{F} = 3x^2(y^2 - 4y)\mathbf{i} + (2x^3y - 4x^3)\mathbf{j}; (-1, 1), (2, 3)$
12. $\mathbf{F} = e^x\cos(y)\mathbf{i} - e^x\sin(y)\mathbf{j}; (0, 0), (2, \pi/4)$

13. $\mathbf{F}=2xy\mathbf{i}+(x^2-1/y)\mathbf{j}$; $(1,3)$, $(2,2)$ (The path cannot cross the x-axis).

14. $\mathbf{F}=\mathbf{i}+(6y+\sin(y))\mathbf{j}$; $(0,0)$, $(1,3)$

15. $\mathbf{F}=(3x^2y^2-6y^3)\mathbf{i}+(2x^3y-18xy^2)\mathbf{j}$; $(0,0)$, $(1,1)$

16. $\mathbf{F}=(y\cos(xz)-xyz\sin(xz))\mathbf{i}+x\cos(xz)\mathbf{j}-x^2\sin(xz)\mathbf{k}$; $(1,0,\pi)$, $(1,1,7)$

17. $\mathbf{F}=\mathbf{i}-9y^2z\mathbf{j}-3y^3\mathbf{k}$; $(1,1,1)$, $(0,3,5)$

18. $\mathbf{F}=-8y^2\mathbf{i}-(16xy+4z)\mathbf{j}-4y\mathbf{k}$; $(-2,1,1)$, $(1,3,2)$

19. $\mathbf{F}=6x^2e^{yz}\mathbf{i}+2x^3ze^{yz}\mathbf{j}+2x^3ye^{yz}\mathbf{k}$; $(0,0,0)$, $(1,2,-1)$

20. $\mathbf{F}=(y-4xz)\mathbf{i}+x\mathbf{j}+(3z^2-2x^2)\mathbf{k}$; $(1,1,1)$, $(3,1,4)$

21. Prove the law of conservation of energy, which states that the sum of the kinetic and potential energies of an object acted on by a conservative force is a constant. *Hint*: The kinetic energy is $(m/2)\|\mathbf{R}'(t)\|2$, where m is the mass and $\mathbf{R}(t)$ describes the trajectory of the particle. The potential energy is $-\varphi(x,y,z)$, where $\mathbf{F}=\nabla\varphi$.

11.5   Surface Integrals

Just as there are integrals of vector fields over curves, there are also integrals of vector fields over surfaces. We begin with some facts about surfaces.

A curve in R^3 is given by coordinate functions of one variable and may be thought of as a one-dimensional object (such as a thin wire). A *surface* is defined by coordinate or parametric functions of two variables,

$$x=x(u,v),\ \ y=y(u,v),\ \text{and}\ z=z(u,v)$$

for (u,v) in some specified set in the (u,v) plane. We call u and v parameters for the surface.

EXAMPLE 11.14

Figure 11.10 shows part of the surface having coordinate functions

$$x=u\cos(v),\ \ y=u\sin(v),\ \text{and}\ z=\frac{1}{2}u^2\sin(2v)$$

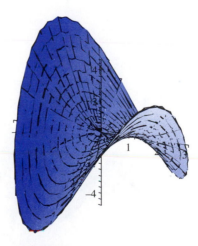

FIGURE 11.10 *The surface* $x=u\cos(v)$, $y=u\sin(v)$, *and* $z=(u^2/2)\sin(2v)$ *in Example 11.14.*

in which u and v can be any real numbers. Since $z = xy$, the surface cuts any plane $z = k$ in a hyperbola $xy = k$. However, the surface intersects a plane $y = \pm x$ in a parabola $z = \pm x^2$. For this reason, the surface is called a *hyperbolic paraboloid*. ◆

Often a surface is defined as a level surface $f(x, y, z) = k$ with f as a given function. For example,

$$f(x, y, z) = (x - 1)^2 + y^2 + (z + 4)^2 = 16$$

has the sphere of radius 4 and center $(1, 0, -4)$ as its graph.

We may also express a surface as a locus of points satisfying an equation $z = f(x, y)$, $y = h(x, z)$, or $x = w(y, z)$. Figure 11.11 shows part of the graph of $z = 6\sin(x - y)/\sqrt{1 + x^2 + y^2}$.

We often write a position vector

$$\mathbf{R}(u, v) = x(u, v)\mathbf{i} + y(u, v)\mathbf{j} + z(u, v)\mathbf{k}$$

for a surface. $\mathbf{R}(u, v)$ can be thought of as an arrow from the origin to the point $(x(u, v), y(u, v), z(u, v))$ on the surface.

Although a surface is different from its graph (the surface is a triple of coordinate functions, the graph is a geometric locus in R^3), often we will informally identify the surface with its graph, just as we sometimes identify a curve with its graph.

A surface is *simple* if it does not fold over and intersect itself. This means that $\mathbf{R}(u_1, v_1) = \mathbf{R}(u_2, v_2)$ can occur only when $u_1 = u_2$ and $v_1 = v_2$.

11.5.1   Normal Vector to a Surface

We would like to define a normal vector to a surface at a point. Previously, this was done for level surfaces.

Let Σ be a surface with coordinate functions $x(u, v)$, $y(u, v)$, and $z(u, v)$. Let P_0 be a point on Σ corresponding to $u = u_0$, and $v = v_0$.

If we fix $v = v_0$, we can define the curve Σ_{v_0} on Σ, having coordinate functions

$$x = x(u, v_0), y = y(u, v_0), z = z(u, v_0).$$

The tangent vector to this curve at P_0 is

$$\mathbf{T}_{u_0} = \frac{\partial x}{\partial u}(u_0, v_0)\mathbf{i} + \frac{\partial y}{\partial u}(u_0, v_0)\mathbf{j} + \frac{\partial z}{\partial u}(u_0, v_0)\mathbf{k}.$$

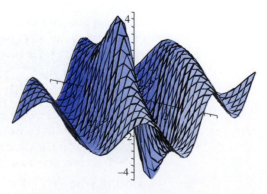

FIGURE 11.11   *The surface* $z = 6\sin(x - y)/$ $\sqrt{1 + x^2 + y^2}$.

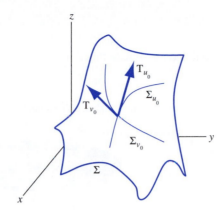

FIGURE 11.12 *Curves* Σ_{u_0} *and* Σ_{v_0} *and tangent vectors* $\mathbf{T}_{u_0}$ *and* $\mathbf{T}_{v_0}$.

Similarly, we can fix $u = u_0$ and form the curve Σ_{u_0} on the surface. The tangent to this curve at P_0 is

$$\mathbf{T}_{v_0} = \frac{\partial x}{\partial v}(u_0, v_0)\mathbf{i} + \frac{\partial y}{\partial v}(u_0, v_0)\mathbf{j} + \frac{\partial z}{\partial v}(u_0, v_0)\mathbf{k}.$$

These two curves and tangent vectors are shown in Figure 11.12.

Assuming that neither of these tangent vectors is the zero vector, they both lie in the tangent plane to the surface at P_0. Their cross product is therefore normal to this tangent plane. This leads us to define the *normal to the surface* at P_0 to be the vector

$$\mathbf{N}(P_0) = \mathbf{T}_{u_0} \times \mathbf{T}_{v_0}$$

$$= \begin{vmatrix} \mathbf{i} & \mathbf{j} & \mathbf{k} \\ \frac{\partial x}{\partial u}(u_0, v_0) & \frac{\partial y}{\partial u}(u_0, v_0) & \frac{\partial z}{\partial u}(u_0, v_0) \\ \frac{\partial x}{\partial v}(u_0, v_0) & \frac{\partial y}{\partial v}(u_0, v_0) & \frac{\partial z}{\partial v}(u_0, v_0) \end{vmatrix}$$

$$= \left(\frac{\partial y}{\partial u}\frac{\partial z}{\partial v} - \frac{\partial z}{\partial u}\frac{\partial y}{\partial v} \right)\mathbf{i} + \left(\frac{\partial z}{\partial u}\frac{\partial x}{\partial v} - \frac{\partial x}{\partial u}\frac{\partial z}{\partial v} \right)\mathbf{j} + \left(\frac{\partial x}{\partial u}\frac{\partial y}{\partial v} - \frac{\partial y}{\partial u}\frac{\partial x}{\partial v} \right)\mathbf{k},$$

in which all partial derivatives are evaluated at (u_0, v_0).

To make this vector easier to write, define the *Jacobian* of two functions f and g to be

$$\frac{\partial(f, g)}{\partial(u, v)} = \begin{vmatrix} \partial f/\partial u & \partial f/\partial v \\ \partial g/\partial u & \partial g/\partial v \end{vmatrix} = \frac{\partial f}{\partial u}\frac{\partial g}{\partial v} - \frac{\partial g}{\partial u}\frac{\partial f}{\partial v}.$$

Then

$$\mathbf{N}(P_0) = \frac{\partial(y, z)}{\partial(u, v)}\mathbf{i} + \frac{\partial(z, x)}{\partial(u, v)}\mathbf{j} + \frac{\partial(x, y)}{\partial(u, v)}\mathbf{k},$$

with all of the partial derivatives evaluated at (u_0, v_0). This expression is easy to remember with an observation. Write (x, y, z) in this order. For the $\mathbf{i}$ component of $\mathbf{N}$, delete x, leaving (y, z), in this order in the numerator of the Jacobian. For the $\mathbf{j}$ component, delete y from (x, y, z), but move to the right, getting (z, x) in the Jacobian. For the $\mathbf{k}$ component, delete z, leaving (x, y) in this order.

EXAMPLE 11.15

The *elliptical cone* has coordinate functions

$$x = au\cos(v), \quad y = au\sin(v), \quad \text{and } z = u$$

with a and b as positive constants. Part of this surface is shown in Figure 11.13.

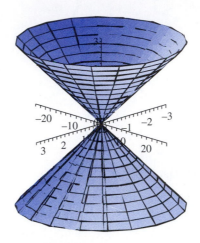

FIGURE 11.13 *Elliptical cone $z = au\cos(v),\ y = bu\sin(v),\ and\ z = u$.*

Since

$$z^2 = \left(\frac{x}{a}\right)^2 + \left(\frac{y}{b}\right)^2,$$

this surface is a "cone" with its major axis the z-axis. Planes $z = k$ parallel to the (x, y) plane intersect this surface in ellipses. We will write the normal vector $\mathbf{P}_0$ at $P_0 = (a\sqrt{3}/4, b/4, 1/2)$ obtained when $u = u_0 = 1/2$ and $v = v_0 = \pi/6$. Compute the Jacobian components:

$$\frac{\partial(y, z)}{\partial(u, v)} = \left[\frac{\partial y}{\partial u}\frac{\partial z}{\partial v} - \frac{\partial z}{\partial u}\frac{\partial y}{\partial v}\right]_{(1/2, \pi/6)}$$

$$= [b\sin(v)(0) - bu\cos(v)]_{(1/2, \pi/6)} = -\sqrt{3}b/4,$$

$$\frac{\partial(z, x)}{\partial(u, v)} = \left[\frac{\partial z}{\partial u}\frac{\partial x}{\partial v} - \frac{\partial x}{\partial u}\frac{\partial z}{\partial v}\right]_{(1/2, \pi/6)}$$

$$= [-au\sin(v) - a\cos(v)(0)]_{(1/2, \pi/6)} = -a/4,$$

and

$$\frac{\partial(x, y)}{\partial(u, v)} = \left[\frac{\partial x}{\partial u}\frac{\partial y}{\partial v} - \frac{\partial y}{\partial u}\frac{\partial x}{\partial v}\right]_{(1/2, \pi/6)}$$

$$= [a\cos(v)bu\cos(v) - b\sin(v)(-au\sin(v))]_{(1/2, \pi/6)} = ab/2.$$

Then

$$\mathbf{N}(P_0) = -\frac{\sqrt{3}b}{4}\mathbf{i} - \frac{a}{4}\mathbf{j} + \frac{ab}{2}\mathbf{k}. \quad \blacklozenge$$

We frequently encounter the case that a surface is given by an equation $z = S(x, y)$ with $u = x$ and $v = y$ as parameters. In this case,

$$\frac{\partial(y, z)}{\partial(u, v)} = \frac{\partial(y, z)}{\partial(x, y)} = \begin{vmatrix} 0 & 1 \\ \partial S/\partial x & \partial S/\partial y \end{vmatrix} = -\frac{\partial S}{\partial x},$$

$$\frac{\partial(z, x)}{\partial u, v} = \frac{\partial(z, x)}{\partial(x, y)} = \begin{vmatrix} \partial S/\partial x & \partial S/\partial y \\ 1 & 0 \end{vmatrix} = -\frac{\partial S}{\partial y},$$

and

$$\frac{\partial(x, y)}{\partial(u, v)} = \frac{\partial(x, y)}{\partial(x, y)} = \begin{vmatrix} 1 & 0 \\ 0 & 1 \end{vmatrix} = 1.$$

Now the normal vector at $P_0 : (x_0, y_0)$ is

$$\mathbf{N}(P_0) = -\frac{\partial S}{\partial x}(x_0, y_0)\mathbf{i} - \frac{\partial S}{\partial y}(x_0, y_0)\mathbf{j} + \mathbf{k}$$

$$= -\frac{\partial z}{\partial x}(x_0, y_0)\mathbf{i} - \frac{\partial z}{\partial y}(x_0, y_0)\mathbf{j} + \mathbf{k}.$$

EXAMPLE 11.16

Let Σ be the hemisphere given by

$$z = \sqrt{4 - x^2 - y^2}.$$

We will find the normal vector at $P_0 : (1, \sqrt{2}, 1)$. Compute

$$\frac{\partial z}{\partial x}\bigg|_{(1,\sqrt{2})} = -1 \text{ and } \frac{\partial z}{\partial y}\bigg|_{(1,\sqrt{2})} = -\sqrt{2}.$$

Then

$$\mathbf{N}(P_0) = \mathbf{i} + \sqrt{2}\mathbf{j} + \mathbf{k}.$$

This result is consistent with the fact that a line from the origin through a point on this hemisphere is normal to the hemisphere at that point. ◆

11.5.2 Tangent Plane to a Surface

If a surface Σ has a normal vector $\mathbf{N}(P_0)$ at a point then it has a tangent plane at P_0. This is the plane through $P_0 : (x_0, y_0, z_0)$ having normal vector $\mathbf{N}(P_0)$. The equation of this tangent plane is

$$\mathbf{N}(P_0) \cdot [(x - x_0)\mathbf{i} + (y - y_0)\mathbf{j} + (z - z_0)\mathbf{k}] = 0$$

or

$$\left[\frac{\partial(y, z)}{\partial(u, v)}\right]_{(u_0, v_0)}(x - x_0) + \left[\frac{\partial(z, x)}{\partial(u, v)}\right]_{(u_0, v_0)}(y - y_0) + \left[\frac{\partial(x, y)}{\partial(u, v)}\right]_{(u_0, v_0)}(z - z_0) = 0.$$

If Σ is given by $z = S(x, y)$, this tangent plane has the equation

$$-\left(\frac{\partial S}{\partial x}\right)_{(x_0, y_0)}(x - x_0) - \left(\frac{\partial S}{\partial y}\right)_{(x_0, y_0)}(y - y_0) + z - z_0 = 0.$$

EXAMPLE 11.17

For the elliptical cone of Example 11.15, the tangent plane at $(\sqrt{3}a/4, b/4, 1/2)$ has the equation

$$-\frac{\sqrt{3}b}{4}\left(x - \frac{\sqrt{3}a}{4}\right) - \frac{a}{4}\left(x - \frac{b}{4}\right) + \frac{ab}{2}\left(z - \frac{1}{2}\right) = 0. \quad \blacklozenge$$

EXAMPLE 11.18

For the hemisphere of Example 11.16, the tangent plane at $(1, \sqrt{2}, 1)$ has the equation

$$(x - 1) + \sqrt{2}(y - \sqrt{2}) + (z - 1) = 0$$

or

$$x + \sqrt{2}y + z = 4. \quad \blacklozenge$$

11.5.3   Piecewise Smooth Surfaces

A curve is smooth if it has a continuous tangent. A *smooth surface* is one that has a continuous normal. A *piecewise smooth surface* is one that consists of a finite number of smooth surfaces. For example, a sphere is smooth and the surface of a cube is piecewise smooth, consisting of six smooth faces. The cube does not have a normal vector (or tangent plane) along an edge.

In calculus, it is shown that the area of a smooth surface Σ given by $z = S(x, y)$ is

$$\text{area of } \Sigma = \iint_D \sqrt{1 + \left(\frac{\partial S}{\partial x}\right)^2 + \left(\frac{\partial S}{\partial y}\right)^2}\, dA$$

where D is the set of points in the (x, y) plane over which the surface is defined. We now recognize that this area is actually the integral of the length of the normal vector:

$$\text{area of } \Sigma = \iint_D \|\mathbf{N}(x, y)\|\, dx\, dy.$$

This is analogous to the formula for the length of a curve as the integral of the length of the tangent vector. More generally, if Σ is given by coordinate functions $x(u, v)$, $y(u, v)$, and $z(u, v)$ for (u, v) varying over some set D in the (u, v) plane, then

$$\text{area of } \Sigma = \iint_D \|\mathbf{N}(u, v)\|\, du\, dv. \tag{11.9}$$

11.5.4   Surface Integrals

The line integral of $f(x, y, z)$ over C with respect to arc length is

$$\int_C f(x, y, z)\, ds = \int_a^b f(x(t), y(t), z(t))\sqrt{x'(t)^2 + y'(t)^2 + z'(t)^2}\, dt.$$

We want to lift this idea up one dimension to integrate a function over a surface instead of over a curve. Now the coordinate functions are functions of two variables u and v, so $\int_a^b \ldots dt$ will be replaced by $\int\int_D, \ldots, du\, dv$. The differential element of arc length ds for C will be replaced by the differential element of surface area on Σ, which by equation (11.9) is $d\sigma = \|\mathbf{N}(u, v)\|\, du\, dv$.

Let Σ be a smooth surface with coordinate functions $x(u, v)$, $y(u, v)$, and $z(u, v)$ for (u, v) in D. Let f be continuous on Σ. Then *the surface integral of* f *over* Σ is denoted $\int\int_{\Sigma} f(x, y, z)d\sigma$ and is defined by

$$\iint_{\Sigma} f(x, y, z)d\sigma = \iint_{D} f(x(u, v), y(u, v), z(u, v)) \| \mathbf{N}(u, v) \| \, dudv.$$

If Σ is piecewise smooth, then the line integral of f over Σ is the sum of the line integrals over the smooth pieces.

If Σ is given by $z = S(x, y)$ for (x, y) in D, then

$$\iint_{\Sigma} f(x, y, z)d\sigma = \iint_{D} f(x, y, S(x, y))\sqrt{1 + \left(\frac{\partial S}{\partial x}\right)^2 + \left(\frac{\partial S}{\partial y}\right)^2} \, dydx.$$

EXAMPLE 11.19

We will compute the surface integral $\int\int_{\Sigma} xyzd\sigma$ over the part of the surface as

$$x = u\cos(v), \, y = u\sin(v), \, z = \frac{1}{2}u^2\sin(2v),$$

corresponding to (u, v) in $D: 1 \le u \le 2, 0 \le v \le \pi$.

First we need the normal vector. The components of $\mathbf{N}(u, v)$ are

$$\frac{\partial(y, z)}{\partial(u, v)} = \begin{vmatrix} \sin(v) & u\cos(v) \\ u\sin(2v) & u^2\cos(2v) \end{vmatrix} = u^2[\sin(v)\cos(2v) - \cos(v)\sin(2v)],$$

$$\frac{\partial(z, x)}{\partial(u, v)} = \begin{vmatrix} u\sin(2v) & u^2\cos(2v) \\ \cos(v) & -u\sin(v) \end{vmatrix} = -u^2[\sin(v)\sin(2v) + \cos(v)\cos(2v)],$$

and

$$\frac{\partial(x, y)}{\partial(u, v)} = \begin{vmatrix} \cos(v) & -u\sin(v) \\ \sin(v) & u\cos(v) \end{vmatrix} = u.$$

Then

$$\| \mathbf{N}(u, v) \|^2 = u^4[\sin(v)\cos(2v) - \cos(v)\sin(2v)]^2$$
$$+ u^4[\sin(v)\sin(2v) + \cos(v)\cos(2v)]^2 + u^2$$
$$= u^2(1 + u^2),$$

so

$$\| \mathbf{N}(u, v) \| = u\sqrt{1 + u^2}.$$

The surface integral is

$$\iint_{\Sigma} xyz\,d\sigma = \iint_{D} [u\cos(v)][u\sin(v)]\left[\frac{1}{2}u^2\sin(2v)\right]u\sqrt{1+u^2}\,dA$$

$$= \int_{0}^{\pi} \cos(v)\sin(v)\sin(2v)\,dv \int_{1}^{2} u^5\sqrt{1+u^2}\,du$$

$$= \frac{\pi}{4}\left(\frac{100}{21}\sqrt{21} - \frac{11}{105}\sqrt{2}\right). \quad \blacklozenge$$

EXAMPLE 11.20

We will evaluate $\int\int_{\Sigma} z\,d\sigma$ over the part of the plane $x+y+z=4$ lying above the rectangle $D: 0 \le x \le 2, 0 \le y \le 1$. This surface is shown in Figure 11.14.

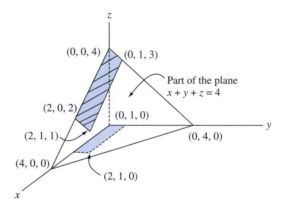

FIGURE 11.14 *Part of the plane $x+y+z=4$ in Example 11.20.*

With $z = S(x,y) = 4 - x - y$, we have

$$\iint_{\Sigma} z\,d\sigma = \iint_{D} z\sqrt{1+(-1)^2+(-1)^2}\,dy\,dx$$

$$= \sqrt{3}\int_{0}^{2}\int_{0}^{1}(4-x-y)\,dy\,dx.$$

First compute

$$\int_{0}^{1}(4-x-y)\,dy = \left[(4-x)y - \frac{1}{2}y^2\right]_{0}^{1}$$

$$= 4 - x - \frac{1}{2} = \frac{7}{2} - x.$$

Then

$$\iint_{\Sigma} z\,d\sigma = \sqrt{3}\int_{0}^{2}\left(\frac{7}{2} - x\right)dx = 5\sqrt{3}. \quad \blacklozenge$$

SECTION 11.5   *PROBLEMS*

In each of Problems 1 through 10, evaluate $\int\int_\Sigma f(x, y, z)\,d\sigma$.

1. $f(x, y, z) = x$, Σ is the part of the plane $x + 4y + z = 10$ in the first octant.

2. $f(x, y, z) = y^2$, Σ is the part of the plane $z = x$ for $0 \le x \le 2, 0 \le y \le 4$.

3. $f(x, y, z) = 1$, Σ is the part of the paraboloid $z = x^2 + y^2$ lying between the planes $z = 2$ and $z = 7$.

4. $f(x, y, z) = x + y$, Σ is the part of the plane $4x + 8y + 10z = 25$ lying above the triangle in the (x, y) plane having vertices $(0, 0)$, $(1, 0)$, and $(1, 1)$.

5. $f(x, y, z) = z$, Σ is the part of the cone $z = \sqrt{x^2 + y^2}$ in the first octant and between the planes $z = 2$ and $z = 4$.

6. $f(x, y, z) = xyz$, Σ is the part of the plane $z = x + y$ with (x, y) in the square with vertices $(0, 0)$, $(1, 0)$, $(0, 1)$, and $(1, 1)$.

7. $f(x, y, z) = y$, Σ is the part of the cylinder $z = x^2$ for $0 \le x \le 2, 0 \le y \le 3$.

8. $f(x, y, z) = x^2$, Σ is the part of the paraboloid $z = 4 - x^2 - y^2$ lying above the (x, y) plane.

9. $f(x, y, z) = z$, Σ is the part of the plane $z = x - y$ for $0 \le x \le 1$ and $0 \le y \le 5$.

10. $f(x, y, z) = xyz$, Σ is the part of the cylinder $z = 1 + y^2$ for $0 \le x \le 1$ and $0 \le y \le 1$.

11.6   Applications of Surface Integrals

11.6.1   Surface Area

If Σ is a piecewise smooth surface, then

$$\iint_\Sigma d\sigma = \iint_D \| \mathbf{N}(u, v) \|\, du\, dv = \text{area of } \Sigma.$$

(This assumes a bounded surface having finite area.) Clearly, we do not need surface integrals to compute areas of surfaces. However, we mention this result because it is in the same spirit as other familiar mensuration formulas:

$$\int_C ds = \text{length of } C,$$

$$\iint_D dA = \text{area of } D,$$

and

$$\iiint_M dV = \text{volume of } M.$$

11.6.2   Mass and Center of Mass of a Shell

Imagine a shell of negligible thickness in the shape of a piecewise smooth surface Σ. Let $\delta(x, y, z)$ be the density of the material of the shell at point (x, y, z). We want to compute the mass of the shell.

Let Σ have coordinate functions $x(u, v)$, $y(u, v)$, $z(u, v)$ for (u, v) in D. Form a grid of lines over D, as in Figure 11.15, by drawing vertical lines Δu units apart and horizontal lines Δv apart. These lines form rectangles $R_1, \ldots, R_n$ that cover D. Each R_j corresponds to a patch of surface Σ_j, as in Figure 11.16. Let (u_j, v_j) be a point in R_j. This corresponds to a point $P_j = (x(u_j, v_j), y(u_j, v_j), z(u_j, v_j))$ on Σ_j. Approximate the mass of Σ_j by the density at P_j

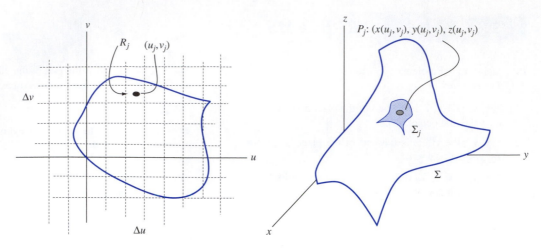

FIGURE 11.15   *Forming a grid over D.*

FIGURE 11.16   *Grid rectangle R_j maps to a patch of surface Σ_j.*

times the area of Σ_j. The mass of the shell is approximately the sum of the approximate masses of these patches of surface:

$$
\text{mass of the shell} \approx \sum_{j=1}^{n} \delta(P_j) \left(\text{area of } \Sigma_j\right).
$$

But the area of Σ_j can be approximated as the length of the normal at P_j times the area of R_j:

$$
\text{area of } \Sigma_j \approx \| \mathbf{N}(P_j) \| \, \Delta u \, \Delta v.
$$

Therefore,

$$
\text{mass of } \Sigma \approx \sum_{j=1}^{n} \delta(P_j) \mathbf{N}(P_j) \| \, \Delta u \, \Delta v.
$$

In the limit as $\Delta u \to 0$ and $\Delta v \to 0$, we obtain

$$
\text{mass of } \Sigma = \iint_{\Sigma} \delta(x, y, z) d\sigma.
$$

The center of mass of the shell is $(\overline{x}, \overline{y}, \overline{z})$, where

$$
\overline{x} = \frac{1}{m} \iint_{\Sigma} x \delta(x, y, z) d\sigma, \; \overline{y} = \frac{1}{m} \iint_{\Sigma} y \delta(x, y, z) d\sigma,
$$

and

$$
\overline{z} = \frac{1}{m} \iint_{\Sigma} z \delta(x, y, z) d\sigma,
$$

in which m is the mass.

If the surface is given as $z = S(x, y)$ for (x, y) in D, then the mass is

$$
m = \iint_{D} \delta(x, y, z) \sqrt{1 + \left(\frac{\partial S}{\partial x}\right)^2 + \left(\frac{\partial S}{\partial y}\right)^2} \, dy \, dx.
$$

EXAMPLE 11.21

We will find the mass and center of mass of the cone $z = \sqrt{x^2 + y^2}$ for $x^2 + y^2 \le 4$ if $\delta(x, y, z) = x^2 + y^2$.

Let D be the disk of radius 2 about the origin. Compute

$$\frac{\partial z}{\partial x} = \frac{x}{z} \text{ and } \frac{\partial z}{\partial y} = \frac{y}{z}.$$

The mass is

$$m = \iint_\Sigma (x^2 + y^2) d\sigma$$

$$= \iint_D (x^2 + y^2) \sqrt{1 + \frac{x^2}{z^2} + \frac{y^2}{z^2}} \, dy \, dx$$

$$= \int_0^{2\pi} \int_0^2 r^2 \sqrt{2} r \, dr \, d\theta$$

$$= 2\sqrt{2}\pi \frac{1}{4} r^4 \Big]_0^2 = 8\sqrt{2}\pi.$$

By symmetry of the surface and of the density function, we expect the center of mass to lie on the z-axis, so $\overline{x} = \overline{y} = 0$. This can be verified by computation. Finally,

$$\overline{z} = \frac{1}{8\sqrt{2}\pi} \iint_\Sigma z(x^2 + y^2) d\sigma$$

$$= \frac{1}{8\sqrt{2}\pi} \iint_D \sqrt{x^2 + y^2}(x^2 + y^2) \sqrt{1 + \frac{x^2}{z^2} + \frac{y^2}{z^2}} \, dy \, dx$$

$$= \frac{1}{8\pi} \int_0^{2\pi} \int_0^2 r(r^2) r \, dr \, d\theta$$

$$= \frac{1}{8\pi} (2\pi) \left[\frac{1}{5} r^5 \right]_0^2 = \frac{8}{5}.$$

The center of mass is $(0, 0, 8/5)$. ◆

11.6.3 Flux of a Fluid Across a Surface

Suppose a fluid moves in some region of 3-space with velocity $\mathbf{V}(x, y, z, t)$. In studying the flow, it is often useful to place an imaginary surface Σ in the fluid and analyze the net volume of fluid flowing across the surface per unit time. This is the *flux* of the fluid across the surface.

Let $\mathbf{n}(u, v, t)$ be the unit normal vector to the surface at time t. If we are thinking of flow out of the surface from its interior, then choose $\mathbf{n}$ to be an outer normal oriented from a point of the surface outward away from the interior.

In a time interval Δt, the volume of fluid flowing across a small piece Σ_j of Σ approximately equals the volume of the cylinder with base Σ_j and altitude $V_n \Delta t$, where V_n is the component of $\mathbf{V}$ in the direction of $\mathbf{n}$ evaluated at some point of Σ_j. This volume (Figure 11.17) is $(V_n \Delta t) A_j$, where A_j is the area of Σ_j. Because $\| \mathbf{n} \| = 1$, $V_n = \mathbf{V} \cdot \mathbf{n}$. The volume of fluid across Σ_j per unit time is

$$\frac{(V_n \Delta t) A_j}{\Delta t} = V_n A_j = \mathbf{V} \cdot \mathbf{n} \Delta t.$$

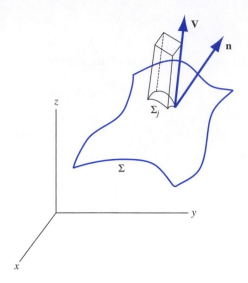

FIGURE 11.17 *Cylinder with base Σ_j and height $V_{\mathbf{n}}\Delta t$.*

Sum these quantities over the entire surface and take a limit as the surface elements are chosen smaller, as we did for the mass of a shell. We get

$$\text{flux of } \mathbf{V} \text{ across } \Sigma \text{ in the direction of } \mathbf{n} = \iint_{\Sigma} \mathbf{V} \cdot \mathbf{n} \, d\sigma.$$

The flux of the fluid (or any vector field) across a surface is therefore computed as the surface integral of the normal component of the field to the surface.

EXAMPLE 11.22

We will calculate the flux of $\mathbf{F} = x\mathbf{i} + y\mathbf{j} + z\mathbf{k}$ across the part of the sphere $x^2 + y^2 + z^2 = 4$ between the planes $z = 1$ and $z = 2$.

The plane $z = 1$ intersects the sphere in the circle $x^2 + y^2 = 3$, $z = 1$. This circle projects onto the circle $x^2 + y^2 = 3$ in the x, y-plane. The plane $z = 2$ hits the sphere at $(0, 0, 2)$ only. Think of Σ as specified by $z = S(x, y) = \sqrt{4 - x^2 + y^2}$ for (x, y) in D, which is the disk $0 \le x^2 + y^2 \le 3$ (Figure 11.18).

To compute the partial derivatives $\partial z / \partial x$ and $\partial z / \partial y$, we can implicitly differentiate the equation of the sphere to get

$$2x + 2z \frac{\partial z}{\partial x} = 0,$$

so

$$\frac{\partial z}{\partial x} = -\frac{x}{z}.$$

Similarly,

$$\frac{\partial z}{\partial y} = -\frac{y}{z}.$$

The vector

$$\frac{x}{z}\mathbf{i} + \frac{y}{z}\mathbf{j} + \mathbf{k}$$

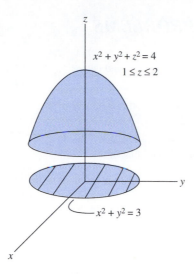

$$x^2 + y^2 + z^2 = 4$$
$$1 \le z \le 2$$

$$x^2 + y^2 = 3$$

FIGURE 11.18 *Surface in Example 11.22.*

is normal to the sphere and oriented outward. This vector has a magnitude of $2/z$, so a unit outer normal is

$$\mathbf{n} = \frac{z}{2}\left(\frac{x}{z}\mathbf{i} + \frac{y}{z}\mathbf{j} + \mathbf{k}\right) = \frac{1}{2}(x\mathbf{i} + y\mathbf{j} + z\mathbf{k}).$$

We need

$$\mathbf{F} \cdot \mathbf{n} = \frac{1}{2}(x^2 + y^2 + z^2).$$

Then

$$
\begin{aligned}
\text{flux} &= \iint_\Sigma \frac{1}{2}(x^2 + y^2 + z^2)\, d\sigma \\
&= \frac{1}{2}\iint_D (x^2 + y^2 + z^2)\sqrt{1 + \frac{x^2}{z^2} + \frac{y^2}{z^2}}\, dA \\
&= \frac{1}{2}\iint_D (x^2 + y^2 + z^2)\sqrt{\frac{x^2 + y^2 + z^2}{z^2}}\, dA \\
&= \frac{1}{2}\iint_D (x^2 + y^2 + z^2)^{3/2}\frac{1}{\sqrt{4 - x^2 - y^2}}\, dA \\
&= 4\iint_D \frac{1}{\sqrt{4 - x^2 - y^2}}\, dA.
\end{aligned}
$$

Here we used the fact that $x^2 + y^2 + z^2 = 4$ on Σ. Converting to polar coordinates, we have

$$
\begin{aligned}
\text{flux} &= 4\int_0^{2\pi}\int_0^{\sqrt{3}} \frac{1}{\sqrt{4 - r^2}}\, r\, dr\, d\theta \\
&= 8\pi[-(4 - r^2)^{1/2}]_0^{\sqrt{3}} = 8\pi. \quad \blacklozenge
\end{aligned}
$$

In each of Problems 1 through 6, find the mass and center of mass of the shell Σ.

1. Σ is a triangle with vertices $(1, 0, 0)$, $(0, 3, 0)$, and $(0, 0, 2)$ with $\delta(x, y, z) = xz + 1$.

2. Σ is the part of the sphere $x^2 + y^2 + z^2 = 9$ above the plane $z = 1$, and the density function is constant.

3. Σ is the cone $z = \sqrt{x^2 + y^2}$ for $x^2 + y^2 \leq 9$, $\delta = $ constant.

4. Σ is the part of the paraboloid $z = 16 - x^2 - y^2$ in the first octant and between the cylinders $x^2 + y^2 = 1$ and $x^2 + y^2 = 9$ with $\delta(x, y, z) = xy/\sqrt{1 + 4x^2 + 4y^2}$.

5. Σ is the paraboloid $z = 6 - x^2 - y^2$ for $z \geq 0$ with $\delta(x, y, z) = \sqrt{1 + 4x^2 + 4y^2}$.

6. Σ is the part of the sphere of radius 1 about the origin, lying in the first octant. The density is constant.

7. Find the flux of $\mathbf{F} = x\mathbf{i} + y\mathbf{j} - z\mathbf{k}$ across the part of the plane in the first octant.

8. Find the flux of $\mathbf{F} = xz\mathbf{i} - y\mathbf{k}$ across the part of the sphere $x^2 + y^2 + z^2 = 4$ above the plane $z = 1$.

11.7   The Divergence Theorem of Gauss

In this and the next section, we will develop the fundamental theorems of vector integral calculus. To get some insight into the divergence theorem of Gauss, begin with Green's theorem. Under appropriate conditions on f and g,

$$\oint_C f\, dx + g\, dy = \iint_D \left(\frac{\partial g}{\partial x} - \frac{\partial f}{\partial y} \right) dA,$$

where C is a simple closed path enclosing the bounded region D. Define a vector field in the plane by

$$\mathbf{F} = g\mathbf{i} - f\mathbf{j}$$

and observe that

$$\nabla \cdot \mathbf{F} = \frac{\partial g}{\partial x} - \frac{\partial f}{\partial y}.$$

Now parametrize C by arc length, so the coordinate functions are $x = x(s)$, $y = y(s)$ for $0 \leq s \leq L$. The unit tangent to C is $\mathbf{T}(s) = x'(s)\mathbf{i} + y'(s)\mathbf{j}$, and the unit normal is $\mathbf{n}(s) = y'(s)\mathbf{i} - x'(s)\mathbf{j}$. These vectors are shown in Figure 11.19. $\mathbf{n}$ is an outer normal to C.

Now

$$\mathbf{F} \cdot \mathbf{n} = g\frac{dy}{ds} + f\frac{dx}{ds},$$

so

$$\oint_C f\, dx + g\, dy = \oint_C \left[f\frac{dx}{ds} + g\frac{dy}{ds} \right] ds$$

$$= \oint_C \mathbf{F} \cdot \mathbf{n}\, ds.$$

We may therefore write a vector version of the conclusion of Green's theorem:

$$\oint_C \mathbf{F} \cdot \mathbf{n}\, ds = \iint_D \nabla \cdot \mathbf{F}\, dA. \tag{11.10}$$

In 3-space, we will call a surface Σ *closed* if it bounds a volume. A sphere is closed, as is a cube, while a hemisphere is not. A normal vector to Σ is an *outer normal* if, when represented as an arrow from a point of Σ, it points away from the volume bounded by Σ.

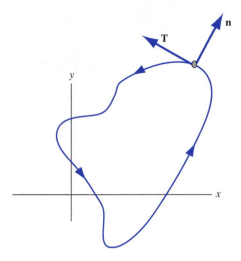

FIGURE 11.19 *Unit tangent and normal vectors to C.*

We can now attempt a generalization of equation (11.10) to 3-space as follows. Replace the closed curve C bounding D with a closed surface Σ bounding a solid region M. Replace the line integral over C with a surface integral over Σ and replace the double integral over D with the triple integral over M. This suggests the following.

THEOREM 11.4   *The Divergence Theorem of Gauss*

Let Σ be a piecewise smooth closed surface bounding a region M of 3-space. Let Σ have a unit outer normal $\mathbf{n}$. Let $\mathbf{F}$ be a vector field with continuous first and second partial derivatives on Σ and throughout M. Then

$$\iint_{\Sigma} \mathbf{F} \cdot \mathbf{n} \, d\sigma = \iiint_{M} \nabla \cdot \mathbf{F} \, dV. \qquad (11.11)$$

The theorem is named for the great nineteenth-century German mathematician and scientist Carl Friedrich Gauss and is actually a conservation of mass equation. Recall that the divergence of a vector field at a point is a measure of the flow of the field away from that point. Equation (11.11) states that the flux of the vector field outward from M across Σ exactly balances the flow of the field from each point in M. Whatever crosses the surface and leaves M must be accounted for by flow out of points of M (in the absence of sources or sinks in M).

We will look at two computational examples to get some feeling for equation (11.11) and then consider some substantial applications.

EXAMPLE 11.23

Let Σ be the piecewise smooth surface consisting of the surface Σ_1 of the cone $z = \sqrt{x^2 + y^2}$ for $x^2 + y^2 \leq 1$ together with the flat cap Σ_2 consisting of the disk $x^2 + y^2 \leq 1$ in the plane $z = 1$. Σ is shown in Figure 11.20. Let $\mathbf{F}(x, y, z) = x\mathbf{i} + y\mathbf{j} + z\mathbf{k}$. We will calculate both sides of equation (11.11).

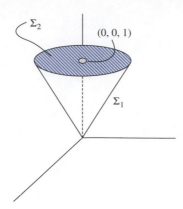

FIGURE 11.20 Σ *in Example 11.23.*

The unit outer normal to Σ_1 is

$$\mathbf{n}_1 = \frac{1}{\sqrt{2}}\left(\frac{x}{z}\mathbf{i} + \frac{y}{z}\mathbf{j} - \mathbf{k}\right).$$

Then

$$\mathbf{F} \cdot \mathbf{n}_1 = \frac{1}{\sqrt{2}}\left(\frac{x^2}{z} + \frac{y^2}{z} - z\right) = 0$$

because on Σ_1, $z^2 = x^2 + y^2$. Therefore,

$$\iint_{\Sigma_1} \mathbf{F} \cdot \mathbf{n}_1 \, d\sigma = 0.$$

The unit outer normal to Σ_2 is $\mathbf{n}_2 = \mathbf{k}$, so $\mathbf{F} \cdot \mathbf{n}_2 = z$. Since $z = 1$ on Σ_2, then

$$\iint_{\Sigma_2} \mathbf{F} \cdot \mathbf{n}_2 \, d\sigma = \iint_{\Sigma_2} z \, d\sigma = \iint_{\Sigma_2} d\sigma$$

$$= \text{area of } \Sigma_2 = \pi.$$

Therefore,

$$\iint_{\Sigma} \mathbf{F} \cdot \mathbf{n} \, d\sigma = \iint_{\Sigma_1} \mathbf{F} \cdot \mathbf{n}_1 \, d\sigma + \iint_{\Sigma_2} \mathbf{F} \cdot \mathbf{n}_2 \, d\sigma = \pi.$$

Now compute the triple integral. The divergence of $\mathbf{F}$ is

$$\nabla \cdot \mathbf{F} = \frac{\partial}{\partial x}x + \frac{\partial}{\partial y}y + \frac{\partial}{\partial z}z = 3,$$

so

$$\iiint_{M} \nabla \cdot \mathbf{F} \, dV = \iiint_{M} 3 \, dV$$

$$= 3[\text{volume of the cone of height 1, radius 1}]$$

$$= 3\left(\frac{1}{3}\right)\pi = \pi. \quad \blacklozenge$$

EXAMPLE 11.24

Let Σ be the piecewise smooth surface of the cube having vertices

$$(0, 0, 0), (1, 0, 0), (0, 1, 0), (0, 0, 1)$$

$$(1, 1, 0), (0, 1, 1), (1, 0, 1), (1, 1, 1).$$

Let $\mathbf{F}(x, y, z) = x^2\mathbf{i} + y^2\mathbf{j} + z^2\mathbf{k}$. We want to compute the flux of $\mathbf{F}$ across Σ. This flux is $\int\int_\Sigma \mathbf{F} \cdot \mathbf{n}\, d\sigma$. We can certainly evaluate this integral, but this will be tedious because Σ has six smooth faces. It is easier to use the triple integral of the divergence theorem. Compute

$$\nabla \cdot \mathbf{F} = 2x + 2y + 2z.$$

Then

$$\text{flux} = \iint_\Sigma \mathbf{F} \cdot \mathbf{N}\, d\sigma$$

$$= \iint_M \nabla \cdot \mathbf{F}\, dV = \iiint_M (2x + 2y + 2z)\, dV$$

$$= \int_0^1 \int_0^1 \int_0^1 (2x + 2y + 2z)\, dz\, dy\, dx$$

$$= \int_0^1 \int_0^1 (2x + 2y + 1)\, dy\, dx$$

$$= \int_0^1 (2x + 2)\, dx = 3. \quad \blacklozenge$$

11.7.1 Archimedes's Principle

Archimedes's principle is that the buoyant force a fluid exerts on a solid object immersed in it is equal to the weight of the fluid displaced. An aircraft carrier floats in the ocean if it displaces a volume of water whose weight at least equals that of the carrier. We will use the divergence theorem to derive this principle.

Imagine a solid object M bounded by a piecewise smooth surface Σ. Let ρ be the constant density of the fluid. Draw a coordinate system, as in Figure 11.21, with M below the surface. Using the fact that pressure equals depth multiplied by density, the pressure $p(x, y, z)$ at a point on Σ is $p(x, y, z) = -\rho z$. The negative sign is given because z is negative in the downward direction, and we want pressure to be positive. Now consider a piece Σ_j of Σ. The force of the pressure on Σ_j is approximately $-\rho z$ multiplied by the area A_j of Σ_j. If $\mathbf{n}$ is the unit outer normal to Σ_j, then the force caused by the pressure on Σ_j is approximately $\rho z \mathbf{n} A_j$. The vertical component of this force is the magnitude of the buoyant force acting upward on Σ_j. This vertical component is $\rho z \mathbf{n} \cdot \mathbf{k} A_j$. Sum these vertical components over Σ to obtain approximately the buoyant force on the object, then take the limit as the surface elements are chosen smaller. We obtain

$$\text{net buoyant force on} \Sigma = \iint_\Sigma \rho z \mathbf{n} \cdot \mathbf{k}\, d\sigma.$$

Write this integral as $\int\int_\Sigma \rho z \mathbf{k} \cdot \mathbf{n}\, d\sigma$ and apply the divergence theorem to obtain

$$\text{net buoyant force on} \Sigma = \iiint_M \nabla \cdot (\rho z \mathbf{k})\, dV.$$

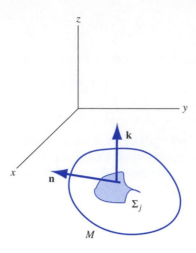

FIGURE 11.21   *Archimedes's Principle.*

But $\nabla \cdot (\rho z \mathbf{k}) = \rho$, so

$$\text{net buoyant force on} \, \Sigma = \iiint\limits_M \rho \, dV = \rho[\text{volume of } M],$$

and ρ multiplied by the volume of M is the weight of the object.

11.7.2   The Heat Equation

We will use the divergence theorem to derive a partial differential equation that models heat conduction and diffusion processes. Suppose some medium (such as a bar of metal or water in a pool) has density $\rho(x, y, z)$, specific heat $\mu(x, y, z)$, and coefficient of thermal conductivity $K(x, y, z)$. Let $u(x, y, z, t)$ be the temperature of the medium at (x, y, z) and time t. We want an equation for u.

We will exploit an idea frequently used in constructing mathematical models. Consider an imaginary smooth closed surface Σ in the medium, bounding a solid region M. The amount of heat energy leaving M across Σ in a time interval Δt is

$$\left(\iint\limits_\Sigma (K \nabla u) \cdot \mathbf{n} \, d\sigma \right) \Delta t.$$

This is the flux of $K \nabla u$ across Σ multiplied by the length of the time interval. But, the change in temperature at (x, y, z) over Δt is approximately $(\partial u / \partial t) \Delta t$, so the resulting heat loss in M is

$$\left(\iiint\limits_M \mu \rho \frac{\partial u}{\partial t} \, dV \right) \Delta t.$$

Assuming no heat sources or sinks in M (which occur, for example, during chemical reactions or radioactive decay), the change in heat energy in M over Δt must equal the heat exchange across Σ:

$$\left(\iint\limits_\Sigma (K \nabla u) \cdot \mathbf{n} \, d\sigma \right) \Delta t = \left(\iiint\limits_M \mu \rho \frac{\partial u}{\partial t} \, dV \right) \Delta t.$$

Therefore,

$$\iint\limits_\Sigma (K \nabla u) \cdot \mathbf{n} \, d\sigma = \iiint\limits_M \mu \rho \frac{\partial u}{\partial t} \, dV.$$

Now use the divergence theorem to convert the surface integral to a triple integral:

$$\iint_{\Sigma} (K\nabla u) \cdot \mathbf{n}\, d\sigma = \iiint_{M} \nabla \cdot (K\nabla u)\, dV.$$

Substitute this into the preceding equation to obtain

$$\iiint_{M} \left(\mu\rho \frac{\partial u}{\partial t} - \nabla \cdot (K\nabla u) \right) dV = 0.$$

Now Σ is an arbitrary closed surface within the medium. If the integrand in the last equation were, say, positive at some point P_0, then it would be positive throughout some (perhaps very small) sphere about P_0, and we could choose Σ as this sphere. But then the triple integral over M of a positive quantity would be nonzero, which is a contradiction. The same conclusion follows if this integrand were negative at some P_0.

Vanishing of the last integral for *every* closed surface Σ in the medium therefore forces the integrand to be identically zero:

$$\mu\rho \frac{\partial u}{\partial t} - \nabla \cdot (K\nabla u) = 0.$$

This is the partial differential equation

$$\mu\rho \frac{\partial u}{\partial t} = \nabla \cdot (K\nabla u)$$

for the temperature function. This equation is called the *heat equation*.

We can expand

$$\nabla \cdot (K\nabla u) = \nabla \cdot \left(K\frac{\partial u}{\partial x}\mathbf{i} + K\frac{\partial u}{\partial y}\mathbf{j} + K\frac{\partial u}{\partial z}\mathbf{k} \right)$$

$$= \frac{\partial}{\partial x}\left(K\frac{\partial u}{\partial x} \right) + \frac{\partial}{\partial y}\left(K\frac{\partial u}{\partial y} \right) + \frac{\partial}{\partial z}\left(K\frac{\partial u}{\partial z} \right)$$

$$= \frac{\partial K}{\partial x}\frac{\partial u}{\partial x} + \frac{\partial K}{\partial y}\frac{\partial u}{\partial y} + \frac{\partial K}{\partial z}\frac{\partial u}{\partial z} + K\left(\frac{\partial^2 u}{\partial x^2} + \frac{\partial^2 u}{\partial y^2} + \frac{\partial^2 u}{\partial z^2} \right)$$

$$= \nabla K \cdot \nabla u + K\nabla^2 u.$$

Here we have introduced the symbol ∇^2, which is defined by

$$\nabla^2 u = \frac{\partial^2 u}{\partial x^2} + \frac{\partial^2 u}{\partial y^2} + \frac{\partial^2 u}{\partial z^2}.$$

∇^2 is called the *Laplacian*, or *Laplacian operator*, and is read "del squared." Now the heat equation is

$$\mu\rho \frac{\partial u}{\partial t} = \nabla K \cdot \nabla u + K\nabla^2 u.$$

If K is constant, then its gradient vector is zero, and this equation simplifies to

$$\frac{\partial u}{\partial t} = \frac{K}{\mu\rho}\nabla^2 u.$$

In the case of one space dimension, $u = u(x, t)$, and we often write this equation as

$$\frac{\partial u}{\partial t} = k\frac{\partial^2 u}{\partial x^2}$$

in which $k = K/\mu\rho$.

The *steady-state heat equation* occurs when $\partial u/\partial t = 0$. In this case, we get $\nabla^2 u = 0$, which is called *Laplace's equation*.

In each of Problems 1 through 8, evaluate either $\int\int_{\Sigma} \mathbf{F} \cdot \mathbf{n}\, d\sigma$ or $\int\int\int_{M} \operatorname{div}(\mathbf{F})\, dV$, whichever is easier.

1. $\mathbf{F} = x\mathbf{i} + y\mathbf{j} - z\mathbf{k}$, Σ is the sphere of radius 4 about $(1, 1, 1)$.

2. $\mathbf{F} = 4x\mathbf{i} - 6y\mathbf{j} + z\mathbf{k}$, Σ is the surface of the solid cylinder $x^2 + y^2 \leq 4$, $0 \leq z \leq 2$, including the end caps of the cylinder.

3. $\mathbf{F} = 2yz\mathbf{i} - 4xz\mathbf{j} + xy\mathbf{k}$, Σ is the sphere of radius 5 about $(-1, 3, 1)$

4. $\mathbf{F} = x^3\mathbf{i} + y^3\mathbf{j} + z^3\mathbf{k}$, Σ is the sphere of radius 1 about the origin.

5. $\mathbf{F} = 4x\mathbf{i} - z\mathbf{j} + x\mathbf{k}$, Σ is the hemisphere $x^2 + y^2 + z^2 = 1$, $z \geq 0$, including the base consisting of points $(x, y, 0)$ with $x^2 + y^2 \leq 1$.

6. $\mathbf{F} = (x - y)\mathbf{i} + (y - 4xz)\mathbf{j} + xz\mathbf{k}$, Σ is the surface of the rectangular box bounded by the coordinate

planes $x = 0, y = 0, z = 0$ and the planes $x = 4$, $y = 2, z = 3$.

7. $\mathbf{F} = x^2\mathbf{i} + y^2\mathbf{j} + z^2\mathbf{k}$, Σ is the cone $z = \sqrt{x^2 + y^2}$ for $x^2 + y^2 \leq 2$, together with the top cap consisting of the points $(x, y, \sqrt{2})$ with $x^2 + y^2 \leq 2$.

8. $\mathbf{F} = x^2\mathbf{i} - e^z\mathbf{j} + z\mathbf{k}$, Σ is the surface bounding the cylinder $x^2 + y^2 \leq 4$, $0 \leq z \leq 2$, including the top and bottom caps of the cylinder.

9. Let Σ be a smooth closed surface and $\mathbf{F}$ a vector field that is continuous with continuous first and second partial derivatives throughout Σ and the region it bounds. Evaluate $\int\int_{\Sigma}(\nabla \times \mathbf{F}) \cdot \mathbf{n}\, d\sigma$.

10. Let Σ be a piecewise smooth closed surface bounding a region M. Show that

$$\text{volume of } M = \frac{1}{3} \iiint_{\Sigma} \mathbf{R} \cdot \mathbf{n}\, d\sigma$$

where $\mathbf{R} = x\mathbf{i} + y\mathbf{j} + z\mathbf{k}$.

11.8   Stokes's Theorem

The divergence theorem can be thought of as a lifting of Green's theorem to 3-space. We can approach Stokes's theorem in the same way. Begin again with

$$\oint_{C} f\, dx + g\, dy = \iint_{D} \left(\frac{\partial g}{\partial x} - \frac{\partial f}{\partial y} \right) dA.$$

Let

$$\mathbf{F}(x, y, z) = f(x, y)\mathbf{i} + g(x, y)\mathbf{j} + 0\mathbf{k}.$$

By thinking of $\mathbf{F}$ as a vector in 3-space, we can take the curl:

$$\nabla \times \mathbf{F} = \begin{vmatrix} \mathbf{i} & \mathbf{j} & \mathbf{k} \\ \partial/\partial x & \partial/\partial y & \partial/\partial z \\ f & g & 0 \end{vmatrix} = \left(\frac{\partial g}{\partial x} - \frac{\partial f}{\partial y} \right) \mathbf{k}.$$

Then

$$(\nabla \times \mathbf{F}) \cdot \mathbf{k} = \frac{\partial g}{\partial x} - \frac{\partial f}{\partial y}.$$

We can therefore write the conclusion of Green's theorem as

$$\oint_{C} \mathbf{F} \cdot d\mathbf{R} = \iint_{D} (\nabla \times \mathbf{F}) \cdot \mathbf{k}\, dA. \tag{11.12}$$

Now think of D as a flat surface in the (x, y) plane with a unit normal $\mathbf{k}$ and bounded by the closed curve C. To lift this up one dimension, allow C to be a curve in 3-space bounded by a surface Σ having a unit outer normal $\mathbf{n}$, as shown in Figure 11.22. Equation (11.12) suggests that

$$\oint_{C} \mathbf{F} \cdot \mathbf{T}\, ds = \iint_{\Sigma} (\nabla \times \mathbf{F}) \cdot \mathbf{N}\, d\sigma.$$

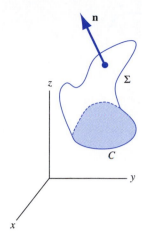

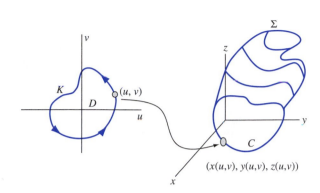

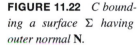

FIGURE 11.22 *C bounding a surface* Σ *having outer normal* **N**.

FIGURE 11.23 *Boundary curve of a surface.*

With some care in confronting subtleties and defining terms, this will be the conclusion of Stokes's theorem.

First we need to explore the idea of a surface and a bounding curve. Suppose Σ is a surface defined by $x = x(u, v)$, $y = y(u, v)$, and $z = z(u, v)$ for (u, v) in some bounded region D of the (u, v) plane. As (u, v) varies over D, the point $(x(u, v), y(u, v), z(u, v))$ traces out Σ. We will assume that D is bounded by a piecewise smooth curve K. As (u, v) traverses K, the corresponding point $(x(u, v), y(u, v), z(u, v))$ traverses a curve C on Σ. This is the curve we call the boundary curve of Σ. This is shown in Figure 11.23.

EXAMPLE 11.25

Let Σ be given by $z = x^2 + y^2$ for $0 \leq x^2 + y^2 \leq 4$. Here x and y are the parameters and vary over the disk of radius 2 in the (x, y) plane. Figure 11.24 shows D and a graph of the surface. The boundary K of D is the circle $x^2 + y^2 = 4$ in the (x, y) plane. This circle maps to the circle $x^2 + y^2 = 4, z = 4$ on Σ. This is the boundary C of Σ, and it is the circle at the top of the bowl-shaped surface. ◆

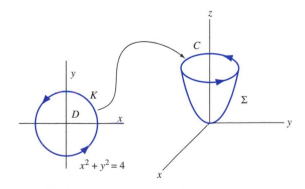

FIGURE 11.24 *The surface in Example 11.25.*

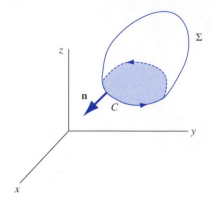

FIGURE 11.25 *Orienting the boundary curve coherently with the normal vector.*

We need a rule for choosing a normal to the surface at each point. We can use the standard normal vector

$$\frac{\partial(y, z)}{\partial(u, v)}\mathbf{i} + \frac{\partial(z, x)}{\partial(u, v)}\mathbf{j} + \frac{\partial(x, y)}{\partial(u, v)}\mathbf{k},$$

dividing this by its length to obtain a unit normal vector. The negative of this is also a unit normal vector. Whichever we use, call it $\mathbf{n}$ and use it consistently throughout the surface. We cannot use $\mathbf{n}$ at some points and $-\mathbf{n}$ at others.

This choice of the normal vector $\mathbf{n}$ is used to determine an orientation on the boundary curve C of Σ. Referring to Figure 11.25, at any point on C, if you stand along $\mathbf{n}$ with your head at the tip of this normal, then the positive direction of C is the one in which you have to walk to have the surface over your left shoulder. This is admittedly informal, but a more rigorous treatment involves topological subtleties we do not wish to engage here. When this direction is chosen on C, we say that C has been *oriented coherently* with $\mathbf{n}$. The choice of normal determines the orientation on the boundary curve. There is no intrinsic positive or negative orientation on this curve in 3-space, simply an orientation coherent with the chosen normal.

With these conventions, we can state the theorem.

THEOREM 11.5   Stokes's Theorem

Let Σ be a piecewise smooth surface bounded by a piecewise smooth curve C. Suppose a unit normal $\mathbf{n}$ has been chosen on Σ, and that C is oriented coherently with this normal. Let $\mathbf{F}(x, y, z)$ be a vector field that is continuous with continuous first and second partial derivatives on Σ. Then,

$$\oint_C \mathbf{F} \cdot d\mathbf{R} = \iint_\Sigma (\nabla \times \mathbf{F}) \cdot \mathbf{n}\, d\sigma. \quad \blacklozenge$$

We will illustrate the theorem with a computational example.

EXAMPLE 11.26

Let $\mathbf{F}(x, y, z) = -y\mathbf{i} + xy\mathbf{j} - xyz\mathbf{k}$. Let Σ consist of the part of the cone $z = \sqrt{x^2 + y^2}$ for $0 \le x^2 + y^2 \le 9$. We will compute both sides of the equation in Stokes's theorem.

The bounding curve C of Σ is the circle $x^2 + y^2 = 3$ and $z = 3$ at the top of the cone. This is similar to Example 11.25. A routine computation yields the normal vector

$$\mathbf{N} = -\frac{x}{z}\mathbf{i} - \frac{y}{z}\mathbf{j} + \mathbf{k}.$$

There is no normal at the origin, where the cone has a sharp point.

For Stokes's theorem, we need a unit normal vector, so divide $\mathbf{N}$ by $\|\mathbf{N}\|$ to get

$$\mathbf{n} = \frac{1}{\sqrt{2}z}(-x\mathbf{i} - y\mathbf{j} + \mathbf{k}).$$

Notice that this vector, if represented as an arrow from a point on the cone, points into the region bounded by the cone. Figure 11.26 shows the orientation on C coherent with $\mathbf{n}$.

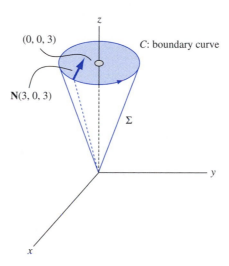

FIGURE 11.26 *The surface and boundary curve in Example 11.26.*

We are now ready to evaluate $\oint_C \mathbf{F} \cdot d\mathbf{R}$. Parametrize C by $x = 3\cos(t)$, $y = 3\sin(t)$, and $z = 3$ for $0 \le t \le 2\pi$. The point $(3\cos(t), 3\sin(t), 3)$ traverses C in the positive direction at t increases from 0 to 2π. Therefore,

$$\oint_C \mathbf{F} \cdot d\mathbf{R} = \oint_C -y\,dx + x\,dy - xyz\,dz$$
$$= \int_0^{2\pi} [-3\sin(t)(-3\cos(t)) + 3\cos(t)(3\cos(t))]dt$$
$$= \int_0^{2\pi} 9\,dt = 18\pi.$$

For the surface integral, compute

$$\nabla \times \mathbf{F} = -xz\mathbf{i} + yz\mathbf{j} + 2\mathbf{k}$$

and

$$(\nabla \times \mathbf{F}) \cdot \mathbf{n} = \frac{1}{\sqrt{2}}(x^2 - y^2 + 2).$$

Then

$$\iint_{\Sigma} (\nabla \times \mathbf{F}) \cdot \mathbf{n} \, d\sigma = \iint_{D} (\nabla \times \mathbf{F}) \cdot \mathbf{n} \parallel \mathbf{N} \parallel dx \, dy$$

$$= \iint_{D} \frac{1}{\sqrt{2}}(x^2 - y^2 + 2)\sqrt{2} dx \, dy$$

$$= \iint_{D} (x^2 - y^2 + 2) \, dx \, dy$$

$$= \int_{0}^{2\pi} \int_{0}^{3} [r^2 \cos^2(\theta) - r^2 \sin^2(\theta)] r \, dr \, d\theta$$

$$= \int_{0}^{2\pi} \int_{0}^{3} r^3 \cos(2\theta) \, dr \, d\theta + \int_{0}^{2\pi} \int_{0}^{3} 2r \, dr \, d\theta$$

$$= \left[\frac{1}{2} \sin(2\theta) \right]_{0}^{2\pi} \left[\frac{1}{4} r^4 \right]_{0}^{3} + 2\pi \left[r^2 \right]_{0}^{3} = 18\pi. \quad \blacklozenge$$

SECTION 11.8   *PROBLEMS*

In each of Problems 1 through 5, use Stokes's theorem to evaluate $\oint_C \mathbf{F} \cdot d\mathbf{R}$ or $\int \int_{\Sigma} (\nabla \times \mathbf{F}) \cdot \mathbf{n} \, d\sigma$, whichever appears easier.

1. $\mathbf{F} = yx^2\mathbf{i} - xy^2\mathbf{j} + z^2\mathbf{k}$ with Σ the hemisphere $x^2 + y^2 + z^2 = 4, z \geq 0$.

2. $\mathbf{F} = xy\mathbf{i} + yz\mathbf{j} + xz\mathbf{k}$ with Σ the paraboloid $z = x^2 + y^2$ for $x^2 + y^2 \leq 9$.

3. $\mathbf{F} = z\mathbf{i} + x\mathbf{j} + y\mathbf{k}$ with Σ the cone $z = \sqrt{x^2 + y^2}$ for $0 \leq z \leq 4$.

4. $\mathbf{F} = z^2\mathbf{i} + x^2\mathbf{j} + y^2\mathbf{k}$ with Σ the part of the paraboloid $z = 6 - x^2 - y^2$ above the (x, y) plane.

5. $\mathbf{F} = xy\mathbf{i} + yz\mathbf{j} + xy\mathbf{k}$ with Σ the part of the plane $2x + 4y + z = 8$ in the first octant.

6. Calculate the circulation of $\mathbf{F} = (x - y)\mathbf{i} + x^2y\mathbf{j} + axz\mathbf{k}$ counterclockwise about the circle $x^2 + y^2 = 1$. Here a is any positive number. *Hint*: Use Stokes's theorem with Σ as any smooth surface having the circle as boundary.

PART 4

Fourier
Analysis and
Eigenfunction
Expansions

$$f](s) = \int_0^\infty e^{-st} t^{-1/2} \, dt = 2 \int_0^\infty e^{-sx^2} \, dx \ (\text{set } x = t^{1/2}) = \frac{2}{\sqrt{s}} \int_0^\infty e^{-z} \, dz \ (\text{set } z = x$$

$$= \sqrt{\frac{\pi}{s}}$$

$$\mathcal{L}[f](s) = \int_0^\infty e$$

$$= 2 \int_0^\infty e^{-s} \, dx$$

$$= \sqrt{\frac{\pi}{s}}$$

$$(s) = \int_0^\infty e^{-st} t^{-1/2} \, dt$$

$$= 2 \int_0^\infty e^{-sx^2} \, dx \ (\text{set } x = t^{1/2})$$

$$= \frac{2}{\sqrt{s}} \int_0^\infty e^{-z^2} \, dz \ (\text{set } z = x\sqrt{s})$$

$$[f](s) = \int_0^\infty e^{-st} t^{-1/2}$$

CHAPTER 12

Fourier Series

In 1807, the French mathematician Joseph Fourier submitted a paper to the French Academy of Sciences in competition for a prize offered for the best mathematical treatment of heat conduction. In the course of this work, Fourier shocked his contemporaries by asserting that "arbitrary" functions (specifying initial temperatures) could be expanded in series of sines and cosines. Consequences of Fourier's work have had an enormous impact on mathematics, science, engineering, medicine, analysis and storage of data, and other areas.

12.1 The Fourier Series of a Function

Let $f(x)$ be defined on $[-L, L]$. We want to choose numbers $a_0, a_1, a_2, \cdots$ and $b_1, b_2, \cdots$, so that

$$f(x) = \frac{1}{2}a_0 + \sum_{k=1}^{\infty}[a_k \cos(k\pi x/L) + b_k \sin(k\pi x/L)]. \tag{12.1}$$

This is a decomposition of the function into a sum of terms, each representing the influence of a different fundamental frequency on the behavior of the function.

To find a_0, integrate equation (12.1) term by term to get

$$\int_{-L}^{L} f(x)dx = \frac{1}{2}\int_{-L}^{L} a_0 dx$$

$$+ \sum_{k=1}^{\infty} a_k \int_{-L}^{L} \cos(k\pi x/L)\,dx + b_k \int_{-L}^{L} \sin(k\pi x/L)\,dx$$

$$= \frac{1}{2}a_0(2L) = \pi a_0,$$

because all of the integrals in the summation are zero. Then

$$a_0 = \frac{1}{L}\int_{-L}^{L} f(x)\,dx. \tag{12.2}$$

269

To solve for the other coefficients in the proposed equation (12.1), we will use the following three facts, which follow by routine integrations.

1. Let m and n be integers. Then

$$\int_{-L}^{L} \cos\left(\frac{n\pi x}{L}\right) \sin\left(\frac{m\pi x}{L}\right) dx = 0. \tag{12.3}$$

2. If $n \neq m$, then

$$\int_{-L}^{L} \cos\left(\frac{n\pi x}{L}\right) \cos\left(\frac{m\pi x}{L}\right) dx = \int_{-L}^{L} \sin\left(\frac{n\pi x}{L}\right) \sin\left(\frac{m\pi x}{L}\right) dx = 0. \tag{12.4}$$

3. If $n \neq 0$, then

$$\int_{-L}^{L} \cos^2\left(\frac{n\pi x}{L}\right) dx = \int_{-L}^{L} \sin^2\left(\frac{n\pi x}{L}\right) dx = L. \tag{12.5}$$

Now let n be any positive integer. To solve for a_n, multiply equation (12.1) by $\cos(n\pi x/L)$ and integrate the resulting equation to get

$$\int_{-L}^{L} f(x) \cos\left(\frac{n\pi x}{L}\right) dx = \frac{1}{2}a_0 \int_{-L}^{L} \cos\left(\frac{n\pi x}{L}\right) dx$$

$$+ \sum_{k=1}^{\infty} \left[a_k \int_{-L}^{L} \cos\left(\frac{k\pi x}{L}\right) \cos\left(\frac{n\pi x}{L}\right) dx + b_k \int_{-L}^{L} \sin\left(\frac{k\pi x}{L}\right) \cos\left(\frac{n\pi x}{L}\right) dx \right].$$

By equations (12.3) and (12.4), all of the terms on the right are zero except the coefficient of a_n, which occurs in the summation when $k = n$. Using equation (12.5), the last equation reduces to

$$\int_{-L}^{L} f(x) \cos\left(\frac{n\pi x}{L}\right) dx = a_n \int_{-L}^{L} \cos^2\left(\frac{n\pi x}{L}\right) dx = a_n L$$

Therefore,

$$a_n = \frac{1}{L} \int_{-L}^{L} f(x) \cos\left(\frac{n\pi x}{L}\right) dx. \tag{12.6}$$

This expression contains a_0 if we let $n = 0$.

Similarly, if we multiply equation (12.1) by $\sin(n\pi x/L)$ instead of $\cos(n\pi x/L)$ and integrate, we obtain

$$b_n = \frac{1}{L} \int_{-L}^{L} f(x) \sin\left(\frac{n\pi x}{L}\right) dx. \tag{12.7}$$

Fourier Coefficients and Series

The numbers

$$a_n = \frac{1}{L} \int_{-L}^{L} f(x) \cos\left(\frac{n\pi x}{L}\right) dx \quad \text{for} \quad n = 0, 1, 2, \cdots \tag{12.8}$$

and

$$b_n = \frac{1}{L} \int_{-L}^{L} f(x) \sin\left(\frac{n\pi x}{L}\right) dx \quad \text{for} \quad n = 1, 2, \cdots \tag{12.9}$$

are called the *Fourier coefficients of f on* $[L, L]$. When these numbers are used, the series in equation (12.1) is called the *Fourier series of f on* $[L, L]$.

EXAMPLE 12.1

Let $f(x) = x - x^2$ for $-\pi \leq x \leq \pi$. Here $L = \pi$. Compute

$$a_0 = \frac{1}{\pi} \int_{-\pi}^{\pi} (x - x^2)\, dx = -\frac{2}{3}\pi^2,$$

$$
\begin{aligned}
a_n &= \frac{1}{\pi} \int_{-\pi}^{\pi} (x - x^2)\cos(nx)\, dx \\
&= \frac{4\sin(n\pi) - 4n\pi\cos(n\pi) - 2n^2\pi^2\sin(n\pi)}{\pi n^3} \\
&= -\frac{4}{n^2}\cos(n\pi) = -\frac{4}{n^2}(-1)^n \\
&= \frac{4(-1)^{n+1}}{n^2},
\end{aligned}
$$

and

$$
\begin{aligned}
b_n &= \frac{1}{\pi} \int_{-\pi}^{\pi} (x - x^2)\sin(nx)\, dx \\
&= \frac{2\sin(n\pi) - 2n\pi\cos(n\pi)}{\pi n^2} \\
&= -\frac{2}{n}\cos(n\pi) = -\frac{2}{n}(-1)^n = \frac{2(-1)^{n+1}}{n},
\end{aligned}
$$

since $\sin(n\pi) = 0$ and $\cos(n\pi) = (-1)^n$ if n is an integer.

The Fourier series of $f(x) = x - x^2$ on $[-\pi, \pi]$ is

$$-\frac{1}{3}\pi^2 + \sum_{n=1}^{\infty}\left[\frac{4(-1)^{n+1}}{n^2}\cos(nx) + \frac{2(-1)^{n+1}}{n}\sin(nx)\right]. \quad \blacklozenge$$

This example illustrates a fundamental issue. We do not know what this Fourier series converges to. We must establish a relationship between the function and its Fourier series on an interval. This will require some assumptions about the function.

Recall that f is *piecewise continuous* on $[a, b]$ if f is continuous at all but perhaps finitely many points of this interval, and, at a point where the function is not continuous, f has a jump discontinuity, which appears as a finite gap in the graph. Figure 12.1 shows a typical piecewise continuous function.

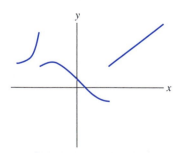

FIGURE 12.1 *A piecewise continuous function.*

If $a < x_0 < b$, denote the left limit of $f(x)$ at x_0 as $f(x_0-)$ and the right limit of $f(x)$ at x_0 as $f(x_0+)$:

$$f(x_0-) = \lim_{h \to 0-} f(x_0 + h) \quad \text{and} f(x_0+) = \lim_{h \to 0+} f(x_0 + h).$$

If f is continuous at x_0, then these left and right limits both equal $f(x_0)$.

EXAMPLE 12.2

Let

$$f(x) = \begin{cases} x & \text{for } -3 \le x < 2 \\ \sin(6x) & \text{for } 2 \le x \le 4. \end{cases}$$

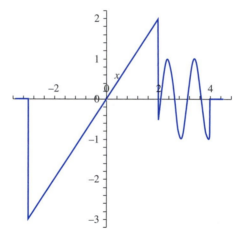

FIGURE 12.2   *f in Example 12.2.*

f is piecewise continuous on $[-3, 4]$, having a single discontinuity at $x = 2$. Furthermore, $f(2-) = 2$, and $f(2+) = 1/2$. A graph of f is shown in Figure 12.2. ◆

> f is *piecewise smooth* on $[a, b]$ if f is piecewise continuous and f' exists and is continuous at all but perhaps finitely many points of (a, b).

EXAMPLE 12.3

The function f of Example 12.2 is differentiable on $(-3, 4)$ except at $x = 2$:

$$f'(x) = \begin{cases} 1 & \text{for } -3 < x < 2 \\ 6\cos(6x) & \text{for } 2 < x < 4. \end{cases}$$

This derivative is itself piecewise continuous. Therefore, f is piecewise smooth on $[-3, 4]$. ◆

We can now state a convergence theorem.

THEOREM 12.1   *Convergence of Fourier Series*

Let f be piecewise smooth on $[-L, L]$. Then, for each x in $(-L, L)$, the Fourier series of f on $[-L, L]$ converges to

$$\frac{1}{2}(f(x+) + f(x-)).$$

At both $-L$ and L this Fourier series converges to

$$\frac{1}{2}(f(L-) + f(-L+)). \quad \blacklozenge$$

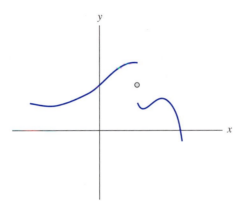

FIGURE 12.3 *Convergence of a Fourier series at a jump discontinuity.*

At any point in $(-L, L)$ at which $f(x)$ is continuous, the Fourier series converges to $f(x)$, because then the right and left limits at x are both equal to $f(x)$. At a point interior to the interval where f has a jump discontinuity, the Fourier series converges to the average of the left and right limits there. This is the point midway in the gap of the graph at the jump discontinuity (Figure 12.3). The Fourier series has the same sum at both ends of the interval.

EXAMPLE 12.4

Let $f(x) = x - x^2$ for $-\pi \leq x \leq \pi$. In Example 12.1, we found the Fourier series of f on $[-\pi, \pi]$. Now we can examine the relationship between this series and $f(x)$.

$f'(x) = 1 - 2x$ is continuous for all x, hence f is piecewise smooth on $[-\pi, \pi]$. For $-\pi < x < \pi$, the Fourier series converges to $x - x^2$. At both π and $-\pi$, the Fourier series converges to

$$\frac{1}{2}(f(\pi-) + f(-\pi+)) = \frac{1}{2}((\pi - \pi^2) + (-\pi - (-\pi)^2))$$

$$= \frac{1}{2}(-2\pi^2) = -\pi^2.$$

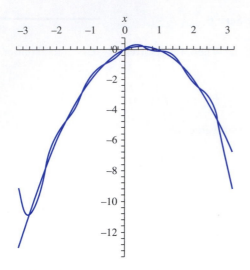

FIGURE 12.4 *Fifth partial sum of the Fourier series in Example 12.4.*

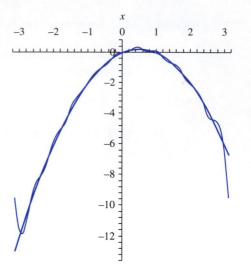

FIGURE 12.5 *Tenth partial sum in Example 12.4.*

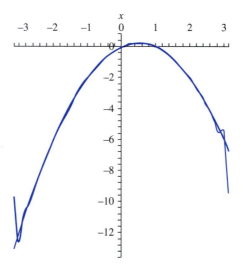

FIGURE 12.6 *Twentieth partial sum in Example 12.4.*

Figures 12.4, 12.5, and 12.6 show the fifth, tenth, and twentieth partial sums of this Fourier series together with a graph of f for comparison. The partial sums are seen to approach the function as more terms are included. ◆

EXAMPLE 12.5

Let $f(x) = e^x$. The Fourier coefficients of f on $[-1, 1]$ are

$$a_0 = \int_{-1}^{1} e^x \, dx = e - e^{-1},$$

$$a_n = \int_{-1}^{1} e^x \cos(n\pi x)\, dx = \frac{(e - e^{-1})(-1)^n}{1 + n^2 \pi^2},$$

and

$$b_n = \int_{-1}^{1} e^x \sin(n\pi x)\, dx = -\frac{(e - e^{-1})(-1)^n (n\pi)}{1 + n^2 \pi^2}.$$

The Fourier series of e^x on $[-1, 1]$ is

$$\frac{1}{2}(e - e^{-1}) + (e - e^{-1}) \sum_{n=1}^{\infty} \left(\frac{(-1)^n}{1 + n^2 \pi^2} \right) (\cos(n\pi x) - n\pi \sin(n\pi x)).$$

Because e^x is continuous with a continuous derivative for all x, this series converges to

$$\begin{cases} e^x & \text{for } -1 < x < 1 \\ \frac{1}{2}(e + e^{-1}) & \text{for } x = 1 \text{ and for } x = -1. \end{cases}$$

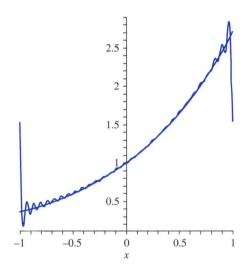

FIGURE 12.7 *Thirtieth partial sum of the Fourier series in Example 12.5.*

Figure 12.7 shows the thirtieth partial sum of this series, suggesting its convergence to the function except at the endpoints -1 and 1. ◆

EXAMPLE 12.6

Let

$$f(x) = \begin{cases} 5\sin(x) & \text{for } -2\pi \leq x < -\pi/2 \\ 4 & \text{for } x = -\pi/2 \\ x^2 & \text{for } -\pi/2 < x < 2 \\ 8\cos(x) & \text{for } 2 \leq x < \pi \\ 4x & \text{for } \pi \leq x \leq 2\pi. \end{cases}$$

f is piecewise smooth on $[-2\pi, 2\pi]$. The Fourier series of f on $[-2\pi, 2\pi]$ converges to

$$\text{Fourier series convergence} \begin{cases} 5\sin(x) & \text{for } -2\pi < x < -\pi/2 \\ \dfrac{1}{2}\left(\dfrac{\pi^2}{4} - 5\right) & \text{for } x = -\pi/2 \\ x^2 & \text{for } -\pi/2 < x < 2 \\ \dfrac{1}{2}(4 + 8\cos(2)) & \text{for } x = 2 \\ 8\cos(x) & \text{for } 2 < x < \pi \\ \dfrac{1}{2}(4\pi - 8) & \text{for } x = \pi \\ 4x & \text{for } -\pi < x < 2\pi \\ 4\pi & \text{for } x = 2\pi \text{ and } x = -2\pi. \end{cases}$$

This conclusion does not require that we write the Fourier series. ◆

12.1.1   Even and Odd Functions

A function f is *even* on $[-L, L]$ if the graph on $[-L, 0]$ is obtained by folding the graph on $[0, L]$ across the vertical axis. This happens when $f(-x) = f(x)$ for $0 < x \le L$. For example, x^{2n} and $\cos(n\pi x/L)$ are even on any $[-L, L]$ for any positive integer n.

A function f is *odd* on $[-L, L]$ if its graph on $[-L, 0)$ is the reflection of the graph on $(0, L]$ through the origin. This occurs when $f(-x) = -f(x)$ for $0 < x \le L$. For example, x^{2n+1} and $\sin(n\pi x/L)$ are odd on $[-L, L]$ for any positive integer n.

Figures 12.8 and 12.9 show examples of even and odd functions, respectively.

A product of two even functions is even, a product of two odd functions is even, and a product of an odd function with an even function is odd.

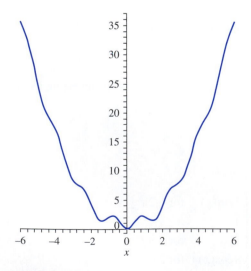

FIGURE 12.8 *An even function.*

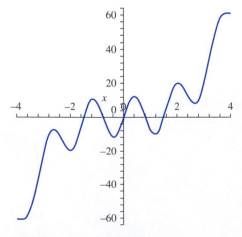

FIGURE 12.9 *An odd function.*

If f is even on $[-L, L]$, then

$$\int_{-L}^{L} f(x)\,dx = 2\int_{0}^{L} f(x)\,dx,$$

and if f is odd on $[-L, L]$, then

$$\int_{-L}^{L} f(x)\,dx = 0.$$

These facts are sometimes useful in computing Fourier coefficients. If f is even, only the cosine terms and possibly the constant term will appear in the Fourier series, because $f(x)\sin(n\pi x/L)$ is odd and the integrals defining the sine coefficients will be zero. If the function is odd, then $f(x)\cos(n\pi x/L)$ is odd and the Fourier series will contain only the sine terms, since the integrals defining the constant term and the coefficients of the cosine terms will be zero.

EXAMPLE 12.7

We will compute the Fourier series of $f(x) = x$ on $[-\pi, \pi]$.

Because $x\cos(nx)$ is an odd function on $[-\pi, \pi]$ for $n = 0, 1, 2, \cdots$, each $a_n = 0$. We need only compute the b_n's:

$$b_n = \frac{1}{\pi}\int_{-\pi}^{\pi} x\sin(nx)\,dx = \frac{2}{\pi}\int_{0}^{\pi} x\sin(nx)\,dx$$

$$= \left[\frac{2}{n^2\pi}\sin(nx) - \frac{2x}{n\pi}\cos(nx)\right]_{0}^{\pi}$$

$$= -\frac{2}{n}\cos(n\pi) = \frac{2}{n}(-1)^{n+1}.$$

The Fourier series of x on $[-\pi, \pi]$ is

$$\sum_{n=1}^{\infty}\frac{2}{n}(-1)^{n+1}\sin(nx).$$

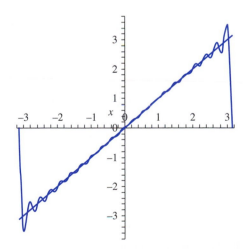

FIGURE 12.10 *Twentieth partial sum of the series in Example 12.7.*

This converges to x for $-\pi < x < \pi$ and to 0 at $x = \pm\pi$. Figure 12.10 shows the twentieth partial sum of this Fourier series compared to the function. ◆

EXAMPLE 12.8

We will write the Fourier series of x^4 on $[-1, 1]$. Since $x^4 \sin(n\pi x)$ is an odd function on $[-1, 1]$ for $n = 1, 2, \cdots$, each $b_n = 0$. Compute

$$a_0 = 2 \int_0^1 x^4 dx = \frac{2}{5}$$

and

$$a_n = 2 \int_0^1 x^4 \cos(n\pi x) dx$$

$$= 2 \left[\frac{(n\pi x)^4 \sin(n\pi x) + 4(n\pi x)^3 \cos(n\pi x)}{(n\pi)^5} \right]_0^1$$

$$+ 2 \left[\frac{-12(n\pi x)^2 \sin(n\pi x) - 24 n\pi x \cos(n\pi x) + 24 \sin(n\pi x)}{(n\pi)^5} \right]_0^1$$

$$= 8 \frac{n^2 \pi^2 - 6}{n^4 \pi^4} (-1)^n.$$

The Fourier series is

$$\frac{1}{5} + \sum_{n=1}^{\infty} \left(\frac{8(-1)^n}{n^4 \pi^4} (n^2 \pi^2 - 6) \right) \cos(n\pi x).$$

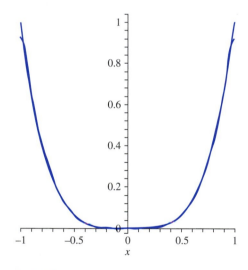

FIGURE 12.11 *Tenth partial sum of the series of Example 12.8.*

This series converges to x^4 for $-1 \le x \le 1$. Figure 12.11 shows the twentieth partial sum of this series compared with the function. ◆

12.1.2   The Gibbs Phenomenon

A.A. Michelson was a physicist who teamed with Morley to show that the "aether," a hypothetical fluid which supposedly permeated all of space, had no effect on the speed of light. Michelson also built a mechanical device for constructing a function from its Fourier coefficients. In one test, he used eighty coefficients for the series of $f(x) = x$ on $[-\pi, \pi]$ and noticed unexpected

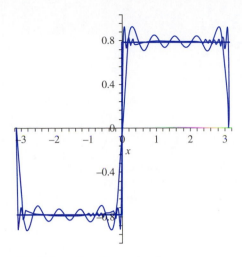

FIGURE 12.12 *The Gibbs phenomenon.*

jumps in the graph near the endpoints. At first he thought this was a problem with his machine. It was subsequently found that this behavior is characteristic of the Fourier series of a function at a point of discontinuity. In the early twentieth century, the Yale mathematician Josiah Willard Gibbs finally explained this behavior.

To illustrate the Gibbs phenomenon, expand f in a Fourier series on $[-\pi, \pi]$ where

$$f(x) = \begin{cases} -\pi/4 & \text{for } -\pi \le x < 0 \\ 0 & \text{for } x = 0 \\ \pi/4 & \text{for } 0 < x \le \pi. \end{cases}$$

This function has a jump discontinuity at 0, but its Fourier series on $[-\pi, \pi]$ converges at 0 to

$$\frac{1}{2}(f(0+) + f(0-)) = \frac{1}{2}\left(\frac{\pi}{4} - \frac{\pi}{4}\right) = 0 = f(0).$$

The Fourier series therefore converges to the function on $(-\pi, \pi)$. This series is

$$\sum_{n=1}^{\infty} \frac{1}{2n-1} \sin((2n-1)x).$$

Figure 12.12 shows the fifth and twenty-fifth partial sums of this series, compared to the function. Notice that both of these partial sums show a peak near 0, which is the point of discontinuity of f. Since the partial sums S_N approach the function as $N \to \infty$, we might expect these peaks to flatten out, but they do not. Instead they remain roughly the same height but move closer to the y-axis as N increases. This is the Gibbs phenomenon.

SECTION 12.1   PROBLEMS

In each of Problems 1 through 10, write the Fourier series of the function on the interval and determine the sum of the Fourier series. Graph some partial sums of the series compared with the graph of the function.

1. $f(x) = 4$ for $-3 \le x \le 3$

2. $f(x) = -x$ for $-1 \le x \le 1$

3. $f(x) = \cosh(\pi x)$ for $-1 \le x \le 1$

4. $f(x) = 1 - |x|$ for $-2 \le x \le 2$

5. $f(x) = \begin{cases} -4 & \text{for } -\pi \le x \le 0 \\ 4 & \text{for } 0 < x \le \pi \end{cases}$

6. $f(x) = \sin(2x)$ for $-\pi \le x \le \pi$

7. $f(x) = x^2 - x + 3$ for $-2 \le x \le 2$

8. $f(x) = \begin{cases} -x & \text{for } -5 \le x < 0 \\ 1+x^2 & \text{for } 0 \le x \le 5 \end{cases}$

9. $f(x) = \begin{cases} 1 & \text{for } -\pi \le x < 0 \\ 2 & \text{for } 0 \le x \le \pi \end{cases}$

10. $f(x) = \cos(x/2) - \sin(x)$ for $-\pi \le x \le \pi$

In each of Problems 11 through 15, use the convergence theorem to determine the sum of the Fourier series of the function on the interval. It is not necessary to write the series to do this.

11. $f(x) = \begin{cases} 2x & \text{for } -3 \le x < -2 \\ 0 & \text{for } -2 \le x < 1 \\ x^2 & \text{for } 1 \le x \le 3 \end{cases}$

12. $f(x) = \begin{cases} 2x - 2 & \text{for } -\pi \le x \le 1 \\ 3 & \text{for } 1 < x \le \pi \end{cases}$

13. $f(x) = \begin{cases} x^2 & \text{for } -\pi \le x \le 0 \\ 2 & \text{for } 0 < x \le \pi \end{cases}$

14. $f(x) = \begin{cases} \cos(x) & \text{for } -2 \le x < 0 \\ \sin(x) & \text{for } 0 \le x \le 2 \end{cases}$

15. $f(x) = \begin{cases} -1 & \text{for } -4 \le x < 0 \\ 1 & \text{for } 0 \le x \le 4 \end{cases}$

16. In Problem 12 write the Fourier series of the function and plot some partial sums, pointing out the occurrence of the Gibbs phenomenon at the points of discontinuity of the function.

17. Carry out the program of Problem 16 for the function of Problem 14.

12.2 Sine and Cosine Series

If $f(x)$ is defined on $[-L, L]$, we may be able to write its Fourier series on this interval. This series may contain just sine terms, just cosine terms, or both sine and cosine terms. We have no control over this, given the function and the interval.

If $f(x)$ is defined on the half-interval $[0, L]$, then we can write a Fourier cosine series (containing just cosine terms) and a Fourier sine series (containing just sine terms) for $f(x)$ on $[0, L]$.

12.2.1 Cosine Series

Suppose $f(x)$ is defined for $0 \le x \le L$. To get a pure cosine series on this interval, imagine reflecting the graph of f across the vertical axis to obtain an even function g defined on $[-L, L]$ (see Figure 12.13). Then

$$g(x) = \begin{cases} f(x) & \text{for } 0 \le x \le L \\ f(-x) & \text{for } -L \le x < 0. \end{cases}$$

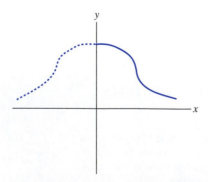

FIGURE 12.13 *Even extension of a function defined on* $[0, L]$.

Because g is even, its Fourier series on $[-L, L]$ has only cosine terms and perhaps the constant term. But $g(x) = f(x)$ for $0 \leq x \leq L$, so this gives a cosine series for f on $[0, L]$. Furthermore, because g is even, the coefficients in the Fourier series of g on $[-L, L]$ are

$$a_n = \frac{1}{L}\int_{-L}^{L} g(x)\cos\left(\frac{n\pi x}{L}\right) dx$$

$$= \frac{2}{L}\int_{0}^{L} g(x)\cos\left(\frac{n\pi x}{L}\right) dx$$

$$= \frac{2}{L}\int_{0}^{L} f(x)\cos\left(\frac{n\pi x}{L}\right) dx.$$

for $n = 0, 1, 2, \cdots$. We can compute a_n strictly in terms of $f(x)$ on $[0, L]$. The construction of g showed us how to obtain this cosine series for f, but we do not need g to compute the coefficients of this series.

This leads us to define the *Fourier cosine coefficients of f on $[0, L]$* to be the numbers

$$a_n = \frac{2}{L}\int_{0}^{L} f(x)\cos(n\pi x/L)\,dx \tag{12.10}$$

for $n = 0, 1, 2, \cdots$. The *Fourier cosine series for f on $[0, L]$* is the series

$$\frac{1}{2}a_0 + \sum_{n=1}^{\infty} a_n \cos\left(\frac{n\pi x}{L}\right) \tag{12.11}$$

in which the a_n's are the Fourier cosine coefficients of f on $[0, L]$.

By applying Theorem 12.1 to g, we obtain the following convergence theorem for cosine series on $[0, L]$.

THEOREM 12.2 Convergence of Fourier Cosine Series

Let f be piecewise smooth on $[0, L]$. Then

1. If $0 < x < L$, the Fourier cosine series for f on $[0, L]$ converges to
$$\frac{1}{2}(f(x+) + f(x-)).$$

2. At 0 this cosine series converges to $f(0+)$.

3. At L this cosine series converges to $f(L-)$. ◆

EXAMPLE 12.9

Let $f(x) = e^{2x}$ for $0 \leq x \leq 1$. We will write the cosine expansion of $f(x)$ on $[0, 1]$. The coefficients are

$$a_0 = 2\int_0^1 e^{2x}\,dx = e^2 - 1.$$

For $n = 1, 2, \cdots$, they are

$$a_n = 2 \int_0^1 e^{2x} \cos(n\pi x)\, dx$$

$$= \left[\frac{4e^{2x} \cos(n\pi x) + 2n\pi e^{2x} \sin(n\pi x)}{4 + n^2\pi^2} \right]_0^1$$

$$= 4 \frac{e^2(-1)^n - 1}{4 + n^2\pi^2}.$$

The cosine series for e^{2x} on $[0, 1]$ is

$$\frac{1}{2}(e^2 - 1) + \sum_{n=1}^{\infty} 4 \frac{e^2(-1)^n - 1}{4 + n^2\pi^2} \cos(n\pi x).$$

This series converges to

$$\begin{cases} e^{2x} & \text{for } 0 < x < 1 \\ 1 & \text{for } x = 0 \\ e^2 & \text{for } x = 1. \end{cases}$$

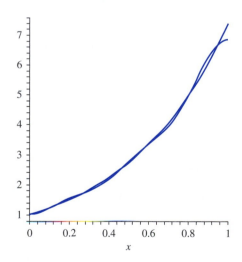

FIGURE 12.14 *Fifth partial sum of the cosine series in Example 12.9.*

This Fourier cosine series converges to e^{2x} for $0 \le x \le 1$. Figure 12.14 shows e^{2x} and the fifth partial sum of this cosine series. ◆

12.2.2 Sine Series

We can also write an expansion of $f(x)$ on $[0, L]$ that contains only sine terms. Now reflect the graph of $f(x)$ on $[0, L]$ through the origin to create an odd function h on $[-L, L]$, with $h(x) = f(x)$ for $0 \le x \le L$ (Figure 12.15).

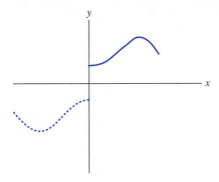

FIGURE 12.15 *Odd extension of a function defined on* $[0, L]$.

The Fourier expansion of h on $[-L, L]$ has only sine terms because h is odd on $[-L, L]$. But $h(x) = f(x)$ on $[0, L]$, so this gives a sine expansion of f on $[0, L]$. This suggests the following definitions.

The *Fourier sine coefficients of f on* $[0, L]$ are

$$b_n = \frac{2}{L} \int_0^L f(x) \sin\left(\frac{n\pi x}{L}\right) dx \tag{12.12}$$

for $n = 1, 2, \cdots$. With these coefficients, the series

$$\sum_{n=1}^{\infty} b_n \sin\left(\frac{n\pi x}{L}\right) \tag{12.13}$$

is the *Fourier sine series for f on* $[0, L]$.

Again, we have a convergence theorem for the sine series directly from the convergence theorem for Fourier series.

THEOREM 12.3 *Convergence of Fourier Sine Series*

Let f be piecewise smooth on $[0, L]$. Then

1. If $0 < x < L$, the Fourier sine series for f on $[0, L]$ converges to

$$\frac{1}{2}(f(x+) + f(x-)).$$

2. At $x = 0$ and $x = L$, this sine series converges to 0. ♦

Conclusion 2 is obvious because each sine term in the series vanishes at $x = 0$ and at $x = L$, regardless of the values of the function there.

EXAMPLE 12.10

Let $f(x) = e^{2x}$ for $0 \leq x \leq 1$. We will write the Fourier sine series of f on $[0, 1]$. The coefficients are

$$b_n = 2 \int_0^1 e^{2x} \sin(n\pi x)\, dx$$

$$= \left[\frac{-2n\pi e^{2x} \cos(n\pi x) + 4e^{2x} \sin(n\pi x)}{4 + n^2 \pi^2} \right]_0^1$$

$$= 2 \frac{n\pi(1 - (-1)^n e^2)}{4 + n^2 \pi^2}.$$

The sine expansion of e^{2x} on $[0, 1]$ is

$$\sum_{n=1}^{\infty} 2 \frac{n\pi(1 - (-1)^n e^2)}{4 + n^2 \pi^2} \sin(n\pi x).$$

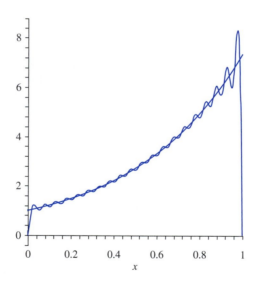

FIGURE 12.16   *Fortieth partial sum of the sine series of Example 12.10.*

This series converges to e^{2x} for $0 < x < 1$ and to 0 at $x = 0$ and $x = 1$. Figure 12.16 shows the function and the fortieth partial sum of this sine expansion. ◆

SECTION 12.2   *PROBLEMS*

In each of Problems 1 through 8, write the Fourier cosine and sine series for f on the interval. Determine the sum of each series. Graph some of the partial sums of these series.

1. $f(x) = 4$ for $0 \leq x \leq 3$

2. $f(x) = \begin{cases} 1 & \text{for } 0 \leq x \leq 1 \\ -1 & \text{for } 1 < x \leq 2 \end{cases}$

3. $f(x) = \begin{cases} 0 & \text{for } 0 \leq x \leq \pi \\ \cos(x) & \text{for } \pi < x \leq 2\pi \end{cases}$

4. $f(x) = 2x$ for $0 \leq x \leq 1$

5. $f(x) = x^2$ for $0 \leq x \leq 2$

6. $f(x) = e^{-x}$ for $0 \leq x \leq 1$

7. $f(x) = \begin{cases} x & \text{for } 0 \leq x \leq 2 \\ 2-x & \text{for } 2 < x \leq 3 \end{cases}$

8. $f(x) = \begin{cases} 1 & \text{for } 0 \leq x < 1 \\ 0 & \text{for } 1 \leq x \leq 3 \\ -1 & \text{for } 3 < x \leq 5 \end{cases}$

9. Sum the series $\sum_{n=1}^{\infty} (-1)^n / (4n^2 - 1)$. *Hint*: Expand $\sin(x)$ in a cosine series on $[0, \pi]$ and choose an appropriate value of x.

12.3   Derivatives and Integrals of Fourier Series

Term by term differentiation of a Fourier series may lead to nonsense.

EXAMPLE 12.11

Let $f(x) = x$ for $-\pi \leq x \leq \pi$. The Fourier series is

$$f(x) = x = \sum_{n=1}^{\infty} \frac{2}{n}(-1)^{n+1} \sin(nx)$$

for $-\pi < x < \pi$. Differentiate this series term by term to get

$$\sum_{n=1}^{\infty} 2(-1)^{n+1} \cos(nx).$$

This series does not converge on $(-\pi, \pi)$. ◆

Sufficient conditions to be able to differentiate a Fourier series are fairly strong.

THEOREM 12.4   *Differentiation of Fourier Series*

Let f be continuous on $[-L, L]$ and suppose that $f(-L) = f(L)$. Let f' be piecewise continuous on $[-L, L]$. Then the Fourier series of f on $[-L, L]$ converges to $f(x)$ on $[-L, L]$:

$$f(x) = \frac{1}{2}a_0 + \sum_{n=1}^{\infty} \left(a_n \cos\left(\frac{n\pi x}{L}\right) + b_n \sin\left(\frac{n\pi x}{L}\right) \right)$$

for $-L \leq x \leq L$. Further, at each x in $(-L, L)$ at which $f''(x)$ exists, the term-by-term derivative of the Fourier series converges to the derivative of the function:

$$f'(x) = \sum_{n=1}^{\infty} \frac{n\pi}{L} \left(-na_n \sin\left(\frac{n\pi x}{L}\right) + b_n \cos\left(\frac{n\pi x}{L}\right) \right). ◆$$

EXAMPLE 12.12

Let $f(x) = x^2$ for $-2 \leq x \leq 2$. By the Fourier convergence theorem,

$$x^2 = \frac{4}{3} + \frac{16}{\pi^2} \sum_{n=1}^{\infty} \frac{(-1)^{n+1}}{n^2} \cos\left(\frac{n\pi x}{2}\right)$$

for $-2 \leq x \leq 2$. Only cosine terms appear in this series because x^2 is an even function. Now, $f'(x) = 2x$ is continuous and f is twice differentiable for all x. Therefore, for $-2 < x < 2$,

$$f'(x) = 2x = \frac{8}{\pi} \sum_{n=1}^{\infty} \frac{(-1)^{n+1}}{n} \sin\left(\frac{n\pi x}{2}\right).$$

This can be verified by expanding $2x$ in a Fourier series on $[-2, 2]$. ◆

Conditions under which a Fourier series can be integrated term by term are much less stringent.

THEOREM 12.5   Integration of Fourier Series

Let f be piecewise continuous on $[-L, L]$ with the Fourier series

$$\frac{1}{2}a_0 + \sum_{n=1}^{\infty} \left(a_n \cos\left(\frac{n\pi x}{L}\right) + b_n \sin\left(\frac{n\pi x}{L}\right) \right).$$

Then, for any x with $-L \leq x \leq L$,

$$\int_{-L}^{x} f(t)\,dt = \frac{1}{2}a_0(x + L)$$

$$+ \frac{L}{\pi} \sum_{n=1}^{\infty} \frac{1}{n} \left[a_n \sin\left(\frac{n\pi x}{L}\right) - b_n \left(\cos\left(\frac{n\pi x}{L}\right) - (-1)^n\right) \right]. ◆$$

The expression on the right in this equation is exactly what we obtain by integrating the Fourier series term by term, from $-L$ to x. This means that we can integrate any piecewise continuous function f from $-L$ to x by integrating its Fourier series term by term. This is true even if the function has jump discontinuities and its Fourier series does not converge to $f(x)$ for all x in $[-L, L]$. Notice, however, that the result of this integration is not a Fourier series.

EXAMPLE 12.13

We have seen in Example 12.11 that

$$f(x) = x = \sum_{n=1}^{\infty} \frac{2}{n}(-1)^{n+1} \sin(nx)$$

on $(-\pi, \pi)$. Term-by-term differentiation results in a series that does not converge on the interval. However, we can integrate this series term by term:

$$\int_{-\pi}^{x} t\,dt = \frac{1}{2}(x^2 - \pi^2)$$

$$= \sum_{n=1}^{\infty} \frac{2}{n}(-1)^{n+1} \int_{-\pi}^{x} \sin(nt)\,dt$$

$$= \sum_{n=1}^{\infty} \frac{2}{n}(-1)^{n+1} \left[-\frac{1}{n}\cos(nx) + \frac{1}{n}\cos(n\pi) \right]$$

$$= \sum_{n=1}^{\infty} \frac{2}{n}(-1)^{n+1}(\cos(nx) - (-1)^n). ◆$$

By integrating a Fourier series, we can derive an important equation satisfied by Fourier coefficients.

THEOREM 12.6 *Parseval's Equation*

Let f be continuous on $[-L, L]$ and let f' be piecewise continuous. Suppose that $f(-L) = f(L)$. Then the Fourier coefficients of f on $[-L, L]$ satisfy

$$\frac{1}{2}a_0^2 + \sum_{n=1}^{\infty}(a_n^2 + b_n^2) = \frac{1}{L}\int_{-L}^{L} f(x)^2\, dx. \quad \blacklozenge$$

In particular, the series of the squares of the Fourier coefficients converges.

To see why Parseval's equation holds, begin with the fact that, from the Fourier convergence theorem,

$$f(x) = \frac{1}{2}a_0 + \sum_{n=1}^{\infty}\left(a_n \cos\left(\frac{n\pi x}{L}\right) + b_n \sin\left(\frac{n\pi x}{L}\right)\right).$$

Multiply this series by $f(x)$ to get

$$f(x)^2 = \frac{1}{2}a_0 f(x) + \sum_{n=1}^{\infty}\left(a_n f(x)\cos\left(\frac{n\pi x}{L}\right) + b_n f(x)\sin\left(n\pi x/L\right)\right).$$

Now derive Parseval's equation by integrating this series term by term and observing that the integrals on the right are Fourier coefficients.

SECTION 12.3   **PROBLEMS**

---

1. Let $f(x) = \begin{cases} 0 & \text{for } -\pi \le x \le 0 \\ x & \text{for } 0 < x \le \pi. \end{cases}$

   (a) Write the Fourier series of $f(x)$ on $[-\pi, \pi]$ and show that this series converges to $f(x)$ on $(-\pi, \pi)$.

   (b) Use Theorem 12.5 to show that this series can be integrated term by term.

   (c) Use the results of (a) and (b) to obtain a trigonometric series expansion of $\int_{-\pi}^{x} f(t)\,dt$ on $[-\pi, \pi]$.

2. Let $f(x) = |x|$ for $-1 \le x \le 1$.

   (a) Write the Fourier series for f on $[-1, 1]$.

   (b) Show that this series can be differentiated term by term to yield the Fourier expansion of $f'(x)$ on $[-\pi, \pi]$.

   (c) Determine $f'(x)$ and expand this function in a Fourier series on $[-\pi, \pi]$. Compare this result with that of (b).

3. Let $f(x) = x\sin(x)$ for $-\pi \le x \le \pi$.

   (a) Write the Fourier series for f on $[-\pi, \pi]$.

   (b) Show that this series can be differentiated term by term and use this fact to obtain the Fourier expansion of $\sin(x) + x\cos(x)$ on $[-\pi, \pi]$.

   (c) Write the Fourier series for $\sin(x) + x\cos(x)$ on $[-\pi, \pi]$ and compare this result with that of (b).

4. Let $f(x) = x^2$ for $-3 \le x \le 3$.

   (a) Write the Fourier series for f on $[-3, 3]$.

   (b) Show that this series can be differentiated term by term and use this to obtain the Fourier expansion of $2x$ on $[-3, 3]$.

   (c) Expand $2x$ in a Fourier series on $[-3, 3]$ and compare this result with that of (b).

12.4 Complex Fourier Series

There is a complex form of Fourier series that is often used. First recall some facts about complex numbers. The *conjugate* of $z = a + ib$ is $\bar{z} = a - ib$. The magnitude of z is

$$|z| = \sqrt{a^2 + b^2}.$$

Magnitude and conjugation are related by

$$z\bar{z} = |z|^2.$$

In polar coordinates, $z = x + iy = r\cos(\theta) + ir\sin(\theta)$, where $r = |z|$. The polar angle θ is called an *argument* of z. If θ is any argument of z, so is $\theta + 2n\pi$ for any integer n. Using Euler's formula, we obtain the *polar form* of z:

$$z = r[\cos(\theta) + i\sin(\theta)] = re^{i\theta}.$$

With $r = 1$, this is

$$e^{i\theta} = \cos(\theta) + i\sin(\theta)$$

and replacing θ with $-\theta$ yields

$$e^{-i\theta} = \cos(\theta) - i\sin(\theta).$$

Solve these equations for $\cos(\theta)$ and $\sin(\theta)$ to obtain the complex exponential forms of the trigonometric functions:

$$\cos(\theta) = \frac{1}{2}\left(e^{i\theta} + e^{-i\theta}\right), \sin(\theta) = \frac{1}{2i}\left(e^{i\theta} - e^{-i\theta}\right).$$

Finally, we will use the fact that, if x is a real number, then $\overline{e^{ix}} = e^{-ix}$. This follows from Euler's formula, since

$$\overline{e^{ix}} = \overline{\cos(x) + i\sin(x)} = \cos(x) - i\sin(x) = e^{-ix}.$$

Now let f be a piecewise smooth periodic function with fundamental period $2L$. This means that $f(x + 2L) = f(x)$ for all x, and L is the smallest positive number with this property. For example, $\sin(2x)$ has fundamental period π. To derive the complex Fourier expansion of $f(x)$, begin with the Fourier series of f on $[-L, L]$. With $\omega_0 = \pi/L$, this series is

$$\frac{1}{2}a_0 + \sum_{n=1}^{\infty}(a_n\cos(n\omega_0 x) + b_n\sin(n\omega_0 x)).$$

Put the complex forms of $\cos(n\omega_0 x)$ and $\sin(n\omega_0 x)$ into this expansion:

$$\frac{1}{2}a_0 + \sum_{n=1}^{\infty}\left[a_n\frac{1}{2}\left(e^{in\omega_0 x} + e^{-in\omega_0 x}\right) + b_n\frac{1}{2i}\left(e^{in\omega_0 x} - e^{-in\omega_0 x}\right)\right]$$

$$= \frac{1}{2}a_0 + \sum_{n=1}^{\infty}\left[\frac{1}{2}(a_n - ib_n)e^{in\omega_0 x} + \frac{1}{2}(a_n + ib_n)e^{-in\omega_0 x}\right].$$

In this series, let

$$d_0 = \frac{1}{2}a_0$$

and, for $n = 1, 2, \ldots,$

$$d_n = \frac{1}{2}(a_n - ib_n).$$

The Fourier series of $f(x)$ on $[-L, L]$ becomes

$$d_0 + \sum_{n=1}^{\infty} d_n e^{in\omega_0 x} + \sum_{n=1}^{\infty} \overline{d_n} e^{-in\omega_0 x} \tag{12.14}$$

Now

$$d_0 = \frac{1}{2} a_0 = \frac{1}{L} \int_{-L}^{L} f(x)\, dx$$

and for $n = 1, 2, \cdots,$

$$d_n = \frac{1}{2}(a_n - ib_n)$$

$$= \frac{1}{2L} \int_{-L}^{L} f(x) \cos(n\omega_0 x)\, dx - \frac{i}{2L} \int_{-L}^{L} f(x) \sin(n\omega_0 x)\, dx$$

$$= \frac{1}{2L} \int_{-L}^{L} f(x)[\cos(n\omega_0 x) - i \sin(n\omega_0 x)]\, dx$$

$$= \frac{1}{2L} \int_{-L}^{L} f(x) e^{-in\omega_0 x}\, dx.$$

Then

$$\overline{d_n} = \frac{1}{2L} \int_{-L}^{L} f(x)\overline{e^{-in\omega_0 x}}\, dx = \frac{1}{2L} \int_{-L}^{L} f(x) e^{in\omega_0 x}\, dx = d_{-n}.$$

Now the expansion of equation (12.14) becomes

$$d_0 + \sum_{n=1}^{\infty} d_n e^{in\omega_0 x} + \sum_{n=1}^{\infty} d_{-n} e^{-in\omega_0 x}$$

or

$$\sum_{n=-\infty}^{\infty} d_n e^{in\omega_0 x}.$$

Complex Fourier Series

This leads us to define the *complex Fourier series of f on $[-L, L]$* to be

$$\sum_{n=-\infty}^{\infty} d_n e^{in\omega_0 x}$$

with coefficients

$$d_n = \frac{1}{2L} \int_{-L}^{L} f(x) e^{-in\omega_0 x}\, dx$$

for $n = 0, \pm 1, \pm 2, \cdots.$

Because of the periodicity of f, the integral defining the coefficients can be carried out over any interval $[\alpha, \alpha + 2L]$ of length $2L$. The Fourier convergence theorem applies to this complex Fourier expansion, since it is just the Fourier series in complex form.

EXAMPLE 12.14

Let $f(x) = x$ for $-1 \leq x < 1$ and suppose f has fundamental period 2, so $f(x+2) = f(x)$ for all x. Figure 12.17 is part of a graph of f. Now $\omega_0 = \pi$.

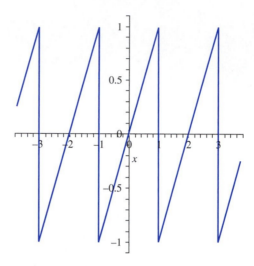

FIGURE 12.17   *Graph of f in Example 12.14.*

Immediately, $d_0 = 0$ because f is an odd function. For $n \neq 0$,

$$d_n = \frac{1}{2} \int_{-1}^{1} x e^{-in\pi x} \, dx$$

$$= \frac{1}{2n^2 \pi^2} \left[in\pi e^{in\pi} - e^{in\pi} + in\pi e^{-in\pi} + e^{-in\pi} \right]$$

$$= \frac{1}{2n^2 \pi^2} \left[in\pi \left(e^{in\pi} + e^{-in\pi} \right) - \left(e^{in\pi} - e^{-in\pi} \right) \right].$$

The complex Fourier series of $f(x)$ is

$$\sum_{n=-\infty, n \neq 0}^{\infty} \frac{1}{2n^2 \pi^2} \left[in\pi \left(e^{in\pi} + e^{-in\pi} \right) - \left(e^{in\pi} - e^{-in\pi} \right) \right] e^{in\pi x}.$$

This converges to x for $-1 < x < 1$. In this example, we can simplify the series. For $n \neq 0$,

$$d_n = \frac{1}{2n^2 \pi^2} [2in\pi \cos(n\pi) - 2i \sin(n\pi)]$$

$$= \frac{i}{n\pi} (-1)^n.$$

Each term $\sin(n\pi)$ is zero. Furthermore in $\sum_{n=-\infty}^{-1}$, replace n with $-n$, sum from $n = 1$ to ∞, and then combine the two summations from 1 to ∞ to write

$$\sum_{n=-\infty, n \neq 0}^{\infty} \frac{i}{n\pi} (-1)^n e^{in\pi x} = \sum_{n=1}^{\infty} \left(\frac{i}{n\pi} (-1)^n e^{in\pi x} + \frac{i}{-n\pi} (-1)^{-n} e^{-in\pi x} \right)$$

$$= \sum_{n=1}^{\infty} \frac{i}{n\pi} (-1)^n \left(e^{in\pi x} - e^{-in\pi x} \right) = \sum_{n=1}^{\infty} \frac{2}{n\pi} (-1)^{n+1} \sin(n\pi x).$$

This is the Fourier series for $f(x) = x$ on $[-1, 1]$. ◆

In each of Problems 1 through 6, write the complex Fourier series of $f(x)$ and determine the sum of the series.

1. $f(x) = 2x$ for $0 \leq x < 3$, period 3

2. $f(x) = x^2$ for $0 \leq x < 2$, period 2

3. $f(x) = \begin{cases} 0 & \text{for } 0 \leq x < 1, \\ 1 & \text{for } 1 \leq x < 4, \text{ period 4} \end{cases}$

4. $f(x) = 1 - x$ for $0 \leq x < 6$, period 6

5. $f(x) = \begin{cases} -1 & \text{for } 0 \leq x < 2, \\ 2 & \text{for } 2 \leq x < 4, \text{ period 4} \end{cases}$

6. $f(x) = e^{-x}$ for $0 \leq x < 5$, period 5

CHAPTER 13

The Fourier Integral and Transforms

13.1   The Fourier Integral

Fourier series are tied to intervals of finite length. If $f(x)$ is defined over the entire line and is not periodic, then the Fourier series representation is replaced with the Fourier integral representation in which $\sum_{n=0}^{\infty}$ is replaced by $\int_{0}^{\infty} \ldots d\omega$.

After some computations which we omit, the *Fourier integral of* $f(x)$ is defined to be

$$\int_{0}^{\infty} (A_{\omega}\cos(\omega x) + B_{\omega}\sin(\omega x))\, d\omega \tag{13.1}$$

in which the *Fourier integral coefficients of* $f(x)$ are

$$A_{\omega} = \frac{1}{\pi} \int_{-\infty}^{\infty} f(\xi)\cos(\omega\xi)\, d\xi \tag{13.2}$$

and

$$B_{\omega} = \frac{1}{\pi} \int_{-\infty}^{\infty} f(\xi)\sin(\omega\xi)\, d\xi. \tag{13.3}$$

The integration variable ω replaces the summation index n in this integral representation.

As with Fourier series, the relationship between the integral (13.1) and $f(x)$ must be determined. This is done by the Fourier integral convergence theorem.

THEOREM 13.1   *Convergence of the Fourier Integral*

Suppose $f(x)$ is defined for all real x and that $\int_{-\infty}^{\infty} |f(x)|\, dx$ converges. Suppose f is piecewise smooth on every interval $[-L, L]$ for $L > 0$. Then at any x the Fourier integral (13.1) of f converges to

$$\frac{1}{2}(f(x+) + f(x-)).$$

293

In particular, if f is continuous at x, then the Fourier integral converges at x to $f(x)$. ◆

When $\int_{-\infty}^{\infty} |f(x)|\,dx$ converges, we say that f is *absolutely integrable* on the real line.

EXAMPLE 13.1

Let

$$f(x) = \begin{cases} 1 & \text{for } -1 \le x \le 1 \\ 0 & \text{for } |x| > 1. \end{cases}$$

Certainly f is absolutely integrable. The Fourier integral coefficients of f are

$$A_\omega = \frac{1}{\pi} \int_{-1}^{1} \cos(\omega\xi)\,d\xi = \frac{2\sin(\omega)}{\pi\omega}$$

and

$$B_\omega = \frac{1}{\pi} \int_{-1}^{1} \sin(\omega\xi)\,d\xi = 0.$$

The Fourier integral of f is

$$\int_{0}^{\infty} \frac{2\sin(\omega)}{\pi\omega} \cos(\omega x)\,d\omega.$$

Because f is continuous for $x \ne \pm 1$, the integral converges to $f(x)$ for $x \ne \pm 1$. At $x = 1$ the integral converges to

$$\frac{1}{2}(f(1+) + f(1-)) = \frac{1}{2}(1 + 0) = \frac{1}{2}.$$

Similarly, the integral converges to $1/2$ at $x = -1$. This Fourier integral is a faithful representation of the function except at 1 and -1, where it averages the ends of the jump discontinuities there.

In view of this convergence, we have

$$\frac{1}{\pi} \int_{0}^{\infty} \frac{2\sin(\omega)}{\omega} \cos(\omega x)\,d\omega = \begin{cases} 1 & \text{for } -1 < x < 1 \\ 1/2 & \text{for } x = \pm 1 \\ 0 & \text{for } |x| > 1. \end{cases} ◆$$

There is another expression for the Fourier integral of a function that is sometimes convenient to use. Insert the coefficients into the Fourier integral:

$$\int_{0}^{\infty} [A_\omega \cos(\omega x) + B_\omega \sin(\omega x)] =$$

$$\int_{0}^{\infty} \left[\left(\frac{1}{\pi} \int_{-\infty}^{\infty} f(\xi) \cos(\omega\xi)\,d\xi \right) \cos(\omega x) + \left(\frac{1}{\pi} \int_{-\infty}^{\infty} f(\xi) \sin(\omega\xi)\,d\xi \right) \sin(\omega x) \right] d\omega$$

$$= \frac{1}{\pi} \int_{0}^{\infty} \int_{-\infty}^{\infty} f(\xi)[\cos(\omega\xi)\cos(\omega x) + \sin(\omega\xi)\sin(\omega x)]\,d\xi\,d\omega$$

$$= \frac{1}{\pi} \int_{0}^{\infty} \int_{-\infty}^{\infty} f(\xi) \cos(\omega(\xi - x))\,d\xi\,d\omega.$$

This gives us the equivalent Fourier integral representation

$$\frac{1}{\pi} \int_0^\infty \int_{-\infty}^\infty f(\xi) \cos(\omega(\xi - x)) \, d\xi \, d\omega \tag{13.4}$$

of f on the real line.

SECTION 13.1   PROBLEMS

In each of Problems 1 through 8, write the Fourier integral representation (13.1) of the function and determine what this integral converges to.

1. $f(x) = \begin{cases} x & \text{for } -\pi \le x \le \pi \\ 0 & \text{for } |x| > \pi \end{cases}$

2. $f(x) = \begin{cases} k & \text{for } -10 \le x \le 10 \\ 0 & \text{for } |x| > 10 \end{cases}$

3. $f(x) = \begin{cases} -1 & \text{for } -\pi \le x \le 0 \\ 1 & \text{for } 0 < x \le \pi \\ 0 & \text{for } |x| > \pi \end{cases}$

4. $f(x) = \begin{cases} \sin(x) & \text{for } -4 \le x \le 0 \\ \cos(x) & \text{for } 0 < x \le 4 \\ 0 & \text{for } |x| > 4 \end{cases}$

5. $f(x) = \begin{cases} x^2 & \text{for } -100 \le x \le 100 \\ 0 & \text{for } |x| > 100 \end{cases}$

6. $f(x) = \begin{cases} |x| & \text{for } -\pi \le x \le 2\pi \\ 0 & \text{for } x < -\pi \text{ and for } x > 2\pi \end{cases}$

7. $f(x) = \begin{cases} \sin(x) & \text{for } -3\pi \le x \le \pi \\ 0 & \text{for } x < -3\pi \text{ and for } x > \pi \end{cases}$

8. $f(x) = \begin{cases} 1/2 & \text{for } -5 \le x < 1 \\ 1 & \text{for } 1 \le x \le 5 \\ 0 & \text{for } |x| > 5 \end{cases}$

13.2   Fourier Cosine and Sine Integrals

We can define Fourier cosine and sine integral expansions for functions defined on the half-line in a manner completely analogous to Fourier cosine and sine expansions of functions defined on a half interval.

Suppose $f(x)$ is defined for $x \ge 0$. If we extend f to an even function on the real line, its Fourier integral contains only cosine terms and is a cosine expansion. If we extend f to an odd function on the entire line, the Fourier integral contains only sine terms. We will summarize these conclusions.

The *Fourier cosine integral of* f on $x \ge 0$ is

$$\int_0^\infty A_\omega \cos(\omega x) \, d\omega \tag{13.5}$$

in which

$$A_\omega = \frac{2}{\pi} \int_0^\infty f(\xi) \cos(\omega \xi) \, d\xi \tag{13.6}$$

is the *Fourier integral cosine coefficient.*

The *Fourier sine integral of f* on $x \geq 0$ is

$$\int_0^\infty B_\omega \sin(\omega x)\, d\omega \tag{13.7}$$

in which

$$B_\omega = \frac{2}{\pi} \int_0^\infty f(\xi) \sin(\omega \xi)\, d\xi \tag{13.8}$$

is the *Fourier integral sine coefficient.*

If f is piecewise smooth on every $[0, L]$, and $\int_0^\infty |f(x)|dx$ converges, then these integrals converge to

$$\frac{1}{2}(f(x+) + f(x-))$$

at each $x > 0$. The cosine integral converges to $f(0+)$ at $x = 0$, and the sine integral converges to 0 at $x = 0$.

EXAMPLE 13.2 Laplace's Integrals

Let $f(x) = e^{-kx}$ for $x \geq 0$, with k a positive number. Then f has a continuous derivative and is absolutely integrable on $[0, \infty)$. For the Fourier cosine integral, compute the coefficients

$$A_\omega = \frac{2}{\pi} \int_0^\infty e^{-k\xi} \cos(\omega \xi)\, d\xi = \frac{2}{\pi} \frac{k}{k^2 + \omega^2}.$$

Then, for $x \geq 0$,

$$e^{-kx} = \frac{2k}{\pi} \int_0^\infty \frac{1}{k^2 + \omega^2} \cos(\omega x)\, d\omega.$$

Next compute the sine coefficients

$$B_\omega = \frac{2}{\pi} \int_0^\infty e^{-k\xi} \sin(\omega \xi)d\xi = \frac{2}{\pi} \frac{\omega}{k^2 + \omega^2}.$$

Then, for $x > 0$, we also have

$$e^{-kx} = \frac{2}{\pi} \int_0^\infty \frac{\omega}{k^2 + \omega^2} \sin(\omega x)\, d\omega.$$

This integral is zero for $x = 0$ and so does not represent $f(x)$ there.

These integral representations are called Laplace's integrals, because A_ω is $2/\pi$ times the Laplace transform of $\sin(kx)$, while B_ω is $2/\pi$ times the Laplace transform of $\cos(kx)$. ◆

In each of Problems 1 through 8, find the Fourier cosine and sine integral representations of the function. Determine what each integral representation converges to.

1. $f(x) = \begin{cases} x^2 & \text{for } 0 \leq x \leq 10 \\ 0 & \text{for } x > 10 \end{cases}$

2. $f(x) = \begin{cases} \sin(x) & \text{for } 0 \leq x \leq 2\pi \\ 0 & \text{for } x > 2\pi \end{cases}$

3. $f(x) = \begin{cases} 1 & \text{for } 0 \leq x \leq 1 \\ 2 & \text{for } 1 < x \leq 4 \\ 0 & \text{for } x > 4 \end{cases}$

4. $f(x) = \begin{cases} \cosh(x) & \text{for } 0 \leq x \leq 5 \\ 0 & \text{for } x > 5 \end{cases}$

5. $f(x) = \begin{cases} 2x + 1 & \text{for } 0 \leq x \leq \pi \\ 2 & \text{for } \pi < x \leq 3\pi \\ 0 & \text{for } x > 3\pi. \end{cases}$

6. $f(x) = \begin{cases} x & \text{for } 0 \leq x \leq 1 \\ x + 1 & \text{for } 1 < x \leq 2 \\ 0 & \text{for } x > 2 \end{cases}$

7. $f(x) = e^{-x}\cos(x)$ for $x \geq 0$

8. $f(x) = xe^{-3x}$ for $x \geq 0$

13.3   The Fourier Transform

We will use equation (13.4) to derive a complex form of the Fourier integral representation of a function, and then use this to define the Fourier transform.

Suppose f is absolutely integrable on the real line, and piecewise smooth on each $[-L, L]$. Then, at any x,

$$\frac{1}{2}(f(x+) + f(x-)) = \frac{1}{\pi} \int_0^\infty \int_{-\infty}^\infty f(\xi)\cos(\omega(\xi - x))\,d\xi\,d\omega.$$

Recall that

$$\cos(x) = \frac{1}{2}\left(e^{ix} + e^{-ix}\right)$$

to obtain

$$\frac{1}{2}(f(x+) + f(x-)) = \frac{1}{\pi} \int_0^\infty \int_{-\infty}^\infty f(\xi)\frac{1}{2}\left(e^{i\omega(\xi-x)} + e^{-i\omega(\xi-x)}\right) d\xi\,d\omega$$

$$= \frac{1}{2\pi} \int_0^\infty \int_{-\infty}^\infty f(\xi)e^{i\omega(\xi-x)}\,d\xi\,d\omega$$

$$+ \frac{1}{2\pi} \int_0^\infty \int_{-\infty}^\infty f(\xi)e^{-i\omega(\xi-x)}\,d\xi\,d\omega.$$

In the next-to-last integral, replace ω with $-\omega$ and compensate for this change by replacing $\int_0^\infty \dots d\omega$ with $\int_{-\infty}^0 \dots d\omega$. This enables us to write

$$\frac{1}{2}(f(x+) + f(x-))$$

$$= \frac{1}{2\pi} \int_{-\infty}^0 \int_{-\infty}^\infty f(\xi)e^{-i\omega(\xi-x)}\,d\xi\,d\omega + \frac{1}{2\pi} \int_0^\infty \int_{-\infty}^\infty f(\xi)e^{-i\omega(\xi-x)}\,d\xi\,d\omega.$$

Now we can combine these integrals to obtain

$$\frac{1}{2}(f(x+) + f(x-)) = \frac{1}{2\pi} \int_{-\infty}^\infty \int_{-\infty}^\infty f(\xi)e^{-i\omega\xi}e^{i\omega x}\,d\xi\,d\omega. \tag{13.9}$$

This is the *complex Fourier integral representation of f on the real line*. If we let

$$C_\omega = \int_{-\infty}^{\infty} f(\xi)e^{-i\omega\xi}\,d\xi,$$

then this integral representation is

$$\frac{1}{2}(f(x+) + f(x-)) = \frac{1}{2\pi}\int_{-\infty}^{\infty} C_\omega e^{i\omega x}\,d\omega.$$

We call C_ω the *complex Fourier integral coefficient of f*. Our main interest in this complex Fourier integral is as a springboard to the Fourier transform, the idea of which is contained in equation (13.9). For emphasis in how we want to think of this equation, write it as

$$\frac{1}{2}(f(x+) + f(x-)) = \frac{1}{2\pi}\int_{-\infty}^{\infty}\left(\int_{-\infty}^{\infty} f(\xi)e^{-i\omega\xi}\,d\xi\right)e^{i\omega x}\,d\omega. \qquad (13.10)$$

The term in large parentheses on the right in equation (13.10) is the Fourier transform of f. We summarize this discussion as follows.

Fourier Transform

If f is absolutely integrable on the real line, then the *Fourier transform* $\mathcal{F}[f]$ of f is the function defined by

$$\mathcal{F}[f](\omega) = \int_{-\infty}^{\infty} f(t)e^{-i\omega t}\,dt.$$

Thus, the Fourier transform of f is the coefficient C_ω in the complex Fourier integral representation of f.

Because of the use of the Fourier transform in applications such as signal analysis, we usually use t (for time) as the variable for the function and ω as the variable of the transformed function $\mathcal{F}[f]$. Engineers refer to ω as the *frequency* of the signal f.

We often denote $\mathcal{F}[f]$ as $\hat{f}$. In this notation,

$$\mathcal{F}[f](\omega) = \hat{f}(\omega).$$

EXAMPLE 13.3

We will determine the transform of $e^{-c|t|}$ with c a positive number. First, write

$$f(t) = e^{-c|t|} = \begin{cases} e^{-ct} & \text{for } t \geq 0 \\ e^{ct} & \text{for } t < 0. \end{cases}$$

Then

$$\mathcal{F}[f](\omega) = \int_{-\infty}^{\infty} e^{-c|t|} e^{-i\omega t}\, dt$$

$$= \int_{-\infty}^{0} e^{ct} e^{-i\omega t}\, dt + \int_{0}^{\infty} e^{-ct} e^{-i\omega t}\, dt$$

$$= \int_{-\infty}^{0} e^{(c-i\omega)t}\, dt + \int_{0}^{\infty} e^{-(c+i\omega)t}\, dt$$

$$= \left[\frac{1}{c-i\omega} e^{(c-i\omega)t} \right]_{-\infty}^{0} + \left[\frac{-1}{c+i\omega} e^{-(c+i\omega)t} \right]_{0}^{\infty}$$

$$= \left(\frac{1}{c+i\omega} + \frac{1}{c-i\omega} \right) = \frac{2c}{c^2 + \omega^2}.$$

We can also write

$$\hat{f}(\omega) = \frac{2c}{c^2 + \omega^2}. \quad \blacklozenge$$

EXAMPLE 13.4

Let $H(t)$ be the Heaviside function, defined by

$$H(t) = \begin{cases} 1 & \text{for } t \geq 0 \\ 0 & \text{for } t < 0. \end{cases}$$

We will compute the Fourier transform of $f(t) = H(t)e^{-5t}$. This is the function

$$f(t) = \begin{cases} e^{-5t} & \text{for } t \geq 0 \\ 0 & \text{for } t < 0. \end{cases}$$

From the definition of $\mathcal{F}$,

$$\hat{f}(\omega) = \int_{-\infty}^{\infty} H(t)e^{-5t} e^{-i\omega t}\, dt$$

$$\int_{0}^{\infty} e^{-5t} e^{-i\omega t}\, dt = \int_{0}^{\infty} e^{-(5+i\omega)t}\, dt$$

$$= -\frac{1}{5+i\omega} \left[e^{-(5+i\omega)t} \right]_{0}^{\infty} = \frac{1}{5+i\omega}. \quad \blacklozenge$$

EXAMPLE 13.5

Let a and k be positive numbers. We will determine $\hat{f}(t)$, where

$$f(t) = \begin{cases} k & \text{for } -a \leq t < a \\ 0 & \text{for } t < -a \text{ and } t \geq a. \end{cases}$$

This is the pulse

$$f(t) = k[H(t+a) - H(t-a)].$$

Then

$$\hat{f}(t) = \int_{-\infty}^{\infty} f(t)e^{-i\omega t}\, dt$$

$$= \int_{-a}^{a} ke^{-i\omega t}\, dt = \left[\frac{-k}{i\omega}e^{-i\omega t}\right]_{-a}^{a}$$

$$= -\frac{k}{i\omega}[e^{-i\omega a} - e^{i\omega a}] = \frac{2k}{\omega}\sin(a\omega). \; \blacklozenge$$

These examples were done by integration. Usually the Fourier transform of a function is computed using tables or software.

Now suppose that f is continuous and f' is piecewise smooth on every interval $[-L, L]$. Because $\hat{f}(\omega)$ is the coefficient in the complex Fourier integral representation of f, we have

$$f(t) = \frac{1}{2\pi}\int_{-\infty}^{\infty}\hat{f}(\omega)e^{i\omega t}\, d\omega. \tag{13.11}$$

Equation (13.11) defines the *inverse Fourier transform*. Given f satisfying certain conditions, we can compute its Fourier transform $\hat{f}$, and conversely, given this transform, we can recover f from equation (13.11).

For this reason, we call the equations

$$\hat{f}(\omega) = \int_{-\infty}^{\infty} f(t)e^{-i\omega t}\, dt \text{ and } f(t) = \frac{1}{2\pi}\int_{-\infty}^{\infty}\hat{f}(\omega)e^{i\omega t}\, d\omega$$

a *transform pair*. We also denote the inverse Fourier transform as $\mathcal{F}^{-1}$:

$$\mathcal{F}^{-1}[\hat{f}] = f \text{ exactly when } \mathcal{F}[f] = \hat{f}.$$

EXAMPLE 13.6

Let

$$f(t) = \begin{cases} 1 - |t| & \text{for } -1 \leq t \leq 1 \\ 0 & \text{for } |t| > 1. \end{cases}$$

Then f is continuous and absolutely integrable, and f' is piecewise continuous. A routine integral gives us the Fourier transform of f:

$$\hat{f}(\omega) = \int_{-\infty}^{\infty} f(t)e^{-i\omega t}\, dt$$

$$= \int_{-1}^{1} (1 - |t|)e^{-i\omega t} = \frac{2(1 - \cos(\omega))}{\omega^2}.$$

As an illustration, we will compute the inverse of this Fourier transform. By equation (13.11),

$$\mathcal{F}^{-1}[\hat{f}](t) = \frac{1}{2\pi}\int_{-\infty}^{\infty}\hat{f}(\omega)e^{i\omega t}\, d\omega$$

$$= \frac{1}{\pi}\int_{-\infty}^{\infty}\frac{1 - \cos(\omega)}{\omega^2}e^{i\omega t}\, d\omega$$

$$= \pi(t + 1)\text{sgn}(t + 1) + \pi(t - 1)\text{sgn}(t - 1) - 2\text{sgn}(t).$$

This integration was obtained using a software package, in which

$$\text{sgn}(t) = \begin{cases} 1 & \text{for } t > 0 \\ -1 & \text{for } t < 0 \\ 0 & \text{for } t = 0. \end{cases}$$

By considering these cases, it is routine to verify that indeed $\mathcal{F}^{-1}[\hat{f}](t) = f(t)$. ◆

The *amplitude spectrum* of a signal $f(t)$ is the graph of $|\hat{f}(\omega)|$.

EXAMPLE 13.7

Let a and k be positive numbers and let

$$f(t) = \begin{cases} k & \text{for } -a \leq t \leq a \\ 0 & \text{for } t < -a \text{ and for } t > a. \end{cases}$$

By Example 13.5,

$$\hat{f}(\omega) = \int_{-\infty}^{\infty} f(t)e^{-i\omega t}\, dt$$

$$= \int_{-a}^{a} ke^{-i\omega t}\, dt = -\frac{k}{i\omega}(e^{-i\omega t} - e^{i\omega t})$$

$$= \frac{2k}{\omega}\sin(a\omega).$$

The amplitude spectrum of f is the graph of

$$|\hat{f}(\omega)| = 2k\left|\frac{\sin(a\omega)}{\omega}\right|$$

shown in Figure 13.1 for $k = 1$, and $a = 2$. ◆

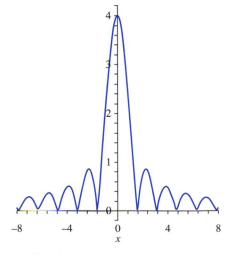

FIGURE 13.1 *Graph of* $|\hat{f}(\omega)|$ *in Example 13.7 for* $k = 1$ *and* $a = 2$.

We will list some properties and computational rules for the Fourier transform.

Linearity

$$\mathcal{F}[f+g]=\mathcal{F}[f]+\mathcal{F}[g]$$

and, for any number k,

$$\mathcal{F}[kf]=k\mathcal{F}[f].$$

Time Shifting
If t_0 is a real number, then

$$\mathcal{F}[f(t-t_0)](\omega)=e^{-i\omega t_0}\hat{f}(\omega).$$

The Fourier transform of a shifted function $f(t-t_0)$ is the Fourier transform of f multiplied by $e^{-i\omega t_0}$.

The inverse version of the time shifting theorem is

$$\mathcal{F}^{-1}[e^{-i\omega t_0}\hat{f}(\omega)](t)=f(t-t_0). \tag{13.12}$$

EXAMPLE 13.8

We will compute

$$\mathcal{F}^{-1}\left[\frac{e^{2i\omega}}{5+i\omega}\right].$$

The presence of the exponential factor $e^{2i\omega}$ suggests use of the inverse version of the time shifting theorem. Put $t_0=-2$ and

$$\hat{f}(\omega)=\frac{1}{5+i\omega}$$

into equation (13.12) to get

$$\mathcal{F}^{-1}[e^{2i\omega}\hat{f}(\omega)](t)=f(t-(-2))=f(t+2),$$

where

$$f(t)=\mathcal{F}^{-1}\left[\frac{1}{5+i\omega}\right]=H(t)e^{-5t}$$

from Example 13.4. Then, by time shifting,

$$\mathcal{F}^{-1}\left[\frac{e^{2i\omega}}{5+i\omega}\right]=f(t+2)=H(t+2)e^{-5(t+2)}. \quad \blacklozenge$$

Frequency Shifting

If ω_0 is any real number, then

$$\mathcal{F}[e^{i\omega_0 t} f(t)] = \hat{f}(\omega - \omega_0).$$

The Fourier transform of a function multiplied by $e^{i\omega_0 t}$ is the Fourier transform of f shifted right by ω_0.

Scaling

If c is any nonzero real number, then

$$\mathcal{F}[f(ct)](\omega) = \frac{1}{|c|} \hat{f}(\omega/c).$$

The inverse version of this is

$$\mathcal{F}^{-1}[\hat{f}(\omega/c)] = |c| f(ct).$$

Time Reversal

$$\mathcal{F}[f(-t)](\omega) = \hat{f}(-\omega).$$

Symmetry

$$\mathcal{F}[\hat{f}(t)](\omega) = 2\pi f(-\omega).$$

If we replace ω by t in the transformed function $\hat{f}$ and then take the transform of this function of t, then we obtain 2π times the original function f with t replaced by $-\omega$.

Modulation

If ω_0 is a real number, then

$$\mathcal{F}[f(t)\cos(\omega_0 t)](\omega) = \frac{1}{2}\left(\hat{f}(\omega + \omega_0) + \hat{f}(\omega - \omega_0)\right)$$

and

$$\mathcal{F}[f(t)\sin(\omega_0 t)](\omega) = \frac{i}{2}\left(\hat{f}(\omega + \omega_0) - \hat{f}(\omega - \omega_0)\right)$$

Operational Formula

To apply the Fourier transform to a differential equation, we must be able to transform a derivative. This is called an *operational rule*. Recall that the kth derivative of f is denoted $f^{(k)}$. As a convenience, we let $f^{(0)} = f$; the "zero-order derivative" of a function is just the function.

Now let n be any positive integer and suppose that $f^{(n-1)}$ is continuous and $f^{(n)}$ is piecewise continuous on each interval $[-L, L]$. Suppose also that $\int_{-\infty}^{\infty} |f^{(n-1)}| \, dt$ converges and that

$$\lim_{t \to \infty} f^{(k)}(t) = \lim_{t \to -\infty} f^{(k)}(t) = 0$$

for $k = 0, 1, 2, \ldots, n - 1$. Then

$$\mathcal{F}[f^{(n)}(t)](\omega) = (i\omega)^n \hat{f}(\omega).$$

In other words, under the given conditions, the Fourier transform of the nth derivative of f is the nth power of $i\omega$ times the Fourier transform of f.

EXAMPLE 13.9

We will solve the differential equation

$$y' - 4y = H(t)e^{-4t}.$$

Apply the Fourier transform to the differential equation to get

$$\mathcal{F}[y'(\omega)] - 4\hat{y}(\omega) = \mathcal{F}[H(t)e^{-4t}](\omega).$$

From the operational rule with $n = 1$,

$$\mathcal{F}[y'](\omega) = i\omega\hat{y}(\omega).$$

Furthermore, from Example 13.4, with 4 in place of 5 we have

$$\mathcal{F}[H(t)e^{-4t}](\omega) = \frac{1}{4+i\omega}.$$

Therefore,

$$i\omega\hat{y} - 4\hat{y} = \frac{1}{4+i\omega}.$$

Solve for $\hat{y}$ to get

$$\hat{y}(\omega) = \frac{-1}{16+\omega^2}.$$

From Example 13.3,

$$y(t) = \mathcal{F}^{-1}\left[\frac{-1}{16+\omega^2}\right] = -\frac{1}{8}e^{-4|t|}. \; \blacklozenge$$

Frequency Differentiation
Let n be a positive integer. Let f be piecewise continuous on $[-L, L]$ for every positive L and assume that $\int_{-\infty}^{\infty} |t^n f(t)|\, dt$ converges. Then

$$\frac{d^n}{d\omega^n}\hat{f}(\omega) = i^{-n}\mathcal{F}[t^n f(t)](\omega).$$

This means that the nth derivative of the Fourier transform of f is i^{-n} times the transform of $t^n f(t)$.

The Fourier Transform of an Integral
Let f be piecewise continuous on every interval $[-L, L]$. Suppose $\int_{-\infty}^{\infty} |f(t)|\, dt$ converges and that $f(0) = 0$. Then

$$\mathcal{F}\left[\int_{-\infty}^{t} f(\tau)d\tau\right](\omega) = \frac{1}{i\omega}\hat{f}(\omega).$$

Convolution

Integral transforms usually have some kind of convolution operation. We have seen a convolution for the Laplace transform. For the Fourier transform, we define the *convolution* of f with g to be the function $f * g$ given by

$$(f * g)(t) = \int_{-\infty}^{\infty} f(t - \tau)g(\tau)\, d\tau.$$

In making this definition, we assume that $\int_a^b f(t)\, dt$ and $\int_a^b g(t)\, dt$ exist for every interval $[a, b]$ and that, for every real number t, $\int_{-\infty}^{\infty} |f(t - \tau)g(t)|\, d\tau$ converges.

Convolution has the following properties.

Commutativity   If $f * g$ is defined, so is $g * f$ and

$$f * g = g * f.$$

Convolution is Linear   This means that for numbers α and β and functions f, g, and h we have

$$(\alpha f + \beta g) * h = \alpha (f * h) + \beta (g * h)$$

provided that all these convolutions are defined.

For the next three properties of convolution, suppose that f and g are bounded and continuous on the real line and that f and g are both absolutely integrable. Then

$$\int_{-\infty}^{\infty} (f * g)(t)\, dt = \int_{-\infty}^{\infty} f(t)\, dt \int_{-\infty}^{\infty} g(t)\, dt.$$

Time Convolution

$$\mathcal{F}[f * g] = \hat{f}\hat{g}.$$

This states that the Fourier transform of the convolution of two functions is the product of the Fourier transforms of the functions. This is known as the *convolution theorem*. The ramification for the inverse Fourier transform is that

$$\mathcal{F}^{-1}[\hat{f}(\omega)\hat{g}(\omega)](t) = (f * g)(t).$$

That is, the inverse Fourier transform of a product of two transformed functions is the convolution of the functions.

Frequency Convolution

$$\mathcal{F}[fg](\omega) = \frac{1}{2\pi}(\hat{f} * \hat{g})(\omega).$$

EXAMPLE 13.10

We will compute

$$\mathcal{F}^{-1}\left[\frac{1}{(4 + \omega^2)(9 + \omega^2)}\right].$$

We want the inverse transform of a product, knowing the inverse of each factor:

$$\mathcal{F}^{-1}\left(\frac{1}{4+\omega^2}\right) = f(t) = \frac{1}{4}e^{-2|t|}$$

and

$$\mathcal{F}^{-1}\left(\frac{1}{9+\omega^2}\right) = g(t) = \frac{1}{6}e^{-3|t|}.$$

The inverse version of the convolution theorem tells us that

$$\mathcal{F}^{-1}\left[\frac{1}{(4+\omega^2)(9+\omega^2)}\right](t) = (f*g)(t) = \frac{1}{24}\int_{-\infty}^{\infty} e^{-2|t-\tau|}e^{-3|\tau|}d\tau.$$

To evaluate this integral, we must consider three cases.

Case 1 If $t > 0$ then

$$24(f*g)(t) = \int_{-\infty}^{0} e^{-2|t-\tau|}e^{-3|\tau|}d\tau + \int_{0}^{t} e^{-2|t-\tau|}e^{-3|\tau|}d\tau + \int_{t}^{\infty} e^{-2|t-\tau|}e^{-3|\tau|}d\tau$$

$$= \int_{-\infty}^{0} e^{-2(t-\tau)}e^{3\tau}d\tau + \int_{0}^{t} e^{-2(t-\tau)}e^{-3\tau}d\tau + \int_{t}^{\infty} e^{-2(t-\tau)}e^{-3\tau}d\tau$$

$$= \frac{6}{5}e^{-2t} - \frac{4}{5}e^{-3t}.$$

Case 2 If $t < 0$, then

$$24(f*g)(t) = \int_{-\infty}^{t} e^{-2|t-\tau|}e^{-3|\tau|}d\tau + \int_{t}^{0} e^{-2|t-\tau|}e^{-3|\tau|}d\tau + \int_{0}^{\infty} e^{-2|t-\tau|}e^{-3|\tau|}d\tau$$

$$= \int_{-\infty}^{t} e^{-2(t-\tau)}e^{3\tau}d\tau + \int_{t}^{0} e^{2(t-\tau)}e^{3\tau}d\tau + \int_{0}^{\infty} e^{2(t-\tau)}e^{-3\tau}d\tau$$

$$= -\frac{4}{5}e^{3t} + \frac{6}{5}e^{2t}.$$

Case 3 Finally, if $t = 0$,

$$24(f*g)(0) = \int_{-\infty}^{\infty} e^{-2|\tau|}e^{-3|\tau|}d\tau = \frac{2}{5}.$$

Therefore,

$$\mathcal{F}^{-1}\left[\frac{1}{(4+\omega^2)(9+\omega^2)}\right](t) = \frac{1}{24}\left(\frac{6}{5}e^{-2|t|} - \frac{4}{5}e^{-3|t|}\right)$$

$$= \frac{1}{20}e^{-2|t|} - \frac{1}{30}e^{-3|t|}. \quad \blacklozenge$$

    PROBLEMS

In each of Problems 1 through 10, find the Fourier transform of the function and graph the amplitude spectrum. Wherever k appears, it is a positive constant. Use can be made of the following transforms:

$$\mathcal{F}[e^{-kt^2}](\omega) = \sqrt{\frac{\pi}{k}} e^{-\omega^2/4k}$$

and

$$\mathcal{F}\left[\frac{1}{k^2+t^2}\right](\omega) = \frac{\pi}{k} e^{-k|\omega|}$$

1. $f(t) = \begin{cases} 1 & \text{for } 0 \le t \le 1 \\ -1 & \text{for } -1 \le t < 0 \\ 0 & \text{for } |t| > 1 \end{cases}$

2. $f(t) = \begin{cases} \sin(t) & \text{for } -k \le t \le k \\ 0 & \text{for } |t| > k \end{cases}$

3. $f(t) = 5[H(t-3) - H(t-11)]$

4. $f(t) = 5e^{-3(t-5)^2}$

5. $f(t) = H(t-k)e^{-t/4}$

6. $f(t) = H(t-k)t^2$

7. $f(t) = 1/(1+t^2)$

8. $f(t) = 3H(t-2)e^{-3t}$

9. $f(t) = 3e^{-4|t+2|}$

10. $f(t) = H(t-3)e^{-2t}$

In each of Problems 11 through 15, find the inverse Fourier transform of the function.

11. $9e^{-(\omega+4)^2/32}$

12. $e^{(20-4\omega)i}/(3 - (5-\omega)i)$

13. $e^{(2\omega-6)i}/(5 - (3-\omega)i)$

14. $10\sin(3\omega)/(\omega+\pi)$

15. $(1+i\omega)/(6-\omega^2+5i\omega)$ *Hint*: Factor the denominator and use partial fractions.

In each of Problems 16, 17, and 18, use convolution to find the inverse Fourier transform of the function.

16. $1/((1+i\omega)(2+i\omega))$

17. $1/(1+i\omega)^2$

18. $\sin(3\omega)/\omega(2+i\omega)$

13.4   Fourier Cosine and Sine Transforms

If f is piecewise smooth on each interval $[0, L]$ and $\int_0^\infty |f(t)|\,dt$ converges, then at each t where f is continuous, the Fourier cosine integral for f is

$$f(t) = \int_0^\infty a_\omega \cos(\omega t)\,d\omega,$$

where

$$a_\omega = \frac{2}{\pi} \int_0^\infty f(t) \cos(\omega t)\,dt.$$

These suggest that we define the *Fourier cosine transform of f* by

$$\mathcal{F}_C[f](\omega) = \int_0^\infty f(t) \cos(\omega t)\,dt. \qquad (13.13)$$

Often we denote $\mathcal{F}_C[f](\omega) = \hat{f}_C(\omega)$.

Notice that

$$\hat{f}_C(\omega) = \frac{\pi}{2} a_\omega$$

and that

$$f(t) = \frac{2}{\pi} \int_0^\infty \hat{f}_C(\omega) \cos(\omega t)\,d\omega. \qquad (13.14)$$

Equations (13.13) and (13.14) form a transform pair for the Fourier cosine transform. Equation (13.14) is the inverse Fourier cosine transform, reproducing f from $\hat{f}_c$. This inverse is denoted $\hat{f}_C^{-1}$.

EXAMPLE 13.11

Let K be a positive number and let

$$f(t) = \begin{cases} 1 & \text{for } 0 \leq t \leq K \\ 0 & \text{for } t > K. \end{cases}$$

Then

$$\hat{f}_C(\omega) = \int_0^\infty f(t) \cos(\omega t)\, dt = \int_0^K \cos(\omega t)\, dt = \frac{\sin(K\omega)}{\omega}. \quad \blacklozenge$$

Using the Fourier sine integral instead of the cosine integral, define the *Fourier sine transform* of f by

$$\mathcal{F}_S[f](\omega) = \int_0^\infty f(t) \sin(\omega t)\, dt.$$

We also denote this as $\hat{f}_S(\omega)$.

If f is continuous at $t > 0$ then the Fourier sine integral representation is

$$f(t) = \int_0^\infty b_\omega \sin(\omega t)\, d\omega,$$

where

$$b_\omega = \frac{2}{\pi} \int_0^\infty f(t) \sin(\omega t)\, dt = \frac{2}{\pi} \hat{f}_S(\omega).$$

This means that

$$f(t) = \frac{2}{\pi} \int_0^\infty \hat{f}_S(\omega) \sin(\omega t)\, d\omega$$

which provides a way of retrieving f from $\hat{f}_S$. This integral is the inverse Fourier sine transform $\hat{f}_S^{-1}$.

EXAMPLE 13.12

With f as in Example 13.11,

$$\hat{f}_S(\omega) = \int_0^\infty f(t) \sin(\omega t)dt = \int_0^K \sin(\omega t)dt = \frac{1}{\omega}[1 - \cos(K\omega)]. \quad \blacklozenge$$

The following operational formulas are needed when these transforms are used to solve differential equations.

Operational Formulas

Let f and f' be continuous on every interval $[0, L]$ and let $\int_0^\infty |f(t)|\,dt$ converge. Suppose $f(t) \to 0$ and $f'(t) \to 0$ as $t \to \infty$. Suppose f'' is piecewise continuous on every $[0, L]$. Then

1. $\mathcal{F}_C[f''(t)](\omega) = -\omega^2 \hat{f}_C(\omega) - f'(0)$.

2. $\mathcal{F}_S[f''(t)](\omega) = -\omega^2 \hat{f}_S(\omega) + \omega f(0)$. ◆

SECTION 13.4   PROBLEMS

In each of Problems 1 through 6, determine the Fourier cosine transform and the Fourier sine transform of the function.

1. $f(t) = e^{-t}$

2. $f(t) = te^{-at}$ with a any positive number

3. $f(t) = \begin{cases} \cos(t) & \text{for } 0 \le t \le K \\ 0 & \text{for } t > K \end{cases}$ with K any positive number

4. $f(t) = \begin{cases} 1 & \text{for } 0 \le t < K \\ -1 & \text{for } K \le t < 2K \\ 0 & \text{for } t \ge 2K \end{cases}$

5. $f(t) = e^{-t}\cos(t)$

6. $f(t) = \begin{cases} \sinh(t) & \text{for } K \le t < 2K \\ 0 & \text{for } 0 \le t < K \text{ and for } t \ge 2K \end{cases}$

CHAPTER 14

Eigenfunction Expansions

The differential equation $y'' + c^2 y = 0$ has solutions $\cos(cx)$ and $\sin(cx)$, which form the basis for Fourier series. Certain other differential equations also have solutions, called special functions, in terms of which we can expand "arbitrary" functions in series, called eigenfunction expansions. This chapter explores these ideas as a prelude to using them in the solution of partial differential equations.

14.1 General Eigenfunction Expansions

Begin with the differential equation

$$y'' + R(x)y' + (Q(x) + \lambda P(x))y = 0$$

on some interval (a, b) with λ as a constant to be determined along with y. Assume that the coefficient functions are continuous on this interval. We will manipulate this differential equation into a special form. Multiply it by

$$r(x) = e^{\int R(x)\,dx}$$

to obtain

$$y'' e^{\int R(x)\,dx} + R(x)y' e^{\int R(x)\,dx} + (Q(x) + \lambda P(x))e^{\int R(x)\,dx} y = 0,$$

which we can write as

$$\left(y' e^{\int R(x)\,dx}\right)' + \left(Q(x)e^{\int R(x)\,dx} + \lambda P(x)e^{\int R(x)\,dx}\right) y = 0.$$

This equation has the general form

$$(ry')' + (q + \lambda p)y = 0. \tag{14.1}$$

When p, q, r, and r' are continuous on (a, b) and $r(x) > 0$ and $p(x) > 0$ on (a, b), we call equation (14.1) the *Sturm-Liouville differential equation.*

This differential equation contains a quantity λ. We want to determine values of λ, called *eigenvalues*, such that there are nontrivial (not identically zero) solutions y of the differential equation satisfying specified conditions, called *boundary conditions* at a and b. For a given eigenvalue λ, such a solution is called an *eigenfunction* associated with λ. There are three kinds of problems for the eigenvalues and eigenfunctions, depending on the form of the boundary conditions.

The Regular Sturm-Liouville Problem
In this problem, $r(x) > 0$ for $a \leq x \leq b$. We want numbers λ for which there are nontrivial solutions of equation (14.1) on $[a, b]$, satisfying *regular boundary conditions*

$$A_1 y(a) + A_2 y'(a) = 0 \text{ and } B_1 y(b) + B_2 y'(b) = 0,$$

in which A_1 and A_2 are given numbers, not both zero, and B_1 and B_2 are also given numbers, not both zero.

The Periodic Sturm-Liouville Problem
Now $r(a) = r(b) \neq 0$, and we want numbers λ for which there are nontrivial solutions of equation (14.1) on $[a, b]$, satisfying *periodic boundary conditions*

$$y(a) = y(b) \text{ and } y'(a) = y'(b).$$

The Singular Sturm-Liouville Problem
For this problem, we want numbers λ for which there are nontrivial solutions of equation (14.1) satisfying one of the following two kinds of boundary conditions.

Type 1 Exactly one of $r(a)$ or $r(b)$ is zero.
 If $r(a) = 0$, then there is no boundary condition specified at a, while at b the boundary condition is

$$B_1 y(b) + B_2 y'(b) = 0$$

with B_1 and B_2 not both zero.

If $r(b) = 0$, there is no boundary condition at b, while at a solutions must satisfy

$$A_1 y(a) + A_2 y'(a) = 0$$

with A_1 and A_2 given and not both zero.

Type 2 $r(a) = r(b) = 0$, although we still assume that $r(x) > 0$ for $a < x < b$.
  We want numbers λ for which there are nontrivial solutions of equation (14.1) that are bounded on $[a, b]$. There are no boundary conditions in this case.

It is possible to show that the eigenvalues associated with all three types of Sturm-Liouville problems must be real numbers.

EXAMPLE 14.1   A Regular Sturm-Liouville Problem

The problem

$$y'' + \lambda y = 0; \; y(0) = y(L) = 0$$

is regular on $[0, L]$. We will solve it for later use. Consider cases on λ.

Case 1 $\lambda = 0$. Then $y(x) = cx + d$ for some constants c and d. But $y(0) = 0 = d$, and then $y(L) = cL = 0$ requires that $c = 0$, so all solutions are trivial when $\lambda = 0$. Therefore, 0 is not an eigenvalue of this problem.

Case 2 $\lambda < 0$, say $\lambda = -k^2$ for $k > 0$. The general solution of $y'' - k^2 y = 0$ is

$$y(x) = c_1 e^{kx} + c_2 e^{-kx}.$$

But $y(0) = c_1 + c_2 = 0$ means that $c_2 = -c_1$, so

$$y(x) = c_1 \left(e^{kx} - e^{-kx} \right).$$

Then

$$y(L) = c_1 \left(e^{kL} - e^{-kL} \right) = 0.$$

If $e^{kL} - e^{-kL} = 0$, we would have $e^{2kL} = 1$ and then $2kL = 0$, which is impossible if $k > 0$ and $L > 0$. Therefore $c_1 = 0$, so $c_2 = 0$ also, and y is the trivial solution. This problem has no negative eigenvalue.

Case 3 $\lambda > 0$, say $\lambda = k^2$ for $k > 0$. Now $y'' + k^2 y = 0$ has general solution

$$y(x) = c_1 \cos(kx) + c_2 \sin(kx).$$

Since $y(0) = c_1 = 0$, then $y = c_2 \sin(kx)$. Then

$$y(L) = c_2 \sin(kL) = 0.$$

To have a nontrivial solution, we cannot have c_2 vanish, so we must choose k so that $\sin(kL) = 0$. Then $kL = n\pi$ for n any positive integer. Using n as an index,

$$\lambda_n = k^2 = \left(\frac{n\pi}{L}\right)^2.$$

These are the eigenvalues of this problem for each positive integer n. Corresponding to each such eigenvalue is the eigenfunction

$$y(x) = \varphi_n(x) = \sin\left(\frac{n\pi x}{L}\right).$$

Any nonzero constant multiple of φ_n is also an eigenfunction corresponding to λ_n. ◆

EXAMPLE 14.2 A Periodic Sturm-Liouville Problem

The problem

$$y'' + \lambda y = 0; \quad y(-L) = y(L), \, y'(-L) = y'(L)$$

is periodic on $[-L, L]$. If we compare $y'' + \lambda y = 0$ with the Sturm-Liouville equation, we have $r(x) = 1$, so $r(-L) = r(L) \neq 0$, as is required for a periodic problem on $[-L, L]$. We will also solve this problem by taking cases on λ.

Case 1 $\lambda = 0$. Then $y(x) = cx + d$. Now

$$y(-L) = -cL + d = y(L) = cL + d$$

implies that $c = 0$. The constant function $y(x) = d$ satisfies both boundary conditions. Thus, 0 is an eigenvalue of this problem with constant (nonzero) functions as eigenfunctions.

Case 2 $\lambda < 0$, say $\lambda = -k^2$.
Now $y(x) = c_1 e^{kx} + c_2 e^{-kx}$. The condition $y(L) = y(-L)$ gives us

$$c_1 e^{-kL} + c_2 e^{kL} = c_1 e^{kL} + c_2 e^{-kL}.$$

Write this as

$$c_1(e^{-kL} - e^{kL}) = c_2(e^{-kL} - e^{kL}).$$

This implies that $c_1 = c_2$, so $y(x) = c_1(e^{kx} + e^{-kx})$. Now $y'(-L) = y'(L)$ gives us

$$c_1 k(e^{-kL} - e^{kL}) = c_1 k(e^{kL} - e^{-kL}).$$

This implies that $c_1 = -c_1$, hence $c_1 = 0$, so $c_2 = 0$ also. This problem has only the trivial solution, so there is no negative eigenvalue.

Case 3 $\lambda > 0$, say $\lambda = k^2$. The general solution of $y'' + k^2 y = 0$ is

$$y(x) = c_1 \cos(kx) + c_2 \sin(kx).$$

Now

$$y(-L) = c_1 \cos(kL) - c_2 \sin(kL) = y(L) = c_1 \cos(kL) + c_2 \sin(kL).$$

This implies that $c_2 \sin(kL) = 0$. Next,

$$y'(-L) = kc_1 \sin(kL) + kc_2 \cos(kL) = y'(L) = -kc_1 \sin(kL) + kc_2 \cos(kL),$$

implying that $kc_1 \sin(kL) = 0$. If $\sin(kL) \neq 0$, then $c_1 = c_2 = 0$, and we have only the trivial solution. For a nontrivial solution, we must have $\sin(kL) = 0$, so that at least one of the constants c_1 and c_2 can be chosen nonzero. But $\sin(kL) = 0$ is satisfied if $kL = n\pi$, with n a positive integer (positive because we chose $k > 0$). Since $\lambda = k^2$, the eigenvalues of this problem, indexed by n, are

$$\lambda_n = \left(\frac{n\pi}{L}\right)^2 \text{ for } n = 1, 2, \cdots$$

and corresponding eigenfunctions are

$$y(x) = \varphi_n(x) = c_1 \cos\left(\frac{n\pi x}{L}\right) + c_2 \sin\left(\frac{n\pi x}{L}\right),$$

with c_1 and c_2 any constants, not both zero.

By choosing $n = 0$ and $c_2 = 0$ but $c_1 \neq 0$, we obtain a constant eigenfunction corresponding to the eigenvalue 0, consolidating cases 1 and 3. ◆

A Fourier sine series on $[0, L]$ is an expansion in the eigenfunctions of Example 14.1. A Fourier series on $[-L, L]$ is an expansion in the eigenfunctions of Example 14.2. This raises an intriguing question. If a Sturm-Liouville problem on $[a, b]$ has eigenfunctions $\varphi_1(x), \varphi_2(x), \ldots,$ we ask whether it might be possible to expand an "arbitrary" function $f(x)$ on $[a, b]$ in a series of these eigenfunctions,

$$f(x) = \sum_{n=1}^{\infty} c_n \varphi_n(x).$$

The key lies in the choice of the coefficients c_k, which in turn hinges on an orthogonality property of the eigenfunctions. Recall that in seeking the coefficients a_n and b_n in the Fourier expansion

$$f(x) = \frac{1}{2}a_0 + \sum_{n=1}^{\infty}(a_n \cos(n\pi x/L) + b_n \sin(n\pi x/L)),$$

we multiplied the series by a particular eigenfunction $\cos(m\pi x/L)$ or $\sin(m\pi x/L)$ and integrated. Because of equations (12.3), (12.4), and (12.5), the integral of a product of any two distinct eigenfunctions vanished, leaving simple integral expressions for a_m and b_m.

We will show that a similar property holds for the eigenfunctions of any Sturm-Liouville problem suggesting a similar strategy for finding the coefficients in a series of eigenfunctions $\sum_{n=1}^{\infty} c_n \varphi_n(x)$.

THEOREM 14.1

Let λ_n and λ_m be distinct eigenvalues of a Sturm-Liouville problem on $[a, b]$. Let φ_n and φ_m be corresponding eigenfunctions. Then

$$\int_a^b p(x)\varphi_n(x)\varphi_m(x)dx = 0, \tag{14.2}$$

where p is the coefficient of λ in the Sturm-Liouville differential equation $(ry')' + (q + \lambda p)y = 0$. ◆

To establish this conclusion, begin with

$$(r\varphi_n')' + (q + \lambda_n p)\varphi_n = 0$$

and

$$(r\varphi_m')' + (q + \lambda_m p)\varphi_m = 0.$$

Multiply the first equation by φ_m and the second by φ_n and subtract to get

$$(r\varphi_n')'\varphi_m - (r\varphi_m')'\varphi_n = (\lambda_m - \lambda_n)p\varphi_n\varphi_m.$$

Integrate this equation from a to b to get

$$(\lambda_m - \lambda_n) \int_a^b p(x)\varphi_n(x)\varphi_m(x)\,dx$$

$$= \int_a^b [(r(x)\varphi_n'(x))'\varphi_m(x) - (r(x)\varphi_m'(x))'\varphi_n(x)]\,dx.$$

Integrate by parts to obtain

$$(\lambda_m - \lambda_n) \int_a^b p(x)\varphi_n(x)\varphi_m(x)\,dx$$

$$= r(b)[\varphi_m(b)\varphi_n'(b) - \varphi_n(b)\varphi_m'(b)] - r(a)[\varphi_m(a)\varphi_n'(a) - \varphi_n(a)\varphi_m'(a)].$$

A systematic application of the different kinds of boundary conditions shows that in each case the right side of this equation is zero. Therefore,

$$(\lambda_m - \lambda_n) \int_a^b p(x)\varphi_n(x)\varphi_m(x)\,dx = 0.$$

But $\lambda_n \neq \lambda_m$, so the integral is zero, establishing equation (14.2).

The relationship (14.2) between eigenfunctions associated with distinct eigenvalues is called *orthogonality of the eigenfunctions with respect to the weight function p*. This terminology derives from the fact that $\int_a^b p(x)f(x)g(x)\,dx$ behaves like a dot product for vectors, and two vectors are called orthogonal when their dot product is zero. In Example 14.1, $p(x) = 1$ and the orthogonality relationship is the familiar

$$\int_0^L \sin(n\pi x/L)\sin(m\pi x/L)\,dx = 0$$

for $n \neq m$.

Using this notion of weighted orthogonality of eigenfunctions, we can solve for the coefficients in a proposed series of eigenfunctions

$$f(x) = \sum_{k=1}^{\infty} c_k \varphi_k(x).$$

Multiply this equation by $p(x)\varphi_n(x)$ and integrate to get

$$\int_a^b p(x)f(x)\varphi_n(x)\,dx = \sum_{k=1}^{\infty} \int_a^b p(x)\varphi_k(x)\varphi_n(x)\,dx.$$

By equation (14.2), all of the integrals in the summation are zero except when $k = n$, yielding

$$\int_a^b p(x)f(x)\varphi_n(x)\,dx = c_n \int_a^b p(x)\varphi_n^2(x)\,dx$$

or

$$c_n = \frac{\int_a^b p(x)f(x)\varphi_n(x)\,dx}{\int_a^b p(x)\varphi_n^2(x)}\,dx. \tag{14.3}$$

By analogy with Fourier series, we call the numbers defined by equation (14.3) the *generalized Fourier coefficients of f with respect to the Sturm-Liouville problem*. With this choice of coefficients, we call

$$\sum_{k=1}^{\infty} c_k \varphi_k(x)$$

the *eigenfunction expansion of f with respect to the eigenfunctions of the Sturm-Liouville problem.*

As with Fourier (trigonometric) series, the question now is the relationship between the function and its eigenfunction expansion. The convergence theorem we will state is like that for Fourier series. This should not be surprising, since Fourier series and Fourier sine and cosine series are themselves eigenfunction expansions.

THEOREM 14.2 *Convergence of Eigenfunction Expansions*

Suppose φ_n, for $n = 1, 2, \cdots$, are the eigenfunctions of a Sturm-Liouville problem on $[a, b]$. Let f be piecewise smooth on $[a, b]$ and let c_n be given by equation (14.3). Then, for $a < x < b$, the eigenfunction expansion $\sum_{n=1}^{\infty} c_n \varphi_n(x)$ converges to

$$\frac{1}{2}(f(x+) + f(x-)).$$

If f is continuous at x, then this eigenfunction expansion converges to $f(x)$. ◆

EXAMPLE 14.3

We will expand $f(x) = x$ in a series of the eigenfunctions of the regular Sturm-Liouville problem

$$y'' + 4y' + (3 + \lambda)y = 0; \quad y(0) = y(1) = 0$$

on $[0, 1]$. First we must generate the eigenvalues and eigenfunctions. If we set $y = e^{rx}$ into the differential equation, we obtain the characteristic equation

$$r^2 + 4r + (3 + \lambda) = 0,$$

with roots

$$r = -2 \pm \sqrt{1 - \lambda}.$$

Take cases on λ.

Case 1: $\lambda = 1$. Then

$$y(x) = c_1 e^{-2x} + c_2 x e^{-2x}.$$

Now

$$y(0) = c_1 = 0 \text{ and } y(1) = c_2 e^{-2} = 0$$

so $c_2 = 0$ and y is the trivial solution. Therefore, 1 is not an eigenvalue of this problem.

Case 2: $1 - \lambda > 0$. Write $1 - \lambda = \alpha^2$ with $\alpha > 0$ to obtain

$$y(x) = c_1 e^{(-2+\alpha)x} + c_2 e^{(-2-\alpha)x}.$$

Then

$$y(0) = c_1 + c_2 = 0,$$

so $c_2 = -c_1$. Also

$$y(1) = c_1 e^{-2}(e^\alpha - e^{-\alpha}) = -2c_1 e^{-2} \sinh(\alpha).$$

Since $\alpha > 0$, $\sinh(\alpha) > 0$, so $c_2 = c_1 = 0$ and again y is the trivial solution. There is no eigenvalue $\lambda < 1$.

Case 3: $1 - \lambda < 0$.   Write $1 - \lambda = -\alpha^2$ with $\alpha > 0$. Now

$$y(x) = c_1 e^{-2x} \cos(\alpha x) + c_2 e^{-2x} \sin(\alpha x).$$

Then

$$y(0) = c_1 = 0.$$

This leaves us with $y(x) = c_2 e^{-2x} \sin(\alpha x)$. Now

$$y(1) = c_2 e^{-2} \sin(\alpha) = 0.$$

To have a nontrivial solution, we need $c_2 \neq 0$, so we must have $\sin(\alpha) = 0$. Then $\alpha = n\pi$ for n a positive integer, and the eigenvalues are

$$\lambda = 1 + \alpha^2 = 1 + n^2\pi^2.$$

The eigenfunctions are

$$\varphi_n(x) = e^{-2x} \sin(n\pi x)$$

for $n = 1, 2, \ldots$.

To compute the coefficients in the expansion of $f(x) = x$ in a series of these eigenfunctions, we need to know the weight function p. For this, write the differential equation in Sturm-Liouville form. Multiply it by

$$e^{\int 4\,dx} = e^{4x}$$

to obtain

$$e^{4x} y'' + 4 e^{4x} y' + (3 + \lambda) e^{4x} y = 0.$$

This is

$$\left(e^{4x} y'\right)' + (3 e^{4x} + \lambda e^{4x}) y = 0.$$

This is in Sturm-Liouville form with $r(x) = p(x) = e^{4x}$ and $q(x) = 3 e^{4x}$. To write $f(x) = x = \sum_{n=1}^{\infty} c_n \varphi_n(x)$, choose

$$c_n = \frac{\int_0^1 p(x) e^{-2x} \sin(n\pi x)\,dx}{\int_0^1 p(x) e^{-4x} \sin^2(n\pi x)\,dx}$$

$$= \frac{\int_0^1 e^{2x} \sin(n\pi x)\,dx}{\int_0^1 \sin^2(n\pi x)\,dx}$$

$$= \frac{-8n\pi - 2e^2 n^3 \pi^3 (-1)^n}{(4 + (n\pi)^2)^2}.$$

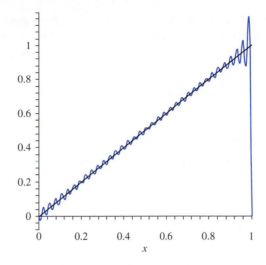

FIGURE 14.1 *Eigenfunction expansion in Example 14.3.*

Figure 14.1 compares graphs of $f(x) = x$ with the 70th partial sum of this expansion. This particular expansion converges fairly slowly to $f(x)$. Note also the Gibbs phenomenon-like behavior at $x = 1$. ◆

SECTION 14.1   PROBLEMS

In each of Problems 1 through 10, classify the Sturm-Liouville problem as regular, periodic, or singular. State the relevant interval and find the eigenvalues and eigenfunctions. In some cases, the eigenvalues may be defined implicitly by a transcendental equation.

1. $y'' + \lambda y = 0; \ y(0) = y'(L) = 0$

2. $y'' + \lambda y = 0; \ y'(0) = y'(L) = 0$

3. $y'' + \lambda y = 0; \ y'(0) = y(4) = 0$

4. $y'' + \lambda y = 0; \ y(0) = y(\pi), \ y'(0) = y'(\pi)$

5. $y'' + \lambda y = 0; \ y(-3\pi) = y(3\pi), \ y'(-3\pi) = y'(3\pi)$

6. $y'' + \lambda y = 0; \ y(0) = 0, \ y(\pi) + 2y'(\pi) = 0$

7. $y'' + \lambda y = 0; \ y(0) - 2y'(0) = 0, \ y'(1) = 0$

8. $y'' + 2y' + (1 + \lambda)y = 0; \ y(0) = y(1) = 0$

9. $(e^{2x}y')' + \lambda e^{2x}y = 0; \ y(0) = y(\pi) = 0$

10. $(e^{-6x}y')' + (1 + \lambda)e^{-6x}y = 0; \ y(0) = y(8) = 0$

In each of Problems 11 through 16, find the eigenfunction expansion of the given function in the eigenfunctions of the Sturm-Liouville problem. In each case determine what this expansion converges to and graph the Nth partial sum of the expansion and the function on the same set of axes. In Problem 11, do the graph for $L = 1$.

11. $f(x) = 1 - x$ for $0 \le x \le L$
    $y'' + \lambda y = 0; \ y(0) = y(L) = 0; \ N = 40$

12. $f(x) = |x|$ for $0 \le x \le \pi$
    $y'' + \lambda y = 0; \ y(0) = y'(\pi) = 0; \ N = 30$

13. $f(x) = \begin{cases} -1 & \text{for } 0 \le x \le 2 \\ 1 & \text{for } 2 < x \le 4 \end{cases}$
    $y'' + \lambda y = 0; \ y'(0) = y(4) = 0; \ N = 40$

14. $f(x) = \sin(2x)$ for $0 \le x \le \pi$
    $y'' + \lambda y = 0; \ y'(0) = y'(\pi) = 0; \ N = 30$

15. $f(x) = x^2$ for $-3\pi \le x \le 3\pi$
    $y'' + \lambda y = 0; \ y(-3\pi) = y(3\pi), \ y'(-3\pi) = y'(3\pi); \ N = 10$

16. $f(x) = \begin{cases} 0 & \text{for } 0 \le x \le 1/2 \\ 1 & \text{for } 1/2 < x \le 1 \end{cases}$
    $y'' + 2y' + (1 + \lambda)y = 0; \ y(0) = y(1) = 0; \ N = 30$

17. Let $\varphi_1, \varphi_2, \ldots$ be the eigenfunctions of a Sturm-Liouville problem on $[a, b]$. Let f be piece-wise smooth on $[a, b]$ with eigenfunction expansion

$\sum_{n=1}^{\infty} c_n \varphi_n(x)$. We call the partial sum $\sum_{n=1}^{N} k_n \varphi_k(x)$ the *best mean approximation to f on $[a, b]$* if the k_n's minimize

$$I_N(f) = \int_a^b p(x) \left(f(x) - \sum_{n=1}^{N} k_n \varphi_n(x) \right)^2 dx.$$

Prove that the best mean approximation is obtained by choosing each $k_n = c_n$,

18. Prove *Bessel's inequality*:

$$\sum_{n=1}^{\infty} c_n^2 \le \int_1^b p(x) f(x)^2 dx.$$

14.2   Fourier-Legendre Expansions

A function is called *special* when it has some distinctive characteristics that make it worthwhile determining its properties and behavior. Special functions often arise as solutions of Sturm-Liouville problems that occur in important contexts. In particular, we will encounter Legendre's equation and Bessel's equation in modeling wave and heat radiation phenomena. We will discuss solutions of Legendre's equation in this section and Bessel's equation in the next.

Legendre's differential equation in Sturm-Liouville form is

$$((1 - x^2)y')' + \lambda y = 0 \text{ for } -1 \le x \le 1.$$

Here $r(x) = 1 - x^2$. Since $r(-1) = r(1) = 0$, there are no boundary conditions, but we seek solutions that are bounded on $[-1, 1]$. To find such solutions, attempt a power series solution $y = \sum_{n=0}^{\infty} a_n x^n$ expanded about 0. Substitute this into Legendre's equation to obtain

$$\sum_{n=2}^{\infty} n(n-1)a_n x^{n-2} - \sum_{n=2}^{\infty} n(n-1)a_n x^n - \sum_{n=0}^{\infty} 2n a_n x^n + \sum_{n=0}^{\infty} \lambda a_n x^n = 0.$$

Rewrite the first summation as

$$\sum_{n=2}^{\infty} n(n-1)a_n x^{n-2} = \sum_{n=0}^{\infty} (n+2)(n+1)a_{n+2} x^n$$

to obtain

$$\sum_{n=0}^{\infty} (n+2)(n+1)a_{n+2} x^n - \sum_{n=2}^{\infty} n(n-1)a_n x^n - \sum_{n=0}^{\infty} 2n a_n x^n + \sum_{n=0}^{\infty} \lambda a_n x^n = 0.$$

Write this as

$$2a_2 + 6a_3 x - 2a_1 x + \lambda a_0 + \lambda a_1 x$$

$$+ \sum_{n=2}^{\infty} [(n+2)(n+1)a_{n+2} - (n^2 + n - \lambda)a_n] x^n = 0.$$

The constant term and the coefficient of each power of x on the left must be zero, so

$$2a_2 + \lambda a_0 = 0$$

$$6a_3 - 2a_1 + \lambda a_1 = 0,$$

and

$$(n+1)(n+2)a_{n+2} - [n(n+1) - \lambda]a_n = 0 \text{ for } n = 2, 3, \dots.$$

From these equations, we obtain, in turn,

$$a_2 = -\frac{\lambda}{2}a_0,$$

$$a_3 = \frac{2-\lambda}{6}a_1 = \frac{2-\lambda}{3!}a_1,$$

and

$$a_{n+2} = \frac{n(n+1)-\lambda}{(n+1)(n+2)}a_n \text{ for } n = 2, 3, \dots.$$

The last equation is a recurrence relation which gives each a_{n+2} in terms of λ and a_n. This enables us to produce the coefficients in turn from previously found coefficients. We already have expressions for a_2 and a_3 in terms of λ, a_0 and a_1. From the recurrence relation with $n = 2$,

$$a_4 = \frac{6-\lambda}{(3)(4)}a_2 = -\frac{\lambda}{2}\frac{6-\lambda}{(3)(4)}a_0 = \frac{-\lambda(6-\lambda)}{4!}a_0$$

$$a_6 = \frac{20-\lambda}{(5)(6)}a_4 = \frac{-\lambda(6-\lambda)(20-\lambda)}{6!}a_0,$$

and so on, with each even-indexed a_{2n} in terms of λ, n, and a_0. Similarly,

$$a_5 = \frac{12-\lambda}{(4)(5)}a_3 = \frac{(2-\lambda)(12-\lambda)}{5!}a_1,$$

$$a_7 = \frac{30-\lambda}{(6)(7)}a_5 = \frac{(2-\lambda)(12-\lambda)(30-\lambda)}{7!}a_1,$$

and so on, with each odd-indexed a_{2n+1} in terms of λ, n, and a_1.

Now we can write the solution

$$y(x) = \sum_{n=0}^{\infty} a_n x^n = a_0\left(1 - \frac{\lambda}{2}x^2 - \frac{\lambda(6-\lambda)}{4!}x^4 - \frac{\lambda(6-\lambda)(20-\lambda)}{6!}x^6 + \dots\right)$$

$$+ a_1\left(x + \frac{2-\lambda}{3!}x^3 + \frac{(2-\lambda)(12-\lambda)}{5!}x^5 + \frac{(2-\lambda)(12-\lambda)(30-\lambda)}{7!}x^7 + \dots\right).$$

These two series solutions in large parentheses are linearly independent, one containing only odd powers of x and the other only even powers. This expression therefore gives the general solution of Legendre's equation with a_0 and a_1 as arbitrary constants.

Now observe that we can obtain polynomial solutions by choosing λ of the form $n(n+1)$ and either a_0 or a_1 as zero. For example:

With $n = 0$ and $a_1 = 0$, we have $\lambda = 0$ and

$$y(x) = a_0 = \text{constant}.$$

With $n = 1$ and $a_0 = 0$, we have $\lambda = 2$ and

$$y(x) = a_1 x.$$

With $n = 2$ and $a_1 = 0$, we have $\lambda = 6$ and

$$y(x) = a_0(1 - 3x^2).$$

With $n = 3$ and $a_0 = 0$, we have $\lambda = 12$ and

$$y(x) = a_1\left(x - \frac{5}{3}x^3\right).$$

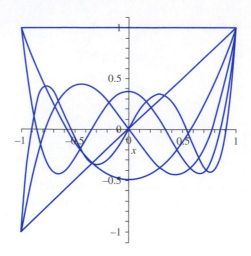

FIGURE 14.2 *The first six Legendre polynomials.*

With $n = 4$ and $a_1 = 0$, we have $\lambda = 20$ and

$$y(x) = a_0 \left(1 - 10x^2 + \frac{35}{3} x^4 \right).$$

and so on. These polynomial solutions are bounded on $[-1, 1]$. If the constant is always chosen so that $y(1) = 1$, we obtain the Legendre polynomials $P_n(x)$, the first six of which are

$$P_0(x) = 1, \; P_1(x) = x, \; P_2(x) = \frac{1}{2}(3x^2 - 1),$$

$$P_3(x) = \frac{1}{2}(5x^3 - 3x), \; P_4(x) = \frac{1}{8}(35x^4 - 30x^2 + 3),$$

and

$$P_5(x) = \frac{1}{8}(63x^5 - 70x^3 + 15x).$$

Graphs of these first six are shown in Figure 14.2.

In summary, the eigenvalues of Legendre's equation are $\lambda = n(n+1)$ for $n = 0, 1, 2, \ldots$. For each n, an eigenfunction is $\varphi_n(x) = P_n(x)$, which is an odd polynomial if n is odd and an even polynomial if n is even.

There is an extensive literature on Legendre polynomials. Here are two frequently used facts.

Rodrigues's Formula
For $n = 1, 2, \ldots,$

$$P_n(x) = \frac{1}{2^n n!} \frac{d^n}{dx^n} ((x^2 - 1)^n).$$

This formula can be used to systematically generate Legendre polynomials.

For example, with $n = 2$ it gives us

$$\frac{1}{2^2 2!} \frac{d^2}{dx^2}((x^2 - 1)^2) = \frac{1}{8} \frac{d^2}{dx^2}(x^4 - 2x^2 + 1)$$

$$= \frac{1}{8}(12x^2 - 4) = \frac{3}{2}x^2 - \frac{1}{2} = P_2(x).$$

Recurrence Relation

For $n = 1, 2, \ldots$,

$$(n + 1)P_{n+1}(x) - (2n + 1)x P_n(x) + n P_{n-1}(x) = 0.$$

This can be used to derive a variety of useful facts about Legendre polynomials.

For example, one can show that the coefficient of x^n in $P_n(x)$ is

$$\frac{1 \cdot 3 \ldots (2n - 1)}{n!}.$$

This gives us the coefficient of x^5 in $P_5(x)$:

$$\frac{1(3)(5)(7)(9)}{5!} = \frac{63}{8}.$$

We saw this previously. The recurrence relation can also be used to derive the important result that

$$\int_{-1}^{1} P_n^2(x)\, dx = \frac{2}{2n + 1}.$$

We will use this shortly for Fourier-Legendre series.

Because the Legendre polynomials are eigenfunctions of a Sturm-Liouville problem in which the coefficient of λ is $p(x) = 1$, they are orthogonal on $[-1, 1]$ with respect to this weight function:

$$\int_{-1}^{1} P_n(x) P_m(x)\, dx = 0 \text{ if } n \neq m.$$

This is the basis for eigenfunction expansions in series of Legendre polynomials (Fourier-Legendre series), to which Theorem 14.2 applies. The Fourier-Legendre coefficients in the expansion of a function f are

$$c_n = \frac{\int_{-1}^{1} f(x) P_n(x)\, dx}{\int_{-1}^{1} P_n^2(x)\, dx} = \frac{2n + 1}{2} \int_{-1}^{1} f(x) P_n(x)\, dx.$$

Here is an example of a Fourier-Legendre expansion.

EXAMPLE 14.4

We will write the Fourier-Legendre expansion of $f(x) = \cos(\pi x/2)$. Because $\cos(\pi x/2)$ is continuous with a continuous derivative,

$$\cos(\pi x/2) = \sum_{n=0}^{\infty} c_n P_n(x)$$

for $-1 < x < 1$ where

$$c_n = \frac{2n+1}{2} \int_{-1}^{1} \cos(\pi x/2) P_n(x) \, dx.$$

There is no simple expression for c_n for arbitrary n, so we will approximate this eigenfunction expansion with the sum of the first six terms. We must therefore compute $c_0, \ldots, c_5$.

Because $\cos(\pi x/2)$ is even on $[-1, 1]$, $\cos(\pi x/2) P_n(x)$ is an odd function for n odd. Then

$$\int_{-1}^{1} \cos(\pi x/2) P_n dx = 0 \text{ for } n \text{ odd},$$

so $c_n = 0$ if n is odd. In particular, $c_1 = c_3 = c_5 = 0$.

For c_0, c_2, and c_4, we can do the integrations exactly because we know $P_0(x)$, $P_2(x)$, and $P_4(x)$. However, it is more efficient to use a software package. If MAPLE is used, the nth Legendre polynomial is denoted *LegendreP(n,x)*. To compute c_n, use the MAPLE code:

```
((2*n + 1)/2)*int(cos(Pi*x/2)*LegendreP(n,x),x=-1..1);
```

With, respectively, $n = 0, 2, 4$, this yields

$$c_0 \approx 0.6366197722, \quad c_2 \approx -0.6870852690, \quad \text{and } c_4 \approx 0.05177890435.$$

If we retain four decimal places, then the sixth partial sum of the Fourier-Legendre expansion of $\cos(\pi x/2)$ is approximately

$$\cos(\pi x/2) \approx 0.6366 - 0.3435(3x^2 - 1) + 0.0065(35x^4 - 30x^2 + 3)$$

for $-1 < x < 1$.

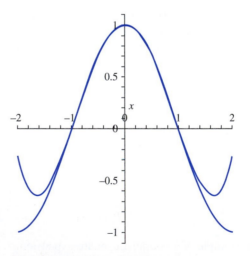

FIGURE 14.3 *Fourier-Legendre expansion of* $\cos(\pi x/2)$ *in Example 14.4.*

Figure 14.3 shows a graph of $\cos(\pi x/2)$ and this partial sum on an interval slightly larger than $[-1, 1]$. Even with this small number of terms, this partial sum is nearly indistinguishable from the function on $[-1, 1]$, within the scale of the graph. This is not to be expected in general with only six terms of the series. Outside $[-1, 1]$, the polynomial approximation rapidly diverges from $\cos(\pi x/2)$. ◆

There are many different (weighted) orthogonal polynomials arising as eigenfunctions of Sturm-Liouville problems. These include Hermite, Laguerre, and Tchebyshev polynomials, and the literature contains many others.

1. Use the recurrence relation to derive $P_6(x)$, $P_7(x)$, and $P_8(x)$.

2. Use Rodrigues's formula to derive $P_2(x)$ through $P_5(x)$.

3. Let $q(x)$ be a polynomial of degree m. Show that $\int_{-1}^{1} q(x) P_n(x)\, dx = 0$ for $n > m$.

For each of Problems 4 through 9, find the first five coefficients in the Fourier-Legendre expansion of the function. Graph the function and this partial sum on the same set of axes for $-1 \le x \le 1$.

4. e^{-x}

5. $\sin(\pi x / 2)$

6. $\cos(x) - \sin(x)$

7. $\sin^2(x)$

8. $(x + 1)\cos(x)$

9. $f(x) = \begin{cases} -1 & \text{for } -1 < x \le 0 \\ 1 & \text{for } 0 < x \le 1. \end{cases}$

14.3   Fourier-Bessel Expansions

The differential equation

$$x^2 y'' + xy' + (x^2 - v^2)y = 0$$

is called *Bessel's equation of order* v. The term *order* as used here refers to the value of the parameter v, not to the order of the equation. v can be any nonnegative number. Bessel's equation arises in many important contexts, for example, in analyzing heat radiation from a cylindrical tank, or vibrations in a circular fixed-frame plate.

Referring to Section 4.2, Bessel's equation has a regular singular point at 0. We therefore seek a Frobenius solution

$$y(x) = \sum_{n=0}^{\infty} a_n x^{n+r}.$$

The idea is to substitute this series into Bessel's equation and solve for r and the a_n's. Complete details of this calculation are carried out in Example 4.5, resulting in $r = \pm v$. Corresponding to $r = v$, we found there the solution

$$y_1(x) = c \sum_{n=0}^{\infty} \frac{(-1)^n}{2^{2n} n! (1 + v)(2 + v) \dots (n + v)} x^{2n+v}, \tag{14.4}$$

in which c can be any number.

The Gamma Function

We will take a brief diversion to develop the *gamma function*, which is used to write solutions of Bessel's equation. For $x > 0$, $\Gamma(x)$ is defined by

$$\Gamma(x) = \int_0^{\infty} t^{x-1} e^{-t}\, dt.$$

For our purpose, the most useful property of the gamma function is that, for $x > 0$,

$$\Gamma(x + 1) = x \Gamma(x).$$

This is easily established by an integration by parts. If $0 < a < b$, then

$$\int_a^b t^x e^{-t}\, dt = \left[t^x(-e^{-t})\right]_a^b - \int_a^b x t^{x-1}(-1)e^{-t}\, dt$$

$$= -b^x e^{-b} + a^x e^{-a} + x \int_a^b t^{x-1} e^{-t}\, dt.$$

In the limit as $a \to 0+$ and $b \to \infty$, this gives us $\Gamma(x+1) = x\Gamma(x)$. This relationship is known as the *factorial property* of the gamma function because, if $x = n$ is a positive integer, then repeated application yields

$$\Gamma(n+1) = n\Gamma(n) = n\Gamma((n-1)+1)$$

$$= n(n-1)\Gamma(n-1) = n(n-1)\Gamma((n-2)+1)$$

$$= n(n-1)(n-2)\Gamma(n-2)$$

$$= \cdots = n(n-1)(n-2)\cdots(2)(1)\Gamma(1).$$

But

$$\Gamma(1) = \int_0^\infty e^{-t}\, dt = 1,$$

so $\Gamma(n+1) = n!$ for n any positive integer.

Using the factorial property, we can extend $\Gamma(x)$ to be defined for negative, non-integer numbers. To begin, if $-1 < x < 0$, then $0 < x+1 < 1$, so $\Gamma(x+1) = x\Gamma(x)$, suggesting that we define

$$\Gamma(x) = \frac{1}{x}\Gamma(x+1).$$

For example,

$$\Gamma(-1/2) = \frac{1}{-1/2}\Gamma(1/2) = -2\Gamma(1/2).$$

Once $\Gamma(x)$ is defined for $-1 < x < 0$, we can in the same way use the factorial property to define $\Gamma(x)$ for $-2 < x < -1$. For example,

$$\Gamma(-3/2) = \frac{1}{-3/2}\Gamma(-1/2) = -\frac{2}{3}\Gamma(-1/2) = \frac{4}{3}\Gamma(1/2).$$

We now return to Bessel functions.

Bessel Functions of the First Kind

Using the factorial property of the gamma function, write

$$\Gamma(n+v+1) = (n+v)\Gamma(n+v)$$

$$= (n+v)(n+v-1)\Gamma(n+v-1)$$

$$= \cdots = (1+v)(2+v)\cdots(n-1+v)(n+v)\Gamma(v+1).$$

Then

$$(1+v)(2+v)\cdots(n+v) = \frac{\Gamma(n+v+1)}{\Gamma(v+1)}.$$

It is customary to choose

$$c = \frac{1}{2^v \Gamma(v+1)}$$

in the solution of equation (14.4) to Bessel's equation. The resulting solution is written $J_\nu(x)$, and is called a *Bessel function of the first kind of order* ν:

$$J_\nu(x) = \sum_{n=0}^{\infty} \frac{(-1)^n}{2^{2n+\nu} n! \Gamma(n+\nu+1)} x^{2n+\nu}.$$

This series converges for all x. This standardization of a particular solution of Bessel's equation of order ν allows for the tabulation of values of the solution, zeros, and other information used in applications. Figure 14.4 shows graphs of $J_0(x)$, $J_{1/2}(x)$, $J_2(x)$, and $J_3(x)$, which increase in amplitude at 0 as ν is chosen larger.

Because Bessel's equation is of second order, a second solution, linearly independent from J_ν, is needed to write the general solution. Theorem 4.1 is our guide to producing a second solution of Bessel's equation of order ν. The following cases occur.

Case 1 If 2ν is not an integer, then J_ν and $J_{-\nu}$ are linearly independent, and the general solution of Bessel's equation of order ν is

$$y(x) = c_1 J_\nu(x) + c_2 J_{-\nu}(x).$$

Case 2 If 2ν is an odd positive integer, then J_ν and $J_{-\nu}$ are again linearly independent. In this case, $2\nu = 2n + 1$ for some positive integer n, so $\nu = n + 1/2$, and the general solution of Bessel's equation of order ν is

$$y(x) = c_1 J_{n+1/2}(x) + c_2 J_{-n-1/2}(x).$$

It also can be shown by manipulating the series for $J_\nu(x)$ that, in this case, $J_\nu(x)$ can be written in closed form as a finite sum of terms involving square roots, sines, and cosines. For example,

$$J_{1/2}(x) = \sqrt{\frac{2}{\pi x}} \sin(x), \quad J_{-1/2}(x) = \sqrt{\frac{2}{\pi x}} \cos(x),$$

$$J_{3/2}(x) = \sqrt{\frac{2}{\pi x}} \left(\frac{1}{x} \sin(x) - \cos(x) \right)$$

and so on.

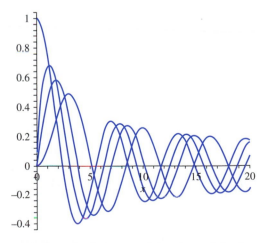

FIGURE 14.4 *Some Bessel functions of the first kind.*

Bessel Functions of the Second Kind

Case 3 If 2ν is an integer but is not of the form $n + 1/2$ for any positive integer n, then it is routine to check that

$$J_{-\nu}(x) = (-1)^{\nu} J_{\nu}(x).$$

In this case, $J_{\nu}(x)$ and $J_{-\nu}(x)$ are linearly dependent. The Frobenius theorem (see Theorem 4.1) tells us to look for a second solution of the form

$$y_2(x) = k J_{\nu}(x) \ln(x) + \sum_{n=1}^{\infty} c_n^* x^{n+r_2},$$

in which r_2 is the second solution of the indicial equation. For Bessel's equation of order ν, $r = -\nu$, so we are attempting a second solution

$$y_2(x) = k J_{\nu}(x) \ln(x) + \sum n = 1^{\infty} c_n^* x^{n+-\nu}.$$

Again, the strategy is to substitute this into Bessel's equation and solve for k and the c_n^*'s. These details are carried out in Example 4.6 for the case $\nu = 0$. In general, for $\nu = 0, 1, 2, \ldots$, a long computation yields the second solution

$$Y_n(x) = \frac{2}{\pi} \left[J_n(x)[\ln(x/2 + \gamma)] + \sum_{j=1}^{\infty} \frac{(-1)^{j+1}[\phi(j) + \phi(j+1)]}{2^{2j+n+1} j!(j+n)!} x^{2j+n} \right]$$

$$- \frac{2}{\pi} \sum_{j=0}^{n-1} \frac{(n-j-1)!}{2^{2j-n+1} j!} x^{2j-n},$$

in which the last summation (from $j = 0$ to $j = n-1$) does not appear in $Y_n(x)$, where $n = 0$;

$$\phi(j) = 1 + \frac{1}{2} + \ldots + \frac{1}{j};$$

and γ is the *Euler-Mascheroni constant*, defined by

$$\gamma = \lim_{j \to \infty} (\phi(j) - \ln(j)).$$

$Y_n(x)$ is the *Bessel function of the second kind of order n*. The general solution of Bessel's equation of integer order n is

$$y(x) = c_1 J_n(x) + c_2 Y_n(x).$$

Figure 14.5 shows graphs of $Y_0(x)$, $Y_1(x)$, and $Y_2(x)$.

It is possible to extend this definition of $Y_n(x)$ to $Y_{\nu}(x)$ for any real ν by setting

$$Y_{\nu}(x) = \frac{1}{\sin(\nu\pi)} [J_{\nu}(x) \cos(\nu\pi) - J_{-\nu}(x)].$$

This is called *Neumann's Bessel function of order ν*.

In some models of physical systems, Bessel's equation occurs in disguised form, requiring a change of variables to write the solution in terms of Bessel functions. In particular, if a, b, and c are real numbers and $\nu = n$ is a nonnegative integer, then the general solution of

$$y'' - \left(\frac{2a-1}{x} \right) y' + \left(b^2 c^2 x^{2c-2} + \frac{a^2 - n^2 c^2}{x^2} \right) y = 0 \tag{14.5}$$

is

$$y(x) = d_1 x^a J_n(bx^c) + d_2 x^a Y_n(bx^c),$$

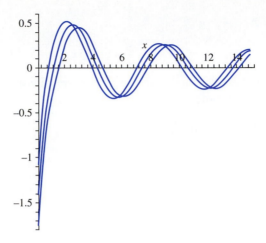

FIGURE 14.5 *Some Bessel functions of the second kind.*

in which d_1 and d_2 are arbitrary constants.

To illustrate, consider the differential equation

$$y'' - \left(\frac{2\sqrt{3} - 1}{x} \right) y' + \left(784x^6 - \frac{61}{x^2} \right) y = 0.$$

To fit this into the template of equation (14.5), we must clearly choose $a = \sqrt{3}$. Because of the x^6 term, we need $2c - 2 = 6$, so $c = 4$. Now we must choose b so that

$$b^2 c^2 = 16b^2 = 784,$$

so $b = 7$. And v must satisfy

$$a^2 - v^2 c^2 = 3 - 16v^2 = -61,$$

so $v = 2$. The general solution is

$$y(x) = d_1 x^{\sqrt{3}} J_2(7x^4) + d_2 x^{\sqrt{3}} Y_2(7x^4).$$

Some Properties of $J_v(x)$

We will derive some important relationships satisfied by Bessel functions of the first kind, $J_v(x)$, for v any nonnegative number.

Property 1

$$\frac{d}{dx}(x^v J_v(x)) = x^v J_{v-1}(x). \tag{14.6}$$

This is a routine exercise in differentiation:

$$\frac{d}{dx}(x^v J_v(x)) = \frac{d}{dx} \sum_{n=0}^{\infty} \frac{(-1)^n}{2^{2n+v} n! \Gamma(n+v+1)} x^{2n+2v}$$

$$= \sum_{n=0}^{\infty} \frac{(-1)^n 2(n+v)}{2^{2n+v} n! (n+v) \Gamma(n+v)} x^{2n+2v-1}$$

$$= x^v \sum_{n=0}^{\infty} \frac{(-1)^n}{2^{2n+v-1} n! \Gamma(n+v)} x^{2n+v-1} = x^v J_{v-1}(x).$$

By a similar argument,

Property 2

$$\frac{d}{dx}(x^{-\nu}J_\nu(x)) = -x^{-\nu}J_{\nu+1}(x). \tag{14.7}$$

Using these equations, we can derive a relationship between Bessel functions of the first kind of different orders.

Property 3   For $x > 0$,

$$\frac{2\nu}{x}J_\nu(x) = J_{\nu+1}(x) + J_{\nu-1}(x). \tag{14.8}$$

To see why this is true, begin by carrying out the differentiations in equations (14.6) and (14.7) to obtain

$$x^\nu J_\nu'(x) + \nu x^{\nu-1}J_\nu(x) = x^\nu J_{\nu-1}(x)$$

and

$$x^{-\nu}J_\nu'(x) - \nu x^{-\nu-1}J_\nu(x) = -x^\nu J_{\nu+1}(x).$$

Multiply the first of these equations by $x^{-\nu}$ and the second by x^ν to obtain

$$J_\nu'(x) + \frac{\nu}{x}J_\nu(x) = J_{\nu-1}(x)$$

and

$$J_\nu'(x) - \frac{\nu}{x}J_\nu(x) = -J_{\nu+1}(x).$$

Now subtract the second of these two equations from the first to obtain equation (14.8).

It is often necessary to be able to solve the equation $J_\nu(x) = 0$. Solutions are called *zeros* of $J_\nu(x)$. The graphs in Figure 14.4 suggest that these functions have positive zeros. We will explore this question more analytically.

To begin, from equation (14.5), $y = J_\nu(kx)$ is a solution of

$$x^2 y'' + xy' + (k^2x^2 - \nu^2)y = 0.$$

Suppose $k > 1$ and set $y(x) = (kx)^{-1/2}u(x)$ to obtain

$$u''(x) + \left(k^2 - \frac{\nu^2 - \frac{1}{4}}{x^2}\right)u(x) = 0. \tag{14.9}$$

Intuitively, this differential equation for u is approximated more closely by $u'' + k^2 u = 0$ as x is chosen larger. Since this equation has solutions $\cos(kx)$ and $\sin(kx)$, we suspect that $J_\nu(x)$ may be approximated by $\cos(kx)/\sqrt{kx}$ and $\sin(kx)/\sqrt{kx}$ as x tends to infinity. Since these functions have infinitely many positive zeros (cross the horizontal axis infinitely many times), so must $J_\nu(kx)$. This suggests that $J_\nu(kx)$ has infinitely many positive zeros.

To firm up this intuition, notice that $v(x) = \sin(x - \alpha)$ is a solution of $v'' + v = 0$ for any number α. We are interested in having $\alpha > 0$. Multiply this equation for v by u and equation (14.9) by v and subtract to obtain

$$uv'' - vu'' = \left(k^2 - \frac{\nu^2 - \frac{1}{4}}{x^2}\right)uv - uv,$$

which we write as

$$(uv' - vu')' = \left(k^2 - 1 - \frac{\nu^2 - \frac{1}{4}}{x^2}\right)uv.$$

Then

$$\int_{\alpha}^{\alpha+\pi} (uv' - vu')' \, dx$$

$$= u(\alpha + \pi)v'(\alpha + \pi) - u(\alpha)v'(\alpha) - v(\alpha + \pi)u'(\alpha + \pi) + v(\alpha)u'(\alpha)$$

$$= -u(\alpha + \pi) - u(\alpha)$$

$$= \int_{\alpha}^{\alpha} \left(k^2 - 1 - \frac{v^2 - \frac{1}{4}}{x^2} \right) u(x)\sin(x - \alpha) \, dx.$$

By the mean value theorem for integrals, there is a number τ between α and $\alpha + \pi$ such that

$$-u(\alpha + \pi) - u(\alpha) = u(\tau) \int_{\alpha}^{\alpha+\pi} \left(k^2 - 1 - \frac{v^2 - \frac{1}{4}}{x^2} \right) \sin(x - \alpha) \, dx.$$

Now $\sin(x - \alpha) > 0$ for $\alpha < x < \alpha + \pi$. Furthermore, we can choose α large enough that

$$k^2 - 1 - \frac{v^2 - \frac{1}{4}}{x^2} > 0$$

for $\alpha \le x \le \alpha + \pi$. This makes the integral on the right positive for such α. But then $u(\alpha)$, $u(\alpha + \pi)$, and $u(\tau)$ cannot all have the same sign, so $u(x)$ must vanish somewhere between α and $\alpha + \pi$. Therefore $J_v(kx)$ must vanish somewhere in this interval.

We summarize this discussion as the fourth property of $J_v(x)$ in our list.

Property 4 $J_v(x)$ has infinitely many positive zeros. Furthermore, if we order these zeros $j_1 < j_2 < \ldots$, then $j_n \to \infty$ as $n \to \infty$.

We claim further that each positive zero of $J_v(x)$ is *simple*. This means that, if $J_v(c) = 0$, then $J_v'(c) \ne 0$. In terms of the graph, $J_v(x)$ would cross the horizontal axis at c with nonzero slope.

Property 5 Every positive zero of $J_v(x)$ is simple.

To see why this is true, suppose $J_v(c) = J_v'(c) = 0$ with $c > 0$. Then $J_v(x)$ is a solution of the initial value problem

$$x^2 y'' + xy' + (k^2 x^2 - v^2)y = 0; \quad y(c) = y'(c) = 0.$$

But $y(x) = 0$ is also a solution of this problem, which has a unique solution in some interval about c. This would make $J_v(x)$ the zero function, which is a contradiction.

Property 6 J_v has no positive zero in common with either J_{v-1} or J_{v+1}.

To prove this, recall that

$$J_v'(x) + \frac{v}{x} J_v(x) = J_{v-1}(x).$$

If $c \ne 0$ and $J_v(c) = J_{v-1}(c) = 0$, then we would have $J_v'(c) = 0$ also, contradicting Property 5. Similarly,

$$J_v'(x) - \frac{v}{x} J_v(x) = J_{v+1}(x)$$

so J_v cannot share a zero with J_{v+1}.

We conclude our list of properties of Bessel functions of the first kind with the *interlacing lemma*.

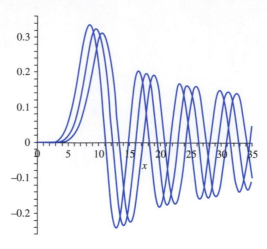

FIGURE 14.6   *The interlacing lemma.*

Property 7   Between any two positive zeros of J_ν, there is a positive zero of $J_{\nu-1}$ and a positive zero of $J_{\nu+1}$. That is, positive zeros of consecutively numbered Bessel functions are intertwined.

This behavior is seen in the graphs of J_7, J_8, and J_9 in Figure 14.6.

To verify this property, suppose a and b are positive zeros of $J_\nu(x)$. First let $f(x) = x^\nu J_\nu(x)$. Then $f(a) = f(b) = 0$. By the mean value theorem, there is some c between a and b such that $f'(c) = 0$. But

$$f'(x) = \frac{d}{dx}(x^\nu J_\nu(x)) = x^\nu J_{\nu-1}(x)$$

so $f'(c) = 0$ implies that $J_{\nu-1}(c) = 0$. Therefore, $J_{\nu-1}$ has a zero between a and b.

A similar argument applied to $g(x) = x^{-\nu} J_\nu(x)$, and using the relation

$$\frac{d}{dx}(x^{-\nu} J_\nu(x)) = -x^{-\nu} J_{\nu+1}(x)$$

shows that $J_{\nu+1}$ also has a zero between a and b.

Fourier-Bessel Expansions

We will cast Bessel's equation in a form to fit within Sturm-Liouville theory. Consider the Bessel equation

$$x^2 y'' + x y' + (j_n^2 x^2 - \nu^2) y = 0,$$

in which j_n is the nth positive zero of J_ν with these zeros ordered so that $0 < j_1 < j_2 < \ldots$. Divide by x to write

$$x y'' + y' + \left(j_n^2 x - \frac{\nu^2}{x^2} \right) y = 0.$$

This is a Sturm-Liouville equation

$$(x y')' + \left(j_n^2 x - \frac{\nu^2}{x^2} \right) y = 0$$

on $(0, 1)$ with $r(x) = x$, $q(x) = -\nu^2/x^2$, $p(x) = x$, and $\lambda = j_n^2$. From equation (14.5) with $a = 0, c = 1$, and $b = j_n$, we know that $y = J_\nu(j_n x)$ is a solution. This differential equation has eigenvalues $\lambda_n = j_n^2$ and eigenfunctions $\varphi_n(x) = J_\nu(j_n x)$.

Since the interval of interest here is $(0, 1)$, there are no boundary conditions. The weight function for the orthogonality of the eigenfunctions is $p(x) = x$:

$$\int_0^1 x J_\nu(j_n x) J_\nu(j_m x)\, dx = 0 \text{ if } n \neq m.$$

If f is a function defined on $[0, 1]$, we can attempt an eigenfunction expansion in terms of the eigenfunctions $J_\nu(j_n x)$. This *Fourier-Bessel* expansion has the form

$$\sum_{n=1}^{\infty} c_n J_\nu(j_n x)$$

where the c_n's are the *Fourier-Bessel coefficients* given as

$$c_n = \frac{\int_0^1 x f(x) J_\nu(j_n x)\, dx}{\int_0^1 x J_\nu^2(j_n x)}\, dx.$$

We can simplify this expression for the coefficients by using the identity

$$\int_0^1 x J_\nu^2(j_n x)\, dx = \frac{1}{2} J_{\nu+1}^2(j_n). \tag{14.10}$$

To derive this, begin with the differential equation

$$x y'' + x y' + (j_n^2 x^2 - \nu^2) y = 0$$

in which $y = J_\nu(j_n x)$. Multiply by $2y'$ to obtain

$$2x^2 y' y'' + 2x(y')^2 + 2(j_n^2 x^2 - \nu^2) y y' = 0,$$

which can be written as

$$[x^2(y')^2 + (j_n^2 x^2 - \nu^2) y^2]' - 2 j_n^2 x y^2 = 0.$$

Integrate, keeping in mind that $y(1) = 0$:

$$0 = [x^2(y')^2 + (j_n^2 x^2 - \nu^2) y^2]_0^1 - 2 j_n^2 \int_0^1 x y^2\, dx$$

$$= (y'(1))^2 - 2 j_n^2 \int_0^1 x y^2\, dx$$

$$= j_n^2 (J_\nu'(j_n x)) - 2 j_n^2 \int_0^1 x (J_\nu(j_n x))^2\, dx.$$

Then

$$\int_0^1 x J_\nu^2(j_n x)\, dx = \frac{1}{2} (J_\nu'(j_n))^2.$$

But in general,

$$J_\nu'(x) - \frac{\nu}{x} J_\nu(x) = -J_{\nu+1}(x).$$

Therefore,

$$J_\nu'(j_k) - \frac{\nu}{j_n} J_\nu(j_n) = -J_{\nu+1}(j_n).$$

Then

$$J_\nu'(j_n) = -J_{\nu+1}(j_n).$$

Therefore,

$$\int_0^1 x J_\nu^2(j_n x)\, dx = \frac{1}{2} J_{\nu+1}^2(j_n).$$

We can therefore write the nth Fourier-Bessel coefficient of f as

$$c_n = \frac{2}{[J_{\nu+1}(j_n)]^2} \int_0^1 x f(x) J_\nu(j_n x)\, dx. \tag{14.11}$$

In general, these coefficients cannot be computed by hand, and a software routine should be used to approximate the positive zeros of $J_\nu(x)$ and to carry out the integrations.

EXAMPLE 14.5

We will compute some terms in the Fourier-Bessel expansion of $f(x) = x(1-x)$ on $[0, 1]$ with $\nu = 1$. In this case, the eigenfunctions are $J_1(j_n x)$ and the j_n's are the positive zeros of $J_1(x)$. The coefficients are

$$c_n = \frac{2 \int_0^1 x^2(1-x) J_1(j_n x)\, dx}{J_2^2(j_n)}.$$

Because $x(1-x)$ is twice differentiable on $[0, 1]$, we will have

$$x(1-x) = \sum_{n=1}^\infty c_n J_\nu(j_n x)\, dx$$

for $0 < x < 1$. We will compute the first four coefficients to illustrate the ideas involved. First we need $j_1, \ldots, j_4$. These can be obtained from tables or from MAPLE by the command

```
evalf(BesselJZeros(ν,n));
```

with $\nu = 1$ and n successively chosen to be 1, 2, 3, and 4. The output is

$$j_1 \approx 3.83170597, \quad j_2 \approx 7.01558667, \quad j_3 \approx 10.17346814, \quad \text{and } j_4 \approx 13.32369194.$$

Using this value for j_1, c_1 is approximately

$$c_1 \approx \frac{2}{J_2(3.83170597)^2} \int_0^1 x^2(1-x) J_1(3.83170597 x)\, dx.$$

To carry out this computation, first numerically approximate the denominator $J_2^2(j_1)$. If MAPLE is used, the denominator is the square of the number computed as

```
evalf (BesselJ(2,3.83170597));
```

The integral in the numerator is computed as

```
evalf (int(x*(1-x)*BesselJ(1,x)(3.83170597*x),x=0..1));
```

This computation yields

$$c_1 \approx 0.45221702.$$

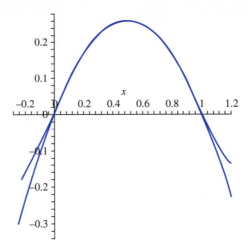

FIGURE 14.7 *Fourier-Bessel expansion in Example 14.6.*

By repeating this calculation for $n = 2, 3, 4$, in turn, with the appropriate j_n inserted, we approximate

$$c_2 \approx -0.03151859, \quad c_3 \approx 0.03201789, \quad \text{and} \quad c_4 \approx -0.00768864.$$

Using the first four terms of the partial sum of the Fourier-Bessel expansion, we have

$$x(1-x) \approx 0.45221702 J_1(3.83170597x) - 0.03151859 J_1(7.01558667x)$$
$$+ 0.03201794 J_1(10.17346814x) - 0.00768864 J_1(13.32369194x).$$

Figure 14.7 compares a graph of $x(1-x)$ with this sum of four terms on $[-1/4, 6/5]$. In the scale of the graph, the approximation appears to be quite good on $[0, 1]$. Outside of $[0, 1]$, the partial sum and the function are unrelated. ◆

SECTION 14.3   *PROBLEMS*

For each of the following functions approximately the first five terms in the Fourier-Bessel expansion of f on $[0, 1]$ in a series of the functions $J_1(j_n x)$, where j_n is the nth positive zero of $J_1(x)$. Compare a graph of this partial sum with f.

1. $f(x) = x$

2. $f(x) = e^{-x}$

3. $f(x) = xe^{-x}$

4. $f(x) = x^2 e^{-x}$

5. $f(x) = \sin(\pi x)$

6. $f(x) = x\cos(\pi x)$

Problems 7 through 12 are defined as follows. For $n = 7, \ldots, 12$, Problem n is to carry out the instructions for Problem $n - 6$, except now expand in a series of the functions $J_2(j_n x)$. In this case, j_n is the nth positive zero of $J_2(x)$.

PART 5

Partial Differential Equations

CHAPTER 15

The Wave Equation

15.1   Derivation of the Equation

Vibrations in a membrane or steel plate, or oscillations along a guitar string, are all modeled by the wave equation and appropriate initial and boundary conditions. We will begin with a derivation of the wave equation under simple conditions.

Suppose an elastic string has its ends fastened by two pegs. The string is displaced, released, and allowed to vibrate in a plane.

Place the string along the x axis from 0 to L and assume that it vibrates in the (x, y) plane. We want a function $y(x, t)$ such that, at time t, the graph of $y(x, t)$ is the shape of the string at that time. We call $y(x, t)$ the *position function* for the string.

Neglect damping forces such as the weight of the string and assume that the tension $\mathbf{T}(x, t)$ acts tangent to the string and that individual particles of the string move only vertically. Also assume that the mass ρ per unit length is constant. Consider a segment of string between x and $x + \Delta x$. By Newton's second law of motion, the net force on this segment due to the tension is equal to the acceleration of the center of mass of this segment multiplied by its mass. This is a vector equation. Its vertical component (Figure 15.1) gives us approximately

$$T(x + \Delta x, t) \sin(\theta + \Delta \theta) - T(x, t) \sin(\theta) = \rho \Delta x \frac{\partial^2 y}{\partial t^2}(\overline{x}, t),$$

where $\overline{x}$ is the center of mass of this segment and $T(x, t) = \| \mathbf{T}(x, t) \|$. Then

$$\frac{T(x + \Delta x, t) \sin(\theta + \Delta \theta) - T(x, t) \sin(\theta)}{\Delta x} = \rho \frac{\partial^2 y}{\partial t^2}(\overline{x}, t).$$

Now $v(x, t) = T(x, t) \sin(\theta)$ is the vertical component of the tension, so this equation becomes

$$\frac{v(x + \Delta x, t) - v(x, t)}{\Delta x} = \rho \frac{\partial^2 y}{\partial t^2}(\overline{x}, t).$$

339

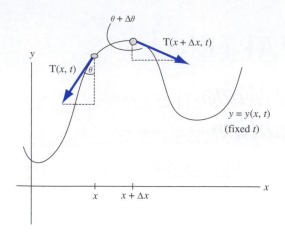

FIGURE 15.1 *Deriving the wave equation.*

As $\Delta x \to 0$, $\overline{x} \to x$, and we obtain

$$\frac{\partial v}{\partial x} = \rho \frac{\partial^2 y}{\partial t^2}.$$

The horizontal component of the tension is $h(x, t) = T(x, t) \cos(\theta)$, so

$$v(x, t) = h(x, t) \tan(\theta) = h(x, t) \frac{\partial y}{\partial x}.$$

Then

$$\frac{\partial}{\partial x} \left(h \frac{\partial y}{\partial x} \right) = \rho \frac{\partial^2 y}{\partial t^2}.$$

To compute the left side of this equation, use the fact that the horizontal component of the tension of the segment is zero, so

$$h(x + \Delta x, t) = h(x, t).$$

This means that h is independent of x, so

$$h \frac{\partial^2 y}{\partial x^2} = \rho \frac{\partial^2 y}{\partial t^2}.$$

Let $c^2 = h/\rho$ to get the *one-dimensional wave equation*

$$\frac{\partial^2 y}{\partial t^2} = c^2 \frac{\partial^2 y}{\partial x^2}. \tag{15.1}$$

The fact that the ends are held fixed is reflected in the *boundary conditions*

$$y(0, t) = y(L, t) = 0 \text{ for } t \geq 0.$$

The initial displacement of the string and the velocity with which it is released at time 0 are the *initial conditions* given by

$$y(x, 0) = f(x) \text{ for } 0 \leq x \leq L.$$

and

$$\frac{\partial y}{\partial t}(x, 0) = g(x).$$

The wave equation, together with initial and boundary conditions, is called an *initial-boundary value problem* for $y(x, t)$.

If the string is released from rest (no initial velocity), then $g(x)$ is identically zero. Furthermore, if the string has its ends fixed at the same level, then the initial position function f must satisfy the compatibility condition $f(0) = f(L) = 0$.

Variations on this problem can allow for moving ends with conditions such as

$$y(0, t) = \alpha(t) \text{ and } y(L, t) = \beta(t).$$

We also can have a forcing term in which an external force drives the motion. In this case, the wave equation is

$$\frac{\partial^2 y}{\partial t^2} = c^2 \frac{\partial^2 y}{\partial x^2} + F(x, t).$$

The wave equation in two space dimensions is

$$\frac{\partial^2 z}{\partial t^2} = c^2 \left(\frac{\partial^2 z}{\partial x^2} + \frac{\partial^2 z}{\partial y^2} \right), \tag{15.2}$$

in which $z(x, y, t)$ is the displacement function.

If the displacement function is $u(r, \theta, t)$ in polar coordinates, this wave equation is

$$\frac{\partial^2 u}{\partial t^2} = c^2 \left(\frac{\partial^2 u}{\partial r^2} + \frac{1}{r} \frac{\partial u}{\partial r} + \frac{1}{r^2} \frac{\partial^2 u}{\partial \theta^2} \right). \tag{15.3}$$

SECTION 15.1   PROBLEMS

1. Let $y(x, t) = \sin(n\pi x/L) \cos(n\pi ct/L)$ for $0 \le x \le L$. Show that y satisfies the one-dimensional wave equation for any positive integer n.

2. Show that $z(x, y, t) = \sin(nx) \cos(my) \cos(\sqrt{n^2 + m^2}t)$ satisfies the two-dimensional wave equation for any positive integers n and m.

3. Let f be a twice-differentiable function of one variable. Show that

$$y(x, t) = \frac{1}{2}[f(x + ct) + f(x - ct)]$$

satisfies the one-dimensional wave equation.

4. Show that

$$y(x, t) = \sin(x) \cos(ct) + \frac{1}{c} \cos(x) \sin(ct)$$

satisfies the one-dimensional wave equation together with the boundary conditions

$$y(0, t) = y(2\pi, t) = \frac{1}{c} \sin(ct) \text{ for } t > 0$$

and the initial conditions

$$y(x, 0) = \sin(x) \text{ and } \frac{\partial y}{\partial t}(x, 0) = \cos(x) \text{ for } 0 < x < \pi.$$

5. Formulate an initial-boundary value problem for vibrations of a rectangular membrane occupying $0 \le x \le a, 0 \le y \le b$ if the initial position is the graph of $z = f(x, y)$ and the initial velocity is $g(x, y)$. The membrane is fastened to a frame along the rectangular boundary of the region.

6. Formulate an initial-boundary value problem for the motion of an elastic string of length L fastened at both ends and released from rest with an initial position given by $f(x)$. The motion is opposed by air resistance, which has a force at each point of magnitude proportional to the square of the velocity at that point.

15.2   Wave Motion on an Interval

We will solve initial-boundary value problems for the wave equation on a closed interval $[0, L]$, first considering the case of zero initial velocity, then zero initial position.

15.2.1   Zero Initial Velocity

An elastic string of length L with fixed ends is released from rest from an initial position given as the graph of $y = f(x)$. The initial-boundary value problem for the position function $y(x, t)$ is

$$\frac{\partial^2 y}{\partial t^2} = c^2 \frac{\partial^2 y}{\partial x^2} \text{ for } 0 < x < L, t > 0,$$

$$y(0, t) = y(L, t) = 0 \text{ for } t \geq 0,$$

$$y(x, 0) = f(x) \text{ for } 0 \leq x \leq L,$$

and

$$\frac{\partial y}{\partial t}(x, 0) = 0.$$

The *Fourier method*, or *method of separation of variables*, is to attempt a solution of the form $y(x, t) = X(x)T(t)$. Substitute this into the wave equation to get

$$XT'' = c^2 X''T,$$

where $X' = dX/dx$ and $T' = dT/dt$. Then

$$\frac{X''}{X} = \frac{T''}{c^2 T}.$$

The left side depends only on x. We could fix x and then the right side, which depends only on t, would be constant. But then the left side must equal the same constant. Therefore, for some number λ, called the *separation constant*,

$$\frac{X''}{X} = \frac{T''}{c^2 T} = -\lambda.$$

Calling the constant $-\lambda$ is common practice. Then

$$X'' + \lambda X = 0 \text{ and } T'' + \lambda c^2 T = 0,$$

which are two ordinary differential equations for X and T. Next use the boundary conditions. First,

$$y(0, t) = X(0)T(t) = 0 \text{ for } t \geq 0$$

implies that $X(0) = 0$. Similarly, $y(x, L) = X(L)T(t) = 0$ implies that $X(L) = 0$. This gives us a Sturm-Liouville problem for X:

$$X'' + \lambda X = 0 \text{ and } X(0) = X(L) = 0.$$

In Example 14.1, we solved this problem for the values of λ (eigenvalues) and corresponding solutions for X (eigenfunctions):

$$\lambda_n = \frac{n^2 \pi^2}{L^2} \text{ and } X_n(x) = \sin\left(\frac{n\pi x}{L}\right) \text{ for } n = 1, 2, \ldots.$$

Next focus on T. Since $\lambda_n = n^2 \pi^2 / L^2$, the differential equation for T is

$$T'' + \frac{n^2 \pi^2 c^2}{L^2} T = 0.$$

Because the string is released from rest,

$$\frac{\partial y}{\partial t}(x, 0) = X(x)T'(0) = 0,$$

so

$$T'(0) = 0.$$

Solutions for $T(t)$ subject to this condition are constant multiples of $\cos(n\pi ct/L)$.

So far, for $n = 1, 2, \ldots$, we have a function

$$y_n(x, t) = b_n \sin(n\pi x/L) \cos(n\pi ct/L)$$

that satisfies the wave equation, both boundary conditions, and the initial condition of zero initial velocity. We have yet to satisfy $y(x, 0) = f(x)$. If $f(x) = K \sin(m\pi x/L)$ for some integer m, then $y(x, t) = K \sin(m\pi x/L) \cos(m\pi ct/L)$ is the solution. But $f(x)$ need not look like this. For example, if the string is picked up at its midpoint, say

$$f(x) = \begin{cases} x & \text{for } 0 \leq x \leq L/2 \\ L - x & \text{for } L/2 \leq x \leq L \end{cases} \tag{15.4}$$

then not even a finite sum $y(x, t) = \sum_{n=0}^{N} y_n(x, t)$ can satisfy $y(x, 0) = f(x)$. It was Fourier's brilliant insight to attempt an infinite superposition

$$y(x, t) = \sum_{n=1}^{\infty} b_n \sin\left(\frac{n\pi x}{L}\right) \cos\left(\frac{n\pi ct}{L}\right).$$

The condition $y(x, 0) = f(x)$ is satisfied if we can choose the coefficients so that

$$y(x, 0) = f(x) = \sum_{n=0}^{\infty} b_n \sin\left(\frac{n\pi x}{L}\right).$$

This is the Fourier sine expansion of $f(x)$ on $[0, L]$, hence choose

$$b_n = \frac{2}{L} \int_0^L f(\xi) \sin\left(\frac{n\pi \xi}{L}\right) d\xi.$$

The solution of this problem, with zero initial velocity and initial position given by f, is

$$y(x, t) = \frac{2}{L} \sum_{n=1}^{\infty} \left(\int_0^L f(\xi) \sin(n\pi \xi/L) \, d\xi \right) \sin\left(\frac{n\pi x}{L}\right) \cos\left(\frac{n\pi ct}{L}\right). \tag{15.5}$$

EXAMPLE 15.1

With f given by equation (15.4) and $L = \pi$, the coefficients are

$$b_n = \frac{2}{\pi} \int_0^\pi f(\xi) \sin(n\xi) d\xi$$

$$= \frac{2}{\pi} \int_0^{\pi/2} \xi \sin(n\xi) \, d\xi + \frac{2}{\pi} \int_{\pi/2}^\pi (\pi - \xi) \sin(n\xi) \, d\xi$$

$$= \frac{4 \sin(n\pi/2)}{\pi n^2}.$$

The solution of this problem is

$$y(x, t) = \sum_{n=1}^{\infty} \frac{4 \sin(n\pi/2)}{\pi n^2} \sin(nx) \cos(nct).$$

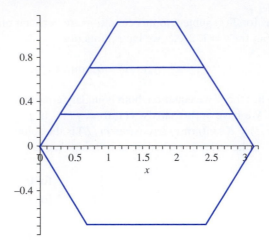

FIGURE 15.2   *Wave motion in Example 15.1.*

Figure 15.2 shows the string profile for $c = 2$ at times $t = 0.3, 0.6, 0.9,$ and 1.2 starting at the top and moving downward in this time frame. ◆

EXAMPLE 15.2

We will solve for $y(x, t)$ on $[0, \pi]$ with zero initial velocity and initial position $f(x) = x \cos(5x/2)$. The coefficients are

$$b_n = \frac{2}{\pi} \int_0^\pi \xi \cos(5\xi/2) \sin(n\xi) \, d\xi = \frac{8}{\pi} \frac{n(-1)^{n+1}}{(5+2n)^2(5-2n)^2}.$$

The solution is

$$y(x, t) = \sum_{n=1}^\infty \frac{8}{\pi} \frac{n(-1)^{n+1}}{(5+2n)^2(5-2n)^2} \sin(nx) \cos(nct).$$

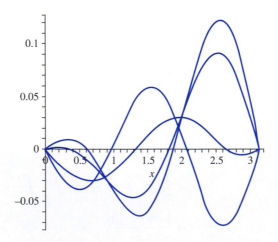

FIGURE 15.3   *Wave profiles in Example 15.2.*

Figure 15.3 shows wave profiles for $c = 1$ at times $t = 0.1, 0.3, 0.6,$ and 0.9 moving downward in this time frame. ◆

15.2.2   Zero Initial Displacement

We will solve the initial-boundary value problem for the wave motion if there is an initial velocity, but no initial displacement. The problem is

$$\frac{\partial^2 y}{\partial t^2} = c^2 \frac{\partial^2 y}{\partial x^2} \text{ for } 0 < x < L, t > 0,$$

$$y(0, t) = y(L, t) = 0 \text{ for } t \geq 0,$$

$$y(x, 0) = 0,$$

$$\text{and } \frac{\partial y}{\partial t}(x, 0) = g(x) \text{ for } 0 < x < L.$$

Again let $y(x, t) = X(x)T(t)$. The problem for X is the same as before,

$$X'' + \lambda X = 0 \text{ and } X(0) = X(L) = 0$$

with eigenvalues $\lambda_n = n^2 \pi^2 / L^2$ and eigenfunctions $\sin(n\pi x/L)$. The problem for T, however, is different. The differential equation is still

$$T'' + \lambda c^2 T = T'' + \left(\frac{n^2 \pi^2 c^2}{L^2} \right) T = 0$$

but now the zero initial displacement gives us $y(x, 0) = X(x)T(0) = 0$, so $T(0) = 0$. Solutions of this problem for $T(t)$ are constant multiples of

$$T_n(t) = \sin \left(\frac{n\pi ct}{L} \right).$$

Now we have functions

$$y_n(x, t) = X_n(x)T_n(t) = \sin \left(\frac{n\pi x}{L} \right) \sin \left(\frac{n\pi ct}{L} \right)$$

that satisfy the wave equation, the boundary conditions, and the initial condition $y(x, 0) = 0$. To satisfy the initial velocity condition, we will generally (depending on g) need a superposition

$$y(x, t) = \sum_{n=1}^{\infty} b_n \sin \left(\frac{n\pi x}{L} \right) \sin \left(\frac{n\pi ct}{L} \right).$$

We must choose the b_n's to satisfy

$$\left. \frac{\partial y}{\partial t} \right]_{t=0} = \sum_{n=1}^{\infty} \left(\frac{n\pi c}{L} \right) b_n \sin \left(\frac{n\pi x}{L} \right) = g(x).$$

Then

$$\frac{L}{n\pi c} g(x) = \sum_{n=1}^{\infty} b_n \sin \left(\frac{n\pi x}{L} \right).$$

This is the Fourier sine expansion of $\frac{L}{n\pi c} g(x)$. Therefore, choose the coefficients

$$b_n = \frac{2}{L} \frac{L}{n\pi c} \int_0^L g(\xi) \sin \left(\frac{n\pi \xi}{L} \right) d\xi$$

or

$$b_n = \frac{2}{n\pi c} \int_0^L g(\xi) \sin \left(\frac{n\pi \xi}{L} \right) d\xi.$$

With this choice of the coefficients, the solution is

$$y(x,t) = \frac{2}{\pi c} \sum_{n=1}^{\infty} \frac{1}{n} \left(\int_0^L g(\xi) \sin(n\pi\xi/L)\,d\xi \right) \sin\left(\frac{n\pi x}{L}\right) \sin\left(\frac{n\pi ct}{L}\right). \qquad (15.6)$$

EXAMPLE 15.3

Suppose the string is released from its horizontal position with an initial velocity given by $g(x) = x(1 + \cos(\pi x/L))$. Equation (15.6) is the solution. First compute the integral

$$\int_0^L g(\xi) \sin(n\pi\xi/L)\,d\xi = \int_0^L \xi\,(1 + \cos(\pi\xi/L)) \sin(n\pi\xi/L)\,d\xi$$

$$= \begin{cases} \dfrac{3L^2}{4\pi} & \text{for } n = 1 \\[2ex] \dfrac{L^2(-1)^n}{n\pi(n^2 - 1)} & \text{for } n = 2, 3, \cdots. \end{cases}$$

The solution is

$$y(x,t) = \frac{2}{\pi c}\left(\frac{3L^2}{4\pi}\right) \sin\left(\frac{\pi x}{L}\right) \sin\left(\frac{\pi ct}{L}\right)$$

$$+ \frac{2}{\pi c} \sum_{n=2}^{\infty} \frac{L^2(-1)^n}{n^2\pi(n^2-1)} \sin\left(\frac{n\pi x}{L}\right) \sin\left(\frac{n\pi ct}{L}\right).$$

With $c = 1$ and $L = \pi$, this solution is

$$y(x,t) = \frac{3}{2}\sin(x)\sin(t) + \sum_{n=2}^{\infty} \frac{2(-1)^n}{n^2(n^2-1)} \sin(nx)\sin(nt).$$

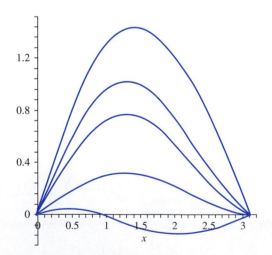

FIGURE 15.4 *Wave moving downward for increasing times in Example 15.3.*

Figure 15.4 shows the wave moving downward at times $t = 0.2, 0.5, 0.7, 1.3$, and 3.4. ◆

15.2.3 Nonzero Initial Displacement and Velocity

Suppose the string has an initial displacement $f(x)$ and an initial velocity $g(x)$. Write the solution $y_f(x, t)$ for the problem with initial displacement f and zero initial velocity and the solution $y_g(x, t)$ for the problem with zero initial displacement and initial velocity $g(x)$. Then the solution for the problem with initial position f and initial velocity g is

$$y(x, t) = y_f(x, t) + y_g(x, t).$$

To see this, observe that this function satisfies the wave equation and boundary conditions, because both $y_f(x, t)$ and $y_g(x, t)$ do. Furthermore,

$$y(x, 0) = y_f(x, 0) + y_g(x, 0) = y_f(x, 0) + 0 = y_f(x, 0) = f(x)$$

and

$$\frac{\partial y}{\partial t}(x, 0) = \frac{\partial y_f}{\partial t}(x, 0) + \frac{\partial y_g}{\partial t}(x, 0) = 0 + g(x) = g(x).$$

EXAMPLE 15.4

We will solve the wave equation on $[0, L]$ with $y(0, t) = y(L, t) = 0$, subject to initial position

$$f(x) = \begin{cases} x & \text{for } 0 \le x \le L/2 \\ L - x & \text{for } L/2 \le x \le L. \end{cases}$$

and initial velocity $g(x) = x(1 + \cos(\pi x/L))$. The solution $y(x, t)$ is the sum of the solutions of the problems solved in Examples 15.1 and 15.3. If $c = 1$ and $L = \pi$, this solution is

$$y(x, t) = \sum_{n=1}^{\infty} \frac{4}{n^2 \pi} \sin(n\pi/2) \sin(nx) \cos(nt)$$

$$+ \frac{3}{2} \sin(x) \sin(t) + \sum_{n=2}^{\infty} \frac{2(-1)^n}{n^2(n^2 - 1)} \sin(nx) \sin(nt).$$

In Figure 15.5, the string's position at $t = 0.2$ is the fourth graph from the top. The wave moves upward to the next graph at $t = 0.6$ then upward again at $t = 1.2$. The highest wave is at

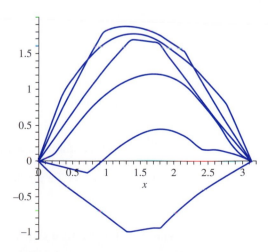

FIGURE 15.5 *Waves in Example 15.4.*

$t = 1.8$. Following this, the wave is partly below the horizontal axis at $t = 2.3$ and completely below this axis at $t = 2.9$. ◆

15.2.4   Influence of Constants and Initial Conditions

It is interesting to observe how the initial conditions and the constant c influence the wave motion.

EXAMPLE 15.5

In Example 15.3, we solved the the problem with zero initial displacement and initial velocity given by $g(x) = x(1 + \cos(\pi x/L))$. If $L = \pi$ this solution is

$$y(x, t) = \frac{3}{2c} \sin(x) \sin(ct) + \sum_{n=2}^{\infty} \frac{2(-1)^n}{cn^2(n^2 - 1)} \sin(nx) \sin(nct).$$

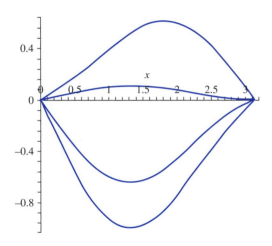

FIGURE 15.6   *Wave profiles for different values of c in Example 15.5.*

Figure 15.3 showed wave profiles at various times for $c = 1$. Figure 15.6 shows wave profiles at time $t = 5.3$ with $c = 1.05$, $c = 1.1$, $c = 1.2$, and $c = 1.65$. These graphs move upward as c increases. ◆

EXAMPLE 15.6

We we will examine the effects of changes in an initial condition on the wave motion with the problem

$$\frac{\partial y^2}{\partial t^2} = 1.44 \frac{\partial y^2}{\partial x^2} \text{ for } 0 < x < \pi, t > 0,$$

$$y(0, t) = y(\pi, t) = 0 \text{ for } t \geq 0,$$

and

$$y(x, 0) = 0, \frac{\partial y}{\partial t}(x, 0) = \sin(\epsilon x) \text{for } 0 < x < \pi,$$

in which ϵ is a positive number that is not an integer.

Use equation (15.6) to write the solution

$$y(x, t) = \frac{5}{3\pi} \sum_{n=1}^{\infty} \frac{\sin(\pi\epsilon)(-1)^{n+1}}{n^2 - \epsilon^2} \sin(nx) \sin(1.2t).$$

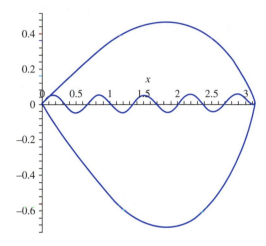

FIGURE 15.7 *Wave profiles in Example 15.6 decreasing as ϵ increases.*

To gauge the effect of ϵ on the motion, compare graphs of this solution for different values of ϵ at given times. Figure 15.7 shows the wave profile at $t = 0.5$ for ϵ equal to 0.7 (wave above the x axis), 1.5 (wave below the axis), and 9.3 (wave oscillating rapidly). ◆

15.2.5 Wave Motion with a Forcing Term

Separation of variables may fail if the partial differential equation contains terms allowing for some type of external forcing, or if the boundary conditions are nonhomogeneous. In such a case, it may be possible to transform the initial-boundary value problem to one that we know how to solve.

EXAMPLE 15.7

We will solve the problem

$$\frac{\partial^2 y}{\partial t^2} = \frac{\partial y^2}{\partial x^2} + Ax \text{ for } 0 < x < L, t > 0,$$

$$y(0, t) = y(L, t) = 0 \text{ for } t \geq 0,$$

and

$$y(x, 0) = 0, \frac{\partial y}{\partial t}(x, 0) = 1 \text{ for } 0 < x < L.$$

A is a positive constant and the term Ax in the wave equation represents an external force having a magnitude of Ax at x. We have let $c = 1$ in this problem.

If we put $y(x, t) = X(x)T(t)$ into the partial differential equation, we obtain

$$XT'' = X''T + Ax,$$

and there is no way to separate the t dependency on one side of an equation and the x dependency on the other. One strategy in such a case is to try to transform this problem into one to which separation of variables applies. Let

$$y(x, t) = Y(x, t) + \psi(x).$$

The idea is to choose ψ to obtain a problem for Y that we can solve. Substitute $y(x, t)$ into the partial differential equation to get

$$\frac{\partial^2 Y}{\partial t^2} = \frac{\partial^2 Y}{\partial x^2} + \psi''(x) + Ax.$$

This will be simplified if we choose ψ so that

$$\psi''(x) + Ax = 0.$$

Integrate twice to get

$$\psi(x) = -A\frac{x^3}{6} + Cx + D,$$

with C and D as constants of integration. To find C and D, look at the boundary conditions. First,

$$y(0, t) = Y(0, t) + \psi(0) = 0.$$

This will be just $y(0, t) = Y(0, t)$ if we make $\psi(0) = 0$, hence choose $D = 0$. Next,

$$y(L, t) = Y(L, t) + \psi(L) = Y(L, t) - A\frac{L^3}{6} + CL = 0.$$

This reduces to $y(L, t) = Y(L, t)$ if we choose C so that

$$\psi(L) = -A\frac{L^3}{6} + CL = 0.$$

With $C = AL^2/6$, we have

$$\psi(x) = \frac{1}{6}Ax(L^2 - x^2).$$

With this ψ,

$$Y(0, t) = Y(L, t) = 0.$$

Next relate the initial conditions for y to the initial conditions for Y. First,

$$Y(x, 0) = y(x, 0) - \psi(x) = -\psi(x) = \frac{1}{6}Ax(x^2 - L^2)$$

and

$$\frac{\partial Y}{\partial t}(x, 0) = \frac{\partial y}{\partial t}(x, 0) = 1.$$

Now we have an initial-boundary value problem for Y:

$$\frac{\partial^2 Y}{\partial t^2} = \frac{\partial Y^2}{\partial x^2} \text{ for } 0 < x < L, t > 0,$$

$$Y(0, t) = Y(L, t) = 0 \text{ for } t \geq 0,$$

and

$$Y(x, 0) = \frac{1}{6}Ax(x^2 - L^2), \frac{\partial Y}{\partial t}(x, 0) = 1 \text{ for } 0 < x < L.$$

We know the solution of this problem. By equations (15.5) and (15.6),

$$Y(x,t) = \frac{2}{L} \sum_{n=1}^{\infty} \left(\int_0^L \frac{1}{6} A\xi(\xi^2 - L^2) \sin(n\pi\xi/L) d\xi \right) \sin(n\pi x/L) \cos(n\pi t/L)$$

$$+ \frac{2}{\pi} \sum_{n=1}^{\infty} \frac{1}{n} \left(\int_0^L \sin(n\pi\xi/L) d\xi \right) \sin(n\pi x/L) \sin(n\pi t/L)$$

$$= \frac{2AL^3}{\pi^3} \sum_{n=1}^{\infty} \frac{(-1)^n}{n^3} \sin(n\pi x/L) \cos(n\pi t/L)$$

$$+ \frac{2L}{\pi^2} \sum_{n=1}^{\infty} \frac{1 - (-1)^n}{n^2} \sin(n\pi x/L) \sin(n\pi t/L).$$

The solution of the original problem is

$$y(x,t) = Y(x,t) + \frac{1}{6}Ax(L^2 - x^2).$$

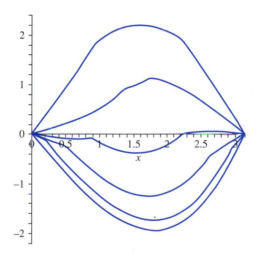

FIGURE 15.8 *Wave profiles in Example 15.7.*

Figure 15.8 shows wave profiles for $c = 1$ and $L = \pi$ at times $t = 0.03, 0.2, 0.5, 0.9, 1.4,$ and 2.2. The waves move upward as t increases in this time range. ◆

15.2.6 Numerical Solutions

It is sometime useful to solve an initial-boundary value problem numerically by determining approximate numerical values of the solution at specific points.

First, we need a scheme for approximating a derivative. If f is a function of one variable, then

$$f'(x_0) = \lim_{h \to 0} \frac{f(x_0 + h) - f(x_0)}{h}.$$

As h is chosen smaller, the difference quotient on the right more closely approximates the derivative. This suggests we look at the approximations

$$f'(x_0) \approx \frac{f(x_0 + h) - f(x_0)}{h}$$

and

$$f'(x_0) \approx \frac{f(x_0 - h) - f(x_0)}{h}$$

with $h > 0$. These are, respectively, the *forward difference approximation* and the *backward difference approximation* to $f'(x_0)$. Their average

$$f'(x_0) \approx \frac{f(x_0 + h) - f(x_0 - h)}{2h}$$

is the *centered difference approximation* of $f'(x_0)$. If we apply this approximation to the second derivative, we obtain

$$f''(x_0) \approx \frac{f(x_0 + h) - 2f(x_0) + f(x_0 - h)}{h^2}.$$

This is the *centered difference approximation of the second derivative.*

Apply this approximation to $y(x, t)$ with increments Δx in x and Δt in t to get

$$\frac{\partial^2 y}{\partial x^2}(x, t) \approx \frac{y(x + \Delta x, t) - 2y(x, t) + y(x - \Delta x, t)}{(\Delta x)^2}$$

and

$$\frac{\partial^2 y}{\partial t^2}(x, t) \approx \frac{y(x, t + \Delta t) - 2y(x, t) + y(x, t - \Delta t)}{(\Delta t)^2}.$$

We will use these to approximate solution values for the following problem at selected points:

$$\frac{\partial y^2}{\partial t^2} = c^2 \frac{\partial y^2}{\partial x^2} \text{ for } 0 < x < L, t > 0,$$

$$y(0, t) = y(L, t) = 0 \text{ for } t \geq 0,$$

and

$$y(x, 0) = f(x), \frac{\partial y}{\partial t}(x, 0) = g(x) \text{ for } 0 \leq x \leq L.$$

The (x, t) region of definition for $y(x, t)$ is the strip $0 \leq x \leq L$ and $t \geq 0$. Choose a positive integer N and let $\Delta x = L/N$. Partition $[0, L]$ by points $x_j = j\Delta x$, which are equally spaced from 0 to L for $j = 0, 1, \ldots, N$. Also choose a positive increment Δt and let $t_k = k\Delta t$. This forms a grid of points (x_j, t_k), called *lattice points*, over the (x, t) strip. It is convenient to write

$$y_{j,k} = y(x_j, t_k) = y(j\Delta x, k\Delta t).$$

Now replace the partial derivatives in the wave equation by their centered difference approximations:

$$\frac{y_{j,k+1} - 2y_{j,k} + y_{j,k-1}}{(\Delta t)^2} = c^2 \frac{y_{j+1,k} - 2y_{j,k} + y_{j-1,k}}{(\Delta x)^2}$$

at (x_j, y_k). Solve this for $y_{j,k+1}$ to get

$$y_{j,k+1} = \left(\frac{c\Delta t}{\Delta x}\right)^2 (y_{j+1,k} - 2y_{j,k} + y_{j-1,k}) + 2y_{j,k} - y_{j,k-1}. \tag{15.7}$$

In Figure 15.9, the points (x_j, t_{k+1}), (x_{j-1}, t_k), (x_j, t_k), (x_{j+1}, t_k), and (x_j, t_{k-1}) form a diamond-shaped configuration with the middle three points at the t_k level, the last point on the lower t_{k-1} level, and the first at the higher t_{k+1} level. From equation (15.7), we know an approximate value of $y_{j,k+1}$ at the top level if we know the approximate values at these four points on the two levels below. Now we can work our way up the strip, using this diamond configuration and equation (15.7) to approximate values at grid points for increasing values of t moving up one level at a time from the two levels immediately below.

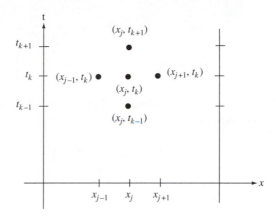

FIGURE 15.9 *Diamond configuration for numerical approximation scheme.*

This process fails at the edges of the (x, t) strip, because we cannot form this five-point configuration there. Now use the boundary and initial conditions. First, $y(x, 0) = f(x)$ at each point on the bottom edge of the strip (the interval $[a, b]$). Next, $y(0, t) = y(L, t) = 0$ on the left and right vertical sides of the strip. Thus,

$$y_{0,k} = y_{L,k} = 0 \text{ for } k = 0, 1, 2, \cdots$$

and

$$y(x_j, 0) = y_{j,0} = f(j\Delta x) \text{ for } j = 1, 2, \cdots, N - 1.$$

In order to get this process started, we need a "−1" layer one step below the bottom edge of the strip. There is no such layer in a natural sense, so we create one to have a starting point for equation (15.7). This layer is created by using the initial condition

$$\frac{\partial y}{\partial t}(x_j, 0) \approx \frac{y(x_j, -\Delta t) - y(x_j, 0)}{\Delta t}$$

$$= \frac{y_{j,-1} - y_{j,0}}{-\Delta t} = g(x_j) = g(j\Delta x) \text{ for } j = 1, 2, \ldots, N - 1.$$

Solve this to obtain

$$y_{j,-1} = y_{j,0} - g(j\Delta x)\Delta t.$$

From here, work upward in the (x, t) strip to approximate values of the solution at successively higher layers.

In applying this method, Δx and Δt should be chosen so that $(c\Delta t/\Delta x)^2 < 1/2$. Otherwise, the method can be unstable and produce results with increasing error as we move up the t layers.

EXAMPLE 15.8

Consider the problem

$$\frac{\partial y^2}{\partial t^2} = \frac{\partial y^2}{\partial x^2} \text{ for } 0 < x < 1, t > 0,$$

$$y(0, t) = y(1, t) = 0 \text{ for } t \geq 0,$$

and

$$y(x, 0) = x \cos(\pi x/2), \quad \frac{\partial y}{\partial t}(x, 0) = \begin{cases} 1 & \text{for } 0 \leq x \leq 1/2 \\ 0 & \text{for } 1/2 < x \leq 1. \end{cases}$$

We will compute some approximate values of the solution. In this case, we can write the exact solution for comparison:

$$y(x, t) = \frac{16}{\pi^2} \sum_{n=1}^{\infty} \left(\frac{(-1)^n}{(4n^2 - 1)^2} \right) \sin(n\pi x) \cos(n\pi t)$$

$$+ \frac{2}{\pi} \sum_{n=1}^{\infty} \frac{1}{n} \left(\frac{\cos(n\pi/2) - 1}{n\pi} \right) \sin(n\pi x) \sin(n\pi t).$$

Choose $N = 10$, so $\Delta x = 0.1$. Let $\Delta t = 0.025$, so $(c \Delta t / \Delta x)^2 = 0.0625 < 1/2$. The equations for the approximations are

$$y_{j,k+1} = (0.0625)(y_{j+1,k} - 2y_{j,k} + y_{j-1,k}) + 2y_{j,k} - y_{j,k-1} \text{ for } j = 1, \ldots, 9, k = 0, 1, 2, \ldots,$$

$$y_{j,0} = f(0.1j) \text{ for } j = 1, \ldots, 9,$$

and

$$y_{j,-1} = j_{j,0} - g(j\Delta x)\Delta t = f(0.1)j - 0.025g(0.1j) \text{ for } j = 1, \ldots, 9.$$

First compute approximate values $y_{j,-1}$ on the fictitious lowest horizontal level:

$$y_{1,-1} = 0.07377, \quad y_{2,-1} = 0.16521, \quad y_{3,-1} = 024320,$$

$$y_{4,-1} = 0.29861, \quad y_{5,-1} = 0.32855, \quad y_{6,-1} = 0.35267,$$

$$y_{7,-1} = 0.31779, \quad y_{8,-1} = 0.24721, \quad y_{9,-1} = 0.14079.$$

Here we have taken a notational liberty and written equal when all values are approximate.

Next compute approximate values $y_{j,0}$ on the next highest level (points on the bottom of the x, t strip):

$$y_{1,0} = 0.09877, \quad y_{2,0} = 0.19021, \quad y_{3,0} = 0.26730,$$

$$y_{4,0} = 0.32361, \quad y_{5,0} = 0.35355, \quad y_{6,0} = 0.35267,$$

$$y_{7,0} = 0.31779, \quad y_{8,0} = 0.24721, \quad y_{9,0} = 0.14079.$$

Move up to the first t layer above the x axis. For $t = 0.025$, put $k = 0$ to compute values

$$y_{j,1} = (0.0625)(y_{j+1,0} - 2y_{j,0} + y_{j-1,0}) + 2y_{j,0} - y_{j,-1} \text{ for } j = 1, \ldots, 9.$$

The computed values are

$$y_{1,1} = 0.12331, \quad y_{2,1} = 0.21431, \quad y_{3,1} = 0.29100,$$

$$y_{4,1} = 0.34696, \quad y_{5,1} = 0.37662, \quad y_{6,1} = 0.35055,$$

$$y_{7,1} = 0.32556, \quad y_{8,1} = 0.24160, \quad y_{9,1} = 0.13864.$$

In this way we can continue to move up as many time steps as we wish. ◆

In each of Problems 1 through 8, solve the initial-boundary value problem using separation of variables. Graph the fiftieth partial sum of the solution for some values of t with $c = 1$ if c is unspecified in the problem.

1.
$$\frac{\partial^2 y}{\partial t^2} = c^2 \frac{\partial^2 y}{\partial x^2} \text{ for } 0 < x < 2, t > 0$$

$$y(0, t) = y(2, t) = 0 \text{ for } t \geq 0$$

$$y(x, 0) = 0, \frac{\partial y}{\partial t}(x, 0) = g(x) \text{ for } 0 \leq x \leq 2$$

where $g(x) = \begin{cases} 2x & \text{for } 0 \leq x \leq 1 \\ 0 & \text{for } 1 < x \leq 2. \end{cases}$

2.
$$\frac{\partial^2 y}{\partial t^2} = 9 \frac{\partial^2 y}{\partial x^2} \text{ for } 0 < x < 4, t > 0$$

$$y(0, t) = y(4, t) = 0 \text{ for } t \geq 0$$

$$y(x, 0) = 2 \sin(\pi x), \frac{\partial y}{\partial t}(x, 0) = 0 \text{ for } 0 \leq x \leq 4$$

3.
$$\frac{\partial^2 y}{\partial t^2} = 4 \frac{\partial^2 y}{\partial x^2} \text{ for } 0 < x < 3, t > 0$$

$$y(0, t) = y(3, t) = 0 \text{ for } t \geq 0$$

$$y(x, 0) = 0, \frac{\partial y}{\partial t}(x, 0) = x(3 - x) \text{ for } 0 \leq x \leq 3$$

4.
$$\frac{\partial^2 y}{\partial t^2} = 9 \frac{\partial^2 y}{\partial x^2} \text{ for } 0 < x < \pi, t > 0$$

$$y(0, t) = y(\pi, t) = 0 \text{ for } t \geq 0$$

$$y(x, 0) = \sin(x), \frac{\partial y}{\partial t}(x, 0) = 1 \text{ for } 0 \leq x \leq \pi$$

5.
$$\frac{\partial^2 y}{\partial t^2} = 8 \frac{\partial^2 y}{\partial x^2} \text{ for } 0 < x < 2\pi, t > 0$$

$$y(0, t) = y(2\pi, t) = 0 \text{ for } t \geq 0$$

$$y(x, 0) = f(x), \frac{\partial y}{\partial t}(x, 0) = 0 \text{ for } 0 \leq x \leq 2\pi$$

where $f(x) = \begin{cases} 3x & \text{for } 0 \leq x \leq \pi \\ 6\pi - 3x & \text{for } \pi < x \leq 2\pi. \end{cases}$

6.
$$\frac{\partial^2 y}{\partial t^2} = 4 \frac{\partial^2 y}{\partial x^2} \text{ for } 0 < x < 5, t > 0$$

$$y(0, t) = y(5, t) = 0 \text{ for } t \geq 0$$

$$y(x, 0) = 0, \frac{\partial y}{\partial t}(x, 0) = g(x) \text{ for } 0 \leq x \leq 5$$

where $g(x) = \begin{cases} 0 & \text{for } 0 \leq x < 4 \\ 5 - x & \text{for } 4 \leq x \leq 5. \end{cases}$

7.
$$\frac{\partial^2 y}{\partial t^2} = 9 \frac{\partial^2 y}{\partial x^2} \text{ for } 0 < x < 2, t > 0$$

$$y(0, t) = y(2, t) = 0 \text{ for } t \geq 0$$

$$y(x, 0) = x(x - 2), \frac{\partial y}{\partial t}(x, 0) = g(x) \text{ for } 0 \leq x \leq 2$$

where $g(x) = \begin{cases} 0 & \text{for } 0 \leq x < 1/2 \text{ and } 1 < x \leq 2 \\ 3 & \text{for } 1/2 \leq x \leq 1. \end{cases}$

8.
$$\frac{\partial^2 y}{\partial t^2} = 25 \frac{\partial^2 y}{\partial x^2} \text{ for } 0 < x < \pi, t > 0$$

$$y(0, t) = y(\pi, t) = 0 \text{ for } t \geq 0$$

$$y(x, 0) = \sin(2x), \frac{\partial y}{\partial t}(x, 0) = \pi - x \text{ for } 0 \leq x \leq \pi$$

9. Solve the initial-boundary value problem

$$\frac{\partial^2 y}{\partial t^2} = 3 \frac{\partial^2 y}{\partial x^2} + 2x \text{ for } 0 < x < 2, t > 0$$

$$y(0, t) = y(2, t) = 0 \text{ for } t \geq 0$$

$$y(x, 0) = 0, \frac{\partial y}{\partial t}(x, 0) = 0 \text{ for } 0 \leq x \leq 2.$$

Graph the fortieth partial sum of the solution for some values of the time.

10. Solve

$$\frac{\partial^2 y}{\partial t^2} = 9 \frac{\partial^2 y}{\partial x^2} + x^2 \text{ for } 0 < x < 4, t > 0$$

$$y(0, t) = y(4, t) = 0 \text{ for } t \geq 0$$

$$y(x, 0) = 0, \frac{\partial y}{\partial t}(x, 0) = 0 \text{ for } 0 \leq x \leq 4.$$

Graph the fortieth partial sum of the solution for selected values of t,

11. Solve

$$\frac{\partial^2 y}{\partial t^2} = \frac{\partial^2 y}{\partial x^2} - \cos(x) \text{ for } 0 < x < 2\pi, t > 0$$

$$y(0, t) = y(2\pi, t) = 0 \text{ for } t \geq 0$$

$$y(x, 0) = 0, \frac{\partial y}{\partial t}(x, 0) = 0 \text{ for } 0 \leq x \leq 2\pi.$$

Graph the fortieth partial sum for some values of the time.

12. Transverse vibrations in a homogeneous rod of length π are modeled by the partial differential equation

$$a^4 \frac{\partial^4 u}{\partial x^4} + \frac{\partial^2 u}{\partial t^2} = 0 \text{ for } 0 < x < \pi, t > 0.$$

Here $u(x, t)$ is the displacement at time t of the cross section through x perpendicular to the x axis and $a^2 = EI/\rho A$, where E is *Young's modulus*, I is the moment of inertia of this cross section, ρ is the constant density, and A is the constant cross-sectional area.

(a) Let $u(x, t) = X(x)T(t)$ to separate the variables.

(b) Solve for values of the separation constant and for X and T in the case of free ends, in which

$$\frac{\partial^2 u}{\partial x^2}(0, t) = \frac{\partial^2 u}{\partial x^2}(\pi, t) = \frac{\partial^3 u}{\partial x^3}(0, t) = \frac{\partial^3 y}{\partial x^3}(\pi, t) = 0$$

for $t > 0$.

(c) Solve for the separation constant and for X and T in the case of supported ends, in which:

$$u(0, t) = u(\pi, t) = \frac{\partial^2 u}{\partial x^2}(0, t) = \frac{\partial^2 u}{\partial x^2}(\pi, t) = 0.$$

13. Solve the *telegraph equation*

$$\frac{\partial^2 u}{\partial t^2} + A\frac{\partial u}{\partial t} + Bu = c^2 \frac{\partial^2 u}{\partial x^2}$$

for $0 < x < L$ and $t > 0$. A and B are positive constants. The boundary conditions are

$$u(0, t) = u(L, t) = 0 \text{ for } t \geq 0,$$

and the initial conditions are

$$u(x, 0) = f(x), \frac{\partial u}{\partial t}(x, 0) = 0 \text{ for } 0 \leq x \leq L.$$

Assume that $A^2 L^2 < 4(BL^2 + c^2\pi^2)$.

14. (a) Write a series solution for

$$\frac{\partial^2 y}{\partial t^2} = 9\frac{\partial^2 y}{\partial x^2} + 5x^3 \text{ for } 0 < x < 4, t > 0$$

$$y(0, t) = y(4, t) = 0 \text{ for } t \geq 0$$

$$y(x, 0) = \cos(\pi x), \frac{\partial y}{\partial t}(x, 0) = 0 \text{ for } 0 \leq x \leq 4.$$

(b) Write a series solution when the forcing term $5x^3$ is removed.

(c) In order to gauge the effect of the forcing term on the motion, graph the fortieth partial sum of the solutions in (a) and (b) on the same set of axes when $t = 0.4$ seconds. Repeat this for $t = 0.8, 1.4, 2, 2.5, 3,$ and 4.

15. (a) Write a series solution for

$$\frac{\partial^2 y}{\partial t^2} = 9\frac{\partial^2 y}{\partial x^2} + \cos(\pi x) \text{ for } 0 < x < 4, t > 0$$

$$y(0, t) = y(4, t) = 0 \text{ for } t \geq 0$$

$$y(x, 0) = x(4 - x), \frac{\partial y}{\partial t}(x, 0) = 0 \text{ for } 0 \leq x \leq 4.$$

(b) Write a series solution when the forcing term $\cos(\pi x)$ is removed.

(c) In order to gauge the effect of the forcing term on the motion, graph the fortieth partial sum of the solutions in (a) and (b) on the same set of axes when $t = 0.6$ seconds and then for $t = 1, 1.4, 2, 3, 5,$ and 7.

16. (a) Write a series solution for

$$\frac{\partial^2 y}{\partial t^2} = 9\frac{\partial^2 y}{\partial x^2} - e^{-x} \text{ for } 0 < x < 4, t > 0$$

$$y(0, t) = y(4, t) = 0 \text{ for } t \geq 0$$

$$y(x, 0) = \sin(\pi x), \frac{\partial y}{\partial t}(x, 0) = 0 \text{ for } 0 \leq x \leq 4.$$

(b) Write a series solution when the forcing term e^{-x} is removed.

(c) In order to gauge the effect of the forcing term on the motion, graph the fortieth partial sum of the solutions in (a) and (b) on the same set of axes when $t = 0.4$ seconds. Repeat this for $t = 0.8, 1.4, 2, 2.5, 3,$ and 4.

In each of Problems 17 through 21, use $\Delta x = 0.1$ and $\Delta t = 0.025$ to compute approximate values of $y(x, t)$ at lattice points in the (x, t) strip $0 \leq x \leq 1$ and $t \geq 0$ for the layers $t = 0$ through $t = 5(0.025) = 0.125$. Write the series solution for $y(x, t)$ and use the fiftieth partial sum to approximate some values of this solution to compare with the approximated values.

17. $\quad \frac{\partial^2 y}{\partial t^2} = \frac{\partial^2 y}{\partial x^2} \text{ for } 0 < x < 1, t > 0$

$$y(0, t) = y(1, t) = 0 \text{ for } t \geq 0$$

$$y(x, 0) = f(x), \frac{\partial y}{\partial t}(x, 0) = 0 \text{ for } 0 \leq x \leq 1,$$

where

$$f(x) = \begin{cases} x & \text{for } 0 \leq x \leq 1/2 \\ 1 - x & \text{for } 1/2 \leq x \leq 1. \end{cases}$$

18. $\quad \frac{\partial^2 y}{\partial t^2} = \frac{\partial^2 y}{\partial x^2} \text{ for } 0 < x < 2, t > 0$

$$y(0, t) = y(2, t) = 0 \text{ for } t \geq 0$$

$$y(x, 0) = 0, \frac{\partial y}{\partial t}(x, 0) = 1 \text{ for } 0 \leq x \leq 2.$$

19. $\quad \frac{\partial^2 y}{\partial t^2} = \frac{\partial^2 y}{\partial x^2} \text{ for } 0 < x < 2, t > 0$

$$y(0, t) = y(2, t) = 0 \text{ for } t \geq 0$$

$$y(x, 0) = \sin(\pi x), \frac{\partial y}{\partial t}(x, 0) = 0 \text{ for } 0 \leq x \leq 2.$$

20. $\quad \frac{\partial^2 y}{\partial t^2} = \frac{\partial^2 y}{\partial x^2} \text{ for } 0 < x < 1, t > 0$

$$y(0, t) = y(1, t) = 0 \text{ for } t \geq 0$$

$$y(x, 0) = x(1 - x^2), \frac{\partial y}{\partial t}(x, 0) = x^2 \text{ for } 0 \leq x \leq 1.$$

15.3 Wave Motion in an Infinite Medium

If great distances are involved (as with sound waves through the ocean or cosmic background radiation across the universe), wave motion is often modeled by the wave equation on $-\infty < x < \infty$. In this case, there is no boundary, hence there is no boundary condition. However, we seek bounded solutions.

The analysis is similar to that for solutions on a closed interval, except that $\int_{-\infty}^{\infty} \ldots d\omega$ replaces $\sum_{n=1}^{\infty}$. As we did on a bounded interval, we will consider separately the cases of zero initial velocity and no initial displacement.

Zero Initial Velocity The initial-boundary value problem is

$$\frac{\partial^2 y}{\partial t^2} = c^2 \frac{\partial^2 y}{\partial x^2} \text{ for } -\infty < x < \infty, t > 0$$

and

$$y(x, 0) = f(x), \frac{\partial y}{\partial t}(x, 0) = 0 \text{ for } -\infty < x < \infty.$$

Separate variables by putting $y(x, t) = X(x)T(t)$. Exactly as with wave motion on a bounded interval, we obtain

$$X'' + \lambda X = 0, T'' + \lambda c^2 T = 0.$$

There are three cases on λ.

Case 1 If $\lambda = 0$ then $X = ax + b$. This is bounded if $a = 0$, so zero is an eigenvalue with constant eigenfunctions.

Case 2 If $\lambda < 0$, write $\lambda = -\omega^2$ with $\omega > 0$. Then $X(x) = c_1 e^{\omega x} + c_2 e^{-\omega x}$, and this function is unbounded on the entire real line if either constant is nonzero. This problem has no negative eigenvalue.

Case 3 If $\lambda > 0$, write $\lambda = \omega^2$ with $\omega > 0$. Now $X(x) = c_1 \cos(\omega x) + c_2 \sin(\omega x)$, which is a bounded function for any choices of positive ω. Every positive number $\lambda = \omega^2$ is an eigenvalue with eigenfunctions of the form $X_\omega = c_1 \cos(\omega x) + c_2 \sin(\omega x)$.

We can consolidate Cases 1 and 3 by allowing $\lambda = 0$ in Case 3.

The equation for T is $T'' + c^2 \omega^2 T = 0$ with solutions of the form

$$T_\omega(t) = a \cos(\omega c t) + b \sin(\omega c t).$$

But

$$\frac{\partial y}{\partial t}(x, 0) = X(x)T'(0) = X(x)\omega c b = 0,$$

and this is satisfied if we choose $b = 0$. This leaves $T_\omega(t)$ as a constant multiple of $\cos(\omega c t)$.

Thus far, for every $\omega \geq 0$, we have functions

$$y_\omega(x, t) = X_\omega(x)T_\omega(t) = [a_\omega \cos(\omega x) + b_\omega \cos(\omega x)] \cos(\omega c t)$$

which satisfy the wave equation and the initial condition $(\partial y/\partial t)(x, 0) = 0$ for all x. We need to satisfy the initial condition $y(x, 0) = f(x)$. For the similar problem on a bounded interval, we attempted a superposition $\sum_{n=1}^{\infty} y_n(x, t)$. Now the eigenvalues fill out the entire nonnegative real line, so the superposition has the form $\int_0^\infty y_\omega(x, t)d\omega$. Thus, attempt a solution

$$y(x, t) = \int_0^\infty [a_\omega \cos(\omega x) + b_\omega \sin(\omega x)] \cos(\omega c t) \, d\omega.$$

The initial condition requires that

$$y(x, 0) = \int_0^\infty [a_\omega \cos(\omega x) + b_\omega \sin(\omega x)] \, d\omega = f(x). \tag{15.8}$$

This is a Fourier integral expansion of $f(x)$ on the real line, so choose a_ω and b_ω as the Fourier integral coefficients of f:

$$a_\omega = \frac{1}{\pi} \int_{-\infty}^\infty f(\xi) \cos(\omega \xi) \, d\xi$$

and

$$b_\omega = \frac{1}{\pi} \int_{-\infty}^\infty f(\xi) \sin(\omega \xi) \, d\xi.$$

With this choice of coefficients, equation (15.8) is the solution.

EXAMPLE 15.9

We will solve the problem

$$\frac{\partial^2 y}{\partial t^2} = c^2 \frac{\partial^2 y}{\partial x^2} \text{ for } -\infty < x < \infty, t > 0$$

and

$$y(x, 0) = e^{-|x|}, \frac{\partial y}{\partial t}(x, 0) = 0 \text{ for } -\infty < x < \infty.$$

Compute the Fourier coefficients of the initial position function. First,

$$a_\omega = \frac{1}{\pi} \int_{-\infty}^\infty e^{-|\xi|} \cos(\omega \xi) \, d\xi = \frac{2}{\pi(1 + \omega^2)}.$$

Because $e^{-|\xi|} \sin(\omega \xi)$ is an odd function of ξ, $b_\omega = 0$. The solution is

$$y(x, t) = \frac{2}{\pi} \int_0^\infty \frac{1}{1 + \omega^2} \cos(\omega x) \cos(\omega c t) \, d\omega. \quad \blacklozenge$$

The solution from equation (15.8) is sometimes seen in a different form. If we insert the integrals for the coefficients into the solution we have

$$y(x, t) = \int_0^\infty [a_\omega \cos(\omega x) + b_\omega \sin(\omega x)] \cos(\omega c t) \, d\omega$$

$$= \frac{1}{\pi} \int_0^\infty \left[\left(\int_{-\infty}^\infty f(\xi) \cos(\omega \xi) \, d\xi \right) \cos(\omega x) \right.$$

$$\left. + \left(\int_{-\infty}^\infty f(\xi) \sin(\omega \xi) \, d\xi \right) \sin(\omega x) \right] \cos(\omega c t) \, d\omega$$

$$= \frac{1}{\pi} \int_{-\infty}^\infty \int_0^\infty [\cos(\omega \xi) \cos(\omega x) + \sin(\omega \xi) \sin(\omega x)] f(\xi) \cos(\omega c t) \, d\omega \, d\xi.$$

Upon applying a trigonometric identity to the term in square brackets, we have

$$y(x, t) = \frac{1}{\pi} \int_{-\infty}^\infty \int_0^\infty \cos(\omega(\xi - x)) f(\xi) \cos(\omega c t) \, d\omega \, d\xi. \tag{15.9}$$

We will use this form of the solution when we solve this problem using the Fourier transform.

Zero Initial Displacement We will solve

$$\frac{\partial^2 y}{\partial t^2} = c^2 \frac{\partial^2 y}{\partial x^2} \text{ for } -\infty < x < \infty, t > 0,$$

and

$$y(x, 0) = 0, \frac{\partial y}{\partial t}(x, 0) = g(x) \text{ for } -\infty < x < \infty.$$

Letting $y(x, t) = X(x)T(t)$, the analysis proceeds exactly as in the zero initial velocity case, except now we find that $T_\omega(t) = \sin(\omega ct)$ because $T(0) = 0$ instead of $T'(0) = 0$. For $\omega \geq 0$, we have functions

$$y_\omega(x, t) = [a_\omega \cos(\omega x) + b_\omega \sin(\omega x)] \sin(\omega ct),$$

which satisfy the wave equation and the condition $y(x, 0) = 0$. To satisfy $(\partial y / \partial t)(x, 0) = g(x)$, attempt a superposition

$$y(x, t) = \int_0^\infty [a_\omega \cos(\omega x) + b_\omega \sin(\omega x)] \sin(\omega ct) \, d\omega. \qquad (15.10)$$

Compute

$$\frac{\partial y}{\partial t}(x, t) = \int_0^\infty [a_\omega \cos(\omega x) + b_\omega \sin(\omega x)] \omega c \cos(\omega ct) d\omega.$$

We must choose the coefficients so that

$$\frac{\partial y}{\partial t}(x, 0) = \int_0^\infty \omega c [a_\omega \cos(\omega x) + b_\omega \sin(\omega x)] \, d\omega = g(x).$$

We can do this by choosing $\omega c a_\omega$ and $\omega c b_\omega$ as the Fourier coefficients in the integral expansion of g. Thus, let

$$a_\omega = \frac{1}{\pi c \omega} \int_{-\infty}^\infty g(\xi) \cos(\omega \xi) \, d\xi \text{ and } b_\omega = \frac{1}{\pi c \omega} \int_{-\infty}^\infty g(\xi) \sin(\omega \xi) \, d\xi.$$

With these choices, equation (15.10) is the solution of the initial-boundary value problem.

EXAMPLE 15.10

Suppose the initial displacement is zero and the initial velocity is given by

$$g(x) = \begin{cases} e^x & \text{for } 0 \leq x \leq 1 \\ 0 & \text{for } x < 0 \text{ and for } x > 1. \end{cases}$$

Compute the coefficients

$$a_\omega = \frac{1}{\pi c \omega} \int_{-\infty}^\infty g(\xi) \cos(\omega \xi) \, d\xi = \frac{1}{\pi c \omega} \int_0^1 e^\xi \cos(\omega \xi) \, d\xi$$

$$= \frac{1}{\pi c \omega} \frac{e \cos(\omega) + e\omega \sin(\omega) - 1}{1 + \omega^2}$$

and

$$b_\omega = \frac{1}{\pi c \omega} \int_{-\infty}^\infty g(\xi) \sin(\omega \xi) \, d\xi = \frac{1}{\pi c \omega} \int_0^1 e^\xi \sin(\omega \xi) \, d\xi$$

$$= -\frac{1}{\pi c \omega} \frac{e\omega \cos(\omega) - e \sin(\omega) - \omega}{1 + \omega^2}.$$

The solution is

$$y(x,t) = \int_0^\infty \left(\frac{1}{\pi c \omega} \frac{e\cos(\omega) + e\omega\sin(\omega) - 1}{1 + \omega^2} \right) \cos(\omega x)\sin(\omega c t)\, d\omega$$

$$- \int_0^\infty \left(\frac{1}{\pi c \omega} \frac{e\omega\cos(\omega) - e\sin(\omega) - \omega}{1 + \omega^2} \right) \sin(\omega x)\sin(\omega c t)\, d\omega. \quad \blacklozenge$$

As with motion on a bounded interval, the general initial-boundary value problem is solved by adding the solution with zero initial velocity to the solution with zero initial displacement.

15.3.1   Solution by Fourier Transform

We will illustrate the use of the Fourier transform to solve the initial-boundary value problem for the wave equation on the real line. Let the initial displacement be $f(x)$ and the initial velocity, $g(x)$.

Begin by taking the Fourier transform of the wave equation. This transform is taken with respect to x, since $-\infty < x < \infty$, which is the appropriate range of values for this transform. Throughout this transform process, t is carried along as a symbol. We have

$$\mathcal{F}\left[\frac{\partial^2 y}{\partial t^2}\right](\omega) = c^2 \mathcal{F}\left[\frac{\partial^2 y}{\partial x^2}\right](\omega). \tag{15.11}$$

On the right side of equation (15.11), we must transform a second derivative with respect to x, which is the variable of the function being transformed. Use the operational rule for the Fourier transform to write

$$\mathcal{F}\left[\frac{\partial^2 y}{\partial x^2}\right](\omega) = (i\omega)^2 \hat{y}(\omega, t) = -\omega^2 \hat{y}(\omega, t).$$

For the left side of equation (15.11), the derivative with respect to t passes through the transform in the x variable:

$$\mathcal{F}\left[\frac{\partial^2 y}{\partial t^2}\right](\omega) = \int_{-\infty}^\infty \frac{\partial^2 y}{\partial t^2}(x,t) e^{-i\omega x}\, dx$$

$$= \frac{\partial^2}{\partial t^2} \int_{-\infty}^\infty y(x,t) e^{-i\omega x}\, dx = \frac{\partial^2}{\partial t^2} \hat{f}(\omega, t).$$

The transformed wave equation is

$$\frac{\partial^2}{\partial t^2} \hat{f}(\omega, t) = -c^2 \omega^2 \hat{y}(x, t).$$

Write this differential equation as

$$\frac{\partial^2}{\partial t^2} \hat{f}(\omega, t) + c^2 \omega^2 \hat{y}(x, t) = 0.$$

Think of this as an ordinary differential equation for $\hat{y}(\omega, t)$ with t as the variable and ω carried along as a parameter. The general solution is

$$\hat{y}(\omega, t) = a_\omega \cos(\omega c t) + b_\omega \sin(\omega c t).$$

To solve for the coefficients to satisfy the initial conditions, transform the initial data. First,

$$\hat{y}(\omega, 0) = a_\omega = \mathcal{F}[y(x, 0)](\omega) = \mathcal{F}[f(x)](\omega) = \hat{f}(\omega),$$

which is the transform of the initial position function. Next,

$$\frac{\partial \hat{y}}{\partial t}(\omega, 0) = c\omega b_\omega = \mathcal{F}\left[\frac{\partial y}{\partial t}(x, 0)\right](\omega, 0)$$

$$= \mathcal{F}[g(x)](\omega) = \hat{g}(\omega),$$

which is the transform of the initial velocity function. Therefore,

$$b_\omega = \frac{1}{\omega c}\hat{g}(\omega).$$

Now

$$\hat{y}(\omega, t) = \hat{f}(\omega)\cos(\omega ct) + \frac{1}{\omega c}\hat{g}(\omega)\sin(\omega ct).$$

This is the Fourier transform of the solution. Invert this to find the solution

$$y(x, t) = \frac{1}{2\pi}\int_{-\infty}^{\infty}[\hat{f}(\omega)\cos(\omega ct) + \frac{1}{\omega c}\hat{g}(\omega)\sin(\omega ct)]e^{i\omega x}\, d\omega.$$

In the case that the string is released from rest, $g(x) = 0$, and this solution is

$$y(x, t) = \frac{1}{2\pi}\int_{-\infty}^{\infty}\hat{f}(\omega)\cos(\omega ct)e^{i\omega x}\, d\omega. \tag{15.12}$$

We claim that this solution (15.12) obtained by using the Fourier transform agrees with the solution (15.9) obtained by separation of variables and the Fourier integral. For the moment, denote the solution (15.12) by Fourier transform as $y_{tr}(x, t)$. Manipulate $y_{tr}(x, t)$ as follows:

$$y_{tr}(x, t) = \frac{1}{2\pi}\int_{-\infty}^{\infty}\hat{f}(\omega)\cos(\omega ct)e^{i\omega x}\, d\omega$$

$$= \frac{1}{2\pi}\int_{-\infty}^{\infty}\left(\int_{-\infty}^{\infty}f(\xi)e^{-i\omega\xi}\, d\xi\right)\cos(\omega ct)e^{i\omega x}\, d\omega$$

$$= \frac{1}{2\pi}\int_{-\infty}^{\infty}\int_{-\infty}^{\infty}e^{-i\omega(\xi-x)}\cos(\omega ct)f(\xi)\, d\omega\, d\xi$$

$$= \frac{1}{2\pi}\int_{-\infty}^{\infty}\int_{-\infty}^{\infty}[\cos(\omega(\xi - x)) - i\sin(\omega(\xi - x))]\cos(\omega ct)f(\xi)\, d\omega\, d\xi.$$

Since $y(x, t)$ must be real-valued, the solution is actually the real part of this integral. Therefore,

$$y_{tr}(x, t) = \frac{1}{2\pi}\int_{-\infty}^{\infty}\int_{-\infty}^{\infty}\cos(\omega(\xi - x))\cos(\omega ct)f(\xi)\, d\omega\, d\xi.$$

Finally, the integrand is an even function of ω, so

$$y_{tr}(x, t) = \frac{1}{\pi}\int_{-\infty}^{\infty}\int_{0}^{\infty}\cos(\omega(\xi - x))\cos(\omega ct)f(\xi)\, d\omega\, d\xi$$

and this is the solution (15.9).

EXAMPLE 15.11

We will use the Fourier transform to solve the wave equation on the real line subject to the initial conditions $g(x) = 0$ and

$$f(x) = \begin{cases} \cos(x) & \text{for } -\pi/2 \leq x \leq \pi/2 \\ 0 & \text{for } |x| > \pi/2. \end{cases}$$

With zero initial velocity, we can use the solution (15.12). We need the Fourier transform of the initial position function:

$$\hat{f}(\omega) = \int_{-\infty}^{\infty} f(\xi) e^{-i\omega\xi} \, d\xi$$

$$= \int_{-\pi/2}^{\pi/2} \cos(\xi) e^{-i\omega\xi} \, d\xi = \begin{cases} 2\cos(\pi\omega/2)/(1-\omega^2) & \text{for } \omega \neq 1 \\ \pi/2 & \text{for } \omega = 1. \end{cases}$$

$\hat{f}$ is continuous because

$$\lim_{\omega \to 1} \frac{2\cos(\pi\omega/2)}{1-\omega^2} = \frac{\pi}{2}.$$

The solution is

$$y(x,t) = \text{Re}\left(\frac{1}{\pi} \int_{-\infty}^{\infty} \frac{2\cos(\pi\omega/2)}{1-\omega^2} \cos(\omega ct) e^{i\omega x} \, d\omega \right).$$

The integral itself is complex because the Fourier transform is a complex quantity. The real part is the solution of the problem. In this case, we can extract the real part easily by writing

$$e^{i\omega x} = \cos(\omega x) + i \sin(\omega x)$$

obtaining

$$y(x,t) = \frac{1}{\pi} \int_{-\infty}^{\infty} \frac{2\cos(\pi\omega/2)}{1-\omega^2} \cos(\omega x) \cos(\omega ct) \, d\omega. \quad \blacklozenge$$

SECTION 15.3   PROBLEMS

In each of Problems 1 through 6, solve the wave equation on the real line for the given initial position f and initial velocity g, first by separation of variables and the Fourier integral, and then by using the Fourier transform. The same solution should result from both methods.

1. $c = 12$, $f(x) = e^{-5|x|}$, and $g(x) = 0$

2. $c = 8$, $g(x) = 0$, and $f(x) = \begin{cases} 8 - x & \text{for } 0 \leq x \leq 8 \\ 0 & \text{for } x < 0 \text{ and} \\ & \text{for } x > 8. \end{cases}$

3. $c = 4$, $f(x) = 0$, and $g(x) = \begin{cases} \sin(x) & \text{for } -\pi \leq x \leq \pi \\ 0 & \text{for } |x| > \pi \end{cases}$

4. $c = 1$, $g(x) = 0$, and $f(x) = \begin{cases} 2 - |x| & \text{for } -2 \leq x \leq 2 \\ 0 & \text{for } |x| > 2 \end{cases}$

5. $c = 3$, $f(x) = 0$, and $g(x) = \begin{cases} e^{-2x} & \text{for } x \geq 1 \\ 0 & \text{for } x < 1 \end{cases}$

6. $c = 2$, $f(x) = 0$, and $g(x) = \begin{cases} 1 & \text{for } 0 \leq x \leq 2 \\ -1 & \text{for } -2 \leq x < 0 \\ 0 & \text{for } x > 2 \text{ and} \\ & \text{for } x < -2 \end{cases}$

15.4   Wave Motion in a Semi-Infinite Medium

We will solve the wave equation on a half-line $0 \leq x < \infty$. The problem is

$$\frac{\partial^2 y}{\partial t^2} = c^2 \frac{\partial^2 y}{\partial x^2} \text{ for } 0 \leq x < \infty, t > 0,$$

$$y(0, t) = 0 \text{ for } t \geq 0$$

and

$$y(x, 0) = f(x), \frac{\partial y}{\partial t}(x, 0) = g(x) \text{ for } 0 \leq x < \infty.$$

We will seek a bounded solution, considering first the case that $g(x) = 0$. Separate variables by putting $y(x, t) = X(x)T(t)$ to obtain

$$X'' + \lambda X = 0 \text{ and } T'' + \lambda c^2 T = 0.$$

Unlike the problem on the entire line, we have a boundary condition at the fixed left end. This means that $y(0, t) = X(0)T(t) = 0$, so $X(0) = 0$ and the problem for X is

$$X'' + \lambda X = 0 \text{ and } X(0) = 0.$$

Upon considering cases as we have done before, we find the eigenvalues $\lambda \geq 0$ and eigenfunctions

$$X_\omega = b_\omega \sin(\omega x).$$

Because the motion is assumed to begin from rest,

$$\frac{\partial y}{\partial t}(x, 0) = X(x)T'(0) = 0,$$

so $T'(0) = 0$. The problem for T is

$$T'' + c^2 \omega^2 T = 0, T'(0) = 0$$

with solutions that are constant multiples of $\cos(\omega c t)$. For each $\omega \geq 0$, we now have a function

$$y_\omega(x, t) = b_\omega \sin(\omega x) \cos(\omega c t)$$

that satisfies the wave equation, the boundary condition, and the initial condition that the string starts from rest. To satisfy the condition $y(x, 0) = f(x)$, we generally need the superposition

$$y(x, t) = \int_0^\infty b_\omega \sin(\omega x) \cos(\omega c t) \, d\omega.$$

Then

$$y(x, 0) = \int_0^\infty b_\omega \sin(\omega x) \, d\omega = f(x),$$

so we must choose

$$b_\omega = \frac{2}{\pi} \int_0^\infty f(\xi) \sin(\omega \xi) \, d\xi,$$

which is the coefficient in the Fourier sine integral representation of f on $[0, \infty)$. This solves the problem when $g(x) = 0$. A similar analysis allows us to write an integral solution when $f(x) = 0$, and the string has an initial velocity $(\partial y / \partial t)(x, 0) = g(x)$.

EXAMPLE 15.12

We will solve the problem

$$\frac{\partial^2 y}{\partial t^2} = 16 \frac{\partial^2 y}{\partial x^2} \text{ for } 0 \leq x < \infty, t > 0,$$

$$y(0, t) = 0 \text{ for } t \geq 0$$

$$\frac{\partial y}{\partial t}(x, 0) = 0 \text{ for } 0 \leq x < \infty$$

and

$$y(x,0) = f(x) = \begin{cases} \sin(\pi x) & \text{for } 0 \le x \le 4 \\ 0 & \text{for } x > 4. \end{cases}$$

To write the solution we need only compute the coefficients

$$b_\omega = \frac{2}{\pi} \int_0^\infty f(\xi) \sin(\omega\xi)\, d\xi$$

$$= \frac{2}{\pi} \int_0^4 \sin(\pi\xi) \sin(\omega\xi)\, d\xi = 8 \sin(\omega) \cos(\omega) \frac{2\cos^2(\omega) - 1}{\omega^2 - \pi^2}.$$

The solution is

$$y(x,t) = \int_0^\infty 8 \sin(\omega) \cos(\omega) \frac{2\cos^2(\omega) - 1}{\omega^2 - \pi^2} \sin(\omega x) \cos(4\omega t)\, d\omega. \quad \blacklozenge$$

15.4.1   Solution by Fourier Sine or Cosine Transform

We will illustrate the use of a Fourier transform to solve the problem on a half-line. Having considered the case of zero initial velocity for the separation of variables solution, we will assume here an initial velocity $g(x)$ but zero initial position $f(x) = 0$. The problem is

$$\frac{\partial^2 y}{\partial t^2} = c^2 \frac{\partial^2 y}{\partial x^2} \text{ for } 0 \le x < \infty, t > 0,$$

$$y(0,t) = 0 \text{ for } t \ge 0$$

and

$$y(x,0) = 0, \frac{\partial y}{\partial t}(x,0) = g(x) \text{ for } 0 \le x < \infty.$$

On the half-line, we can try a Fourier sine or a Fourier cosine transform of $y(x,t)$, thinking of x as the variable and t as a parameter. The choice depends on the operational rules for these transforms. The cosine transform requires that we know something about the derivative (with respect to x) of y at $x = 0$, while the sine transform requires information about the function itself at $x = 0$. Since we are given that $y(0,t) = 0$, this is the kind of information we have, so choose the sine transform. Apply $\mathcal{F}_S$ to the wave equation and use the fact that the derivative with respect to t goes through the transform to get

$$\frac{\partial^2}{\partial t^2} \hat{y}_S = c^2 \mathcal{F}_S \left[\frac{\partial^2 y}{\partial x^2} \right] = -c^2 \omega^2 \hat{y}_S(\omega,t) + \omega c^2 y(0,t) = -c^2 \omega^2 \hat{y}_S(\omega,t).$$

This is a differential equation for $\hat{y}_S(\omega,t)$ with t as the variable:

$$\frac{\partial^2 y}{\partial t^2} \hat{y}_S + c^2 \omega^2 \hat{y}_S(\omega,t) = 0.$$

The general solution is

$$\hat{y}_S(\omega,t) = a_\omega \cos(\omega c t) + b_\omega \sin(\omega c t).$$

Since

$$a_\omega = \hat{y}_S(\omega,0) = \mathcal{F}_S[y(x,0)](\omega) = \mathcal{F}_S[0](\omega) = 0,$$

and

$$\frac{\partial \hat{y}_S}{\partial t}(\omega,0) = \omega c b_\omega = \hat{g}_S(\omega),$$

then

$$b_\omega = \frac{1}{\omega c}\hat{g}_s(\omega).$$

This gives us

$$\hat{y}_s(\omega, t) = \frac{1}{\omega c}\hat{g}(\omega)\sin(\omega c t),$$

which is the sine transform of the solution. We obtain the solution by inverting:

$$y(x, t) = \frac{2}{\pi}\int_0^\infty \frac{1}{\omega c}\hat{g}_s(\omega)\sin(\omega x)\sin(\omega c t)\,d\omega.$$

If we have a wave equation on the half-line with zero initial velocity and initial position given by f, then we can proceed as we have just done, but using the Fourier cosine transform instead of the sine transform. As usual, a problem with initial displacement and velocity can be solved as the sum of the solution with zero initial velocity and the solution with zero initial position.

SECTION 15.4   *PROBLEMS*

In each of Problems 1 through 5, solve the problem for wave equation on the half-line for the given c, initial position f, and initial velocity g, first by using a separation of variables, and then by using an appropriate Fourier transform.

1. $c = 3$, $g(x) = 0$, $f(x) = \begin{cases} x(1-x) & \text{for } 0 \le x \le 1 \\ 0 & \text{for } x > 1 \end{cases}$

2. $c = 3$, $f(x) = 0$, $g(x) = \begin{cases} 0 & \text{for } 0 \le x < 4 \\ 2 & \text{for } 4 < x \le 11 \\ 0 & \text{for } x > 11 \end{cases}$

3. $c = 2$, $f(x) = 0$, $g(x) = \begin{cases} \cos(x) & \text{for } \pi/2 \le x \le 5\pi/2 \\ 0 & \text{for } 0 \le x < \pi/2 \\ & \text{and for } x > 5\pi/2 \end{cases}$

4. $c = 6$, $f(x) = -2e^{-x}$, $g(x) = 0$

5. $c = 14$, $f(x) = 0$, $g(x) = \begin{cases} x^2(3-x) & \text{for } 0 \le x \le 3 \\ 0 & \text{for } x > 3 \end{cases}$

15.5   d'Alembert's Solution

In this section, we will derive d'Alembert's solution of a wave problem on the real line. It will be convenient to denote partial derivatives by subscripts, such as $\partial u/\partial t = u_t$ and $\partial u/\partial x = u_x$. The problem we will solve is

$$u_{tt} = c^2 u_{xx} \text{ for } -\infty < x < \infty, t > 0,$$

and

$$u(x, 0) = f(x), u_t(x, 0) = g(x) \text{ for } -\infty < x < \infty.$$

Here we are using $u(x, t)$ for the position function of the wave. A graph of the wave's profile at time t is the graph of $y = u(x, t)$ in the (x, y) plane for that value of t.

This initial-boundary value problem is called the *Cauchy problem for the wave equation*. The lines $x - ct = k_1$ and $x + ct = k_2$ in the (x, t) plane are called *characteristics* of the wave equation. These are straight lines of slope $1/c$ and $-1/c$ in the (x, t) plane. Exploiting these characteristics, make the change of variables

$$\xi = x - ct, \eta = x + ct.$$

The transformation is invertible, since we can solve for x and t to get

$$x = \frac{1}{2}(\xi + \eta), t = \frac{1}{2c}(-\xi + \eta).$$

Define

$$U(\xi, \eta) = u((\xi + \eta)/2, (-\xi + \eta)/2c).$$

We must compute chain rule derivatives:

$$u_x = U_\xi \xi_x + U_\eta \eta_x = U_\xi + U_\eta$$

and

$$u_{xx} = U_{\xi\xi}\xi_x + U_{\xi\eta}\eta_x + U_{\eta\xi}\xi_x + U_{\eta\eta}\eta_x$$
$$= U_{\xi\xi} + 2U_{\xi\eta} + U_{\eta\eta}.$$

Similarly,

$$u_{tt} = c^2 U_{\xi\xi} - 2c^2 U_{\xi\eta} + c^2 U_{\eta\eta}.$$

The wave equation transforms to

$$u_{tt} - c^2 u_{xx} = 0 = [c^2 U_{\xi\xi} - 2c^2 U_{\xi\eta} + c^2 U_{\eta\eta}] - c^2[U_{\xi\xi} + 2U_{\xi\eta} + U_{\eta\eta}]$$

or

$$U_{\xi\eta} = 0.$$

Now we see the rationale for this change of variables. The transformed equation $U_{\xi\eta} = 0$ is easy to solve. First, $(U_\eta)_\xi = 0$ means that U_η is independent of ξ, so for some function h,

$$U_\eta = h(\eta).$$

Then

$$U(\xi, \eta) = \int h(\eta)d\eta + F(\xi)$$

in which this integration with respect to η may have ξ in its "constant" of integration. Now $\int h(\eta)\,d\eta$ is just another function of η, so we conclude that $U(\xi, \eta)$ must be a sum of a function just of ξ and a function just of η:

$$U(\xi, \eta) = F(\xi) + G(\eta).$$

This function satisfies the transformed wave equation for any twice-differentiable functions F and G of one variable, and conversely, every solution of the transformed wave equation has this form. In terms of x and t, this means that every solution of the one-dimensional (unforced) wave equation has the form

$$u(x, t) = F(x - ct) + G(x + ct). \tag{15.13}$$

Thus far we have dealt with just the partial differential equation. The idea now is to choose F and G to obtain a solution satisfying the initial conditions $y(x, 0) = f(x)$ and $y_t(x, 0) = g(x)$. First we need

$$u(x, 0) = F(x) + G(x) = f(x) \tag{15.14}$$

and

$$u_t(x,0) = -cF'(x) + cG'(x) = g(x). \tag{15.15}$$

Integrate equation (15.15) and rearrange terms to get

$$-F(x) + G(x) = \frac{1}{c} \int_0^x g(w)\,dw - F(0) + G(0).$$

Add this to equation (15.14) to get

$$2G(x) = f(x) + \frac{1}{c} \int_0^x g(w)\,dw - F(0) + G(0).$$

Then

$$G(x) = \frac{1}{2} f(x) + \frac{1}{2c} \int_0^x g(w)\,dw - \frac{1}{2} F(0) + \frac{1}{2} G(0). \tag{15.16}$$

But then, from equation (15.14),

$$F(x) = f(x) - G(x) = \frac{1}{2} f(x) - \frac{1}{2c} \int_0^x g(w)\,dw + \frac{1}{2} F(0) - \frac{1}{2} G(0). \tag{15.17}$$

Finally, use equations (15.16) and (15.17) to obtain the solution

$$u(x,t) = F(x - ct) + G(x + ct)$$

$$= \frac{1}{2} f(x - ct) - \frac{1}{2c} \int_0^{x-ct} g(w)\,dw + \frac{1}{2} F(0) - \frac{1}{2} G(0)$$

$$+ \frac{1}{2} f(x + ct) + \frac{1}{2c} \int_0^{x+ct} g(w)\,dw - \frac{1}{2} F(0) + \frac{1}{2} G(0).$$

After combining the integrals, we have the solution

$$u(x,t) = \frac{1}{2} (f(x - ct) + f(x + ct)) + \frac{1}{2c} \int_{x-ct}^{x+ct} g(w)\,dw. \tag{15.18}$$

This is *d'Alembert's solution* of the Cauchy problem for the wave equation on the real line. It gives an explicit solution in terms of the initial position and velocity functions.

EXAMPLE 15.13

We will solve the initial-boundary value problem

$$u_{tt} = 4u_{xx} \text{ for } -\infty < x < \infty, t > 0,$$

and

$$u(x,0) = e^{-|x|}, u_t(x,0) = \cos(4x) \text{ for } -\infty < x < \infty.$$

Immediately,

$$u(x,t) = \frac{1}{2} \left(e^{-|x-2t|} + e^{-|x+2t|} \right) + \frac{1}{4} \int_{x-2t}^{x+2t} \cos(4w)\,dw$$

$$= \frac{1}{2} \left(e^{-|x-2t|} + e^{-|x+2t|} \right) + \frac{1}{16} (\sin(4(x + 2t)) - \sin(4(x - 2t)))$$

$$= \frac{1}{2} \left(e^{-|x-2t|} + e^{-|x+2t|} \right) + \frac{1}{8} \sin(4x) \cos(8t). \; \blacklozenge$$

15.5.1   Forward and Backward Waves

Write d'Alembert's solution as

$$u(x,t) = \frac{1}{2}\left(f(x-ct) - \frac{1}{c}\int_0^{x-ct} g(w)\,dw \right)$$

$$+ \frac{1}{2}\left(f(x+ct) + \frac{1}{c}\int_0^{x+ct} g(w)\,dw \right)$$

$$= \varphi(x-ct) + \beta(x+ct),$$

where

$$\varphi(x) = \frac{1}{2}f(x) - \frac{1}{2c}\int_0^x g(w)\,dw$$

and

$$\beta(x) = \frac{1}{2}f(x) + \frac{1}{2c}\int_0^x g(w)\,dw.$$

We call $\varphi(x-ct)$ a forward wave. Its graph is the graph of $\varphi(x)$ translated ct units to the right, and so it may be thought of as a wave moving to the right with speed c. We call $\beta(x+ct)$ a backward wave. Its graph is the graph of $\beta(x)$ translated ct units to the left, and it may be thought of as a wave moving to the left with speed c. This allows us to think of the wave profile $y = u(x,t)$ at any time t as a sum of a wave moving to the right and a wave moving to the left.

EXAMPLE 15.14

Suppose $g(x) = 0$, $c = 1$, and

$$f(x) = \begin{cases} 4 - x^2 & \text{for } -2 \leq x \leq 2 \\ 0 & \text{for } |x| > 2. \end{cases}$$

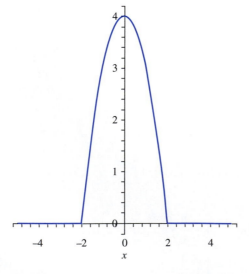

FIGURE 15.10 *Initial position in Example 15.14.*

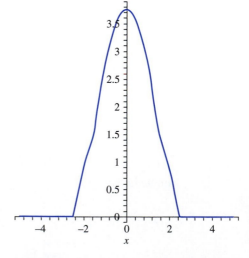

FIGURE 15.11 *Wave in Example 15.14 at time $t = 0.5$.*

The solution is

$$u(x,t) = \varphi(x+ct) + \beta(x+ct) = \frac{1}{2}(f(x-t) + f(x+t)).$$

At any time t, the wave profile consists of the initial position function translated t units to the right and superimposed on the initial position function translated t units to the left. Figures 15.10 through 15.16 show this profile at increasing times.

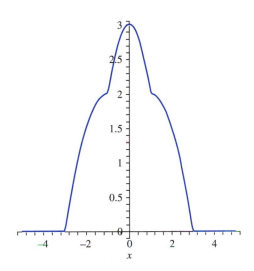

FIGURE 15.12 $t = 1$.

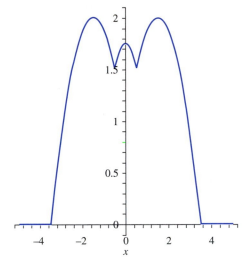

FIGURE 15.13 $t = 1.5$.

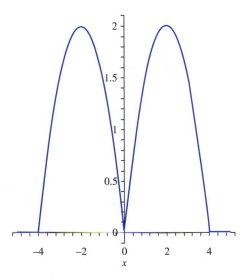

FIGURE 15.14 $t = 2$.

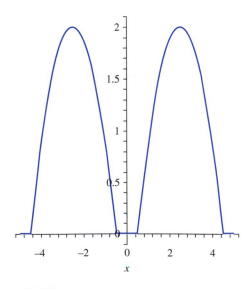

FIGURE 15.15 $t = 2.5$.

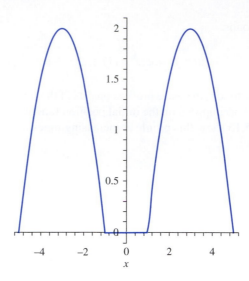

FIGURE 15.16 $t = 3$.

Notice that by time $t = 3$, the wave has split into two disjointed copies of the original ($t = 0$) wave, one continuing to move to the right, the other to the left). This separation is because $f(x)$ in this example is nonzero on only a bounded interval. ♦

15.5.2   Forced Wave Motion

Using the characteristics, we will write a solution for the Cauchy problem on the line with a forcing term:

$$u_{tt} = c^2 u_{xx} + F(x, t) \text{ for } -\infty < x < \infty, t > 0,$$

and

$$u(x, 0) = f(x), u_t(x, 0) = g(x) \text{ for } -\infty < x < \infty.$$

Suppose we want the solution at $P_0 : (x_0, t_0)$. There are two characteristics through P_0, namely the straight lines

$$x - ct = x_0 - ct_0 \text{ and } x + ct = x_0 + ct_0.$$

Use parts of these to form the *characteristic triangle* of Figure 15.17, with vertices $(x_0 - ct_0, 0)$, $(x_0 + ct_0, 0)$, and (x_0, t_0) and sides L, M, and I. Let Δ denote this solid triangle and compute the double integral of $-F$ over Δ.

$$-\iint_\Delta F(x, t) dA = \iint_\Delta (c^2 u_{xx} - u_{tt}) dA = \iint_\Delta \left(\frac{\partial}{\partial x} (c^2 u_x) - \frac{\partial}{\partial t} (u_t) \right) dA.$$

Apply Green's theorem to the last integral to obtain

$$-\iint_\Delta F(x, t) dA = \oint_C u_t \, dx + c^2 u_x \, dt,$$

where C is the boundary of Δ oriented counterclockwise. C consists of the line segments L, M, and I. Compute the line integral over each side.

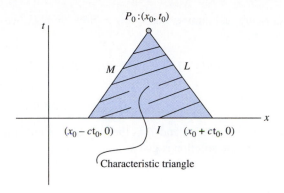

FIGURE 15.17 *The characteristic triangle.*

On I, $t = 0$ and x varies from $x_0 - ct_0$ to $x_0 + ct_0$, so

$$\int_I u_t\, dx + c^2 u_x\, dt = \int_{x_0-ct_0}^{x_0+ct_0} u_t(x, 0)\, dx = \int_{x_0-ct_0}^{x_0+ct_0} g(w)\, dw.$$

On L, $x + ct = x_0 + ct_0$, so $dx = -c\, dt$ and

$$\int_L u_t\, dx + c^2 u_x\, dt = \int_L u_t(-c)\, dt + c^2 u_x \left(-\frac{1}{c}\right) dx = -c \int_L du$$
$$= -c[u(x_0, t_0) - u(x_0 + ct_0, 0)].$$

Finally, on M, $x - ct = x_0 - ct_0$, so $dx = c\, dt$ and

$$\int_M u_t\, dx + c^2 u_x\, dt = \int_M u_t c\, dt + c^2 \left(\frac{1}{c}\right) dx = c \int_M du$$
$$= c[u(x_0 - ct_0, 0) - u(x_0, t_0)].$$

M has an initial point (x_0, t_0) and terminal point $(x_0 - ct_0, 0)$ because of the counterclockwise orientation on C. Upon summing these integrals, we obtain

$$-\iint_\Delta F(x, y)\, dA = \int_{x_0-ct_0}^{x_0+ct_0} g(w)\, dw$$
$$- c[u(x_0, t_0) - u(x_0 + ct_0, 0)] + c[u(x_0 - ct_0, 0) - u(x_0, t_0)].$$

Then

$$-\iint_\Delta F(x, y)\, dA = \int_{x_0-ct_0}^{x_0+ct_0} g(w)\, dw - 2cu(x_0, t_0) + c[f(x_0 + ct_0) + f(x_0 - ct_0)].$$

Solve this equation for $u(x_0, t_0)$ to get

$$u(x_0, t_0) = \frac{1}{2}(f(x_0 - ct_0) + f(x_0 + ct_0))$$
$$+ \frac{1}{2c} \int_{x_0-ct_0}^{x_0+ct_0} g(w)\, dw + \frac{1}{2c} \iint_\Delta F(x, t)\, dA.$$

Since x_0 is any real number and t_0 any positive number, we can drop the subscripts and write the solution $u(x, t)$ as

$$u(x, t) = \frac{1}{2}(f(x - ct) + f(x + ct)) + \frac{1}{2c}\int_{x-ct}^{x+ct} g(w)\,dw$$

$$+ \frac{1}{2c}\iint_\Delta F(X, T)\,dX\,dT$$

in which we have used X and T as the variables of integration to avoid confusion with the point (x, t) at which the solution is given.

EXAMPLE 15.15

We will solve the problem

$$u_{tt} = 25u_{xx} + x^2 t^2 \text{ for } -\infty < x < \infty, t > 0,$$

and

$$u(x, 0) = x\cos(x), u_t(x, 0) = e^{-x} \text{ for } -\infty < x < \infty.$$

The solution at any x and time t is

$$u(x, t) = \frac{1}{2}[(x - 5t)\cos(x - 5t) + (x + 5t)\cos(x + 5t)]$$

$$+ \frac{1}{10}\int_{x-5t}^{x+5t} e^w\,dw + \frac{1}{10}\iint_\Delta X^2 T^2\,dX\,dT.$$

Compute

$$\frac{1}{10}\int_{x-5t}^{x+5t} e^{-w}\,dw = \frac{1}{10}\left(e^{-x+5t} - e^{-x-5t}\right)$$

and, from Figure 15.18,

$$\frac{1}{10}\iint_\Delta X^2 T^2\,dX\,dT = \frac{1}{10}\int_{x-5t+5T}^{x+5t-5T} X^2 T^2\,dX\,dT = \frac{1}{12}x^2 t^4 + \frac{5}{36}t^6.$$

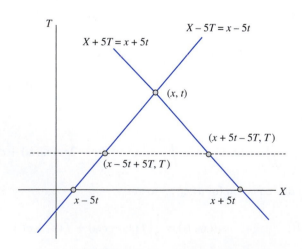

FIGURE 15.18 *Characteristic triangle in Example 15.14*

The solution is

$$u(x,t) = \frac{1}{2}[(x-5t)\cos(x-5t)+(x+5t)\cos(x+5t)]$$

$$+\frac{1}{10}\left(e^{-x+5t}-e^{-x-5t}\right)+\frac{1}{12}x^2t^4+\frac{5}{36}t^6. \quad \blacklozenge$$

SECTION 15.5   PROBLEMS

In each of Problems 1 through 6, write the d'Alembert solution for the problem

$$u_{tt} = c^2 u_{xx} \text{ for } -\infty < x < \infty, t > 0$$

$$u(x,0) = f(x), u_t(x,0) = g(x) \text{ for } -\infty < x < \infty.$$

1. $c=1$, $f(x)=x^2$, $g(x)=-x$

2. $c=4$, $f(x)=x^2-2x$, $g(x)=\cos(x)$

3. $c=7$, $f(x)=\cos(\pi x)$, $g(x)=1-x^2$

4. $c=5$, $f(x)=\sin(2x)$, $g(x)=x^3$

5. $c=14$, $f(x)=e^x$, $g(x)=x$

6. $c=12$, $f(x)=-5x+x^2$, $g(x)=3$

In each of Problems 7 through 12, solve the problem

$$u_{tt} = c^2 u_{xx} + F(x,t) \text{ for } -\infty < x < \infty, t > 0,$$

$$u(x,0) = f(x), u_t(x,0) = g(x) \text{ for } -\infty < x < \infty.$$

7. $c=4$, $f(x)=x$, $g(x)=e^{-x}$, $F(x,t)=x+t$

8. $c=2$, $f(x)=\sin(x)$, $g(x)=2x$, $F(x,t)=2xt$

9. $c=8$, $f(x)=x^2-x$, $g(x)=\cos(2x)$, $F(x,t)=xt^2$

10. $c=4$, $f(x)=x^2$, $g(x)=xe^{-x}$, $F(x,t)=x\sin(t)$

11. $c=3$, $f(x)=\cosh(x)$, $g(x)=1$, $F(x,t)=3xt^3$

12. $c=7$, $f(x)=1+x$, $g(x)=\sin(x)$, $F(x,t)=x-\cos(t)$

In each of Problems 13 through 16, write the solution of the problem

$$u_{tt} = c^2 u_{xx} \text{ for } -\infty < x < \infty, t > 0,$$

$$u(x,0) = f(x), u_t(x,0) = 0 \text{ for } -\infty < x < \infty$$

as a sum of a forward and a backward wave. Graph the initial position function and then graph the solution at selected times, showing the wave as a superposition of a wave moving to the right and a wave moving to the left.

13. $f(x)=\begin{cases}\sin(2x) & \text{for } -\pi \leq x \leq \pi \\ 0 & \text{for } |x| > \pi\end{cases}$

14. $f(x)=\begin{cases}1-|x| & \text{for } -1 \leq x \leq 1 \\ 0 & \text{for } |x| > 1\end{cases}$

15. $f(x)=\begin{cases}\cos(x) & \text{for } -\pi/2 \leq x \leq \pi/2 \\ 0 & \text{for } |x| > \pi/2\end{cases}$

16. $f(x)=\begin{cases}1-x^2 & \text{for } |x| \leq 1 \\ 0 & \text{for } |x| > 1\end{cases}$

15.6   Vibrations in a Circular Membrane

Imagine an elastic membrane of radius R fastened onto a circular frame (such as a drum). The membrane is set in motion from a given initial position and with a given initial velocity. In polar coordinates, the membrane occupies the disk $r \leq R$. Assume that the particle of membrane at (r, θ) vibrates vertical to the (x, y) plane, and let the displacement of this particle at time t be $z(r, \theta, t)$.

The wave equation in polar coordinates is

$$\frac{\partial^2 z}{\partial t^2} = c^2\left(\frac{\partial^2 z}{\partial r^2} + \frac{1}{r}\frac{\partial z}{\partial r} + \frac{1}{r^2}\frac{\partial^2 z}{\partial \theta^2}\right).$$

We will assume axial symmetry, which means that the motion is independent of θ. Then $z = z(r, t)$, and the wave equation is

$$\frac{\partial^2 z}{\partial t^2} = c^2\left(\frac{\partial^2 z}{\partial r^2} + \frac{1}{r}\frac{\partial z}{\partial r}\right).$$

The initial position is $z(r, 0) = f(r)$, and the initial velocity is $(\partial z/\partial t)(r, 0) = g(r)$.

Attempt a solution $z(r, t) = F(r)T(t)$. A routine calculation leads to

$$F'' + \frac{1}{r}F' + \frac{\lambda}{c^2}F = 0 \text{ and } T'' + \lambda T = 0.$$

If $\lambda = \omega^2 > 0$, this equation for F is a zero-order Bessel equation with solutions (bounded on the disk $r < R$) that are multiples of

$$J_0\left(\frac{\omega}{c}r\right).$$

The equation for T is

$$T'' + \omega^2 T = 0$$

with solutions of the form

$$T(t) = a\cos(\omega t) + b\sin(\omega t).$$

For each positive number ω, we now have a function

$$z_\omega(r, t) = a_\omega J_0\left(\frac{\omega}{c}r\right)\cos(\omega t) + b_\omega J_0\left(\frac{\omega}{c}r\right)\sin(\omega t).$$

that satisfies the wave equation.

Because the membrane is fixed on a circular frame,

$$z_\omega(R, t) = a_\omega J_0\left(\frac{\omega}{c}R\right)\cos(\omega t) + b_\omega J_0\left(\frac{\omega}{c}R\right)\sin(\omega t) = 0$$

for $t > 0$. This equation will be satisfied for all $t > 0$ if $J_0(\omega R/c) = 0$. Thus, choose ω so that $\omega R/c$ is a positive zero of $J_0(x)$. Let these zeros be $j_1 < j_2 < \ldots$. For each positive integer n, choose ω_n so that $\omega_n R/c = j_n$. This gives us the eigenvalues

$$\lambda_n = \omega_n^2 = \left(\frac{j_n c}{R}\right)^2$$

and the corresponding eigenfunctions

$$J_0\left(\frac{j_n}{R}r\right).$$

The functions

$$z_n(r, t) = a_n J_0\left(\frac{j_n}{R}r\right)\cos\left(\frac{j_n ct}{R}\right) + b_n J_0\left(\frac{j_n}{R}r\right)\sin\left(\frac{j_n ct}{R}\right)$$

satisfy the wave equation and the boundary condition that $z(R, 0) = 0$ for each positive integer n. To satisfy the initial condition $z(r, 0) = f(r)$, use a superposition

$$z(r, t) = \sum_{n=1}^{\infty}\left[a_n J_0\left(\frac{j_n}{R}r\right)\cos\left(\frac{j_n ct}{R}\right) + b_n J_0\left(\frac{j_n}{R}r\right)\sin\left(\frac{j_n ct}{R}\right)\right]. \tag{15.19}$$

The initial condition gives us

$$z(r, 0) = f(r) = \sum_{n=1}^{\infty} a_n J_0\left(\frac{j_n r}{R}\right).$$

This is like a Fourier-Bessel expansion, but we developed these in Chapter 14 for the interval $[0, 1]$. Therefore, let $s = r/R$ to convert this expansion to

$$f(Rs) = \sum_{n=1}^{\infty} a_n J_0(j_n s)$$

on $0 \leq s \leq 1$. From Sturm-Liouville theory, choose

$$a_n = \frac{2 \int_0^1 s f(Rs) J_0(j_n s)\, ds}{J_1^2(j_n)}$$

for $n = 1, 2, \ldots$.

Next solve for the b_n's. We need

$$\frac{\partial z}{\partial t}(r, 0) = g(r) = \sum_{n=1}^{\infty} b_n \frac{j_n c}{R} J_0\left(\frac{j_n r}{R}\right).$$

Again, we let $s = r/R$ to normalize the interval to $[0, 1]$ and obtain

$$b_n = \frac{2R}{j_n c} \frac{\int_0^1 s g(Rs) J_0(j_n s)\, ds}{J_1^2(j_n)}.$$

With these coefficients, equation (15.19) is the solution for $z(r, t)$.

15.6.1 Normal Modes of Vibration

The numbers $\omega_n = j_n c/R$ are the *frequencies of normal modes of vibration* of the membrane with periods $2\pi/\omega_n = 2\pi R/j_n c$. The *normal modes of vibration* are the functions $z_n(r, t)$, which are often written in phase angle form as

$$z_n(r, t) = A_n J_0\left(\frac{j_n r}{R}\right) \cos(\omega_n t + \delta_n)$$

in which $\omega_n = j_n c/R$, $A_n = \sqrt{a_n^2 + b_n^2}$ and $\delta_n = \arctan(-b_n/a_n)$ if $a_n \neq 0$.

The first normal mode is

$$z_1(r, t) = A_1 J_0\left(\frac{j_1 r}{R}\right) \cos(\omega_1 t + \delta_1).$$

As r varies from 0 to R, $j_1 r/R$ varies from 0 to j_1, which is the first positive zero of J_0. At any time t, a radial section through the membrane takes the shape of the graph of $J_0(x)$ for $0 \leq x \leq j_1$ (Figure 15.19).

The second normal mode is

$$z_2(r, t) = A_2 J_0\left(\frac{j_2 r}{R}\right) \cos(\omega_2 t + \delta_2).$$

As r varies from 0 to R, $j_2 r/R$ varies from 0 to j_2, passing through j_1 along the way. Since $J_0(j_2 r/R) = 0$ when $j_2 r/R = j_1$, this mode has a nodal circle (fixed in the motion) at radius

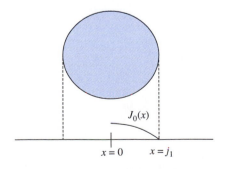

FIGURE 15.19 *First normal mode of vibration.*

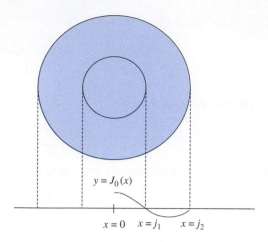

FIGURE 15.20   *Second normal mode.*

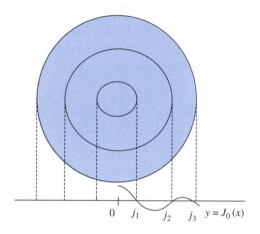

FIGURE 15.21   *Third normal mode.*

$r = j_1 R/j_2$. A section through the membrane takes the shape of the graph of $J_0(x)$ for $0 \leq x \leq j_2$ (Figure 15.20).

Similarly, the third normal mode is

$$z_3(r, t) = A_3 J_0 \left(\frac{j_3 r}{R} \right) \cos(\omega_3 t + \delta_3),$$

and this mode has two nodes: one at $r = j_1 R/j_3$ and the second at $r = j_2 R/j_3$. Now a radial section has the shape of a graph of $J_0(x)$ for $0 \leq x \leq j_3$ (Figure 15.21).

In general the nth normal mode has $n - 1$ nodes (fixed circles in the motion of the membrane), occurring at $j_1 R/j_n, j_2 R/j_n, \ldots, j_{n-1} R/j_n$.

SECTION 15.6   PROBLEMS

1. Let $c = R = 1$, $f(r) = 1 - r$, and $g(r) = 0$. Approximate the coefficients a_1 through a_5 in the solution for the motion of the membrane and graph the fifth partial sum for a selection of different times.

2. Repeat Problem 1 with $f(r) = 1 - r^2$.

3. Repeat Problem 1 with $f(r) = \sin(\pi r)$.

15.7   Vibrations in a Rectangular Membrane

Suppose an elastic membrane is attached to a rectangular frame that bounds the region $0 \leq x \leq L, 0 \leq y \leq K$. The membrane is given an initial displacement and released with a given initial velocity. We want the displacement function $z(x, y, t)$.

The initial-boundary value problem for z is

$$\frac{\partial^2 z}{\partial t^2} = c^2 \left(\frac{\partial^2 z}{\partial x^2} + \frac{\partial^2 z}{\partial y^2} \right) \text{ for } 0 < x < L, 0 < y < K, t > 0,$$

$$z(x, 0, t) = z(x, K, t) = 0 \text{ for } 0 < x < L, t > 0,$$

$$z(0, y, t) = z(L, y, t) = 0 \text{ for } 0 < y < K, t > 0,$$

$$z(x, y, 0) = f(x, y) \text{ for } 0 < x < L, 0 < y < K,$$

$$\frac{\partial z}{\partial t}(x, y, 0) = g(x, y) \text{ for } 0 < x < L, 0 < y < K.$$

We will solve this for the case $g(x, y) = 0$, so the membrane is displaced and released from rest. To attempt a separation of variables, substitute $z(x, y, t) = X(x)Y(y)T(t)$ into the wave equation to get

$$XYT'' = c^2(X''YT + XY''T)$$

or

$$\frac{T''}{c^2T} - \frac{Y''}{Y} = \frac{X''}{X}.$$

The left side depends only on y and t and the right only on x, and these variables are independent, so both sides must be constant:

$$\frac{T''}{c^2T} - \frac{Y''}{Y} = \frac{X''}{X} = -\lambda.$$

Then

$$X'' + \lambda X = 0 \text{ and } \frac{T''}{c^2T} + \lambda = \frac{Y''}{Y}.$$

The equation for T and Y has one side dependent only on t and the other side only on y. Therefore, for some constant μ,

$$\frac{T''}{c^2T} + \lambda = \frac{Y''}{Y} = -\mu.$$

Then

$$Y'' + \mu Y = 0 \text{ and } T'' + c^2(\lambda + \mu)T = 0.$$

Separation of variables has introduced two separation constants. From the boundary conditions,

$$X(0) = X(L) = Y(0) = Y(K) = 0.$$

We have solved these problems for X and Y before, obtaining eigenvalues and eigenfunctions

$$\lambda_n = \frac{n^2\pi^2}{L^2}, \ X_n(x) = \sin\left(\frac{n\pi x}{L}\right)$$

and

$$\mu_m = \frac{m^2\pi^2}{L^2}, \ Y_m(x) = \sin\left(\frac{m\pi y}{K}\right)$$

with n and m varying independently over the positive integers. The problem for T becomes

$$T'' + c^2 \left(\frac{n^2 \pi^2}{L^2} + \frac{m^2 \pi^2}{K^2} \right) T = 0.$$

With zero initial velocity, we have $T'(0) = 0$. Therefore, $T(t)$ must be a constant multiple of $\cos(\alpha_{nm} \pi c t)$, where

$$\alpha_{nm} = \sqrt{\frac{n^2}{L^2} + \frac{m^2}{K^2}}.$$

For each positive integer n and m, we now have functions

$$z_{nm}(x, y, t) = a_{nm} \sin\left(\frac{n\pi x}{L}\right) \sin\left(\frac{m\pi y}{K}\right) \cos(\alpha_{nm} \pi c t).$$

that satisfy the wave equation and the boundary conditions, as well as the condition of zero initial velocity. To satisfy $z(x, y, 0) = f(x, y)$, attempt a superposition, which is now a double sum:

$$z(x, y, t) = \sum_{n=1}^{\infty} \sum_{m=1}^{\infty} z_{nm}(x, y, t).$$

We must choose the coefficients so that

$$z(x, y, 0) = \sum_{n=1}^{\infty} \sum_{m=1}^{\infty} a_{nm} \sin\left(\frac{n\pi x}{L}\right) \sin\left(\frac{m\pi y}{K}\right) = f(x, y).$$

If we think of y as fixed for the moment, then $f(x, y) = h_y(x)$ is a function of x. Now

$$f(x, y) = h_y(x) = \sum_{n=1}^{\infty} \left[\sum_{m=1}^{\infty} a_{nm} \sin\left(\frac{m\pi y}{K}\right) \right] \sin\left(\frac{n\pi x}{L}\right)$$

is the Fourier sine expansion in x of $f(x, y)$ on $[0, L]$. Therefore, the coefficient of $\sin(n\pi x / L)$, which is the entire sum in square brackets, is the Fourier sine coefficient of this function. For a given n,

$$\sum_{m=1}^{\infty} a_{nm} \sin\left(\frac{m\pi y}{K}\right) = \frac{2}{L} \int_0^L h_y(\xi) \sin\left(\frac{n\pi \xi}{L}\right) d\xi$$

$$= \frac{2}{L} \int_0^L f(\xi, y) \sin\left(\frac{n\pi \xi}{L}\right) d\xi$$

The integral on the right is a function of y. On the left is a series that we can think of as its Fourier sine expansion on $[0, K]$ with the a_{nm}'s as coefficients. Therefore,

$$a_{nm} = \frac{2}{K} \int_0^K \left[\frac{2}{L} \int_0^L f(\xi, y) \sin\left(\frac{n\pi \xi}{L}\right) d\xi \right] \sin\left(\frac{m\pi \eta}{K}\right) d\eta$$

$$= \frac{4}{LK} \int_0^L \int_0^K f(\xi, \eta) \sin\left(\frac{n\pi \xi}{L}\right) \sin\left(\frac{m\pi \eta}{K}\right) d\eta \, d\xi.$$

With this choice of constants, we have the solution for the displacement function $z(x, y, t)$.

EXAMPLE 15.16

Suppose the initial position function is

$$z(x, y, 0) = x(L - x)y(K - y),$$

and the initial velocity is zero. Compute

$$a_{nm} = \frac{4}{LK} \int_0^L \int_0^K \xi(L-\xi)\eta(K-\eta)\sin\left(\frac{n\pi\xi}{L}\right)\sin\left(\frac{m\pi\eta}{K}\right) d\eta \, d\xi$$

$$= \frac{16L^2K^2}{(nm\pi^2)^3}[(-1)^n - 1][(-1)^m - 1].$$

The solution is

$$z(x, y, t) =$$

$$\sum_{n=1}^{\infty}\sum_{m=1}^{\infty} \frac{16L^2K^2}{(nm\pi^2)^3}[(-1)^n - 1][(-1)^m - 1]\sin\left(\frac{n\pi x}{L}\right)\sin\left(\frac{m\pi y}{K}\right)\cos(\alpha_{nm}\pi ct). \quad \blacklozenge$$

SECTION 15.7 *PROBLEMS*

In each of Problems 1, 2, and 3, solve the problem for the rectangular membrane with the given c, L, K, $f(x, y)$, and $g(x, y)$.

1. $c = 1, L = K = 2\pi, f(x, y) = x^2 \sin(y), g(x, y) = 0$

2. $c = 3, L = K = \pi, f(x, y) = 0, g(x, y) = xy$

3. $c = 2, L = K = 2\pi, f(x, y) = 0, g(x, y) = 1$

CHAPTER 16

The Heat Equation

16.1   Initial and Boundary Conditions

In Section 11.7, we used Gauss's divergence theorem to derive a partial differential equation modeling heat distribution or diffusion. In the absence of sources or sinks within the medium, the one-dimensional heat equation is

$$\frac{\partial u}{\partial t} = k \frac{\partial^2 u}{\partial x^2} \tag{16.1}$$

in which k depends on the medium and is often assumed constant.

Equation (16.1) can be solved subject to a variety of boundary and initial conditions. For example,

$$\frac{\partial u}{\partial t} = k \frac{\partial^2 u}{\partial x^2} \text{ for } 0 < x < L, t > 0,$$

$$u(0, t) = T_1, u(L, t) = T_2 \text{ for } t \geq 0,$$

and

$$u(x, 0) = f(x) \text{ for } 0 \leq x \leq L$$

models the temperature distribution in a thin, homogeneous bar of length L whose left end is kept at temperature T_1 and right end at temperature T_2, and having initial temperature $f(x)$ in the cross section at x.

The initial-boundary value problem

$$\frac{\partial u}{\partial t} = k \frac{\partial^2 u}{\partial x^2} \text{ for } 0 < x < L, t > 0,$$

$$\frac{\partial u}{\partial x}(0, t) = \frac{\partial u}{\partial x}(L, t) = 0 \text{ for } t \geq 0,$$

and

$$u(x, 0) = f(x) \text{ for } 0 \leq x \leq L$$

models the distribution in a bar of length L having no heat loss across its ends (*insulation conditions*) and an initial temperature function f.

Other kinds of boundary conditions can also be specified. If the left end is kept at constant temperature T and the right end is insulated, then

$$u(0, t) = T \text{ and } \frac{\partial u}{\partial x}(L, t) = 0 \text{ for } t > 0.$$

Free radiation or *convection* occurs when the bar loses energy by radiation from its ends into the surrounding medium, which is assumed to be maintained at constant temperature T. Now the boundary conditions have the form

$$\frac{\partial u}{\partial x}(0, t) = A[u(0, t) - T], \text{ and } \frac{\partial u}{\partial x}(L, t) = -A[u(L, t) - T] \text{ for } t \geq 0,$$

in which A is a positive constant. Notice that, if the bar is kept hotter than the surrounding medium, then the heat flow as measured by $\partial u / \partial x$ must be positive at one end and negative at the other.

Boundary conditions

$$u(0, t) = T_1, \frac{\partial u}{\partial x}(L, t) = -A[u(L, t) - T_2]$$

are used if the left end is kept at constant temperature T_1 while the right end radiates heat energy into a medium of constant temperature T_2.

As with the wave equation, we can also consider the heat equation on the line or half-line, subject to various conditions.

SECTION 16.1   *PROBLEMS*

1. Formulate an initial-boundary value problem modeling heat conduction in a thin, homogeneous bar of length L if the left end is kept at temperature zero and the right end is insulated. The initial temperature function is f.

2. Formulate an initial-boundary value problem modeling heat conduction in a thin, homogeneous bar of length L if the left end is kept at temperature $\alpha(t)$ and the

right end at temperature $\beta(t)$. The initial temperature function in the cross section at x is $f(x)$.

3. Formulate an initial-boundary value problem for the temperature distribution in a thin bar of length L if the left end is insulated and the right end is kept at temperature $\beta(t)$. The initial temperature function is f.

16.2   The Heat Equation on [0, L]

We will solve several initial-boundary value problems on an interval $[0, L]$.

16.2.1   Ends Kept at Temperature Zero

If the initial temperature in the cross section at x is $f(x)$ and the ends of the bar are kept at a temperature of zero, the problem for the temperature distribution function is

$$\frac{\partial u}{\partial t} = k \frac{\partial^2 u}{\partial x^2} \text{ for } 0 < x < L, t > 0,$$

$$u(0, t) = u(L, t) = 0 \text{ for } t \geq 0,$$

and

$$u(x,0)=f(x) \text{ for } 0\le x\le L.$$

Put $u(x,t)=X(x)T(t)$ into the heat equation to obtain

$$\frac{X''}{X}=\frac{T'}{kT}=-\lambda,$$

in which λ is the separation constant. Now,

$$u(0,t)=X(0)T(t)=0=u(L,t)=X(L)T(t)$$

for all $t\ge 0$, so $X(0)=X(L)=0$, and the problem for X is

$$X''+\lambda X=0; X(0)=X(L)=0$$

with eigenvalues $\lambda_n=n^2\pi^2/L^2$ and eigenfunctions $\sin(n\pi x/L)$.

It is in the time dependence that the heat and wave equations differ. The equation for T is

$$T'+\frac{n^2\pi^2 k}{L^2}T=0,$$

which is a first-order differential equation with solutions that are constant multiples of $e^{-n^2\pi^2 kt/L^2}$. For $n=1,2,\ldots$, the functions

$$u_n(x,t)=c_n\sin\left(\frac{n\pi x}{L}\right)e^{-n^2\pi^2 kt/L^2}$$

satisfy the heat equation and the boundary conditions $u(0,t)=u(L,t)=0$. To satisfy the initial condition, we must (depending on f) use a superposition

$$u(x,t)=\sum_{n=1}^{\infty}c_n\sin\left(\frac{n\pi x}{L}\right)e^{-n^2\pi^2 kt/L^2}$$

and choose the coefficients so that

$$u(x,0)=f(x)=\sum_{n=1}^{\infty}c_n\sin\left(\frac{n\pi x}{L}\right).$$

This series is the Fourier sine expansion of f on $[0,L]$, so choose

$$c_n=\frac{2}{L}\int_0^L f(\xi)\sin\left(\frac{n\pi\xi}{L}\right)d\xi.$$

With this choice, the solution is

$$u(x,t)=\frac{2}{L}\sum_{n=1}^{\infty}\left(\int_0^L f(\xi)\sin(n\pi\xi/L)\,d\xi\right)\sin(n\pi x/L)e^{-n^2\pi^2 kt/L^2}. \tag{16.2}$$

EXAMPLE 16.1

Suppose the ends are kept at zero temperature and the initial temperature is $f(x)=A$, which is constant. Compute

$$c_n=\frac{2}{L}\int_0^L A\sin(n\pi\xi/L)\,d\xi=\frac{2A}{n\pi}[1-(-1)^n].$$

The solution is

$$u(x,t)=\frac{2A}{\pi}\sum_{n=1}^{\infty}\frac{1-(-1)^n}{n}\sin\left(\frac{n\pi x}{L}\right)e^{-n^2\pi^2 kt/L^2}.$$

Since

$$1 - (-1)^n = \begin{cases} 2 & \text{if } n \text{ is odd} \\ 0 & \text{if } n \text{ is even,} \end{cases}$$

we can omit the even values of n in the summation to write the solution

$$u(x, t) = \frac{4A}{\pi} \sum_{n=1}^{\infty} \frac{1}{2n-1} \sin\left(\frac{(2n-1)\pi x}{L}\right) e^{-(2n-1)^2 \pi^2 kt/L^2}. \; \blacklozenge$$

16.2.2   Insulated Ends

Suppose the bar has insulated ends, hence there is no energy loss across the ends. The temperature distribution is modeled by the initial-boundary value problem

$$\frac{\partial u}{\partial t} = k \frac{\partial^2 u}{\partial x^2} \text{ for } 0 < x < L, t > 0,$$

$$\frac{\partial u}{\partial x}(0, t) = \frac{\partial u}{\partial x}(L, t) = 0 \text{ for } t \geq 0,$$

and

$$u(x, 0) = f(x) \text{ for } 0 \leq x \leq L.$$

As before, separation of variables yields

$$X'' + \lambda X = 0 \text{ and } T' + \lambda k T = 0.$$

The insulation conditions give us

$$\frac{\partial u}{\partial x}(0, t) = X'(0)T(t) = \frac{\partial u}{\partial x}(L, t) = X'(L)T(t) = 0$$

so $X'(0) = X'(L) = 0$, and the problem for X is

$$X'' + \lambda X = 0; \; X'(0) = X'(L) = 0.$$

Previously, we solved this problem for the eigenvalues $\lambda_n = n^2\pi^2/L^2$ and eigenfunctions $\cos(n\pi x/L)$ for $n = 0, 1, 2, \ldots$.

The equation for T is

$$T' + \frac{n^2\pi^2 k}{L^2} T = 0.$$

For $n = 0$, we get $T_0(t) = $ constant. For $n = 1, 2, \ldots$, we get

$$T_n(t) = e^{-n^2\pi^2 kt/L^2}$$

or constant multiples of this function. We now have a function

$$u_n(x, t) = c_n \cos\left(\frac{n\pi x}{L}\right) e^{-n^2\pi^2 kt/L^2},$$

which satisfies the heat equation and the insulation boundary conditions for $n = 0, 1, 2, \ldots$. To satisfy the initial condition, use the superposition

$$u(x, t) = \frac{1}{2}c_0 + \sum_{n=1}^{\infty} c_n \cos\left(\frac{n\pi x}{L}\right) e^{-n^2\pi^2 kt/L^2}.$$

Then

$$u(x, 0) = f(x) = \frac{1}{2}c_0 + \sum_{n=1}^{\infty} c_n \cos\left(\frac{n\pi x}{L}\right),$$

so choose the c_n's to be the Fourier cosine coefficients of f on $[0, L]$:

$$c_n = \frac{2}{L} \int_0^L f(\xi) \cos\left(\frac{n\pi\xi}{L}\right) d\xi.$$

EXAMPLE 16.2

Suppose the ends of the bar are insulated and the left half of the bar is initially at constant temperature A, while the right half is initially at temperature zero. Then

$$f(x) = \begin{cases} A & \text{for } 0 \le x \le L/2 \\ 0 & \text{for } L/2 < x \le L. \end{cases}$$

Compute

$$c_0 = \frac{2}{L} \int_0^{L/2} A \, d\xi = A$$

and, for $n = 1, 2, \ldots$,

$$c_n = \frac{2}{L} \int_0^{L/2} A \cos\left(\frac{n\pi\xi}{L}\right) d\xi = \frac{2A}{n\pi} \sin(n\pi/2).$$

The solution is

$$u(x, t) = \frac{A}{2} + \frac{2A}{\pi} \sum_{n=1}^{\infty} \frac{1}{n} \sin\left(\frac{n\pi}{2}\right) \cos\left(\frac{n\pi x}{L}\right) e^{-n^2\pi^2 kt/L^2}.$$

Since $\sin(n\pi/2) = 0$ if n is even, we can retain only odd n in this summation to write this solution as

$$u(x, t) = \frac{A}{2} + \frac{2A}{\pi} \sum_{n=1}^{\infty} \frac{1}{2n-1} \cos\left(\frac{(2n-1)\pi x}{L}\right) e^{-(2n-1)^2\pi^2 kt/L^2}. \; \blacklozenge$$

16.2.3 Radiating End

Suppose the left end of the bar is maintained at temperature zero, while the right end radiates energy into the surrounding medium, which is kept at temperature zero. If the initial temperature function is f, then the temperature distribution is modeled by the initial-boundary value problem

$$\frac{\partial u}{\partial t} = k \frac{\partial^2 u}{\partial x^2} \text{ for } 0 < x < L, t > 0,$$

$$u(0, t) = 0, \frac{\partial u}{\partial x}(L, t) = -Au(L, t) \text{ for } t > 0,$$

and

$$u(x, 0) = f(x) \text{ for } 0 \le x \le L.$$

A is a positive constant called the *transfer coefficient*. Let $u(x, t) = X(x)T(t)$ to obtain

$$X'' + \lambda X = 0, T' + \lambda kT = 0.$$

Because $u(0, t) = X(0)T(t) = 0$, then $X(0) = 0$. From the radiation condition at the right end,

$$X'(L)T(t) = -AX(L)T(t) \text{ for } t > 0,$$

so

$$X'(L) + AX(L) = 0.$$

The problem for X is

$$X'' + \lambda X = 0, \ X(0) = 0, \ \text{and} \ X'(L) + AX(L) = 0.$$

This is a regular Sturm-Liouville problem. To solve for the eigenvalues and eigenfunctions, consider cases on λ.

Case 1 $\lambda = 0$. We get $X(x) = cx + d$, so $X(0) = d = 0$. Then $X(x) = cx$, so

$$X'(L) + AX(L) = c + cL = 0,$$

and this forces $c = 0$. This case yields only the trivial solution, so 0 is not an eigenvalue of this problem.

Case 2 $\lambda < 0$. Write $\lambda = -\alpha^2$ with $\alpha > 0$. Then $X'' - \alpha^2 X = 0$ with solutions

$$X(x) = ce^{\alpha x} + de^{-\alpha x}.$$

Now $X(0) = c + d = 0$ implies that $d = -c$, so

$$X(x) = c(e^{\alpha x} - e^{-\alpha x}).$$

From the radiation condition at L,

$$X'(L) = \alpha c(e^{\alpha L} + e^{-\alpha L}) = -AX(L) = -Ac(e^{\alpha L} - e^{-\alpha L}).$$

Now

$$\alpha c(e^{\alpha L} + e^{-\alpha L}) > 0 \ \text{and} \ -Ace^{\alpha L} - e^{-\alpha L} < 0.$$

We conclude that $c = 0$, so this case also has only the trivial solution. This problem has no negative eigenvalue.

Case 3 $\lambda > 0$. Set $\lambda = \alpha^2$ with $\alpha > 0$. Now $X'' + \alpha^2 X = 0$, so

$$X(x) = c\cos(\alpha x) + d\sin(\alpha x).$$

Then $X(0) = c = 0$, so $X(x) = d\sin(\alpha x)$. Next,

$$X'(L) = \alpha d\cos(\alpha L) = -AX(L) = -d\sin(\alpha L).$$

If $d = 0$, we have only the trivial solution. If $d \neq 0$, then $\alpha \cos(\alpha L) = -\sin(\alpha L)$, so

$$\tan(\alpha L) = -\frac{\alpha}{A}. \tag{16.3}$$

We have a nontrivial solution for X only if α is chosen to satisfy this transcendental equation, which we cannot solve algebraically for α. However, let $z = \alpha L$. Then equation (16.3) is $\tan(z) = -z/AL$. Part of the graphs of $y = \tan(z)$ and $y = -z/AL$ are shown in Figure 16.1, and they have infinitely many points of intersection for $z > 0$. The z-coordinates of these points are solutions of $\tan(z) = -z/AL$. Let these z-coordinates be $z_1, z_2, \ldots$ in increasing order. Since $\alpha = z/L$, the eigenvalues of this problem for X are

$$\lambda_n = \alpha_n^2 = \frac{z_n^2}{L^2}.$$

The eigenfunctions are functions $X_n(x) = \sin(\alpha_n x) = \sin(z_n x/L)$.

With these eigenvalues, the problem for T is

$$T' + \frac{z_n^2 k}{L^2} T = 0$$

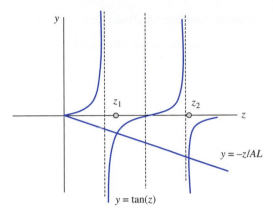

FIGURE 16.1 *Eigenvalues as the problem with radiating end.*

with solutions that are constant multiples of

$$T_n(t) = e^{-z_n^2 kt/L^2}.$$

For each positive integer n, we now have a function

$$u_n(x, t) = X_n(x)T_n(t) = \sin\left(\frac{z_n x}{L}\right)e^{-z_n^2 kt/L^2}$$

that satisfies the heat equation and the boundary conditions. To satisfy the initial condition, write the superposition

$$u(x, t) = \sum_{n=1}^{\infty} c_n \sin\left(\frac{z_n x}{L}\right)e^{-z_n^2 kt/L^2}.$$

We must choose the c_n's to satisfy

$$u(x, 0) = f(x) = \sum_{n=1}^{\infty} c_n \sin\left(\frac{z_n x}{L}\right).$$

This is not a Fourier sine series, but it is an eigenfunction expansion in the eigenfunctions of a Sturm-Liouville problem. The coefficients are

$$c_n = \frac{\int_0^L f(\xi) \sin(z_n \xi/L)\, d\xi}{\int_0^L \sin^2(z_n \xi/L)\, d\xi}.$$

The difficulty in computing these numbers is one we frequently encounter in such problems. We do not have an explicit formula for the z_n's. However, we can compute approximate values of the z_n's for a given L. For the first four values with $L = 1$, we obtain the approximate values

$$z_1 \approx 2.0288, \quad z_2 \approx 4.9132, \quad z_3 \approx 7.9787, \quad \text{and } z_4 \approx 11.0855.$$

Using these numbers, we can perform numerical integrations to approximate

$$c_1 \approx 1.9207, \quad c_2 \approx 2.6593, \quad c_3 \approx 4.1457, \quad \text{and } c_4 \approx 5.6329.$$

With just the first four terms, we have an approximation of

$$u(x, t) \approx 1.9027 \sin(2.0288x)e^{-4.1160kt} + 2.6593 \sin(4.9132x)e^{-24.1395kt}$$

$$+ 4.1457 \sin(7.9787x)e^{-63.6597kt} + 5.6329 \sin(11.0855x)e^{-1,228,883kt}.$$

Depending on k, these exponentials may be decaying so fast that these first four terms would be a good enough approximation for our purpose.

16.2.4 Transformation of Problems

As with the wave equation, it may be impossible to separate variables in a diffusion problem, depending on the partial differential equation and/or the boundary conditions. Sometimes we can transform such a problem into one to which a separation of variables applies.

EXAMPLE 16.3

A thin, homogeneous bar extends from $x = 0$ to $x = L$. The left end is maintained at constant temperature T_1 and the right end at constant temperature T_2. The initial temperature in the cross section at x is $f(x)$.

The initial-boundary value problem modeling this setting is

$$\frac{\partial u}{\partial t} = k\frac{\partial^2 u}{\partial x^2} \text{ for } 0 < x < L, t > 0,$$

$$u(0,t) = T_1, u(L,t) = T_2 \text{ for } t > 0,$$

and

$$u(x,0) = f(x) \text{ for } 0 \le x \le L.$$

If we set $u(x,t) = X(x)T(t)$, we must have $X(0)T(t) = T_1$. If $T_1 = 0$, this is satisfied by requiring that $X(0) = 0$. If $T_1 \ne 0$, then $X(0) \ne 0$, and we must have $T(t) = T_1/X(0) = $ constant, which is not acceptable.

Perturb the temperature distribution function by setting

$$u(x,t) = U(x,t) + \psi(x).$$

The idea is to choose ψ to obtain a problem for U that we know how to solve. Substitute U into the wave equation to get

$$\frac{\partial U}{\partial t} = k\left(\frac{\partial^2 U}{\partial x^2} + \psi''(x)\right).$$

This is the standard heat equation (16.1) if $\psi''(x) = 0$ and, hence, if $\psi(x) = cx + d$. Now

$$u(0,t) = T_1 = U(0,t) + \psi(0),$$

and this is the more friendly condition $U(0,t) = 0$ if $\psi(0) = T_1$. Thus, choose $d = T_1$ to have $\psi(x) = cx + T_1$. Next we need

$$u(L,t) = T_2 = U(L,t) + \psi(L),$$

and this is just $U(L,t) = 0$ if $\psi(L) = T_2$. We need $cL + T_1 = T_2$, so $c = (T_2 - T_1)/L$ and

$$\psi(x) = \frac{1}{L}(T_2 - T_1)x + T_1.$$

Finally,

$$u(x,0) = f(x) = U(x,0) + \psi(x),$$

so

$$U(x,0) = f(x) - \psi(x) = f(x) - \frac{1}{L}(T_2 - T_1)x - T_1.$$

The problem for U is:

$$\frac{\partial U}{\partial t} = k\frac{\partial^2 U}{\partial x^2} \text{ for } 0 < x < L, t > 0,$$

$$U(0,t) = U(L,t) = 0 \text{ for } t > 0,$$

and

$$U(x,0) = f(x) - \frac{1}{L}(T_2 - T_1)x - T_1 \text{ for } 0 \le x \le L.$$

We know the solution of this problem for U:

$$U(x,t) = \frac{2}{L}\sum_{n=1}^{\infty} c_n \sin\left(\frac{n\pi x}{L}\right) e^{-n^2\pi^2 kt/L^2},$$

where

$$c_n = \frac{2}{L}\int_0^L \left[f(\xi) - \frac{1}{L}(T_2 - T_1)\xi - T_1 \right] \sin\left(\frac{n\pi\xi}{L}\right) d\xi.$$

The solution of the original problem is

$$u(x,t) = U(x,t) + \frac{1}{L}(T_2 - T_1)x + T_1. \quad \blacklozenge$$

EXAMPLE 16.3

Suppose $T_1 = 1$, $T_2 = 2$, and $f(x) = 3/2$. Now

$$\psi(x) = \frac{1}{L}x + 1$$

and

$$c_n = \frac{2}{L}\int_0^L \left[\frac{3}{2} - \frac{1}{L}x - 1 \right] \sin\left(\frac{n\pi\xi}{L}\right) d\xi$$

$$= \frac{2}{L}\int_0^L \left(\frac{1}{2} - \frac{1}{L}x \right) \sin\left(\frac{n\pi\xi}{L}\right) d\xi$$

$$= \frac{1 + (-1)^n}{n\pi}.$$

The solution is

$$u(x,t) = \sum_{n=1}^{\infty} \left(\frac{1 + (-1)^n}{n\pi} \right) \sin\left(\frac{n\pi x}{L}\right) e^{-n^2\pi^2 kt/L^2} + \frac{1}{L}x + 1. \quad \blacklozenge$$

Sometimes an initial-boundary value problem involving the heat equation can be transformed into a simpler problem by multiplying by an exponential function $e^{\alpha x + \beta t}$ and making appropriate choices for α and β. We will pursue this idea in the problems.

16.2.5   The Heat Equation with a Source Term

We will determine the solution of the initial-boundary value problem

$$\frac{\partial u}{\partial t} = k\frac{\partial^2 u}{\partial x^2} + F(x,t) \text{ for } 0 < x < L, t > 0,$$

$$u(0,t) = u(L,t) = 0 \text{ for } t > 0,$$

and

$$u(x,0) = f(x) \text{ for } 0 \le x \le L.$$

The term $F(x, t)$ could, for example, account for a source or loss of energy within the medium.

It is routine to try separation of variables and find that it does not apply to this heat equation. A different approach is needed.

As a starting point, we know that, in the case that $F(x, t) = 0$ for all x and t, the solution has the form

$$u(x, t) = \sum_{n=1}^{\infty} c_n \sin\left(\frac{n\pi x}{L}\right) e^{-n^2\pi^2 kt/L^2}.$$

Taking a cue from this, we will attempt a solution of the current problem of the form

$$u(x, t) = \sum_{n=1}^{\infty} T_n(t) \sin\left(\frac{n\pi x}{L}\right). \tag{16.4}$$

We must find the functions $T_n(t)$. If t is fixed, then the left side of equation (16.4) is a function of x, and the right side is its Fourier sine expansion on $[0, L]$. The coefficients must be

$$T_n(t) = \frac{2}{L} \int_0^L u(\xi, t) \sin\left(\frac{n\pi \xi}{L}\right) d\xi. \tag{16.5}$$

Assume that for any $t \geq 0$, $F(x, t)$, thought of as a function of x, also can be expanded in a Fourier sine series on $[0, L]$:

$$F(x, t) = \sum_{n=1}^{\infty} B_n(t) \sin\left(\frac{n\pi \xi}{L}\right), \tag{16.6}$$

where

$$B_n(t) = \frac{2}{L} \int_0^L F(\xi, t) \sin\left(\frac{n\pi \xi}{L}\right) d\xi. \tag{16.7}$$

Differentiate equation (16.5) with respect to t to get

$$T_n'(t) = \frac{2}{L} \int_0^L \frac{\partial u}{\partial t}(\xi, t) \sin\left(\frac{n\pi \xi}{L}\right) d\xi. \tag{16.8}$$

Substitute for $\partial u / \partial t$ from the heat equation to get

$$T_n'(t) = \frac{2k}{L} \int_0^L \frac{\partial^2 u}{\partial x^2}(\xi, t) \sin\left(\frac{n\pi \xi}{L}\right) d\xi + \frac{2}{L} \int_0^L F(\xi, t) \sin\left(\frac{n\pi \xi}{L}\right) d\xi.$$

In view of equation (16.6), this equation becomes

$$T_n'(t) = \frac{2k}{L} \int_0^L \frac{\partial^2 u}{\partial x^2}(\xi, t) \sin\left(\frac{n\pi \xi}{L}\right) d\xi + B_n(t). \tag{16.9}$$

Integrate by parts twice, and at the last step make use of the boundary conditions and equation (16.5). We get

$$\int_0^L \frac{\partial^2 u}{\partial x^2}(\xi, t) \sin\left(\frac{n\pi \xi}{L}\right) d\xi$$

$$= \left[\frac{\partial u}{\partial x}(\xi, t) \sin\left(\frac{n\pi \xi}{L}\right)\right]_0^L - \int_0^L \frac{n\pi}{L} \frac{\partial u}{\partial x}(\xi, t) \cos\left(\frac{n\pi \xi}{L}\right) d\xi$$

$$= -\frac{n\pi}{L}\left[u(\xi, t) \cos\left(\frac{n\pi \xi}{L}\right)\right]_0^L + \frac{n\pi}{L} \int_0^L -\frac{n\pi}{L} u(\xi, t) \sin\left(\frac{n\pi \xi}{L}\right) d\xi$$

$$= -\frac{n^2\pi^2}{L^2} \int_0^L u(\xi, t) \sin\left(\frac{n\pi\xi}{L}\right) d\xi$$

$$= -\frac{n^2\pi^2}{L^2}\frac{L}{2}T_n(t) = -\frac{n^2\pi^2}{2L}T_n(t).$$

Substitute this into equation (16.9) to get

$$T_n' = -\frac{n^2\pi^2 k}{L^2}T_n(t) + B_n(t).$$

For $n = 1, 2, \ldots$, this is a first-order differential equation for $T_n(t)$, which we write as

$$T_n'(t) + \frac{n^2\pi^2 k}{L^2}T_n(t) = B_n(t).$$

Next use equation (16.5) to obtain the condition

$$T_n(0) = \frac{2}{L}\int_0^L u(\xi, 0)\sin\left(\frac{n\pi\xi}{L}\right) d\xi = \frac{2}{L}\int_0^L f(\xi)\sin\left(\frac{n\pi\xi}{L}\right) d\xi = b_n,$$

the nth coefficient in the Fourier sine expansion of f on $[0, L]$. Solve the differential equation for $T_n(t)$ subject to this boundary condition to get

$$T_n(t) = \int_0^t e^{-n^2\pi^2 k(t-\tau)/L^2} B_n(\tau)\, d\tau + b_n e^{-n^2\pi^2 kt/L^2}.$$

Finally, substitute this into equation (16.4) to obtain the solution

$$u(x, t) = \sum_{n=1}^{\infty}\left(\int_0^t e^{-n^2\pi^2 k(t-\tau)/L^2} B_n(\tau)\, d\tau\right)\sin\left(\frac{n\pi x}{L}\right)$$

$$+ \frac{2}{L}\sum_{n=1}^{\infty}\left(\int_0^L f(\xi)\sin\left(\frac{n\pi\xi}{L}\right) d\xi\right)\sin\left(\frac{n\pi x}{L}\right)e^{-n^2\pi^2 kt/L^2}. \qquad (16.10)$$

Notice that the last term is the solution of the problem if $F(x, t) = 0$, while the first term is the effect of this source term on the solution.

EXAMPLE 16.4

We will solve the problem

$$\frac{\partial u}{\partial t} = 4\frac{\partial^2 u}{\partial x^2} + xt \text{ for } 0 < x < \pi, t > 0,$$

$$u(0, t) = u(\pi, t) = 0 \text{ for } t \geq 0,$$

and

$$u(x, 0) = f(x) = \begin{cases} 20 & \text{for } 0 \leq x \leq \pi/4 \\ 0 & \text{for } \pi/4 < x \leq \pi. \end{cases}$$

Since we have a formula (16.10) for the solution, we need only carry out the integrations. First compute

$$B_n(t) = \frac{2}{\pi} \int_0^\pi \xi t \sin(n\xi)\, d\xi = \frac{2(-1)^{n+1}}{n} t.$$

This enables us to evaluate

$$\int_0^t e^{-4n^2(t-\tau)} B_n(\tau)\, d\tau = \int_0^t \frac{2(-1)^{n+1}}{n} \tau e^{-4n^2(t-\tau)}\, d\tau$$

$$= \frac{1}{8}(-1)^{n+1} \frac{-1 + 4n^2 t + e^{-4n^2 t}}{n^5}.$$

Finally, we need

$$b_n = \frac{2}{\pi} \int_0^\pi f(\xi) \sin(n\xi)\, d\xi = \frac{40}{\pi} \int_0^{\pi/4} \sin(n\xi)\, d\xi = \frac{40}{\pi} \frac{1 - \cos(n\pi/4)}{\pi}.$$

With all the components in place, the solution is

$$u(x, t) = \sum_{n=1}^\infty \left(\frac{1}{8}(-1)^{n+1} \frac{-1 + 4n^2 t + e^{-4n^2 t}}{n^5} \right) \sin(nx)$$

$$+ \sum_{n=1}^\infty \frac{40}{\pi} \frac{1 - \cos(n\pi/4)}{n} \sin(nx) e^{-4n^2 t}.$$

The second term on the right is the solution of the problem with the xt term omitted. Denote this "no-source" solution as

$$u_0(x, t) = \sum_{n=1}^\infty \frac{40}{\pi} \frac{1 - \cos(n\pi/4)}{n} \sin(nx) e^{-4n^2 t}.$$

Then the solution with the source term is

$$u(x, t) = u_0(x, t) + \sum_{n=1}^\infty \left(\frac{1}{8}(-1)^{n+1} \frac{-1 + 4n^2 t + e^{-4n^2 t}}{n^5} \right) \sin(nx).$$

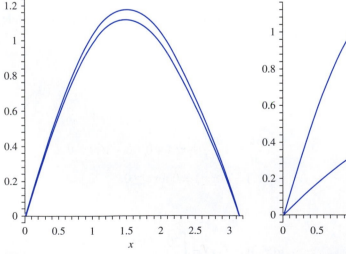

FIGURE 16.2   *u(x, t) and $u_0(x, t)$ at $t = 0.3$.*

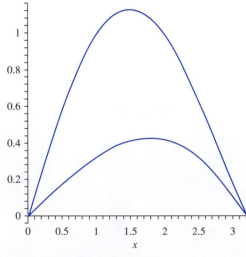

FIGURE 16.3 $t = 0.8$.

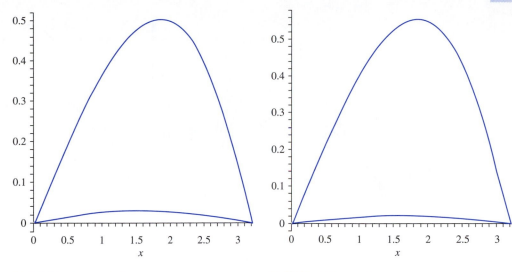

FIGURE 16.4 $t = 1.2$.   **FIGURE 16.5** $t = 1.3$.

Writing the solution in this way enables us to gauge the effect of the xt term on the solution. Figures 16.2 through 16.5 compare graphs of $u(x, t)$ and $u_0(x, t)$ at times $t = 0.3, 0.8, 1.2$, and 1.3. In each figure, $u_0(x, t)$ falls below $u(x, t)$. Both solutions decay to zero as t increases, but the effect of the xt term is to retard this decay. ◆

16.2.6 Effects of Boundary Conditions and Constants

We will investigate how constants and certain terms appearing in diffusion problems influence the behavior of solutions.

EXAMPLE 16.5

A homogeneous bar of length π has initial temperature function $f(x) = x^2 \cos(x/2)$ and ends maintained at a temperature of zero. The temperature distribution function satisfies

$$\frac{\partial u}{\partial t} = k \frac{\partial^2 u}{\partial x^2} \text{ for } 0 < x < \pi, t > 0,$$

$$u(0, t) = u(\pi, t) = 0 \text{ for } t > 0,$$

and

$$u(x, 0) = x^2 \cos(x/2) \text{ for } 0 \leq x \leq \pi.$$

The solution is

$$u(x, t) = \frac{2}{\pi} \sum_{n=1}^{\infty} \left(\int_0^{\pi} \xi^2 \cos(\xi/2) \sin(n\xi) \, d\xi \right) \sin(nx) e^{-n^2 kt}$$

$$= \frac{4}{\pi} \sum_{n=1}^{\infty} \frac{16\pi n(-1)^n - 64\pi n^3(-1)^n - 48n - 64n^3}{64n^6 - 48n^4 + 12n^2 - 1} \sin(nx) e^{-n^2 kt}.$$

To gauge the effect of the diffusivity constant k on the solution, Figure 16.6 shows graphs of $y = u(x, t)$ for $t = 0.2$ and for $k = 0.3, 0.6, 1.1$, and 2.7. Figure 16.7 has the graphs for the same

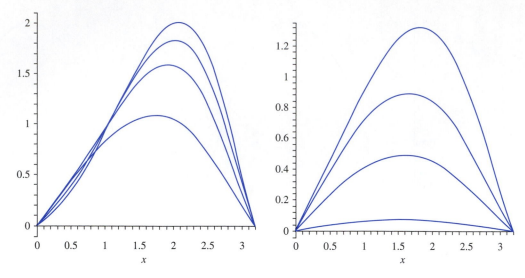

FIGURE 16.6 *The solution in Example 16.5 at* $t = 0.2$ *and* $k = 0.3, 0.6, 1.4,$ *and* 2.7.

FIGURE 16.7 *The solution in Example 16.5 at* $t = 1.2$ *and* $k = 0.3, 0.6, 1.4,$ *and* 2.7.

values of k, but $t = 1.2$. For each k, the temperature function decays with time, as we expect. However, for each time, the temperature function has a smaller maximum as k increases. ◆

EXAMPLE 16.6

We will examine the effects on $u(x, t)$, depending on whether the ends of the bar are kept at a temperature of zero or are insulated. Suppose the initial temperature function is $f(x) = x^2(\pi - x)$, and both $L = \pi$ and $k = 1/4$. For the ends at temperature zero, we find that

$$u_1(x, t) = \sum_{n=1}^{\infty} \left(\frac{8(-1)^{n+1} - 4}{n^3} \right) \sin(nx) e^{-n^2 t/4}.$$

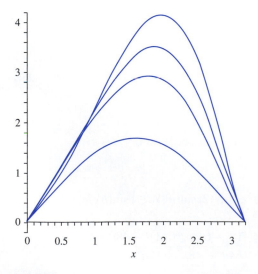

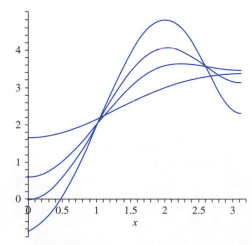

FIGURE 16.8 $u_1(x, t)$ *at times* $t = 0.4, 0.9, 1.5,$ *and* 3.6.

FIGURE 16.9 $u_2(x, t)$ *at times* $t = 0.4, 0.9, 1.5,$ *and* 3.6.

If the ends are insulated, the solution is

$$u_2(x, t) = \frac{1}{12}\pi^3 + \sum_{n=1}^{\infty} \left(\frac{n^2\pi^2(-1)^{n+1} + 6(-1)^n - 6}{n^4} \right) \cos(nx)e^{-n^2t/4}.$$

Figure 16.8 shows $u_1(x, t)$, and Figure 16.9 shows $u_2(x, t)$, at times $t = 0.4, 0.9, 1.5$, and 3.6. Both solutions decrease as t increases. However, as $t \to \infty$, $u_2(x, t) \to \pi^3/12$, as suggested in Figure 16.9. ◆

16.2.7 Numerical Approximation of Solutions

We will outline a strategy for approximating solution values of the diffusion problem

$$\frac{\partial u}{\partial t} = k\frac{\partial^2 u}{\partial x^2} \text{ for } 0 < x < L, t > 0,$$

$$u(0, t) = u(L, t) = 0 \text{ for } t \geq 0,$$

and

$$u(x, 0) = f(x) \text{ for } 0 \leq x \leq L.$$

As with the wave equation, form a grid of points over the (x, t) strip $0 \leq x \leq L$ and $t \geq 0$. Choose a positive integer N, a positive number Δt, let $\Delta x = L/N$, and form lattice points (x_i, t_j) where $x_i = i\Delta x$ and $t_j = j\Delta t$ for $i = 0, 1, \ldots, N$ and $j = 0, 1, 2, \ldots$. Denote $u_{i,j} = u(x_i, t_j)$. Use centered difference approximations for the partial derivatives to replace the heat equation with

$$\frac{u_{i,j+1} - u_{i,j}}{\Delta t} = k\left(\frac{u_{i+1,j} - 2u_{i,j} + u_{i-1,j}}{(\Delta x)^2} \right).$$

Solve this to obtain

$$u_{i,j+1} = \frac{k\Delta t}{(\Delta x)^2}(u_{i+1,j} - 2u_{i,j} + u_{i-1,j}) + u_{i,j}.$$

This enables us to approximate solutions at $x = j + 1$ at the t-level from information at the next lower level, assuming approximations have already been made at that level (Figure 16.10).

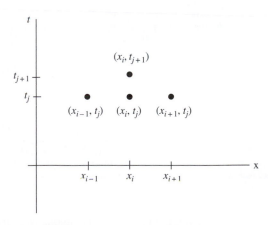

FIGURE 16.10 *Dependence of the approximation at (x_j, t_{k+1}) on three points at level t_k.*

In this way, we can move upward in the (x, t) strip, making new approximations from previous ones. Data for the starting layer is provided by the initial and boundary conditions. First,

$$u_{0,j} = u_{N,j} = 0$$

gives approximate values at lattice points on the left and right sides of the strip, and

$$u_{i,0} = f(x_i) = f(i \Delta x).$$

The quantity $k \Delta t / (\Delta x)^2$ should be less than $1/2$ to ensure stability of the method.

Notice that here we need information only on the t-layer immediately below the layer on which we are computing approximate values. For the wave equation, we needed information on two layers below the layer of current activity. This is due to the difference in the order of the time derivatives in the heat and wave equations.

EXAMPLE 16.7

Let $L = k = 1$ and $f(x) = x(1 - x)$. The solution is

$$u(x, t) = \frac{8}{\pi^3} \sum_{n=1}^{\infty} \frac{1}{(2n-1)^3} \sin((2n-1)\pi x) e^{-(2n-1)^2 \pi^2 t}.$$

To make numerical approximations of solution values, we will use $\Delta x = 0.1$ (so $N = 10$) and $\Delta t = 0.0025$. We know that $u_{0,j} = u_{10,j} = 0$ and

$$u_{i,0} = f(i \Delta x) = i(0.1)(1 - i(0.1)).$$

This initiates the approximation. To move up the t-layers, use

$$u_{i,j+1} = 0.25(u_{i+1,j} - 2u_{i,j} + u_{i-1,j}) + u_{i,j}.$$

TABLE 16.1   *Approximate Values of $u(x, t)$*

i	$u_{i,1}$	$u_{i,2}$	$u_{i,3}$	$u_{i,4}$	$u_{i,5}$
0	0.09	0.16	0.21	0.24	0.25
1	0.085	0.155	0.205	0.235	0.245
2	0.081	0.150	0.200	0.230	0.240
3	0.078	0.145	0.195	0.225	0.235
4	0.075	0.141	0.190	0.220	0.230

i	$u_{i,6}$	$u_{i,7}$	$u_{i,8}$	$u_{i,9}$	$u_{i,10}$
0	0.24	0.21	0.16	0.09	0
1	0.235	0.205	0.155	0.085	0
2	0.230	0.200	0.150	0.081	0
3	0.225	0.195	0.145	0.078	0
4	0.220	0.190	0.141	0.075	0

Go first to the $j = 1$ layer, then to $j = 2$, and so on, each time using the latest approximate values to compute the values at points on the next layer up. Through the first four layers above the bottom of the strip, Table 16.1 gives the approximate values $u_{i,j}$ of $u(x_i, t_j)$. ◆

In each of Problems 1 through 7, write a solution of the initial-boundary value problem. Graph the twentieth partial sum of the series for $u(x, t)$ on the same set of axes for different values of the time.

1. $$\frac{\partial u}{\partial t} = k \frac{\partial^2 u}{\partial x^2} \text{ for } 0 < x < L, t > 0,$$
   $u(0, t) = u(L, t) = 0$ for $t \geq 0$,
   $u(x, 0) = x(L - x)$ for $0 \leq x \leq L$

2. $$\frac{\partial u}{\partial t} = 4 \frac{\partial^2 u}{\partial x^2} \text{ for } 0 < x < L, t > 0,$$
   $u(0, t) = u(L, t) = 0$ for $t \geq 0$,
   $u(x, 0) = x^2(L - x)$ for $0 \leq x \leq L$

3. $$\frac{\partial u}{\partial t} = 3 \frac{\partial^2 u}{\partial x^2} \text{ for } 0 < x < L, t > 0,$$
   $u(0, t) = u(L, t) = 0$ for $t \geq 0$,
   $u(x, 0) = L(1 - \cos(2\pi x/L))$ for $0 \leq x \leq L$

4. $$\frac{\partial u}{\partial t} = \frac{\partial^2 u}{\partial x^2} \text{ for } 0 < x < \pi, t > 0,$$
   $\frac{\partial u}{\partial x}(0, t) = \frac{\partial u}{\partial x}(\pi, t) = 0$ for $t \geq 0$,
   $u(x, 0) = \sin(x)$ for $0 \leq x \leq \pi$

5. $$\frac{\partial u}{\partial t} = 4 \frac{\partial^2 u}{\partial x^2} \text{ for } 0 < x < 2\pi, t > 0,$$
   $\frac{\partial u}{\partial x}(0, t) = \frac{\partial u}{\partial x}(2\pi, t) = 0$ for $t \geq 0$,
   $u(x, 0) = x(2\pi - x)$ for $0 \leq x \leq 2\pi$

6. $$\frac{\partial u}{\partial t} = 4 \frac{\partial^2 u}{\partial x^2} \text{ for } 0 < x < 3, t > 0,$$
   $\frac{\partial u}{\partial x}(0, t) = \frac{\partial u}{\partial x}(3, t) = 0$ for $t \geq 0$,
   $u(x, 0) = x^2$ for $0 \leq x \leq 3$

7. $$\frac{\partial u}{\partial t} = 2 \frac{\partial^2 u}{\partial x^2} \text{ for } 0 < x < 6, t > 0,$$
   $\frac{\partial u}{\partial x}(0, t) = \frac{\partial u}{\partial x}(6, t) = 0$ for $t \geq 0$,
   $u(x, 0) = e^{-x}$ for $0 \leq x \leq 6$

8. A thin, homogeneous bar of length L has insulated ends and initial temperature B, which is a positive constant. Find the temperature distribution in the bar.

9. A thin, homogeneous bar of length L has initial temperature $f(x) = B$. The right end $x = L$ is insulated, while the left end is kept at zero temperature. Find the temperature distribution in the bar.

10. A thin, homogeneous bar has a thermal diffusivity of 9 and a length of 2 cm. It has insulated sides at its left end maintained at zero temperature, while its right end is perfectly insulated. The bar has an initial

temperature $f(x) = x^2$ for $0 \leq x \leq 2$. Determine the temperature distribution $u(x, t)$ and $\lim_{t \to 0} u(x, t)$.

11. Show that the partial differential equation
$$\frac{\partial u}{\partial t} = k \left(\frac{\partial^2 u}{\partial x^2} + A \frac{\partial u}{\partial x} + Bu \right)$$
can be transformed into a standard heat equation for v by letting $u(x, t) = e^{\alpha x + \beta t} v(x, t)$ and choosing α and β appropriately.

12. Use the idea of Problem 11 to solve
$$\frac{\partial u}{\partial t} = \frac{\partial^2 u}{\partial x^2} + 4 \frac{\partial u}{\partial x} + 2u \text{ for } 0 < x < \pi, t > 0,$$
$u(0, t) = u(\pi, t) = 0$ for $t \geq 0$,

and
$$u(x, 0) = x(\pi - x) \text{ for } 0 \leq x \leq \pi.$$

13. Solve
$$\frac{\partial u}{\partial t} = \frac{\partial^2 u}{\partial x^2} + 6 \frac{\partial u}{\partial x} \text{ for } 0 < x < 4, t > 0,$$
$u(0, t) = u(4, t) = 0$ for $t \geq 0$,

and
$$u(x, 0) = 1 \text{ for } 0 \leq x \leq 4.$$

Graph the twentieth partial sum of the solution for selected times.

14. Solve
$$\frac{\partial u}{\partial t} = \frac{\partial^2 u}{\partial x^2} - 6 \frac{\partial u}{\partial x} \text{ for } 0 < x < \pi, t > 0,$$
$u(0, t) = u(\pi, t) = 0$ for $t \geq 0$,

and
$$u(x, 0) = x(\pi - x) \text{ for } 0 \leq x \leq \pi.$$

Graph the twentieth partial sum of the solution for selected times.

15. Solve
$$\frac{\partial u}{\partial t} = 16 \frac{\partial^2 u}{\partial x^2} \text{ for } 0 < x < 1, t > 0,$$
$u(0, t) = 2, u(1, t) = 5$ for $t \geq 0$,

and
$$u(x, 0) = x^2 \text{ for } 0 \leq x \leq 1.$$

16. Solve

$$\frac{\partial u}{\partial t} = k\frac{\partial^2 u}{\partial x^2} \text{ for } 0 < x < L, t > 0,$$

$$u(0, t) = T, u(L, t) = 0 \text{ for } t \geq 0,$$

and

$$u(x, 0) = x(L - x) \text{ for } 0 \leq x \leq L.$$

17. Solve

$$\frac{\partial u}{\partial t} = 4\frac{\partial^2 u}{\partial x^2} - Au \text{ for } 0 < x < 9, t > 0,$$

$$u(0, t) = u(9, t) = 0 \text{ for } t \geq 0,$$

and

$$u(x, 0) = 0 \text{ for } 0 \leq x \leq 9.$$

Graph the twentieth partial sum of the solution for $t = 0.2$ and $A = 1/4$, $1/2$, 1, and 3. Repeat this with $t = 0.7$ and $t = 1.4$. This gives some idea of the effect of the term $-Au$ on the solution.

18. Solve

$$\frac{\partial u}{\partial t} = 9\frac{\partial^2 u}{\partial x^2} \text{ for } 0 < x < L, t > 0,$$

$$u(0, t) = T, u(L, t) = 0 \text{ for } t \geq 0,$$

and

$$u(x, 0) = 0 \text{ for } 0 \leq x \leq L.$$

In each of Problems 19 through 23, solve the problem

$$\frac{\partial u}{\partial t} = k\frac{\partial^2 u}{\partial x^2} + F(x, t) \text{ for } 0 < x < L, t > 0,$$

$$u(0, t) = u(L, t) = 0 \text{ for } t \geq 0,$$

and

$$u(x, 0) = f(x) \text{ for } 0 \leq x \leq L.$$

Choose a value of time, and using the same set of axes, graph the twentieth partial sum of the solution together with the solution of the problem with $F(x, t)$ removed. Repeat this for several times. This gives some sense of the influence of $F(x, t)$ on the temperature distribution.

19. $k = 4$, $F(x, t) = t$, $f(x) = x(\pi - x)$, $L = \pi$

20. $k = 1$, $F(x, t) = x \sin(t)$, $f(x) = 1$, $L = 4$

21. $k = 1$, $F(x, t) = t\cos(x)$, $f(x) = x^2(5 - x)$, $L = 5$

22. $k = 4$, $F(x, t) = \begin{cases} K & \text{for } 0 \leq x \leq 1 \\ 0 & \text{for } 1 < x \leq 2 \end{cases}$
    $f(x) = \sin(\pi x/2)$, $L = 2$

23. $k = 16$, $F(x, t) = xt$, $f(x) = K$, $L = 3$

24. Find approximate solution values for the problem

$$\frac{\partial u}{\partial t} = \frac{\partial^2 u}{\partial x^2} \text{ for } 0 < x < 2, t > 0,$$

$$u(0, t) = u(2, t) = 0 \text{ for } t \geq 0,$$

and

$$u(x, 0) = \sin(\pi x/2) \text{ for } 0 \leq x \leq 2.$$

Let $\Delta x = 0.2$ and $\Delta t = 0.0025$. Carry out calculations for the first four horizontal layers, including the $t = 0$ layer.

25. Find approximate solution values for the problem

$$\frac{\partial u}{\partial t} = \frac{\partial^2 u}{\partial x^2} \text{ for } 0 < x < 1, t > 0,$$

$$u(0, t) = u(1, t) = 0 \text{ for } t \geq 0,$$

and

$$u(x, 0) = x^2(1 - x) \text{ for } 0 \leq x \leq 1.$$

Let $\Delta x = 0.2$ and $\Delta t = 0.0025$. Carry out calculations for the first four horizontal layers, including the $t = 0$ layer.

26. Find approximate solution values for the problem

$$\frac{\partial u}{\partial t} = \frac{\partial^2 u}{\partial x^2} \text{ for } 0 < x < 1, t > 0,$$

$$u(0, t) = u(1, t) = 0 \text{ for } t \geq 0,$$

and

$$u(x, 0) = x\cos(\pi x/2) \text{ for } 0 \leq x \leq 1$$

Let $\Delta x = 0.2$ and $\Delta t = 0.0025$. Carry out calculations for the first four horizontal layers, including the $t = 0$ layer.

16.3   Solutions in an Infinite Medium

We will consider problems involving the heat equation over the entire line or half-line.

16.3.1 Problems on the Real Line

For a setting in which one dimension is very much greater than the others, it is sometimes useful to model heat conduction or a diffusion process by imagining the space variable free to vary over the entire real line. In this case, there is no boundary condition, but we look for bounded solutions. The problem we will solve is

$$\frac{\partial u}{\partial t} = k \frac{\partial^2 u}{\partial x^2} \text{ for } -\infty < x < \infty, t > 0$$

and

$$u(x,0) = f(x) \text{ for } -\infty < x < \infty.$$

Separation of variables yields

$$X'' + \lambda X = 0, \ T' + \lambda kT = 0.$$

As with the wave equation on the line, the eigenvalues $\lambda = \omega^2 \geq 0$ and the eigenfunctions have the form $a_\omega \cos(\omega x) + b_\omega \sin(\omega x)$.

The problem for T is $T' + k\omega^2 T = 0$ with solutions that are constant multiples of $e^{-\omega^2 kt}$.

For $\omega \geq 0$, we now have functions

$$u_\omega(x,t) = [a_\omega \cos(\omega x) + b_\omega \sin(\omega x)]e^{-\omega^2 kt}$$

that satisfy the heat equation and are bounded for all x. To satisfy the initial condition, attempt a superposition

$$u(x,t) = \int_0^\infty [a_\omega \cos(\omega x) + b_\omega \sin(\omega x)]e^{-\omega^2 kt} \, d\omega. \tag{16.11}$$

We must choose the coefficients so that

$$u(x,0) = f(x) = \int_0^\infty [a_\omega \cos(\omega x) + b_\omega \sin(\omega x)] \, d\omega.$$

This is the Fourier integral of f on the real line, so the coefficients should be chosen as the Fourier integral coefficients of f as

$$a_\omega = \frac{1}{\pi} \int_{-\infty}^\infty f(\xi) \cos(\omega\xi) \, d\xi \ \text{ and } \ b_\omega = \frac{1}{\pi} \int_{-\infty}^\infty f(\xi) \sin(\omega\xi) \, d\xi.$$

EXAMPLE 16.8

Suppose the initial temperature function is $f(x) = e^{-|x|}$. Compute the coefficients

$$a_\omega = \frac{1}{\pi} \int_{-\infty}^\infty e^{-|\xi|} \cos(\omega\xi) \, d\xi = \frac{2}{\pi} \frac{1}{1+\omega^2}$$

and $b_\omega = 0$, because $e^{-|x|} \sin(\omega x)$ is an odd function. The solution is

$$u(x,t) = \frac{2}{\pi} \int_0^\infty \frac{1}{1+\omega^2} \cos(\omega x)e^{-\omega^2 kt} \, d\omega. \ \blacklozenge$$

In general, if we put the integrals for a_ω and b_ω into the solution of equation (16.11) and argue as we did for the solution of the wave equation on the line using equation (15.9), we obtain an alternative form of this solution:

$$u(x,t) = \frac{1}{\pi} \int_0^\infty \int_{-\infty}^\infty \cos(\omega(x-\xi))f(\xi)e^{-\omega^2 kt} \, d\xi. \tag{16.12}$$

This solution can be written as a single integral. First write equation (16.12) as

$$u(x,t) = \frac{1}{2\pi} \int_{-\infty}^{\infty} \int_{-\infty}^{\infty} \cos(\omega(x-\xi)) f(\xi) e^{-\omega^2 kt} \, d\xi.$$

Now we need the following integral. If α and β are real numbers with $\beta \neq 0$, then

$$\int_{-\infty}^{\infty} e^{-\zeta^2} \cos\left(\frac{\alpha\zeta}{\beta}\right) d\zeta = \sqrt{\pi} e^{-\alpha^2/4\beta^2}. \tag{16.13}$$

A derivation of this integral is sketched in Problem 11. In this integral, let

$$\xi = \sqrt{kt}\,\omega, \quad \alpha = x - \xi, \text{ and } \beta = \sqrt{kt}$$

to obtain

$$\int_{-\infty}^{\infty} e^{-\omega^2 kt} \cos(\omega(x-\xi)) d\omega = \frac{\sqrt{\pi}}{\sqrt{kt}} e^{-(x-\xi)^2/4kt}.$$

Now the solution (16.12) is

$$u(x,t) = \frac{1}{2\sqrt{\pi kt}} \int_{-\infty}^{\infty} f(\xi) e^{-(x-\xi)^2/4kt} \, d\xi. \tag{16.14}$$

16.3.2   Solution by Fourier Transform

We will illustrate the use of the Fourier transform to solve the problem

$$\frac{\partial u}{\partial t} = k \frac{\partial^2 u}{\partial x^2} \text{ for } -\infty < x < \infty, t > 0$$

and

$$u(x,0) = f(x) \text{ for } -\infty < x < \infty.$$

Take the Fourier transform of the heat equation with respect to x to get

$$\mathcal{F}\left[\frac{\partial u}{\partial t}\right] = k\mathcal{F}\left[\frac{\partial^2 u}{\partial x^2}\right].$$

Because x and t are independent,

$$\mathcal{F}\left[\frac{\partial u}{\partial t}\right] = \frac{\partial}{\partial t} \hat{u}(\omega, t).$$

Next use the operational rule for the Fourier transform to write

$$\mathcal{F}\left[\frac{\partial^2 u}{\partial x^2}\right] = -\omega^2 \hat{u}(\omega, t).$$

The transformed heat equation is

$$\frac{\partial}{\partial t} \hat{u}(\omega, t) + k\omega^2 \hat{u}(\omega, t) = 0$$

with solutions

$$\hat{u}(\omega, t) = a_\omega e^{-\omega^2 kt}.$$

Since $u(x,0) = f(x)$,

$$\hat{u}(\omega, 0) = \hat{f}(\omega) = a_\omega.$$

We now have the Fourier transform of the solution:

$$\hat{u}(\omega, t) = \hat{f}(\omega) e^{-\omega^2 kt}.$$

Apply the inverse Fourier transform. Since this transform is complex valued and the solution for a temperature distribution is real, the solution is the real part of this inverse:

$$u(x,t) = \operatorname{Re} \mathcal{F}^{-1}\left[\hat{f}(\omega)e^{-\omega^2 kt}\right](t)$$

$$= \operatorname{Re} \frac{1}{2\pi}\int_{-\infty}^{\infty} \hat{f}(\omega)e^{-\omega^2 kt}e^{i\omega x}\,d\omega$$

$$= \operatorname{Re} \frac{1}{2\pi}\int_{-\infty}^{\infty}\left(\int_{-\infty}^{\infty} f(\xi)e^{-i\omega\xi}\,d\xi\right)e^{i\omega x}e^{-\omega^2 kt}\,d\omega$$

$$= \operatorname{Re} \frac{1}{2\pi}\int_{-\infty}^{\infty}\int_{-\infty}^{\infty} f(\xi)e^{-i\omega(\xi-x)}e^{-\omega^2 kt}\,d\xi\,d\omega$$

$$= \operatorname{Re} \frac{1}{2\pi}\int_{-\infty}^{\infty}\int_{-\infty}^{\infty} f(\xi)[\cos(\omega(\xi-x)) - i\sin(\omega(\xi-x))]e^{-\omega^2 kt}\,d\xi\,d\omega$$

$$= \frac{1}{2\pi}\int_{-\infty}^{\infty}\int_{-\infty}^{\infty} f(\xi)\cos(\omega(\xi-x))e^{-\omega^2 kt}\,d\xi\,d\omega,$$

which is the same expression obtained by separation of variables.

16.3.3   Problems on the Half-Line

We will solve the problem

$$\frac{\partial u}{\partial t} = k\frac{\partial^2 u}{\partial x^2} \text{ for } 0 < x < \infty, t > 0,$$

$$u(0,t) = 0 \text{ for } t > 0,$$

and

$$u(x,0) = f(x) \text{ for } 0 < x < \infty$$

for the half-line, taking the case that the left end at $x = 0$ is kept at zero temperature. Putting $u(x,t) = X(x)T(t)$ leads in the usual way to

$$X'' + \lambda X = 0; X(0) = 0$$

and

$$T' + k\lambda T = 0.$$

The eigenvalues are $\lambda = \omega^2$ with $\omega > 0$, with eigenfunctions $b_\omega \sin(\omega x)$. With these values of λ, $T(t)$ is a constant multiple of $e^{-\omega^2 kt}$. For each $\omega > 0$, the function

$$u_\omega(x,t) = b_\omega \sin(\omega x)e^{-\omega^2 kt}$$

satisfies the heat equation and the boundary condition. For the initial condition, write

$$u(x,t) = \int_0^\infty b_\omega \sin(\omega x)e^{-\omega^2 kt}\,d\omega.$$

Then

$$u(x,0) = f(x) = \int_0^\infty b_\omega \sin(\omega x)\,d\omega$$

requires that we choose the coefficients as the Fourier integral sine coefficients of f on $[0, \infty)$:

$$b_\omega = \frac{2}{\pi}\int_0^\infty f(\xi)\sin(\omega\xi)\,d\xi.$$

EXAMPLE 16.9

Suppose the initial temperature function is

$$f(x) = \begin{cases} \pi - x & \text{for } 0 \le x \le \pi \\ 0 & \text{for } x > \pi. \end{cases}$$

Compute the coefficients

$$b_\omega = \frac{2}{\pi} \int_0^\pi (\pi - \xi) \sin(\omega\xi)\, d\xi = \frac{2}{\omega} \frac{\pi\omega - \sin(\pi\omega)}{\omega^2}.$$

The solution is

$$u(x, t) = \frac{2}{\pi} \int_0^\infty \frac{\pi\omega - \sin(\pi\omega)}{\omega^2} \sin(\omega x) e^{-\omega^2 kt}\, d\omega. \quad \blacklozenge$$

16.3.4   Solution by Fourier Sine Transform

We can also solve this problem on the half-line using the Fourier sine transform. Take this transform of the heat equation with respect to x to get

$$\frac{\partial u}{\partial t} \hat{u}_S(\omega, t) = -\omega^2 k \hat{u}_S(\omega, t) + \omega k u(0, t) = -\omega^2 k \hat{u}_S(\omega, t).$$

This has general solution

$$\hat{u}_S(\omega, t) = b_\omega e^{-\omega^2 kt}.$$

Furthermore,

$$\hat{u}_S(\omega, 0) = \hat{f}_S(\omega) = b_\omega,$$

so

$$\hat{u}_S(\omega, t) = \hat{f}_S(\omega) e^{-\omega^2 kt}.$$

Upon taking the inverse Fourier sine transform, we obtain the solution

$$u(x, t) = \frac{2}{\pi} \int_0^\infty \hat{f}_S(\omega) e^{-\omega^2 kt} \sin(\omega x)\, d\omega.$$

It is a calculation along lines we have done previously to insert the integral for $\hat{f}_S(\omega)$ into this expression to obtain the solution by separation of variables.

SECTION 16.3   PROBLEMS

In each of Problems 1 through 4, solve the problem

$$\frac{\partial u}{\partial t} = k \frac{\partial^2 u}{\partial x^2} \text{ for } -\infty < x < \infty, t > 0$$

and

$$u(x, 0) = f(x) \text{ for } -\infty < x < \infty.$$

1. $f(x) = e^{-4|x|}$

2. $f(x) = \begin{cases} \sin(x) & \text{for } |x| \le \pi \\ 0 & \text{for } |x| > \pi \end{cases}$

3. $f(x) = \begin{cases} x & \text{for } 0 \le x \le 4 \\ 0 & \text{for } x < 0 \text{ and for } x > 4 \end{cases}$

4. $f(x) = \begin{cases} e^{-x} & \text{for } |x| \le 1 \\ 0 & \text{for } |x| > 1 \end{cases}$

In each of Problems 5 through 8, solve the problem

$$\frac{\partial u}{\partial t} = k \frac{\partial^2 u}{\partial x^2} \text{ for } x > 0, t > 0,$$

$$u(0, t) = 0 \text{ for } t > 0,$$

and

$$u(x, 0) = f(x) \text{ for } x > 0.$$

5. $f(x) = e^{-\alpha x}$ with α any positive constant.

6. $f(x) = xe^{-\alpha x}$ with α any positive constant.

7. $f(x) = \begin{cases} 1 & \text{for } 0 \le x \le h \\ 0 & \text{for } x > h \end{cases}$
    with h any positive number.

8. $f(x) = \begin{cases} x & \text{for } 0 \le x \le 2 \\ 0 & \text{for } x > 2 \end{cases}$

In each of Problems 9 and 10, use a Fourier transform on the half-line to solve the problem.

9.
$$\frac{\partial u}{\partial t} = \frac{\partial^2 u}{\partial x^2} - tu \text{ for } x > 0, t > 0,$$
$$u(0, t) = 0 \text{ for } t > 0,$$
$$u(x, 0) = xe^{-x} \text{ for } x > 0$$

10.
$$\frac{\partial u}{\partial t} = \frac{\partial^2 u}{\partial x^2} - u \text{ for } x > 0, t > 0,$$
$$u(0, t) = 0 \text{ for } t > 0,$$
$$\frac{\partial u}{\partial x}(0, t) = f(t) \text{ for } t > 0$$

11. Derive the integral (16.13). *Hint*: This can be done using complex function theory and a contour integral. Here is another way. Let

$$F(x) = \int_0^\infty e^{-\zeta^2} \cos(x\zeta) \, d\zeta.$$

Compute $F'(x)$ by differentiating under the integral sign and show that $F'(x) = -xF(x)/2$. Solve for $F(x)$ subject to the initial condition that

$$F(0) = \int_0^\infty e^{-\zeta^2} \, d\zeta = \frac{\sqrt{\pi}}{2}.$$

This integral for $F(0)$ is familiar from statistics and can be assumed known. Finally, let $x = \alpha/\beta$.

16.4   Heat Conduction in an Infinite Cylinder

We will determine the temperature distribution in a solid, infinitely long, homogeneous cylinder of radius R with its axis along the z axis in 3-space.

In cylindrical coordinates, the heat equation for the temperature distribution $U(r, \theta, z, t)$ is

$$\frac{\partial U}{\partial t} = k \left(\frac{\partial^2 U}{\partial r^2} + \frac{1}{r} \frac{\partial U}{\partial r} + \frac{1}{r^2} \frac{\partial^2 U}{\partial \theta^2} + \frac{\partial^2 U}{\partial z^2} \right).$$

This is a formidable equation to solve at this stage, and we will restrict it to the special case that the temperature at any point in the cylinder depends only on the time t and the horizontal distance r from the z axis. This symmetry means that $\partial U / \partial \theta = \partial U / \partial z = 0$, and the heat equation becomes

$$\frac{\partial U}{\partial t} = k \left(\frac{\partial^2 U}{\partial r^2} + \frac{1}{r} \frac{\partial U}{\partial r} \right)$$

for $0 \le r < R, t > 0$. We will write $U(r, t)$ with dependence only on r and t and assume the boundary condition

$$U(R, t) = 0 \text{ for } t > 0.$$

The initial condition is

$$U(r, 0) = f(r) \text{ for } 0 \le r < R.$$

Put $U(r, t) = F(r)T(t)$ and separate the variables, obtaining

$$\frac{T'}{kT} = \frac{F'' + (1/r)F'(r)}{F(r)} = -\lambda.$$

Then

$$T' + \lambda T = 0 \text{ and } F'' + \frac{1}{r} F' + \lambda F = 0.$$

Since $U(R,t) = F(R)T(t) = 0$, then $F(R) = 0$. The problem for F is a singular Sturm-Liouville problem $[0, R]$. To determine the eigenvalues and eigenfunctions take cases on λ.

Case 1 $\lambda = 0$.   Let

$$F'' + \frac{1}{r}F' = 0$$

with a general solution of the form $F(r) = c\ln(r) + d$. Since $\ln(r) \to -\infty$ as $r \to 0$, which is the center of the cylinder, we must choose $c = 0$. Then $F(r) = d$. Since $T'(t) = 0$ if $\lambda = 0$, then $T(t) = $ constant also. In this case, $U(r,t) = $ constant, and this constant must be zero because $U(R, 0) = 0$. $U(r, t) = 0$ is indeed the solution if $f(r) = 0$. If $f(r)$ is not identically zero, then $\lambda = 0$ does not contribute to the solution.

Case 2 $\lambda < 0$.   Write $\lambda = -\omega^2$ with $\omega > 0$. Now $T' - k\omega^2 T = 0$ has the general solution

$$T(t) = ce^{\omega^2 kt},$$

and this is unbounded as t increases. To have a bounded solution, we reject this case.

Case 3 $\lambda > 0$.   Write $\lambda = \omega^2$ with $\omega > 0$. Now $T' + k\omega^2 T = 0$ with the general solution

$$T(t) = ce^{-\omega^2 kt}.$$

This is a bounded function. The equation for F becomes

$$F''(r) + \frac{1}{r}F'(r) + \omega^2 F(r) = 0.$$

Write this as

$$r^2 F''(r) + rF'(r) + \omega^2 F(r) = 0.$$

This is Bessel's equation of order zero. Bounded solutions on $[0, R]$ are constant multiples of $J_0(\omega r)$.

Thus far, we have

$$U_\omega(r,t) = a_\omega J_0(\omega r)e^{-\omega^2 kt}.$$

The condition $U(R, 0) = 0$ requires that

$$J_0(\omega R) = 0.$$

Let $j_1 < j_2 < \dots$ be the positive zeros of $J_0(x)$ in ascending order. For $J_0(\omega R) = 0$, there must be some positive integer n such that $\omega R = j_n$. Denote $\omega_n = j_n/R$. This gives us the eigenvalues of this problem as

$$\lambda_n = \omega_n^2 = \frac{j_n^2}{R^2}.$$

The eigenfunctions are constant multiples of $J_0(j_n r/R)$.

Now we have, for $n = 1, 2, \dots$, functions

$$U_n(r,t) = a_n J_0\left(\frac{j_n r}{R}\right)e^{-j_n^2 kt/R^2}$$

satisfying the heat equation and the boundary condition. To satisfy the initial condition, employ a superposition

$$U(r,t) = \sum_{n=1}^{\infty} a_n J_0\left(\frac{j_n r}{R}\right)e^{-j_n^2 kt/R^2}.$$

Now we must choose the coefficients so that

$$U(r,0) = f(r) = \sum_{n=1}^{\infty} a_n J_0\left(\frac{j_n r}{R}\right).$$

Let $\xi = r/R$ to write

$$f(\xi R) = \sum_{n=1}^{\infty} a_n J_0(j_n \xi)$$

for $0 \le \xi \le 1$. In this framework, previous results on eigenfunction expansions apply, and we can write

$$a_n = \frac{2 \int_0^1 r f(\xi R) J_0(j_n \xi) \, d\xi}{J_1^2(j_n)}.$$

With these coefficients we have the solution for $U(r,t)$.

SECTION 16.4   PROBLEMS

1. Suppose $R = 1$, $k = 1$, and $f(r) = r$. Assume that $U(1,t) = 0$ for $t > 0$. Use a numerical integration to approximate the coefficients $a_1, \ldots, a_5$ and use these numbers in the fifth partial sum of the series solution to approximate $U(r,t)$. Graph this partial sum for different values of t.

2. Repeat the calculations of Problem 1 with $k = 16$, $R = 3$, and $f(r) = e^r$.

3. Repeat the calculations of Problem 1 with $k = 1/2$, $R = 3$, and $f(r) = 9 - r^2$.

16.5   Heat Conduction in a Rectangular Plate

We will solve for the temperature distribution $u(x, y, t)$ in a flat, square, homogeneous plate covering the region $0 \le x \le 1$ and $0 \le y \le 1$ in the plane. The sides are kept at a temperature of zero, and the interior temperature at (x, y) at time 0 is $f(x, y)$.

To be specific, we will let $f(x, y) = x(1 - x^2)y(1 - y)$ and solve the initial-boundary value problem

$$\frac{\partial u}{\partial t} = k\left(\frac{\partial^2 u}{\partial x^2} + \frac{\partial^2 u}{\partial y^2}\right) \text{ for } 0 < x < 1, 0 < y < 1, t > 0,$$

$$u(x, 0, t) = u(x, 1, t) = 0 \text{ for } 0 < x < 1, t > 0,$$

$$u(0, y, t) = u(1, y, t) = 0 \text{ for } 0 < y < 1, t > 0,$$

and

$$u(x, y, 0) = f(x, y) = x(1 - x^2)y(1 - y).$$

Let $u(x, y, t) = X(x)Y(y)T(t)$ and separate variables to obtain

$$X'' + \lambda X = 0, \ Y'' + \mu Y = 0, \text{ and } T' + (\lambda + \mu)T = 0$$

as in the analysis of wave motion of a membrane in Chapter 15. The boundary conditions imply that

$$X(0) = X(1) = 0 \text{ and } Y(0) = Y(1) = 0,$$

so the eigenvalues and eigenfunctions are

$$\lambda_n = n^2\pi^2 \text{ and } X_n(x) = \sin(n\pi x)$$

and

$$\mu_m = m^2\pi^2 \text{ and } Y_m(y) = \sin(m\pi y)$$

for $n = 1, 2, \ldots$ and $m = 1, 2, \ldots$. The equation for T becomes

$$T' + (n^2 + m^2)\pi^2 T = 0$$

with solutions that are constant multiples of $e^{-(n^2+m^2)\pi^2 kt}$. For each positive integer m and n, we now have functions

$$u_{nm}(x, y, t) = \sum_{n=1}^{\infty}\sum_{m=1}^{\infty} c_{nm}\sin(n\pi x)\sin(m\pi y)e^{-(n^2+m^2)\pi^2 kt}$$

satisfying the heat equation and the boundary conditions. Reasoning as we did with the two-dimensional wave equation, we get

$$c_{nm} = 4\int_0^1\int_0^1 \xi(1-\xi^2)\eta(1-\eta)\sin(n\pi\xi)\sin(m\pi\eta)\,d\xi\,d\eta$$

$$= 48\left(\frac{(-1)^n}{n^3\pi^3}\right)\left(\frac{(-1)^m - 1}{m^3\pi^3}\right).$$

The solution is

$$u(x, y, t) =$$

$$\frac{48}{\pi^6}\sum_{n=1}^{\infty}\sum_{m=1}^{\infty}\left(\frac{(-1)^n}{n^3}\right)\left(\frac{(-1)^m - 1}{m^3}\right)\sin(n\pi x)\sin(m\pi y)e^{-(n^2+m^2)\pi^2 kt}. \quad\blacklozenge$$

SECTION 16.5   PROBLEMS

1. Write the solution for the general problem

$$\frac{\partial u}{\partial t} = k\left(\frac{\partial^2 u}{\partial x^2} + \frac{\partial^2 u}{\partial y^2}\right) \text{ for } 0 < x < L,$$

$$0 < y < K, t > 0,$$

$$u(x, 0, t) = u(x, K, t) = 0 \text{ for } 0 < x < L, t > 0,$$

$$u(0, y, t) = u(L, y, t) = 0 \text{ for } 0 < y < K, t > 0,$$

and

$$u(x, y, 0) = f(x, y) \text{ for } 0 \le x \le L, 0 \le y \le K.$$

2. Solve this problem when $k = 4, L = 2, K = 3$, and $f(x, y) = x^2(L - x)\sin(y)(K - y)$.

3. Solve this problem when $k = 1, L = \pi, K = \pi$, and $f(x, y) = \sin(x)\cos(y/2)$.

CHAPTER 17

The Potential Equation

17.1 Laplace's Equation

The partial differential equation

$$\frac{\partial^2 u}{\partial x^2} + \frac{\partial^2 u}{\partial y^2} = 0$$

is called *Laplace's equation* in two dimensions. In three dimensions, Laplace's equation is

$$\frac{\partial^2 u}{\partial x^2} + \frac{\partial^2 u}{\partial y^2} + \frac{\partial^2 u}{\partial z^2} = 0.$$

These equations are often written $\nabla^2 u = 0$, which is read "del squared u".

Laplace's equation arises in several contexts. It is the steady-state heat equation, occurring when $\partial u / \partial t = 0$. It is also called the *potential equation*. If a vector field $\mathbf{F}$ has a potential φ, then φ must satisfy Laplace's equation.

A function satisfying Laplace's equation in a region of the plane (or 3-space) is said to be *harmonic* on that region. For example, $x^2 - y^2$ and xy are both harmonic over the entire plane.

A *Dirichlet problem* for a region D consists of finding a function that is harmonic on D and assumes specified values on the boundary of D. We will be primarily concerned with Dirichlet problems in the plane, in which D is a region that is bounded by one or more piecewise smooth curves. It is customary to denote the boundary by ∂D. The Dirichlet problem for D is to solve

$$\nabla^2 u = 0 \text{ on } D$$

and

$$u(x, y) = f(x, y) \text{ for } (x, y) \text{ in } \partial D$$

with $f(x, y)$ as a given function. The function f is called *boundary data* for D.

SECTION 17.1   *PROBLEMS*

1. Show that if f and g are harmonic on D, so are $f + g$ and, for any constant c, cf.

2. Show that the following functions are harmonic (on the entire plane if D is not specified).

   (a) $x^3 - 3xy^2$

   (b) $3x^2y - y^3$

   (c) $x^4 - 6x^2y^2 + y^4$

   (d) $4x^3y - 4xy^3$

   (e) $\sin(x)(e^y + e^{-y})$

   (f) $\cos(x)(e^y - e^{-y})$

   (g) $e^{-x}\cos(y)$

   (h) $\ln(x^2 + y^2)$ if D is the plane with the origin removed.

17.2   Dirichlet Problem for a Rectangle

The region D exerts a great influence on our ability to explicitly solve a Dirichlet problem or even whether a solution exists. Some regions admit solutions by Fourier methods. In this and the next section, we will treat two such cases, that D is a rectangle or disk in the plane.

Let D be the solid rectangle consisting of points (x, y) with $0 \leq x \leq L$ and $0 \leq y \leq K$. We will solve the Dirichlet problem for D.

This problem can be solved by separation of variables if the boundary data is nonzero on only one side of D. We will illustrate this for the case that this is the upper horizontal side of D. The problem in this case is

$$\nabla^2 u = 0 \text{ on } D,$$

$$u(x, 0) = 0 \text{ for } 0 \leq x \leq L,$$

$$u(0, y) = 0 \text{ for } 0 \leq y \leq K,$$

$$u(L, y) = 0 \text{ for } 0 \leq y \leq K,$$

and

$$u(x, K) = f(x) \text{ for } 0 \leq x \leq L.$$

Figure 17.1 shows D and the boundary data. Let $u(x, y) = X(x)Y(y)$ in Laplace's equation to obtain

$$\frac{X''}{X} = -\frac{Y''}{Y} = -\lambda,$$

or

$$X'' + \lambda X = 0 \text{ and } Y'' - \lambda Y = 0.$$

From the boundary conditions, $X(0) = X(L) = Y(0) = 0$, so the problems for X and Y are

$$X'' + \lambda X = 0; \quad X(0) = X(L) = 0$$

and

$$Y'' - \lambda Y = 0; \quad Y(0) = 0.$$

The problem for X has eigenvalues and eigenfunctions

$$\lambda_n = \frac{n^2\pi^2}{L^2} \text{ and } X_n(x) = \sin\left(\frac{n\pi x}{L}\right).$$

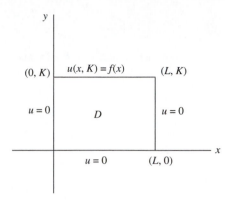

FIGURE 17.1 *D and the boundary data for this problem.*

for $n = 1, 2, \cdots$.

The problem for Y becomes

$$Y'' - \frac{n^2\pi^2}{L^2}Y = 0; \ Y(0) = 0$$

with solutions that are constant multiples of

$$Y_n(y) = \sinh\left(\frac{n\pi y}{L}\right).$$

For each positive integer n, we have a function

$$u_n(x, y) = \sin\left(\frac{n\pi x}{L}\right)\sinh\left(\frac{n\pi y}{L}\right)$$

that is harmonic on D and satisfies the zero boundary conditions on the lower side and vertical sides of D. For the condition on the top side, use a superposition

$$u(x, y) = \sum_{n=1}^{\infty} b_n \sin\left(\frac{n\pi x}{L}\right)\sinh\left(\frac{n\pi y}{L}\right).$$

We need

$$u(x, K) = f(x) = \sum_{n=1}^{\infty} b_n \sin\left(\frac{n\pi x}{L}\right)\sinh\left(\frac{n\pi K}{L}\right).$$

This is a Fourier sine expansion of $f(x)$ on $[0, L]$, with coefficient $b_n \sinh(n\pi K/L)$. Thus, choose $b_n \sinh(n\pi K/L)$ to be the nth Fourier sine coefficient of $f(x)$ on $[0, L]$:

$$b_n \sinh(n\pi K/L) = \frac{2}{L}\int_0^L f(\xi)\sin\left(\frac{n\pi \xi}{L}\right)d\xi.$$

Then

$$b_n = \frac{2}{L\sinh(n\pi K/L)}\int_0^L f(\xi)\sin\left(\frac{n\pi \xi}{L}\right)d\xi.$$

With this choice of coefficients, the solution can be written

$$u(x, y) = \sum_{n=1}^{\infty}\frac{2}{L}\left(\int_0^{\infty} f(\xi)\sin(n\pi \xi/L)\,d\xi\right)\sin(n\pi x/L)\frac{\sinh(n\pi y/L)}{\sinh(n\pi K/L)}.$$

EXAMPLE 17.1

Suppose, in the problem just solved, $L = K = \pi$ and $f(x) = x(\pi - x)$. Compute the numbers

$$\frac{2}{\pi} \int_0^\pi \xi(\pi - \xi) \sin(n\xi)\, d\xi = \frac{4}{\pi n^3}(-1)^n.$$

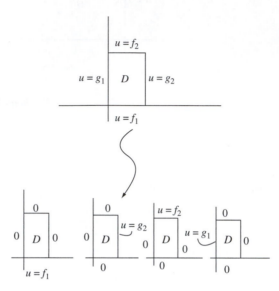

FIGURE 17.2 $u(x, y) = \sum_{j=1}^4 u_j(x, y)$.

The solution is

$$u(x, y) = \sum_{n=1}^\infty \frac{4}{\pi n^3}(-1)^n \sin(nx)\frac{\sinh(ny)}{\sinh(n\pi)}. \quad \blacklozenge$$

If nonzero boundary data is prescribed on all four sides of D, then define four Dirichlet problems, in each of which the data is nonzero on just one side (Figure 17.2). Each of these problems can be solved by separation of variables. The sum of the solutions of these four problems is the solution of the original problem.

SECTION 17.2   PROBLEMS

In each of Problems 1 through 5, solve the Dirichlet problem for the given rectangle and boundary conditions.

1. $u(0, y) = u(1, y) = 0$ for $0 \le y \le \pi$ and $u(x, \pi) = 0, u(x, 0) = \sin(\pi x)$ for $0 \le x \le 1$.

2. $u(0, y) = y(2 - y), u(3, y) = 0$ for $0 \le y \le 2$ and $u(x, 0) = u(x, 2) = 0$ for $0 \le x \le 3$

3. $u(0, y) = u(1, y) = 0$ for $0 \le y \le 4$ and $u(x, 4) = x\cos(\pi x/2), u(x, 0) = 0$ for $0 \le x \le 1$

4. $u(0, y) = \sin(y), u(\pi, y) = 0$ for $0 \le y \le \pi$ and $u(x, 0) = x(\pi - x), u(x, \pi) = 0$ for $0 \le x \le \pi$

5. $u(0, y) = 0, u(2, y) = \sin(y)$ for $0 \le y \le \pi$ and $u(x, 0) = 0, u(x, \pi) = x\sin(\pi x)$ for $0 \le x \le 2$

6. Apply separation of variables to the problem

$$\nabla^2 u = 0 \text{ for } 0 < x < a, 0 < y < b,$$

$$u(x, 0) = \frac{\partial u}{\partial y}(x, b) = 0 \text{ for } 0 \le x \le a,$$

and

$$u(0, y) = 0, u(a, y) = g(y) \text{ for } 0 \le y \le b.$$

7. Use separation of variables to solve

$$\nabla^2 u = 0 \text{ for } 0 < x < a, 0 < y < b,$$

$$u(x, 0) = 0, u(x, b) = f(x) \text{ for } 0 \le x \le a,$$

and

$$u(0, y) = \frac{\partial u}{\partial x}(a, y) = 0 \text{ for } 0 \le y \le b.$$

8. Solve for the steady-state temperature distribution in a homogeneous, thin, flat plate covering the rectangle $0 \le x \le a, 0 \le y \le b$ if the temperature on the vertical and lower sides are kept at zero and the temperature along the top side is $f(x) = x(x - a)^2$.

9. Solve for the steady-state temperature distribution in a thin, flat plate covering the rectangle $0 \le x \le 4, 0 \le y \le 1$ if the temperature on the horizontal sides is zero while the temperature on the left side is $f(y) = \sin(\pi y)$ and on the right side is $g(y) = y(1 - y)$.

17.3   Dirichlet Problem for a Disk

We will solve the Dirichlet problem for a disk of radius R centered at the origin. The boundary of this disk is the circle $x^2 + y^2 = R^2$. Using polar coordinates, the problem for $u(r, \theta)$ is

$$\nabla^2 u = \frac{\partial^2 u}{\partial r^2} + \frac{1}{r} \frac{\partial u}{\partial r} + \frac{1}{r^2} \frac{\partial^2 u}{\partial \theta^2} = 0 \text{ for } 0 \le r < R, -\pi \le \theta \le \pi$$

and

$$u(R, \theta) = f(\theta) \text{ for } -\pi \le \theta \le \pi.$$

It is easy to check that the functions $1, r^n \cos(n\theta)$, and $r^n \sin(n\theta)$ are harmonic on the entire plane. Thinking ahead to the possibility of a Fourier series to satisfy the boundary condition, attempt a solution in a series of these functions:

$$u(r, \theta) = \frac{1}{2} a_0 + \sum_{n=1}^{\infty} (a_n r^n \cos(n\theta) + b_n r^n \sin(n\theta)).$$

This would require that

$$u(R, \theta) = f(\theta) = \frac{1}{2} a_0 + \sum_{n=1}^{\infty} (a_n R^n \cos(n\theta) + b_n R^n \sin(n\theta)).$$

This is a Fourier expansion of $f(x)$ on $[-\pi, \pi]$ if we choose the coefficients $a_n R^n$ and $b_n R^n$ to be the Fourier coefficients of f on $[-\pi, \pi]$. This means that

$$a_0 = \frac{1}{\pi} \int_{-\pi}^{\pi} f(\xi) \, d\xi$$

and, for $n = 1, 2, \cdots$,

$$a_n = \frac{1}{\pi R^n} \int_{-\pi}^{\pi} f(\xi) \cos(n\xi) \, d\xi \text{ and } b_n = \frac{1}{\pi R^n} \int_{-\pi}^{\pi} f(\xi) \sin(n\xi) \, d\xi.$$

The solution is

$$u(r, \theta) = \frac{1}{2\pi} \int_{-\pi}^{\pi} f(\xi) \, d\xi$$

$$+ \frac{1}{\pi} \sum_{n=1}^{\infty} \left(\frac{r}{R}\right)^n \left(\int_{-\pi}^{\pi} f(\xi) \cos(n\xi) \, d\xi \cos(n\theta) + \int_{-\pi}^{\pi} f(\xi) \sin(n\xi) \, d\xi \sin(n\theta) \right).$$

$$(17.1)$$

This can be rearranged to

$$u(r, \theta) = \frac{1}{2\pi} \int_{-\pi}^{\pi} f(\xi)\, d\xi$$

$$+ \frac{1}{\pi} \sum_{n=1}^{\infty} \left(\frac{r}{R}\right)^n \int_{-\pi}^{\pi} f(\xi) \cos(n(\xi - \theta))\, d\xi. \tag{17.2}$$

This can also be written as

$$u(r, \theta) = \frac{1}{2\pi} \int_{-\pi}^{\pi} \left[1 + 2 \sum_{n=1}^{\infty} \left(\frac{r}{R}\right)^n \cos(n(\xi - \theta)) \right] f(\xi)\, d\xi, \tag{17.3}$$

which we will use to derive Poisson's integral formula.

EXAMPLE 17.2

We will solve the Dirichlet problem for the disk of radius 4 about the origin if $u(4, \theta) = f(\theta) = \theta^2$. Using equation (17.1), the solution is

$$u(r, \theta)$$

$$= \frac{1}{2\pi} \int_{-\pi}^{\pi} \xi^2\, d\xi + \frac{1}{\pi} \sum_{n=1}^{\infty} \left(\frac{r}{4}\right)^n \left(\int_{-\pi}^{\pi} \xi^2 \cos(n\xi)\, d\xi \cos(n\theta) + \int_{-\pi}^{\pi} \xi^2 \sin(n\xi)\, d\xi \sin(n\theta) \right)$$

$$= \frac{1}{3}\pi^2 + \sum_{n=1}^{\infty} \frac{4(-1)^n}{n^2} \left(\frac{r}{4}\right)^n \cos(n\theta). \ \blacklozenge$$

EXAMPLE 17.3

To solve the Dirichlet problem

$$\nabla^2 u(x, y) = 0 \text{ for } x^2 + y^2 < 9$$

and

$$u(x, y) = x^2 y^2 \text{ for } x^2 + y^2 = 9,$$

convert the problem to polar coordinates. Let $U(r, \theta) = u(r \cos(\theta), r \sin(\theta))$. On the boundary,

$$U(3, \theta) = 9 \cos^2(\theta) 9 \sin^2(\theta) = 81 \sin^2(\theta) \cos^2(\theta) = f(\theta).$$

The solution in polar coordinates is

$$U(r, \theta) = \frac{1}{2\pi} \int_{\pi}^{\pi} 81 \cos^2(\xi) \sin^2(\xi)\, d\xi$$

$$+ \frac{1}{\pi} \sum_{n=1}^{\infty} \left(\frac{r}{3}\right)^n \left[\int_{-\pi}^{\pi} 81 \cos^2(\xi) \sin^2(\xi) \cos(n\xi)\, d\xi \cos(n\theta) \right.$$

$$\left. + \int_{-\pi}^{\pi} 81 \cos^2(\xi) \sin^2(\xi) \sin(n\xi)\, d\xi \sin(n\theta) \right].$$

There remains to evaluate the integrals:

$$\frac{1}{2\pi} \int_{-\pi}^{\pi} 81 \cos^2(\xi) \sin^2(\xi)\, d\xi = \frac{81}{4}\pi,$$

$$\int_{-\pi}^{\pi} 81 \cos^2(\xi) \sin^2(\xi) \cos(n\xi)\, d\xi = \begin{cases} 0 & \text{if } n \neq 4 \\ -81\pi/8 & \text{if } n = 4, \end{cases}$$

and

$$\int_{-\pi}^{\pi} 81 \cos^2(\xi) \sin^2(\xi) \sin(n\xi)\, d\xi = 0.$$

The solution is

$$U(r,\theta) = \frac{1}{2\pi}\frac{81\pi}{4} - \frac{1}{\pi}\frac{81\pi}{8}\left(\frac{r}{3}\right)^4 \cos(4\theta)$$

$$= \frac{81}{8} - \frac{1}{8}r^4 \cos(4\theta).$$

To convert this solution to rectangular coordinates, use the fact that

$$\cos(4\theta) = 8\cos^4(\theta) - 8\cos^2(\theta) + 1$$

to obtain

$$U(r,\theta) = \frac{81}{8} - \frac{1}{8}(8r^4 \cos^4(\theta) - 8r^4 \cos^2(\theta) + r^4)$$

$$= \frac{81}{8} - \frac{1}{8}(8r^4 \cos^4(\theta) - 8r^2 r^2 \cos^2(\theta) + r^4).$$

Then

$$u(x,y) = \frac{81}{8} - \frac{1}{8}(8x^4 - 8(x^2 + y^2)x^2 + (x^2 + y^2)^2)$$

$$= \frac{81}{8} - \frac{1}{8}(x^4 + y^4 - 6x^2 y^2). \quad \blacklozenge$$

SECTION 17.3   PROBLEMS

In each of Problems 1 through 8, write the solution of the Dirichlet problem for the disk with the given boundary data.

1. $R = 3$, $f(\theta) = 1$

2. $R = 3$, $f(\theta) = 8\cos(4\theta)$

3. $R = 2$, $f(\theta) = \theta^2 - \theta$

4. $R = 5$, $f(\theta) = \theta \cos(\theta)$

5. $R = 4$, $f(\theta) = e^{-\theta}$

6. $R = 1$, $f(\theta) = \sin^2(\theta)$

7. $R = 8$, $f(\theta) = 1 - \theta^2$

8. $R = 4$, $f(\theta) = \theta e^{2\theta}$

In each of Problems 9 through 12, solve the problem by converting it to polar coordinates.

9. $\nabla^2 u(x,y) = 0$ for $x^2 + y^2 < 16$,
   $u(x,y) = x^2$ for $x^2 + y^2 = 16$

10. $\nabla^2 u(x,y) = 0$ for $x^2 + y^2 < 9$,
    $u(x,y) = x - y$ for $x^2 + y^2 = 9$

11. $\nabla^2 u(x,y) = 0$ for $x^2 + y^2 < 4$,
    $u(x,y) = x^2 - y^2$ for $x^2 + y^2 = 4$

12. $\nabla^2 u(x,y) = 0$ for $x^2 + y^2 < 25$,
    $u(x,y) = xy$ for $x^2 + y^2 = 25$

17.4   Poisson's Integral Formula

We will derive an integral formula due to Poisson for the Dirichlet problem for a disk. Suppose the disk is centered at the origin and has radius 1, and that $u(1, \theta) = f(\theta)$. By equation (17.3), the solution with $R = 1$ is

$$u(r, \theta) = \frac{1}{2\pi} \int_{-\pi}^{\pi} \left[1 + 2 \sum_{n=1}^{\infty} r^n \cos(n(\xi - \theta)) \right] f(\xi)\, d\xi.$$

The quantity

$$P(r, \zeta) = \frac{1}{2\pi} \left[1 + 2 \sum_{n=1}^{\infty} r^n \cos(n\zeta) \right]$$

is called the *Poisson kernel*. In terms of this kernel function, the solution is

$$u(r, \theta) = \int_{-\pi}^{\pi} P(r, \xi - \theta) f(\xi)\, d\xi.$$

We will evaluate the sum in the Poisson kernel in closed form, yielding Poisson's integral formula for the solution. Let z be a complex number. In polar form, $z = re^{i\zeta}$ with $r < 1$ inside the unit disk. By Euler's formula,

$$z^n = r^n e^{in\zeta} = r^n \cos(n\zeta) + i r^n \sin(n\zeta).$$

This enables us to recognize $r^n \cos(n\zeta)$, which appears in the Poisson kernel, as the real part of z^n and write

$$1 + 2 \sum_{n=1}^{\infty} r^n \cos(n\zeta) = \text{Re} \left(1 + 2 \sum_{n=1}^{\infty} z^n \right).$$

But $|z| = r < 1$, so this sum is just a geometric series:

$$\sum_{n=1}^{\infty} z^n = \frac{z}{1 - z}.$$

Combining these observations, we have

$$1 + 2 \sum_{n=1}^{\infty} r^n \cos(n\theta) = \text{Re} \left(1 + 2 \sum_{n=1}^{\infty} z^n \right) = \text{Re} \left(1 + 2 \frac{z}{1 - z} \right)$$

$$= \text{Re} \left(\frac{1 + z}{1 - z} \right) = \text{Re} \left(\frac{1 + re^{i\zeta}}{1 - re^{i\zeta}} \right).$$

A simple manipulation gives us

$$\frac{1 + re^{i\zeta}}{1 - re^{i\zeta}} = \frac{1 - r^2 + 2ir \sin(\zeta)}{1 + r^2 - 2r \cos(\zeta)}.$$

The real part of this is easily extracted, yielding

$$1 + 2 \sum_{n=1}^{\infty} r^n \cos(n\zeta) = \frac{1 - r^2}{1 + r^2 - 2r \cos(\zeta)}.$$

Therefore, the solution of the Dirichlet problem for the unit disk is

$$u(r, \theta) = \frac{1}{2\pi} \int_{-\pi}^{\pi} \frac{1 - r^2}{1 + r^2 - 2r \cos(\xi - \theta)} f(\xi)\, d\xi.$$

This is *Poisson's integral formula.* For a disk of radius R, a simple change of variables gives us the solution

$$u(r,\theta) = \frac{1}{2\pi} \int_{-\pi}^{\pi} \frac{R^2 - r^2}{R^2 + r^2 - 2Rr\cos(\xi - \theta)} f(\xi)\, d\xi.$$

EXAMPLE 17.4

The solution of the problem of Example 17.2 also can be written as

$$u(r,\theta) = \frac{1}{2\pi} \int_{-\pi}^{\pi} \frac{16 - r^2}{16 + r^2 - 8r\cos(\xi - \theta)} \xi^2\, d\xi$$

$$= \frac{16 - r^2}{2\pi} \int_{-\pi}^{\pi} \frac{\xi^2}{16 + r^2 - 8r\cos(\xi - \theta)}\, d\xi$$

for $0 \leq r < 4, -\pi \leq \theta \leq \pi$. This integral solution may be more suitable for numerical approximations of values of the solution at specific points than the infinite series solution. ◆

SECTION 17.4   *PROBLEMS*

In each of Problems 1 through 4, find an integral formula for the solution of the Dirichlet problem. Use a numerical integration routine to approximate $u(r,\theta)$ at the given points.

1. $R = 1, f(\theta) = \theta; (1/2, \pi), (3/4, \pi/3), (0.2, \pi/4)$

2. $R = 4, f(\theta) = \sin(4\theta); (1, \pi/6), (3, 7\pi/2), (1, \pi/4), (2.5, \pi/12)$

3. $R = 15, f(\theta) = \theta^3 - \theta; (4, \pi), (12, 3\pi/2), (8, \pi/4), (7, 0)$

4. $R = 6, f(\theta) = e^{-\theta}; (5.5, 3\pi/5), (4, 2\pi/7), (1, \pi), (4, 9\pi/4)$

17.5   Dirichlet Problem for Unbounded Regions

When D is unbounded (has points arbitrarily far from the origin), we may use a Fourier integral or transform to solve a Dirichlet problem on D.

17.5.1   The Upper Half-Plane

We will solve the problem

$$\nabla^2 u(x,y) = 0 \text{ for } -\infty < x < \infty, y > 0$$

and

$$u(x,0) = f(x) \text{ for } -\infty < x < \infty.$$

This is a Dirichlet problem because the horizontal axis is the boundary of the upper half-plane. We seek a bounded solution.

Put $u(x,y) = X(x)Y(y)$ and obtain

$$X'' + \lambda X = 0 \text{ and } Y'' - \lambda Y = 0.$$

The eigenvalues are $\lambda = \omega^2$ with $\omega \geq 0$, and the eigenfunctions are

$$X_\omega(x) = a_\omega \cos(\omega x) + b_\omega \sin(\omega x).$$

The equation for Y is $Y'' - \omega^2 Y = 0$ with constant multiples of $e^{-\omega y}$ as bounded solutions because $y \geq 0$. For each $\omega \geq 0$, we have a solution

$$u_\omega(x, y) = (a_\omega \cos(\omega x) + b_\omega \sin(\omega x))e^{-\omega y}$$

of Laplace's equation. To obtain a solution satisfying the boundary condition, use the superposition

$$u(x, y) = \int_0^\infty (a_\omega \cos(\omega x) + b_\omega \sin(\omega x))e^{-\omega y}\, d\omega.$$

We need

$$u(x, 0) = f(x) = \int_0^\infty (a_\omega \cos(\omega x) + b_\omega \sin(\omega x))\, d\omega.$$

The coefficients are the Fourier integral coefficients of f on the real line:

$$a_\omega = \frac{1}{\pi}\int_{-\infty}^\infty f(\xi)\cos(\omega\xi)\, d\xi \quad \text{and} \quad b_\omega = \frac{1}{\pi}\int_{-\infty}^\infty f(\xi)\sin(\omega\xi)\, d\xi.$$

Insert these coefficients into the integral expression for $u(x, y)$:

$$u(x, y) = \frac{1}{\pi}\int_0^\infty \int_{-\infty}^\infty [f(\xi)\cos(\omega\xi)\cos(\omega x) + f(\xi)\sin(\omega\xi)\sin(\omega x)]e^{-\omega y}\, d\xi\, d\omega$$

$$= \frac{1}{\pi}\int_{-\infty}^\infty \left[\int_0^\infty \cos(\omega(\xi - x))e^{-\omega y}\, d\omega\right] f(\xi)\, d\xi.$$

The inner integral can be evaluated explicitly as

$$\int_0^\infty \cos(\omega(\xi - x))e^{-\omega y}\, d\omega$$

$$= \left[\frac{e^{-\omega y}}{y^2 + (\xi - x)^2}[-y\cos(\omega(\xi - x)) + (\xi - x)\sin(\omega(\xi - x))]\right]_0^\infty$$

$$= \frac{y}{y^2 + (\xi - x)^2}.$$

The solution of the Dirichlet problem for the upper half-plane is therefore

$$u(x, y) = \frac{y}{\pi}\int_{-\infty}^\infty \frac{f(\xi)}{y^2 + (\xi - x)^2}\, d\xi. \tag{17.4}$$

Solution Using the Fourier Transform

We can also solve this problem for the upper half-plane using the Fourier transform. Apply the transform in the x variable to Laplace's equation to obtain

$$\mathcal{F}\left[\frac{\partial^2 u}{\partial x^2}\right] + \mathcal{F}\left[\frac{\partial^2 u}{\partial y^2}\right] = \frac{\partial^2 \hat{u}}{\partial y^2}(\omega, y) - \omega^2 \hat{u}(\omega, y) = 0.$$

This has general solution

$$\hat{u}(\omega, y) = a_\omega e^{\omega y} + b_\omega e^{-\omega y}.$$

We want this to be bounded. But for $\omega > 0$, $e^{\omega y} \to \infty$ as $y \to \infty$, so $a_\omega = 0$ for positive ω. Also, $e^{-\omega y} \to \infty$ as $y \to \infty$ if $\omega < 0$, so for negative ω, we must have $b_\omega = 0$. Therefore,

$$\hat{u}(\omega, y) = \begin{cases} b_\omega e^{-\omega y} & \text{for } \omega > 0 \\ a_\omega e^{\omega y} & \text{for } \omega < 0. \end{cases}$$

We can consolidate these cases by writing

$$\hat{u}(\omega, y) = c_\omega e^{-|\omega|y}.$$

To solve for the constants, take the transform of $u(x, 0) = f(x)$ to get

$$\hat{u}(\omega, 0) = \hat{f}(\omega) = c_\omega.$$

Then

$$\hat{u}(\omega, y) = \hat{f}(\omega)e^{-|\omega|y}.$$

To retrieve the solution u apply the inverse Fourier transform to get

$$u(x, y) = \mathcal{F}^{-1}\left[\hat{f}(\omega)e^{-|\omega|y}\right](x)$$

$$= \frac{1}{2\pi} \int_{-\infty}^{\infty} \hat{f}(\omega)e^{-|\omega|y}e^{i\omega x}\, d\omega$$

$$= \frac{1}{2\pi} \int_{-\infty}^{\infty} \left(\int_{-\infty}^{\infty} f(\xi)e^{-i\omega\xi}\, d\xi\right) e^{-|\omega|y}e^{i\omega x}\, d\omega$$

$$= \frac{1}{2\pi} \int_{-\infty}^{\infty} \left(\int_{-\infty}^{\infty} e^{-|\omega|y}e^{-i\omega(\xi-x)}\, d\omega\right) f(\xi)\, d\xi.$$

A routine integration gives us

$$\int_{-\infty}^{\infty} e^{-|\omega|y}e^{-i\omega(\xi-x)}\, d\omega = \frac{2y}{y^2 + (\xi - x)^2}.$$

Then

$$u(x, y) = \frac{1}{2\pi} \int_{-\infty}^{\infty} \left(\frac{2y}{y^2 + (\xi - x)^2}\right) f(\xi)\, d\xi$$

$$= \frac{y}{\pi} \int_{-\infty}^{\infty} \frac{f(\xi)}{y^2 + (\xi - x)^2}\, d\xi,$$

in agreement with the solution obtained by separation of variables.

17.5.2 The Right Quarter-Plane

The right quarter-plane has the nonnegative horizontal and vertical axes as boundary. The Dirichlet problem for this region is

$$\nabla^2 u(x, y) = 0 \text{ for } x > 0, y > 0,$$

$$u(x, 0) = f(x) \text{ for } x \geq 0,$$

and

$$u(0, y) = g(y) \text{ for } y \geq 0.$$

This problem can be solved by solving separately the cases that $f(x)$ or $g(y)$ is identically zero. Separation of variables applies to both cases, and the solution of the given problem is the sum of the solutions of these simpler problems.

We will demonstrate a different method for the case that $g(y) = 0$. Notice that, if we fold the upper half-plane across the vertical axis, we obtain the right quarter-plane. This suggests that we might be able to use the solution for the upper half-plane to solve the problem for the right quarter-plane. To do this, let

$$w(x) = \begin{cases} f(x) & \text{for } x \geq 0 \\ \text{anything} & \text{for } x < 0. \end{cases}$$

where by "anything" we mean we will fill in this part shortly. We now have a Dirichlet problem for the upper half-plane with data function $u(x, 0) = w(x)$. We know the solution u_{hp} of this problem for the upper half-plane:

$$u_{hp}(x, y) = \frac{y}{\pi} \int_{-\infty}^{\infty} \frac{w(\xi)}{y^2 + (\xi - x)^2} d\xi.$$

Write this as

$$u_{hp}(x, y) = \frac{y}{\pi} \left[\int_{-\infty}^{0} \frac{w(\xi)}{y^2 + (\xi - x)^2} d\xi + \int_{0}^{\infty} \frac{w(\xi)}{y^2 + (\xi - x)^2} d\xi \right].$$

Change variables in the first integral on the right by letting $\zeta = -\xi$:

$$\int_{-\infty}^{0} \frac{w(\xi)}{y^2 + (\xi - x)^2} d\xi = \int_{0}^{\infty} \frac{w(-\zeta)}{y^2 + (\zeta + x)^2} (-1) d\zeta.$$

For uniformity in notation, replace the dummy variable of integration on the right with ξ to write

$$u_{hp}(x, y) = \frac{y}{\pi} \left[\int_{\infty}^{0} \frac{w(-\xi)}{y^2 + (\xi + x)^2} (-1) d\xi + \int_{0}^{\infty} \frac{w(\xi)}{y^2 + (\xi - x)^2} d\xi \right]$$

$$= \frac{y}{\pi} \int_{0}^{\infty} \left(\frac{w(-\xi)}{y^2 + (\xi + x)^2} + \frac{f(\xi)}{y^2 + (\xi - x)^2} \right) d\xi.$$

In the last integral, we used the fact that $w(\xi) = f(\xi)$ for $\xi \geq 0$. Now fill in the "anything" in the definition of w. Notice that the last integral will vanish at points $(0, y)$ on the positive y axis if $f(\xi) + w(-\xi) = 0$ for $\xi \geq 0$. This will occur if $w(-\xi) = -f(\xi)$. Make w the odd extension of f to the entire line. In this way, we obtain the solution for the upper half-plane as

$$u_{hp}(x, y) = \frac{y}{\pi} \int_{0}^{\infty} \left(\frac{1}{y^2 + (\xi - x)^2} - \frac{1}{y^2 + (\xi + x)^2} \right) f(\xi) d\xi.$$

But this function is also harmonic on the right quarter-plane, vanishes when $x = 0$, and equals $f(x)$ if $x \geq 0$ and $y = 0$. Therefore, this function is also the solution of the problem for the right quarter-plane (in this case of zero data along the positive y axis).

EXAMPLE 17.5

We have a formula for the solution of

$$\nabla^2 u(x, y) = 0 \text{ for } x > 0, y > 0,$$

$$u(0, y) = 0 \text{ for } y > 0,$$

and

$$u(x, 0) = 1 \text{ for } x > 0.$$

With $f(x) = 1$, we can write the solution

$$u(x, y) = \frac{y}{\pi} \int_{0}^{\infty} \frac{1}{y^2 + (\xi - x)^2} d\xi - \frac{y}{\pi} \int_{0}^{\infty} \frac{1}{y^2 + (\xi + x)^2} d\xi.$$

A routine integration yields

$$\frac{y}{\pi} \int_{0}^{\infty} \frac{1}{y^2 + (\xi - x)^2} d\xi = \frac{1}{2} + \frac{1}{\pi} \arctan\left(\frac{x}{y}\right)$$

and

$$\frac{y}{\pi} \int_0^\infty \frac{1}{y^2 + (\xi + x)^2} \, d\xi = \frac{1}{2} - \frac{1}{\pi} \arctan\left(\frac{x}{y}\right).$$

The solution is

$$u(x, y) = \frac{2}{\pi} \arctan\left(\frac{x}{y}\right).$$

This function is harmonic on the right quarter-plane and $u(0, y) = 0$ for $y > 0$. Furthermore, if $x > 0$,

$$\lim_{y \to 0+} \frac{2}{\pi} \arctan\left(\frac{x}{y}\right) = \frac{2}{\pi}\frac{\pi}{2} = 1,$$

as required. ◆

SECTION 17.5   PROBLEMS

1. Write an integral solution for the Dirichlet problem for the upper half-plane if

$$u(x, 0) = \begin{cases} -1 & \text{for } -4 \leq x < 0 \\ 1 & \text{for } 0 \leq x < 4 \\ 0 & \text{for } |x| > 4. \end{cases}$$

2. Write an integral solution for the Dirichlet problem for the upper half-plane if $u(x, 0) = e^{-|x|}$.

3. Write an integral solution for the Dirichlet problem for the right quarter-plane if $u(x, 0) = e^{-x} \cos(x)$ for $x > 0$ and $u(0, y) = 0$ for $y > 0$.

4. Write an integral solution for the Dirichlet problem for the right quarter-plane if $u(x, 0) = 0$ for $x > 0$ and $u(0, y) = g(y)$ for $y > 0$. Use separation of variables and then derive a solution using an appropriate Fourier transform.

5. Find an integral solution of the Dirichlet problem for the right quarter-plane if $u(x, 0) = f(x)$ and $u(0, y) = g(y)$.

6. Write an integral solution for the Dirichlet problem for the lower half-plane $y < 0$ if $u(x, 0) = f(x)$.

7. Find the steady-state temperature distribution in a thin, homogeneous, flat plate extending over the right quarter plane if the temperature on the vertical side is e^{-y} and the temperature on the horizontal side is maintained at zero.

8. Solve the Dirichlet problem for the strip $-\infty < x < \infty$, $0 < y < 1$ if $u(x, 0) = 0$ for $x < 0$ and $u(x, 0) = e^{-\alpha x}$ for $x > 0$ with α a positive number.

9. Solve for the steady-state temperature distribution in an infinite, homogeneous, flat plate covering the half-plane $y \geq 0$ if the temperature on the boundary $y = 0$ is kept at zero for $x < 4$, constant A for $4 \leq x \leq 8$, and zero for $x > 8$.

10. Solve the following problem:

$$\nabla^2 u(x, y) = 0 \text{ for } 0 < x < \pi, 0 < y < 2$$

with boundary conditions $u(0, y) = 0$ and $u(\pi, y) = 4$ for $0 < y < 2$, and

$$\frac{\partial u}{\partial y}(x, 0) = u(x, 2) = 0 \text{ for } 0 < x < \pi.$$

11. Solve for the steady-state temperature distribution in a homogeneous, infinite, flat plate covering the half-plane $x \geq 0$ if the temperature on the boundary $x = 0$ is $f(y)$, where

$$f(y) = \begin{cases} 1 & \text{for } |y| \leq 1 \\ 0 & \text{for } |y| > 1. \end{cases}$$

12. Write a general expression for the steady-state temperature distribution in an infinite, homogeneous, flat plate covering the strip $0 \leq y \leq 1$, $x \geq 0$ if the temperature on the left boundary and on the bottom side is zero while the temperature on the top part of the boundary is $f(x)$.

17.6   A Dirichlet Problem for a Cube

To illustrate a Dirichlet problem in three dimensions we will solve:

$$\nabla^2 u(x, y, z) = 0 \text{ for } 0 < x < A, 0 < y < B, 0 < z < C,$$

$$u(x, y, 0) = u(x, y, C) = 0,$$

$$u(0, y, z) = u(A, y, z) = 0,$$

and

$$u(x, 0, z) = 0, u(x, B, z) = f(x, z).$$

Let $u(x, y, z) = X(x)Y(y)Z(z)$ to obtain

$$\frac{X''}{X} = -\frac{Y''}{Y} - \frac{Z''}{Z} = -\lambda.$$

After a second separation (of y and z), we have

$$\frac{Z''}{Z} = \lambda - \frac{Y''}{Y} = -\mu.$$

Then,

$$X'' + \lambda X = 0, \ Z'' + \mu Z = 0, \text{ and } Y'' - (\lambda + \mu)Y = 0.$$

From the boundary conditions,

$$X(0) = X(A) = 0, \ Z(0) = Z(C) = 0, \text{ and } Y(0) = 0.$$

The problems for X and Z have eigenvalues and eigenfunctions

$$\lambda_n = \frac{n^2\pi^2}{A^2} \text{ and } X_n(x) = \sin\left(\frac{n\pi x}{A}\right)$$

and

$$\mu_m = \frac{m^2\pi^2}{C^2} \text{ and } Z_m(z) = \sin\left(\frac{m\pi z}{C}\right),$$

with n and m independently varying over the positive integers. The problem for Y is

$$Y'' - \left(\frac{n^2\pi^2}{A^2} + \frac{m^2\pi^2}{C^2}\right)Y = 0; \ Y(0) = 0$$

with solutions that are constant multiples of

$$Y_{nm}(y) = \sinh(\beta_{nm}y),$$

where

$$\beta_{nm} = \sqrt{\frac{n^2\pi^2}{A^2} + \frac{m^2\pi^2}{C^2}}.$$

Attempt a solution

$$u(x, y, z) = \sum_{n=1}^{\infty}\sum_{m=1}^{\infty} c_{nm} \sin\left(\frac{n\pi x}{A}\right) \sin\left(\frac{m\pi z}{C}\right) \sinh(\beta_{nm}y).$$

We must choose the coefficients so that

$$u(x, B, z) = f(x, z) = \sum_{n=1}^{\infty}\sum_{m=1}^{\infty} c_{nm} \sin\left(\frac{n\pi x}{A}\right) \sin\left(\frac{m\pi z}{C}\right) \sinh(\beta_{nm}B).$$

This is a double Fourier series for $f(x, z)$ on $0 \le x \le A$ and $0 \le z \le C$. We have seen this type of expansion before (Sections 15.7 and 16.5), leading us to choose

$$c_{nm} = \frac{2}{AC \sinh(\beta_{nm} B)} \int_0^A \int_0^C f(\xi, \zeta) \sin\left(\frac{n\pi\xi}{A}\right) \sin\left(\frac{m\pi\zeta}{C}\right) d\zeta \, d\xi.$$

As usual, if nonzero data is prescribed on more than one face, split the Dirichlet problem into a sum of problems, on each of which there is nonzero data on only one face.

SECTION 17.6   PROBLEMS

1. Solve

$$\nabla^2 u(x, y, z) = 0 \text{ for } 0 < x < 1, 0 < y < 1, 0 < z < 1,$$

$$u(0, y, z) = u(1, y, z) = 0,$$

$$u(x, 0, z) = u(x, 1, z) = 0,$$

$$u(x, y, 0) = 0, u(x, y, 1) = xy.$$

2. Solve

$$\nabla^2 u(x, y, z) = 0 \text{ for } 0 < x < 2\pi, 0 < y < 2\pi, 0 < z < 1,$$

$$u(x, y, 0) = u(x, y, 1) = 0,$$

$$u(x, 0, z) = u(x, 2\pi, z) = 0,$$

$$u(0, y, z) = 0, u(2\pi, y, z) = z.$$

3. Solve

$$\nabla^2 u(x, y, z) = 0 \text{ for } 0 < x < 1, 0 < y < 2\pi, 0 < z < \pi,$$

$$u(0, y, z) = u(1, y, z) = 0,$$

$$u(x, 0, z) = u(x, y, 0) = 0,$$

$$u(x, y, \pi) = x(\pi - x)(\pi - y), u(x, 2\pi, z) = 2.$$

4. Solve

$$\nabla^2 u(x, y, z) = 0 \text{ for } 0 < x < 1, 0 < y < 2, 0 < z < \pi,$$

$$u(x, 0, z) = u(x, 2, z) = 0,$$

$$u(0, y, z) = u(x, y, \pi) = 0,$$

$$u(x, y, 0) = x^2(1 - x)y(2 - y), u(1, y, z)$$

$$= \sin(\pi y) \sin(z).$$

17.7   Steady-State Heat Equation for a Sphere

We will solve for the steady-state temperature function in a solid sphere, given the temperature at all times on the surface.

Let the sphere be centered at the origin and have a radius of R. Use spherical coordinates (ρ, θ, φ), in which ρ is the distance from the origin to the point, θ is the polar angle between the positive x axis and the projection onto the (x, y) plane of the line from the origin to the point, and φ is the angle of declination from the positive z axis to this line (Figure 17.3).

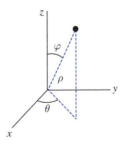

FIGURE 17.3 *Spherical coordinates.*

Assuming symmetry of the temperature function about the z axis, the solution is independent of θ and depends only on ρ and φ. Laplace's equation in spherical coordinates (with independence from θ) is

$$\nabla^2 u(\rho, \varphi) = \frac{\partial^2 u}{\partial \rho^2} + \frac{2}{\rho}\frac{\partial u}{\partial \rho} + \frac{1}{\rho^2}\frac{\partial^2 u}{\partial \varphi^2} + \frac{\cot(\varphi)}{\rho^2}\frac{\partial u}{\partial \varphi} = 0.$$

The temperature on the surface is $u(R, \varphi) = f(\varphi)$ with f given.

Let $u(\rho, \varphi) = X(\rho)\Phi(\varphi)$ to obtain

$$X''\Phi + \frac{2}{\rho}X'\Phi + \frac{1}{\rho^2}X\Phi'' + \frac{\cot(\varphi)}{\rho^2}X\Phi' = 0.$$

Upon dividing this equation by $X\Phi$, we can separate the variables, obtaining

$$\frac{\Phi''}{\Phi} + \cot(\varphi)\frac{\Phi'}{\Phi} = -\rho^2\frac{X''}{X} - 2\rho\frac{X'}{X} = -\lambda.$$

Then

$$\rho^2 X'' + 2\rho X' - \lambda X = 0 \text{ and } \Phi'' + \cot(\varphi)\Phi' + \lambda\Phi = 0.$$

To solve this equation for Φ, write it as

$$\frac{1}{\sin(\varphi)}[\Phi'\sin(\varphi)]' + \lambda\Phi = 0. \tag{17.5}$$

Change variables by putting $x = \cos(\varphi)$. Then $\varphi = \arccos(x)$. Let

$$G(x) = \Phi(\arccos(x)).$$

Since $0 \le \varphi \le \pi$, then $-1 \le x \le 1$. Compute

$$\Phi'(\varphi)\sin(\varphi) = \sin(\varphi)\frac{d\Phi}{dx}\frac{dx}{d\varphi}$$
$$= \sin(\varphi)G'(x)[-\sin(\varphi)]$$
$$= -\sin^2(\varphi)G'(x) = -[1 - \cos^2(x)]G'(x)$$
$$= -(1 - x^2)G'(x).$$

Then

$$\frac{d}{d\varphi}[\Phi'(\varphi)\sin(\varphi)] = -\frac{d}{d\varphi}[(1-x^2)G'(x)]$$
$$= -\frac{d}{dx}[(1-x^2)G'(x)]\frac{dx}{d\varphi}$$
$$= -\frac{d}{dx}[(1-x^2)G'(x)][-\sin(\varphi)].$$

We conclude that

$$\frac{1}{\sin(\varphi)}\frac{d}{d\varphi}[\Phi'(\varphi)\sin(\varphi)] = \frac{d}{dx}[(1-x^2)G'(x)].$$

The point to this calculation is that equation (17.5) transforms to

$$[(1-x^2)G'(x)]' + \lambda G(x) = 0,$$

which is Legendre's differential equation (Section 14.2). For bounded solutions on $[-1, 1]$, the eigenvalues are $\lambda_n = n(n+1)$ for $n = 0, 1, 2, \cdots$, and the eigenfunctions are constant multiples of the Legendre polynomials $P_n(x)$. For nonnegative integer n, we therefore have a solution of the differential equation for Φ:

$$\Phi_n(\varphi) = G(\cos(\varphi)) = P_n(\cos(\varphi)).$$

Now that we know the eigenvalues, the differential equation for X is

$$\rho^2 X'' + 2\rho X' - n(n+1)X = 0.$$

This has the general solution

$$X(\rho) = a\rho^n + b\rho^{-n-1}.$$

Choose $b = 0$ to have a solution that is bounded as $\rho \to 0+$, which is the center of the sphere. For each nonnegative integer n, we now have a function

$$u_n(\rho, \varphi) = a_n \rho^n P_n(\cos(\varphi))$$

that satisfies the steady-state heat equation. To satisfy the boundary condition, write a superposition

$$u(\rho, \varphi) = \sum_{n=0}^{\infty} a_n \rho^n P_n(\cos(\varphi)).$$

We must choose the coefficients to satisfy

$$u(R, \varphi) = \sum_{n=0}^{\infty} a_n R^n P_n \cos(\varphi).$$

To put this into the context of an eigenfunction expansion in terms of Legendre polynomials, recall that $\varphi = \arccos(x)$ to obtain

$$\sum_{n=0}^{\infty} a_n R^n P_n(x) = f(\arccos(x)).$$

Then

$$a_n R^n = \frac{2n+1}{2} \int_{-1}^{1} f(\arccos(x)) P_n(x)\, dx$$

so

$$a_n = \frac{2n+1}{2R^n} \int_{-1}^{1} f(\arccos(x)) P_n(x)\, dx.$$

The steady-state temperature function is

$$u(\rho, \varphi) = \sum_{n=0}^{\infty} \frac{2n+1}{2} \left(\int_{-1}^{1} f(\arccos(x)) P_n(x)\, dx \right) \left(\frac{\rho}{R} \right)^n P_n(\cos(\varphi)).$$

EXAMPLE 17.6

For $f(\varphi) = \varphi$ the solution is

$$u(\rho, \varphi) = \sum_{n=0}^{\infty} \frac{2n+1}{2} \left(\int_{-1}^{1} \arccos(x) P_n(x)\, dx \right) \left(\frac{\rho}{R} \right)^n P_n(\cos(\varphi)).$$

We will use numerical integrations to approximate the first six coefficients. $P_0(x), \cdots, P_5(x)$ are given in Section 14.2. Compute

$$\int_{-1}^{1} \arccos(x)\, dx \approx \pi,$$

$$\int_{-1}^{1} x \arccos(x)\, dx \approx -0.7854,$$

$$\int_{-1}^{1} \frac{1}{2}(3x^2 - 1) \arccos(x)\, dx = 0,$$

$$\int_{-1}^{1} \frac{1}{2}(5x^3 - 3x) \arccos(x)\, dx \approx -.049087,$$

$$\int_{-1}^{1} \frac{1}{8}(35x^4 - 30x^2 + 3) \arccos(x)\, dx = 0,$$

and

$$\int_{-1}^{1} \frac{1}{8}(63x^5 - 70x^3 + 15x) \arccos(x)\, dx \approx -0.012272.$$

Then

$$u(\rho, \varphi) \approx \frac{\pi}{2} - \frac{3}{2}(0.7854)\frac{\rho}{R}\cos(\varphi) - \frac{7}{2}(0.049087)\frac{1}{2}\left(\frac{\rho}{R}\right)^3 (5\cos^3(\varphi) - 3\cos(\varphi))$$

$$- \frac{11}{2}(0.012272)\left(\frac{\rho}{R}\right)^5 \frac{1}{8}(63\cos^5(\varphi) - 70\cos^3(\varphi) + 15\cos(\varphi)). \quad \blacklozenge$$

SECTION 17.7   *PROBLEMS*

In each of Problems 1 through 4, write a solution for the steady-state temperature distribution in the sphere if the boundary data is given by $f(\varphi)$. Use numerical integration to approximate the first six terms in the solution.

1. $f(\varphi) = A\varphi^2$, in which A is a positive number.
2. $f(\varphi) = \sin(\varphi)$

3. $f(\varphi) = \varphi^3$
4. $f(\varphi) = 2 - \varphi^2$
5. Solve for the steady-state temperature distribution in a hollowed-out sphere given in spherical coordinates by $R_1 \leq \rho \leq R_2$. The inner surface is kept at constant temperature T and the outer surface at temperature zero. Assume that u is a function of ρ and φ only.

APPENDIX A

Guide to Notation

This is a brief guide to some of the notation and symbols used. Each symbol is given with a section in which it is defined or used.

$\mathcal{L}[f]$ Laplace transform of f (3.1)

$\mathcal{L}[f](s)$ Laplace transform of f evaluated at s (3.1)

$\mathcal{L}^{-1}[F]$ inverse Laplace transform of F (3.1)

$H(t)$ Heaviside function (3.3.2)

$\delta(t)$ delta function (3.5)

$< a, b, c >$ vector with components a, b, c (5.1)

$a\mathbf{i} + b\mathbf{j} + c\mathbf{k}$ standard form of a vector (5.1)

$\| \mathbf{V} \|$ norm (magnitude) of a vector $\mathbf{V}$ (5.1)

$\mathbf{F} \cdot \mathbf{G}$ dot product of vectors $\mathbf{F}$ and $\mathbf{G}$ (5.2)

$\mathbf{F} \times \mathbf{G}$ cross product of $\mathbf{F}$ and $\mathbf{G}$ (5.3)

R^n n - space, consisting of $n-$ vectors $< x_1, x_2, \cdots, x_n >$ (5.4)

$[a_{ij}]$ matrix whose i, j - element is a_{ij} (6.1)

$\mathbf{O}_{nm}$ $n \times m$ zero matrix (6.1)

$\mathbf{I}_n$ $n \times n$ identity matrix (6.1)

$\mathbf{A}^t$ transpose of $\mathbf{A}$ (6.1)

$\mathbf{A}_R$ reduced form of $\mathbf{A}$ (6.2)

rank$(\mathbf{A})$ rank of a matrix $\mathbf{A}$ (6.2)

$[\mathbf{A} \vdots \mathbf{B}]$ augmented matrix (6.3)

$\mathbf{A}^{-1}$ inverse of the matrix $\mathbf{A}$ (6.4)

$|\mathbf{A}|$ or det $\mathbf{A}$ determinant of $\mathbf{A}$ (7.1)

$p_{\mathbf{A}}(\lambda)$ characteristic polynomial of $\mathbf{A}$ (8.1)

$\mathbf{\Omega}$ often denotes the fundamental matrix of a system $\mathbf{X}' = \mathbf{A}\mathbf{X}$ (9.1)

$\mathbf{T}$ often denotes a tangent vector (10.2)

$\mathbf{N}$ often denotes a normal vector (10.2)

κ curvature (10.2)

∇ del operator (10.4)

$\nabla \varphi$ or grad φ gradient of φ (10.4)

$D_{\mathbf{u}}\varphi(P)$ directional derivative of φ in the direction of $\mathbf{u}$ at P (10.4)

$\int_C f\,dx + g\,dy + h\,dz$ line integral (11.1)

$\int_C \mathbf{F} \cdot d\mathbf{R}$ alternative notation for $\int_C f\,dx + g\,dy + h\,dz$, with $\mathbf{F} = f\mathbf{i} + g\mathbf{j} + h\mathbf{k}$ (11.1)

$\int_C f(x, y, z)\,ds$ line integral of f over C with respect to arc length (11.1.1)

$\frac{\partial(f,g)}{\partial(u,v)}$ Jacobian of f and g with respect to u and v (11.5.1)

$\int\int_\Sigma f(x, y, z)\,d\sigma$ surface integral of f over Σ (11.5.4)

$\mathbf{n}$ often denotes a unit normal vector (11.6)

$f(x_0-)$, $f(x_0+)$ left and right limits, respectively, of $f(x)$ at x_0 (12.1)

$\mathcal{F}[f]$ or $\hat{f}$ Fourier transform of f (13.3)

$\mathcal{F}_C[f]$ or $\hat{f}_C$ Fourier cosine transform of f (13.4)

$\mathcal{F}_S[f]$ or $\hat{f}_S$ Fourier sine transform of f (13.4)

$P_n(x)$ the nth Legendre polynomial (14.2)

$\Gamma(x)$ gamma function (14.3)

$J_\nu(x)$ Bessel function of the first kind of order ν (14.3)

$Y_n(x)$ Bessel function of the second kind of order n (14.3)

$\nabla^2 u$ Laplacian of u (17.1)

A MAPLE Primer

This section is designed to assist students in using MAPLE to carry out some of the calculations encountered in analyzing mathematical models in science and engineering. In most cases, examples are given which can be easily adapted to general use. Additional examples and details can be found in MAPLE's HELP function.

Beginning Computations

Numerical computations are carried out as one might expect, with an asterisk * denoting a product and a wedge $\wedge$ a power. If we type

```
2 ∧ 14;
```

we obtain the fourteenth power of 2, which is $16, 380$. Note the semi-colon ending this MAPLE command. Semi-colons will generally end commands. To multiply this number by 19, type

```
19*2 ∧ 14;
```

Pi is stored in MAPLE as `Pi` (note the upper case P—MAPLE is case sensitive). The exponential function is denoted `exp`, and the number e is obtained as `exp(1)`. If we enter

```
(Pi ∧ 2)*exp(1);
```

this will simply return the symbolic product

$$(\pi^2)e$$

To obtain the (approximate) decimal value of this product, use the `evalf` command:

```
evalf((Pi ∧ 2)*exp(1));
```

This will return the decimal 26.82836630. As another example,

```
evalf(cos(3)+sin(3));
```

will return the decimal value $-.8488724885$ of $\cos(3) + \sin(3)$.

To solve an equation, use the `solve` command. For example,

```
solve(x∧2+2*x-1=0,x);
```

will return the roots $-1 + \sqrt{2}$ and $-1 - \sqrt{2}$ of $x^2 + 2x - 1 = 0$. Note that this `solve` command includes a designation of x as the variable for which to solve. For the solutions in decimal form, use

```
evalf(solve(x∧2+2*x-1=0,x);
```

which will return the decimal values 0.414213562 and -2.414213562 of these roots. As another example,

```
evalf(solve(cos(x)-x=0,x));
```

yields the approximate solution 0.7390851332 of $\cos(x) - x = 0$.

To solve a system of equations, enter the system in curly brackets in the `solve` command. For example

```
solve({x-2*y=4,5*x+y=-3},{x,y});
```

gives the solution $x = -2/11, y = -23/11$. Again, note the designation of x and y as the variables to be solved for.

In order to define a function, say $f(x) = x\sin(5x) - 3x$, enter

```
f:=x→x*sin(5*x)-3*x;
```

This names the function f and the variable x. The arrow must be typed into MAPLE as a dash followed by a "greater than" symbol. If we want to evaluate this function at a point, say $\pi/4$, simply type

```
f(Pi/4);
```

to obtain the value $-\pi\sqrt{2}/8 - 3\pi/4$.

To plot a graph of f, say on the interval $[-1, 3]$, enter

```
plot(f(x),x=-1..3);
```

If we wish, we can enter a function directly into the `plot` command without prior definition. For example,

```
plot(x*cos(3*x)-exp(x),x=-1..1);
```

will graph $x\cos(3x) - e^x$ for $-1 \le x \le 1$.

To plot several graphs of functions that have been defined, enter, for example,

```
plot({f(x),g(x),h(x)},x=a..b);
```

We can guarantee that all the graphs are in one color, say black, by

```
plot({f(x),g(x),h(x)},x=a..b,color=black);
```

Sometimes we want to enter a function having jump discontinuities. This is done by specifying the value of the function on successive intervals. For example, to define

$$s(x) = \begin{cases} x & \text{for } x < -1 \\ \cos(3x) & \text{for } -1 < x < 4 \\ x^2 & \text{for } 4 < x < 9 \\ \sin(4x) & \text{for } x > 9 \end{cases}$$

enter

```
s:=x→piecewise(x<-1,x,x<4,cos(3*x),x<9,x∧2,sin(4*x));
```

We can differentiate $f(x)$ by

```
diff(f(x),x);
```

This can be done with a previously entered $f(x)$, or we can put the function into the command, as with

```
diff(x*cos(3*x)*exp(2*x),x);
```

We can also take a derivative using D(f)(x).
To take an indefinite integral $\int f(x)\, dx$, use the int command:

```
int(f(x),x);
```

Note that the variable must be specified. For a definite integral $\int_a^b f(x)\, dx$, use

```
int(f(x),x=a..b);
```

This can also be used with improper integrals. For example, enter

```
int(exp(-x),x=0..infinity);
```

to evaluate $\int_0^\infty e^{-x}\, dx = 1$.
MAPLE has a sum command. If $a1, \ldots, an$ are n numbers that have been defined in some way, then their sum is computed as

```
sum(aj,j=1..n);
```

It may be necessary to precede this command with eval f to obtain a decimal evaluation. Often it is convenient in a summation to define the sequence as a function. For example, if we define

```
a:=j→j*sin(Pi/j);
```

then

```
sum(a(j),j=4..7);
```

will produce the sum

$$4\sin(\pi/4)+5\sin(\pi/5)+6\sin(\pi/6)+7\sin(\pi/7).$$

Preceding the `sum` command with `evalf` will produce the decimal value of this sum.

Note: If a MAPLE file is saved, then any code that has been entered will be retained. However, if the file is closed and then reopened, commands must be reactivated. For example, if a function `f(x)` was previously entered, then place the cursor at the end of the line defining `f(x)` and hit ENTER again to reactivate this function.

Ordinary Differential Equations

Some operations with differential equations require that a package of differential equation subroutines be opened. This is done by

```
with(DEtools);
```

When this command is ended with a semi-colon, there follows a list of the subroutines. If a colon is used instead, the subroutines are activated but not listed.

For a direction field of $y' = y^2$ on a grid $-2 < x < 2, -2 < y < 2$, type

```
DEplot(diff(y(x),x) =y(x)∧2,x=-2..2,y=-2..2,color=black);
```

Note that y is entered as $y(x)$ in the specification of the differential equation. The instruction to show the direction field in black is optional.

For a field plot with sketches of some integral curves, enter initial conditions specifying these curves, for example

```
DEplot(diff(y(x),x) =y(x)∧2,y(x)=-4..4,y=-3..3,[[y(0)=-1/2],

   [y(0)=1/2],[y(0)=1],[y(0)=-2]],color=black,linecolor

   =[black,black,black,black]);
```

This produces a black direction field over the grid $-4 < x < 4, -3 < y < 3$ with sketches (in black) of the integral curves through $(0, -1/2)$, $(0, 1/2)$, $(0, 1)$ and $(0, -2)$.

We can solve some differential equations using the `dsolve` command. For example, for the general solution of $y' - (1/x)y = -2$, enter

```
dsolve(diff(y(x),x) - (1/x)*y(x) =-2,y(x));
```

This returns the general solution

$$y(x)=C_1x-2x\ln(x).$$

The arbitrary constant in the MAPLE output is denoted $C1$.

As an example of a second-order differential equation, consider

$$y'' - 4y' + y=x^3 - \sin(2x).$$

Enter

```
dsolve(diff(diff(y(x),x),x) - 4*diff(y(x),x) +y(x)
    =x∧3 - sin(2*x),y(x));
```

This gives the general solution

$$y(x) = C_1 e^{(2+\sqrt{3})x} + C_2 e^{(2-\sqrt{3})x} + 90x + 12x^2 + x^3$$

$$+ 336 - \frac{8}{73}\cos(2x) + \frac{3}{73}\sin(2x).$$

For an initial value problem, include the initial condition(s). For example,

```
dsolve(diff(y(x),x) - (1/x)*y(x) = -2,y(1) = 5,y(x));
```

gives the solution $y = (1/2)x^2 + x + 1$ of the initial value problem $y' - (1/x)y = -2$; $y(1) = 5$.
We can also solve some systems of differential equations. For example, to solve

$$y_1' - 4y_2' = 1, \; y_1' + 2y_2' = t$$

enter

```
dsolve({diff(y1(t),t) - 4*diff(y2(t),t) = 1,diff(y1(t),t)
    + 2*diff(y2(t),t) = t},{y1(t),y2(t)});
```

to obtain the general solution

$$y_1(t) = \frac{1}{10}t^2 - \frac{1}{5}t + c_1, \; y_2(t) = \frac{2}{5}t^2 + \frac{1}{5}t + c_2.$$

MAPLE can be used to write the first n terms of the power series solution of a differential equation with analytic coefficients expanded about 0 and with initial condition(s) specified at 0. To illustrate, define the differential equation $y' - \cos(x)y = e^x$ in MAPLE by

```
deq1: = diff(y(x),x) - cos(x)*y(x) = exp(x);
```

Now obtain a power series solution about 0, with $y(0) = 5$, by

```
dsolve(deq1,y(0) = 5,y(x),series);
```

This produces the output

$$y(x) = 5 + 6x + \frac{7}{2}x^2 + \frac{1}{2}x^3 - \frac{7}{12}x^4 - \frac{5}{12}x^5 + O(x^6).$$

The symbol $O(x^6)$ means that terms involving sixth and higher powers of x have not been included in this expression. In the absence of an instruction, MAPLE defaults to giving the first six terms of a power series solution about 0. For terms up to and including x^n, include a value of n in the instruction. For example, to obtain terms up to and including x^{10}, enter

```
dsolve({deq2,y(0) = 5},y(x),series,order = 10);
```

Matrix Manipulations

To work with matrices, load the linear algebra package, using

```
with(linalg);
```

Again, the semi-colon will give a list of the subroutines. To avoid the list, end this command with a colon.

There are many ways to enter a matrix. One way is to enter the dimension and the rows. For example,

```
Matrix(4,2,[[-1,3],[6,1],[6,5],[-1,-2]]);
```

enters the 4×2 matrix

$$\begin{pmatrix} -1 & 3 \\ 6 & 1 \\ 6 & 5 \\ -1 & -2 \end{pmatrix}$$

To give this matrix the name K, enter

```
K:=Matrix(4,2,[[-1,3],[6,1],[6,5],[-1,-2]]);
```

For the inverse of a nonsingular matrix A which has been entered, use

```
inverse(A);
```

For the rank of a matrix L (which need not be square) use

```
rank(L);
```

For the determinant of square A, use

```
det(A);
```

Add two $n \times m$ matrices as we would expect,

```
A+B;
```

Multiply matrices A and B by

```
A.B;
```

Elementary row operations can be carried out very efficiently. Let A be an $n \times m$ matrix that has been entered into MAPLE. To interchange rows k and j and A to form a new matrix B, use

```
B:=swaprow(A,k,j);
```

To form S by multiplying row r of B by x, use

```
S:=mulrow(B,r,x);
```

And to form T from S by adding x times row r_1 to row r_2, use

```
T: = addrow(S,r1,r2,x);
```

Finally, suppose we select an element $a_{k,j}$ of A, and we want to proceed by elementary `row` operations to a new matrix W having zeros above and below this element. W is produced in one command by

```
W: = pivot(A,k,j);
```

The corresponding column operations can be achieved by replacing `row` with `col` in these commands.

The characteristic polynomial of a square matrix A is obtained using

```
charpoly(A,x);
```

in which the variable in which the polynomial will be written (here x) must be specified. The command

```
eigenvals(A);
```

lists the eigenvalues of A (real or complex), giving each eigenvalue according to its multiplicity. And the command

```
eigenvects(A);
```

lists each eigenvalue along with its multiplicity and, for each eigenvalue, as many linearly independent eigenvectors as can be found for that eigenvalue. To illustrate, let

```
A: = Matrix(3,3,[[1,-1,0],[0,1,1],[0,0,-1]]);
```

This is the matrix of Example 8.3. Now

```
eigenvects(A);
```

gives the output

```
[1,2,[1,0,0]],[-1,1,[1,2,-4]]
```

This gives eigenvalue 1 with multiplicity 2, and every eigenvector associated with 1 is a multiple of $(1, 0, 0)$. Eigenvalue -1 has multiplicity 1 and eigenvector $(1, 2, -4)$.

Now let B be the 3×3 matrix of Example 8.5:

```
B: = Matrix(3,3,[[5,-4,4],[12,-11,12],[4,-4,5]]);
```

The command

```
eigenvals(B);
```

gives the list 1, 1, −3 of eigenvalues with 1 having multiplicity 2 and −3 having multiplicity 1. Next use

```
eigenvects(B);
```

to obtain the output

$$[1, 2, [1, 1, 0], [-1, 0, 1]], [-3, 1, [1, 3, 1]]$$

giving two linearly eigenvectors associated with eigenvalue 1.

Integral Transforms

MAPLE has subroutines for several integral transforms. To load these, enter

```
with(inttrans);
```

To take the Laplace transform of $f(t)$, use

```
laplace(f(t),t,s);
```

in which $f(t)$ may have been loaded previously or can be specified in the command. For example,

```
laplace(t*cos(t),t,s);
```

returns the transform $(s^2 - 1)((s^2 + 1)^2)$ of $t\cos(t)$.

For the inverse Laplace transform of $1/((s^2 + 4)^2)$, use

```
invlaplace(1/((s ∧ 2 + 4) ∧ 2),s,t);
```

to obtain $(1/16)\sin(2t) - (1/8)\cos(2t)$.

For the Fourier, Fourier sine, and Fourier cosine transforms, the commands are similar, except replace `laplace` with `fourier`, `fouriersin`, or `fouriercos`. For example, for the Fourier transform of $e^{-|t|}$, use

```
fourier(exp(-abs(t)),t,w);
```

to obtain the transform $2/(1 + w^2)$. To compute the inverse Fourier transform of $1/(1 + w)$, use

```
invfourier((1/(1+w),w,t);
```

to obtain the inverse transform $\frac{1}{2}ie^{-it}(2H(t) - 1)$, where $H(t)$ is the Heaviside function.

Special Functions

In MAPLE, $J_n(x)$ is called `BesselJ(n,x)`, and the Bessel function of the second kind, $Y_n(x)$, is denoted `BesselY(n,x)`. To evaluate, for example, $J_3(1.2)$, use

```
evalf(BesselJ(3,1.2));
```

For integrals involving Bessel functions, use the `int` command. For example,

```
evalf(int(x*BesselJ(1,x)*cos(3*x),x=0..1));
```

will give a decimal evaluation of $\int_0^1 x J_1(x)\cos(3x)\,dx$.

We can find (approximately) the mth (in order of increasing magnitude) positive zeros of $J_n(x)$ by using

```
evalf(BesselJZeros(n,m);
```

For example,

```
evalf(BesselJZeros(3,7));
```

returns the seventh positive zero of $J_3(x)$.

The nth Legendre polynomial $P_n(x)$ is denoted `LegendreP(n,x)` in MAPLE. Since Legendre's differential equation is second order, there is a second, linearly independent solution, often denoted $Q_n(x)$. In MAPLE this function is `LegendreQ(n,x)`. To obtain $P_n(x)$ explicitly, use

```
simplify(LegendreP(n,x));
```

Integrals involving Legendre polynomials can be done using the `int` command. For example,

```
evalf(int(sin(x)*LegendreP(5,x),x=-1..1));
```

will give a decimal evaluation of $\int_{-1}^1 \sin(x)P_5(x)\,dx$.

Legendre polynomials form a special case of a class of special functions called orthogonal polynomials, which include Laguerre polynomials, Hermite polynomials, and many others. The command

```
with(orthopoly);
```

will call up MAPLE's subroutines of orthogonal polynomials.

Answers to Selected Problems

CHAPTER 1

Section 1.1

1. Yes, since

$$2\varphi(x)\varphi'(x) = 2\sqrt{x-1}\left(\frac{1}{2\sqrt{x-1}}\right) = 1 \text{ for } x > 1.$$

3. Yes   5. Yes   7. $2x^2 = y^3 + c$

9. not separable   11. $y = 1/(1-cx)$; also $y = 0$

13. $\sec(y) = Ax$ 15. not separable

17. $\frac{1}{2}y^2 - y + \ln(y+1) = \ln(x) - 2$

19. $(\ln(y))^2 = 3x^2 - 3$ 21. $3y\sin(3y) + \cos(3y) = 9x^2 - 5$

23. $45°$ F   25. 8.57 kg

Section 1.2

1. $y = cx^3 + 2x^3 \ln|x|$ 3. $y = \frac{1}{2}x - \frac{1}{4} + ce^{-2x}$

5. $y = 4x^2 + 4x + 2 + ce^{2x}$ 7. $y = x^2 - x - 2$

9. $y = x + 1 + 4(x+1)^{-2}$ 11. $y = -2x^2 + cx$

13.

$$A_1(t) = 50 - 30e^{-t/20}, \; A_2(t) = 75 + 90e^{-t/20} - 75e^{-t/30}$$

$A_1(t)$ has its minimum value of $5450/81$ pounds at $60\ln(9/5)$ minutes.

Section 1.3

1. $2xy^2 + e^{xy} + y^2 = c$ 3. not exact

5. $y^3 + xy + \ln|x| = c$ 7. $\alpha = -3$; $x^2y^3 - 3xy - 3y^2 = c$

9. $3xy^4 - x = 47$ 11. $x\sin(2y - x) = \pi/24$

Section 1.4

1. The maximum height is 342.25 feet; the object hits the ground at 148 ft/sec, 4.75 seconds after it was dropped.

3.

$$v(t) = \begin{cases} 32 - 32e^{-t} & \text{for } 0 \leq t \leq 4 \\ 8(1 + ke^{-8t})/(1 - ke^{-8t}) & \text{for } t > 4, \end{cases}$$

with $k = e^{32}(3e^4 - 4)/(5e^4 - 4)$. $\lim_{t \to \infty} v(t) = 8$ ft/sec,

$$s(t) = \begin{cases} 32(t + e^{-t} - 1) & \text{for } 0 \le t \le 4 \\ 8t + 2\ln(1 - ke^{-8t}) + 64 \\ +32e^{-4} - 2\ln(2e^4/(5e^4 - 4)) & \text{for } t > 4 \end{cases}$$

5. 17.5 ft   7. $t = 2\sqrt{R}g$ with R as the radius of the circle

9. $y = (-3/4)\ln|x| + C$ 11. $(y - 1)^2 + \frac{1}{2}x^2 = C$

Section 1.5

1. $f(x, y) = \sin(xy)$ and $\partial f/\partial y = x\cos(xy)$ are continuous for all (x, y), hence in any rectangle centered at $(\pi/2, 1)$.

3. $f(x, y) = x^2 - y^2 + 8x/y$ and $\partial f/\partial y = -2y - 8x/y^2$ are continuous on any rectangle centered at $(0, -4)$ that does not intersect the x - axis.

5. Two solutions are

$$y = -\frac{1}{2}\ln\left(e^{-2y_0} + 2(x - x_0)\right)$$

and

$$y = -\frac{1}{2}\ln\left(e^{-2y_0} + 2(x_0 - x)\right).$$

The theorem does not apply, because the differential equation is $y' = \pm 2y$.

Section 1.6

1. Figure A.1   3. Figure A.2   5. Figure A.3

Section 1.7

1. Approximate values are given in Table A.1.

3. See Table A.2. In this case the exact solution $y = 5e^{3x^2/2}$ increases so rapidly that the Euler approximations are increasingly inaccurate as x increases from 0.

5. See Table A.3.

7. See Table A.4.

9. See Table A.5.

11. See Table A.6.

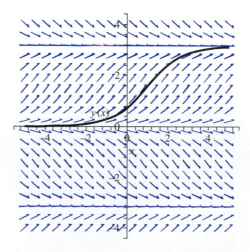

FIGURE A.1 *Direction field and solution for Problem 1, Section 1.6*

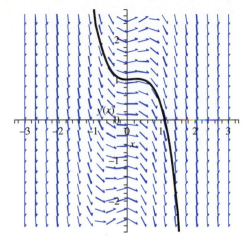

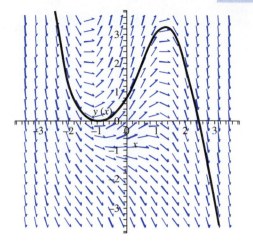

FIGURE A.2 *Direction field and solution for Problem 3 in Section 1.6.*

FIGURE A.3 *Direction field and solution for Problem 5 in Section 1.6.*

TABLE A.1	Euler Approximations in Problem 1 in Section 1.7	
x	$y_{app}(x_k)$	$y(x_k)$
0.0	1	1
0.05	1	1.001250521
0.10	1.002498958	1.001250521
0.15	1.007503103	1.005008335
0.20	1.015031072	1.020133420
0.25	1.025113849	1.031575844
0.30	1.037794710	1.045675942
0.35	1.053129175	1.062502832
0.40	1.071184959	1.082138316
0.45	1.092041913	1.104676904
0.50	1.115791943	1.130225803

TABLE A.2	Euler Approximations in Problem 3 in Section 1.7	
x	$y_{app}(x_k)$	$y(x_k)$
0.00	5	5
0.05	5	5.018785200
0.10	5.0375	5.075565325
0.15	5.1130625	5.171629965
0.20	5.228106406	5.309182735
0.25	5.384949598	5.491425700
0.30	7.404305698	5.722683920
0.35	10.73624326	6.008576785
0.40	16.37277097	6.356245750
0.45	26.19643355	6.774651405
0.50	43.87902620	7.274957075

TABLE A.3	Euler Approximations in Problem 5 in Section 1.7	
x	yapp(x_k)	$y(x_k)$
1	−2	−2
1.05	−2.127015115	−2.129163318
1.10	−2.258244233	−2.262726023
1.15	−2.393836450	−2.400852694
1.20	−2.533952645	−2.543722054
1.25	−2.678768165	−2.691527843
1.30	−2.828472691	−2.844479697
1.35	−2.983271267	−3.002804084
1.40	−3.143385165	−3.166745253
1.45	−3.309052780	−3.336566227
1.50	−3.480530557	−3.512549830

TABLE A.4	Runge-Kutta Approximation in Problem 7 in Section 1.7
x	y_k
0.0	2
0.2	2.16257799
0.4	2.27783452
0.6	2.34198641
0.8	2.35938954
1.0	2.33750216
1.2	2.28392071
1.4	2.20519759
1.6	2.106598
1.8	2.99222519
2.0	2.8652422

TABLE A.5	Runge-Kutta Approximation in Problem 9 in Section 1.7
x	y_k
0	1
0.2	1.26466198
0.4	1.45391187
0.6	1.58485267
0.8	1.67218108
1.0	1.7274517
1.2	1.75945615
1.4	1.77481079
1.6	1.7784748
1.8	1.77415235
2.0	1.76459409

TABLE A.6	*Runge-Kutta Approximation in Problem 11 in Section 1.7*

x	y_k
0	4
0.2	3.43866949
0.4	2.94940876
0.6	2.52453424
0.8	2.15677984
1.0	1.83939807
1.2	1.56621078
1.4	1.33162448
1.6	1.13062448
1.8	0.958734358
2.0	0.812012458

CHAPTER 2

Section 2.1

1. $y(x) = c_1 \sin(6x) + c_2 \cos(6x)$; $y(x) = \frac{1}{3}\sin(6x) - 5\cos(6x)$
3. $y(x) = c_1 e^{-2x} + c_2 e^{-x}$; $y(x) = 4e^{-2x} - 7e^{-x}$
5. $y(x) = c_1 e^x \cos(x) + c_2 e^x \sin(x)$; $y(x) = 6e^x \cos(x) - 5e^x \sin(x)$
7. $y(x) = c_1 e^{4x} + c_2 e^{-4x} - \frac{x^2}{4} + \frac{1}{2}$
9. $y(x) = c_1 e^{3x} \cos(2x) + c_2 e^{3x} \sin(2x) - 8e^x$

Section 2.2

1. $y = c_2 e^{-2x} + c_2 e^{3x}$ 3. $y = e^{-3x}(c_1 + c_2 x)$
5. $y = e^{-5x}(c_1 \cos(x) + c_2 \sin(x))$
7. $y = e^{-3x/2}[c_1 \cos(3\sqrt{7}x/2) + c_2 \sin(3\sqrt{7}x/2)]$
9. $y = e^{7x}(c_1 + c_2 x)$
11. $y = 5 - 2e^{-3x}$ 13. $y(x) = 0$
15. $y = \frac{9}{7}e^{3(x-2)} + \frac{5}{7}e^{-4(x-2)}$
17. $y = e^{x-1}(29 - 17x)$
19. $y = e^{(x+2)/2}\left[\cos(\sqrt{15}(x+2)/2) + \frac{5}{\sqrt{15}}\sin(\sqrt{15}(x+2)/2)\right]$

Section 2.3

1. $y = c_1 \cos(x) + c_2 \sin(x) - \cos(x)\ln|\sec(x) + \tan(x)|$
3. $y = c_1 \cos(3x) + c_2 \sin(3x) + 4x\sin(3x) + \frac{4}{3}\ln|\cos(3x)|\cos(3x)$
5. $y = c_1 e^x + c_2 e^{2x} - e^{2x}\cos(e^{-x})$
7. $y = c_1 e^{2x} + c_2 e^{-x} - x^2 + x - 4$
9. $y = e^x[c_1 \cos(3x) + c_2 \sin(3x)] + 2x^2 + x - 1$
11. $y = c_1 e^{2x} + c_2 e^{4x} + e^x$
13. $y = c_1 e^x + c_2 e^{2x} + +3\cos(x) + \sin(x)$
15. $y = e^{2x}[c_1 \cos(3x) + c_2 \sin(3x)] + \frac{1}{3}e^{2x} - \frac{1}{2}e^{3x}$
17. $y = \frac{7}{4}e^{2x} - \frac{3}{4}e^{-2x} - \frac{7}{4}xe^{2x} - \frac{1}{4}x$
19. $y = \frac{3}{8}e^{-2x} - \frac{19}{120}e^{-6x} + \frac{1}{5}e^{-x} + \frac{7}{12}$
21. $y = 2e^{4x} + 2e^{-2x} - 2e^{-x} - e^{2x}$
23. $y = 4e^{-x} - \sin^2(x) - 2$

Section 2.4

1. With $y(0) = 5$,

$$y = e^{-2t}\left[5\cosh(\sqrt{2}t) + \frac{10}{\sqrt{2}}\sinh(\sqrt{2}t)\right];$$

with $y(0) = 0$,

$$y = \frac{5}{\sqrt{2}}e^{-2t}\sinh(\sqrt{2}t).$$

These functions are graphed in Figure A.4.

3. With $y(0) = 5$,

$$y = \frac{5}{2}e^{-t}[2\cos(2t) + \sin(2t)]$$

and with $y(0) = 0$,

$$y = \frac{5}{2}e^{-t}\sin(2t).$$

Graphs are in Figure A.5.

5. $y = \frac{A}{\sqrt{2}}e^{-2t}\sinh(\sqrt{2}t)$; graphs are in Figure A.6, proceeding from the lowest to the highest graph as A increases.

7. $y = Ate^{-2t}$, with the graphs in Figure A.7 moving from the lowest to the highest as A increases.

9. $y = \frac{A}{2}e^{-t}\sin(2t)$, with the graphs moving from lowest to highest as A increases. Graphs are given in Figure A.8, increasing as A increases.

11. At most once; a condition on $y(0)$ alone is not enough to guarantee this, and a condition on $y'(0)$ is also needed.

13. Increasing C decreases the frequency.

15. (a)

$$y(t) = \frac{1}{373}e^{-3t}\left(2266\cosh(\sqrt{7}t) + \frac{6582}{\sqrt{7}}\sinh(\sqrt{7}t)\right)$$

$$+ \frac{1}{373}(-28\cos(3t) + 72\sin(3t))$$

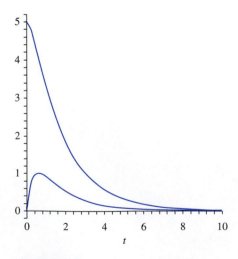

FIGURE A.4 *Graphs in solution 1 in Section 2.4.*

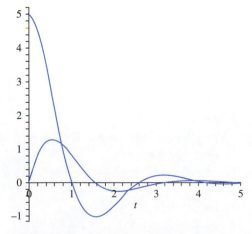

FIGURE A.5 *Graphs in solution 3 in Section 2.4.*

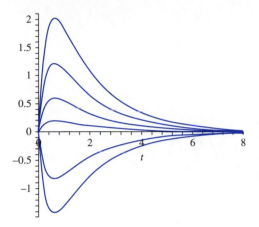

FIGURE A.6 *Graphs in solution 5 in Section 2.4.*

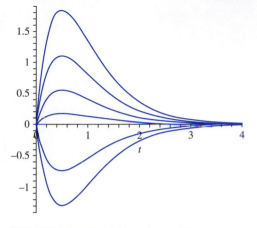

FIGURE A.7 *Graphs in solution 7 in Section 2.4.*

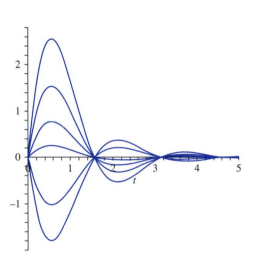

FIGURE A.8 *Graphs in solution 9 in Section 2.4.*

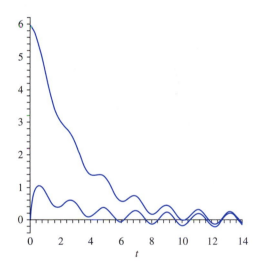

FIGURE A.9 *Graphs in solution 15 in Section 2.4.*

(b)

$$y(t) = \frac{1}{373} e^{-3t} \left(28 \cosh(\sqrt{7}t) + \frac{2106}{\sqrt{7}} \sinh(\sqrt{7}t) \right)$$
$$+ \frac{1}{373} \left(-28 \cos(3t) + 72 \sin(3t) \right)$$

Graphs are shown in Figure A.9.

17. (a)

$$y(t) = \frac{1}{15} e^{-t/2} \left(98 \cos(\sqrt{11}t/2) + \frac{74}{\sqrt{11}} \sin(\sqrt{11}t/2) \right)$$
$$\frac{1}{15} (-8 \cos(3t) + 4 \sin(3t))$$

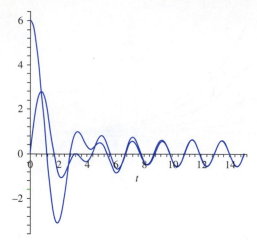

FIGURE A.10   *Graphs in solution 17 in Section 2.4.*

(b)

$$y(t) = \frac{1}{15} e^{-t/2} \left(8\cos(\sqrt{11}t/2) + \frac{164}{\sqrt{11}} \sin(\sqrt{11}t/2) \right)$$
$$+ \frac{1}{15}(-8\cos(3t) + 4\sin(3t))$$

Graphs are given in Figure A.10.

CHAPTER 3

Section 3.1

1. $3\frac{s^2-4}{(s^2+4)^2}$
3. $\frac{14}{s^2} - \frac{7}{s^2+4}$
5. $-10\frac{1}{(s+4)^3} + \frac{3}{s^2+9}$
7. $\cos(8t)$ 9. $e^{-42t} - t^3 e^{-3t}/6$

Section 3.2

1. $y = \frac{1}{4} - \frac{13}{4} e^{-4t}$
3. $y = -\frac{4}{17} e^{-4t} + \frac{4}{17}\cos(t) + \frac{1}{17}\sin(t)$
5. $y = -\frac{1}{4} + \frac{1}{2}t + \frac{17}{4} e^{2t}$
7. $y = \frac{22}{25} e^{2t} - \frac{13}{5} t e^{2t} + \frac{3}{25}\cos(t) - \frac{4}{25}\sin(t)$
9. $y = \frac{1}{16} + \frac{1}{16}t - \frac{33}{16}\cos(4t) + \frac{15}{64}\sin(4t)$

Section 3.3

1. $\frac{6}{(s+2)^4} - \frac{3}{(s+2)^2} + \frac{2}{s+2}$
3. $\frac{1}{s}(1 - e^{-7s}) + \frac{s}{s^2+1}\cos(7)e^{-7s} - \frac{1}{s^2+1}\sin(7)e^{-7s}$
5. $\frac{1}{s^2} - \frac{11}{s} e^{-3s} - \frac{4}{s^2} e^{-3s}$
7. $\frac{1}{s+1} - \frac{2}{(s+1)^3} + \frac{1}{(s+1)^2+1}$

9. $\frac{s}{s^2+1} + \left(\frac{2}{s} - \frac{s}{s^2+1} - \frac{1}{s^2+1}\right)e^{-2\pi s}$

11. $\frac{s^2+4s-5}{(s^2+4s+13)^2}$

13. $e^{2t}\sin(t)$ 15. $\cos(3(t-2))H(t-2)$

17. $\frac{1}{\sqrt{2}}e^{-3t}\sinh(\sqrt{2}t)$

19. $e^{-3t}\cosh(2\sqrt{2}t) - \frac{1}{2\sqrt{2}}e^{-3t}\sinh(2\sqrt{2}t)$

21. $y = \cos(2t) + \frac{3}{4}[1-\cos(2(t-4))]H(t-4)$

23. $y = \left[-\frac{1}{4} + \frac{1}{12}e^{2(t-6)} + \frac{1}{6}e^{-(t-6)}\cos(\sqrt{3}(t-6))\right]H(t-6)$

25. $y = -\frac{1}{4} + \frac{2}{5}e^t - \frac{3}{20}\cos(2t) - \frac{1}{5}\sin(2t) - \left[-\frac{1}{4} + \frac{2}{5}e^{t-5} + \frac{3}{20}\cos(2(t-5)) - \frac{1}{5}\sin(2(t-5))\right]H(t-5)$

27. $E_{\text{out}} = 5e^{-4t} + 10[1-e^{-4(t-5)}]H(t-5)$

29. $i(t) = \frac{k}{R}(1-e^{-Rt/L}) - \frac{k}{R}(1-e^{-R(t-5)/L})H(t-5)$

Section 3.4

1. $\frac{1}{16}[\sinh(2t) - \sin(2t)]$

3. $[\cos(at) - \cos(bt)]/[(b-a)(b+a)]$ if $b^2 \neq a^2$; $t\sin(at)/2a$ if $b^2 = a^2$

5. $\frac{1}{a^4}[1-\cos(at)] - \frac{1}{2a^3}t\sin(at)$

7. $\left(\frac{1}{2} - \frac{1}{2}e^{-2(t-4)}\right)H(t-4)$

9. $y(t) = e^{3t} * f(t) - e^{2t} * f(t)$

11. $y(t) = \frac{1}{4}e^{6t} * f(t) - \frac{1}{4}e^{2t} * f(t) + 2e^{6t} - 5e^{2t}$

13. $y(t) = \frac{1}{3}\sin(3t) * f(t) - \cos(3t) + \frac{1}{3}\sin(3t)$

15. $y(t) = \frac{4}{3}e^t - \frac{1}{4}e^{2t} - \frac{1}{12}e^{-2t} - \frac{1}{3}e^t * f(t) + \frac{1}{4}e^{2t} * f(t) + \frac{1}{12}e^{-2t} * f(t)$

17. $f(t) = \frac{1}{2}e^{-2t} - \frac{3}{2}$ 19. $f(t) = \cosh(t)$

21. $f(t) = 3 + \frac{2}{5}\sqrt{15}e^{t/2}\sin(\sqrt{15}/2)$

Section 3.5

1. $y(t) = 3[e^{-2(t-2)} - e^{-3(t-2)}]H(t-2) - 4[e^{-2(t-5)} - e^{-3(t-5)}]H(t-5)$

3. $y = 6(e^{-2t} - e^{-t} + te^t)$

5. $\varphi(t) = (B+9)e^{-2t} - (B+6)e^{-3t}$; $\varphi(0) = 3, \varphi'(0) = B$

7. $y(t) = \sqrt{\frac{m}{k}}v_0\sin\left(\sqrt{\frac{k}{m}}t\right)$

CHAPTER 4

Section 4.1

1. a_0 is arbitrary, $a_1 = 1, 2a_2 - a_0 = -1$ and $a_n = \frac{1}{n}a_{n-2}$ for $n = 3, 4, \ldots$;

$$y(x) = a_0 + (a_0 - 1)\left[\frac{1}{2}x^2 + \frac{1}{2(4)}x^4 + \frac{1}{2(4)(6)}x^6 + \cdots\right]$$

$$+ x + \frac{1}{3}x^3 + \frac{1}{3(5)}x^5 + \frac{1}{3(5)(7)}x^7 + \cdots.$$

3. a_0 is arbitrary, $a_1 + a_0 = 0, 2a_2 + a_1 = 1$,

$$a_{n+1} = \frac{1}{n+1}(a_{n-2} - a_n) \text{ for } n = 2, 3, \ldots,$$

$$y(x) = a_0 \left[1 - x + \frac{1}{2!}x^2 + \frac{1}{3!}x^3 - \frac{7}{4!}x^4 + \cdots \right]$$

$$+ \frac{1}{2!}x^2 - \frac{1}{3!}x^3 + \frac{1}{4!}x^4 + \frac{11}{5!}x^5 - \frac{31}{6!}x^6 + \cdots$$

5. a_0 and a_1 are arbitrary, $a_2 = \frac{1}{2}(3 - a_0)$,

$$a_{n+2} = \frac{n-1}{(n+1)(n+2)} a_n \text{ for } n = 1, 2, \ldots,$$

$$y(x) = a_0 + a_1 x$$

$$+ (3 - a_0) \left[\frac{1}{2!}x^2 + \frac{1}{4!}x^4 + \frac{3}{6!}x^6 + \frac{3(5)}{8!}x^8 + \frac{3(5)(7)}{10!}x^{10} + \cdots \right]$$

7. a_0, a_1 are arbitrary, $a_2 + a_0 = 0$, $6a_3 + 2a_1 = 1$,

$$a_{n+2} = \frac{(n-1)a_{n-1} - 2a_n}{(n+1)(n+2)} \text{ for } n = 2, 3, \ldots,$$

$$y(x) = a_0 + a_1 x - a_0 x^2 + \left(\frac{1}{6} - \frac{1}{3}a_1 \right) x^3 + \left(\frac{1}{6}a_0 + \frac{1}{12}a_1 \right) x^4 + \cdots$$

9. a_0, a_1 are arbitrary, $2a_2 + a_1 + 2a_0 = 1$, $6a_3 + 2a_2 + a_1 = 0$,

$$a_{n+2} = \frac{(n-2)a_n - (n+1)a_{n+1}}{(n+1)(n+2)} \text{ for } n = 3, 4, \ldots,$$

$$y(x) = a_0 + a_1 x + \left(\frac{1}{2} - a_0 - \frac{1}{2}a_1 \right) x^2$$

$$+ \left(\frac{1}{3}a_0 - \frac{1}{6} \right) x^3 - \left(\frac{1}{24} + \frac{1}{12}a_0 \right) x^4 + \cdots$$

Section 4.2

1. $y_1(x) = c_0(x - 1)$,

$$y_2(x) = c_0^* [(x - 1) \ln(x) - 3x + \frac{1}{4}x^2$$

$$+ \frac{1}{36}x^3 + \frac{1}{288}x^4 + \frac{1}{2400}x^5 + \ldots]$$

3. $y_1(x) = c_0[x^4 + 2x^5 + 3x^6 + 4x^7 + \ldots]$
$$= c_0 \frac{x^4}{(x-1)^2}, \quad y_2(x) = c_0^* \frac{3-4x}{(x-1)^2}$$
$y_2(x) = \frac{3-4x}{(1-x)^2}$

5.

$$y_1(x) = c_0 \left[x^{1/2} - \frac{1}{2(1!)(3)} x^{3/2} + \frac{1}{2^2(2!)(3)(5)} x^{5/2} \right.$$

$$\left. - \frac{1}{2^3(3!)(3)(5)(7)} x^{7/2} + \frac{1}{2^4(4!)(3)(5)(7)(9)} x^{9/2} + \cdots \right],$$

$$y_2(x) = c_0^* \left[1 - \frac{1}{2}x + \frac{1}{2^2(3!)(3)(5)} x^3 \right.$$

$$\left. + \frac{1}{2^4(4!)(3)(5)(7)} x^4 + \cdots \right]$$

7. $y_1(x) = c_0[x^2 + \frac{1}{3!}x^4 + \frac{1}{5!}x^6 + \frac{1}{7!}x^8 + \ldots]$
$y_2(x) = c_0^*[x - x^2 + \frac{1}{2!}x^3 - \frac{1}{3!}x^4 + \frac{1}{4!}x^5 - \ldots]$

9. $y_1(x) = c_0(1 - x)$, $y_2(x) = c_0^* \left((1 - x) \ln \left(\frac{x}{x-2} \right) - 2 \right)$

CHAPTER 5

Section 5.1

1. $(2+\sqrt{2})\mathbf{i}+3\mathbf{j}, (2-\sqrt{2})\mathbf{i}-9\mathbf{j}+10\mathbf{k}$,
   $4\mathbf{i}-6\mathbf{j}+10\mathbf{k}, 3\sqrt{2}\mathbf{i}+18\mathbf{j}-15\mathbf{k}, \sqrt{38}$
3. $3\mathbf{i}-\mathbf{k}, \mathbf{i}-10\mathbf{j}+\mathbf{k}$,
   $4\mathbf{i}-10\mathbf{j}, 3\mathbf{i}+15\mathbf{j}-3\mathbf{k}, \sqrt{29}$
5. $3\mathbf{i}-\mathbf{j}+3\mathbf{k}, -\mathbf{i}-3\mathbf{j}-\mathbf{k}$,
   $2\mathbf{i}+2\mathbf{j}+2\mathbf{k}, 6\mathbf{i}-6\mathbf{j}+6\mathbf{k}, \sqrt{3}$
7. $\frac{3}{\sqrt{5}}(-5\mathbf{i}-4\mathbf{j}+2\mathbf{k})$
9. $\frac{4}{9}(-4\mathbf{i}+7\mathbf{j}+4\mathbf{k})$
11. $x=3+6t, y=-t, z=0, -\infty<t<\infty$
13. $x=0, y=1+t, z=3+2t, -\infty<t<\infty$
15. $x=2+3t, y=-3-9t, z=6+2t, -\infty<t<\infty$

Section 5.2

1. $2, \cos(\theta)=2/\sqrt{14}$, not orthogonal
3. $-23, \cos(\theta)=-23/\sqrt{29}\sqrt{41}$, not orthogonal
5. $-18, \cos(\theta)=-9/\sqrt{10}$, not orthogonal
7. $3x-y+4z=4$ 9. $4x-3y+2z=25$
11. $7x+6y-5z=-26$

Section 5.3

1. $\mathbf{F}\times\mathbf{G}=8\mathbf{i}+2\mathbf{j}+12\mathbf{k}$
3. $\mathbf{F}\times\mathbf{G}=-8\mathbf{i}-12\mathbf{j}-5\mathbf{k}$
5. not collinear, $x-2y+z=3$
7. not collinear, $2x-11y+z=0$
9. not collinear, $29x+37y-12z=30$
11. $\mathbf{i}-\mathbf{j}+2\mathbf{k}$

Section 5.4

1. independent   3. independent   5. dependent
7. dependent   9. independent
11. A basis consists of $<1,0,0,-1>$ and $<0,1,-1,0>$ and the dimension is 2.
13. A basis consists of $<1,0,0,0>, <0,0,1,0>$ and $<0,0,0,1>$. The dimension is 3.
15. The vector $<0,1,0,2,0,3,0>$ forms a basis and the dimension is 1.

CHAPTER 6

Section 6.1

1. $\begin{pmatrix} 14 & -2 & 6 \\ 10 & -5 & -6 \\ -26 & -43 & -8 \end{pmatrix}$

3. $\begin{pmatrix} 2+2x-x^2 & -12x+(1-x)(x+e^x+2\cos(x)) \\ 4+2x+2e^x+2xe^x & -22-2x+e^{2x}+2e^x\cos(x) \end{pmatrix}$

5. $\begin{pmatrix} -36 & 0 & 68 & 196 & 20 \\ 128 & -40 & -36 & -8 & 72 \end{pmatrix}$

7.

$$AB = \begin{pmatrix} -10 & -34 & -16 & -30 & -14 \\ 10 & -2 & -11 & -8 & -45 \\ -5 & 1 & 15 & 61 & -63 \end{pmatrix}$$

BA is not defined.

9. **AB** = (115) and

$$BA = \begin{pmatrix} 3 & -18 & -6 & -42 & 66 \\ -2 & 12 & 4 & 28 & -44 \\ -6 & 36 & 12 & 84 & -132 \\ 0 & 0 & 0 & 0 & 0 \\ 4 & -24 & -8 & -56 & 88 \end{pmatrix}$$

11. **AB** is not defined;

$$BA = \begin{pmatrix} 410 & 36 & -56 & 227 \\ 17 & 253 & 40 & -1 \end{pmatrix}$$

Section 6.2

1. $\alpha \begin{pmatrix} -1 \\ 1 \\ 1 \\ 0 \end{pmatrix} + \beta \begin{pmatrix} 1 \\ -1 \\ 0 \\ 1 \end{pmatrix}$

3. $\alpha \begin{pmatrix} 0 \\ 0 \\ 0 \end{pmatrix}$

(only the trivial solution).

5. $\alpha \begin{pmatrix} -9/4 \\ -7/4 \\ -5/8 \\ 13/8 \\ 1 \end{pmatrix}$

7. $\alpha \begin{pmatrix} -5/6 \\ -2/3 \\ -8/3 \\ -2/3 \\ 1 \\ 0 \end{pmatrix} + \beta \begin{pmatrix} -5/9 \\ -10/9 \\ -13/9 \\ -1/9 \\ 0 \\ 1 \end{pmatrix}$

9. $\alpha \begin{pmatrix} 1 \\ 1 \\ 0 \\ 1 \\ 1 \\ 0 \\ 0 \end{pmatrix} + \beta \begin{pmatrix} -2 \\ -3/2 \\ 2/3 \\ -4/3 \\ 0 \\ 1 \\ 0 \end{pmatrix} + \gamma \begin{pmatrix} 0 \\ 1/2 \\ -3 \\ 0 \\ 0 \\ 0 \\ 1 \end{pmatrix}$

Section 6.3

In each problem, the unique or general solution is given, or it is noted that the system is inconsistent.

1. $\begin{pmatrix} 1 \\ 1/2 \\ 4 \end{pmatrix}$

3. $\alpha \begin{pmatrix} 1 \\ 1 \\ 3/2 \\ 1 \\ 0 \\ 0 \end{pmatrix} + \beta \begin{pmatrix} 0 \\ 0 \\ 1/2 \\ 0 \\ 1 \\ 0 \end{pmatrix} + \gamma \begin{pmatrix} -17/2 \\ -6 \\ -51/4 \\ 0 \\ 0 \\ 1 \end{pmatrix} + \begin{pmatrix} 9/2 \\ 3 \\ 25/4 \\ 0 \\ 0 \\ 0 \end{pmatrix}$

5. $\alpha \begin{pmatrix} 2 \\ 2 \\ 7 \\ 3/2 \\ 1 \\ 0 \end{pmatrix} + \beta \begin{pmatrix} -2 \\ -1 \\ -9/2 \\ -3/4 \\ 0 \\ 1 \end{pmatrix} + \begin{pmatrix} -4 \\ -4 \\ -38 \\ -11/2 \\ 0 \\ 0 \end{pmatrix}$

7. $\alpha \begin{pmatrix} -1/2 \\ -1 \\ 3 \\ 1 \\ 0 \end{pmatrix} + \beta \begin{pmatrix} -3/4 \\ 1 \\ -2 \\ 0 \\ 1 \end{pmatrix} + \begin{pmatrix} 9/8 \\ 2 \\ 0 \\ 0 \\ 0 \end{pmatrix}$

9. $\alpha \begin{pmatrix} -1 \\ -1 \\ 0 \\ 0 \\ 0 \\ 0 \\ 0 \end{pmatrix} + \beta \begin{pmatrix} 1 \\ 0 \\ 0 \\ 1 \\ 0 \\ 0 \\ 0 \end{pmatrix} + \gamma \begin{pmatrix} -3/14 \\ 0 \\ 3/14 \\ 0 \\ 1 \\ 0 \\ 0 \end{pmatrix} + \delta \begin{pmatrix} -1 \\ 0 \\ 0 \\ 0 \\ 0 \\ 1 \\ 0 \end{pmatrix} + \epsilon \begin{pmatrix} 1/14 \\ 0 \\ -1/14 \\ 0 \\ 0 \\ 0 \\ 1 \end{pmatrix} + \begin{pmatrix} -29/7 \\ 0 \\ 1/7 \\ 0 \\ 0 \\ 0 \\ 0 \end{pmatrix}$

Section 6.4

1. $\dfrac{1}{5}\begin{pmatrix} -1 & 2 \\ 2 & 1 \end{pmatrix}$ 3. $\dfrac{1}{12}\begin{pmatrix} -2 & 2 \\ 1 & 5 \end{pmatrix}$ 5. $\dfrac{1}{12}\begin{pmatrix} 3 & -2 \\ -3 & 6 \end{pmatrix}$

7. $\dfrac{1}{31}\begin{pmatrix} -6 & 11 & 2 \\ 3 & 10 & -1 \\ 1 & -7 & 10 \end{pmatrix}$ 9. $\dfrac{1}{11}\begin{pmatrix} -23 \\ -75 \\ -9 \\ 14 \end{pmatrix}$ 11. $\dfrac{1}{7}\begin{pmatrix} 22 \\ 27 \\ 30 \end{pmatrix}$

CHAPTER 7

Section 7.1

1. $\det(\mathbf{B}) = \sum_{p}(-1)^{\mathrm{sgn}(p)} b_{1p(1)} b_{2p(2)} \cdots b_{np(n)}$

$\quad = \sum_{p}(-1)^{\mathrm{sgn}(p)} (\alpha a_{1p(1)})(\alpha a_{2p(2)}) \cdots (\alpha a_{np(n)})$

$\quad = \alpha^n \sum_{p}(-1)^{\mathrm{sgn}(p)} a_{1p(1)} a_{2p(2)} \cdots a_{np(n)} = \alpha^n \det(\mathbf{A}).$

3. From Problem 1,

$$\det(\mathbf{A}) = \det(\mathbf{A}^t) = (-1)^n \det(\mathbf{A}).$$

If n is odd, then $(-1)^n = -1$ and $\det(\mathbf{A}) = -\det(\mathbf{A})$, hence $\det(\mathbf{A}) = 0$.

Section 7.2

1. -22 3. -14 5. -2247 7. -122

Section 7.3

1. 32 3. 3 5. -773 7. -152

9. Subtract row one from rows two and three to obtain

$$\begin{vmatrix} 1 & a & a^2 \\ 1 & b & b^2 \\ 1 & c & c^2 \end{vmatrix} = \begin{vmatrix} 1 & a & a^2 \\ 0 & b-a & b^2-a^2 \\ 0 & c-a & c^2-a^2 \end{vmatrix}$$

$$= (b-a)(c^2-a^2) - (c-a)(b^2-a^2).$$

A routine algebraic manipulation now yields the requested result.

Section 7.4

1. $\dfrac{1}{13}\begin{pmatrix} 6 & 1 \\ -1 & 2 \end{pmatrix}$ 3. $\dfrac{1}{5}\begin{pmatrix} -4 & 1 \\ 1 & 1 \end{pmatrix}$

5. $\dfrac{1}{32}\begin{pmatrix} 5 & 3 & 1 \\ -8 & -24 & 24 \\ -2 & -14 & 6 \end{pmatrix}$ 7. $\dfrac{1}{29}\begin{pmatrix} -1 & 25 & -21 \\ -8 & -3 & 6 \\ -1 & -4 & 8 \end{pmatrix}$

Section 7.5

1. $x_1 = -11/47, x_2 = -100/47$ 3. $x_1 = -1/2, x_2 = -19/22, x_3 = 2/11$
5. $x_1 = 5/6, x_2 = -10/3, x_3 = -5/6$ 7. $x_1 = -86, x_2 = -109/2, x_3 = -43/2, x_4 = 37/2$

CHAPTER 8

Section 8.1

When an eigenvalue is listed more than once, as many linearly independent eigenvectors as are associated with that eigenvalue are given.

1. $1+\sqrt{6}, \begin{pmatrix} \sqrt{6}/2 \\ 1 \end{pmatrix}, 1-\sqrt{6}, \begin{pmatrix} -\sqrt{6}/2 \\ 1 \end{pmatrix}$ 3. $-5, \begin{pmatrix} 7 \\ -1 \end{pmatrix}, 2, \begin{pmatrix} 0 \\ 1 \end{pmatrix}$

5. $\frac{1}{2}(3+\sqrt{47}i), \begin{pmatrix} -1+\sqrt{47}i \\ 4 \end{pmatrix}, \frac{1}{2}(3-\sqrt{47}i), \begin{pmatrix} -1-\sqrt{47}i \\ 4 \end{pmatrix}$

7. $0, \begin{pmatrix} 0 \\ 1 \\ 0 \end{pmatrix}, 2, \begin{pmatrix} 2 \\ 1 \\ 0 \end{pmatrix}, 3, \begin{pmatrix} 0 \\ 2 \\ 3 \end{pmatrix}$ 9. $0, 0, \begin{pmatrix} 1 \\ 0 \\ 3 \end{pmatrix}, -3, \begin{pmatrix} 1 \\ 0 \\ 0 \end{pmatrix}$

11. $1, \begin{pmatrix} -2 \\ -11 \\ 0 \\ 1 \end{pmatrix}, 2, \begin{pmatrix} 0 \\ 0 \\ 1 \\ 0 \end{pmatrix}, \frac{1}{2}(-1+\sqrt{53}), \begin{pmatrix} -7+\sqrt{53} \\ 0 \\ 0 \\ 2 \end{pmatrix} \frac{1}{2}(-1-\sqrt{53}), \begin{pmatrix} -7-\sqrt{53} \\ 0 \\ 0 \\ 2 \end{pmatrix}$

13. $0, \begin{pmatrix} 1 \\ 2 \end{pmatrix}, 5, \begin{pmatrix} -2 \\ 1 \end{pmatrix}$

15. $5+\sqrt{2} \begin{pmatrix} 1+\sqrt{2} \\ 1 \end{pmatrix}, 5-\sqrt{2}, \begin{pmatrix} 1-\sqrt{2} \\ 1 \end{pmatrix}$

17. $3, \begin{pmatrix} 0 \\ 0 \\ 1 \end{pmatrix}, -1+\sqrt{2}, \begin{pmatrix} 1+\sqrt{2} \\ 1 \\ 0 \end{pmatrix}, -1-\sqrt{2}, \begin{pmatrix} 1-\sqrt{2} \\ 0 \\ 4 \end{pmatrix}$

19. If λ is an eigenvalue with eigenvector $\mathbf{E}$, then

$$\mathbf{A}^k\mathbf{E} = \mathbf{A}^{k-1}\mathbf{A}\mathbf{E}$$

$$= \mathbf{A}^{k-1}\lambda\mathbf{E} = \lambda\mathbf{A}^{k-2}\mathbf{A}\mathbf{E}$$

$$= \lambda\mathbf{A}^{k-2}\lambda\mathbf{E} = \lambda^2\mathbf{A}^{k-2}\mathbf{E}$$

$$= \cdots = \lambda^{k-1}\mathbf{A}\mathbf{E} = \lambda^k\mathbf{E},$$

which means that λ^k is an eigenvalue of $\mathbf{A}^k$ with eigenvector $\mathbf{E}$.

Section 8.2

In Problems 1 through 7, P is given

1. $\begin{pmatrix} -3 + \sqrt{7}i & -3 - \sqrt{7}i \\ 8 & 8 \end{pmatrix}$

3. Not diagonalizable

5. $\begin{pmatrix} 0 & 5 & 0 \\ 1 & 1 & -3/2 \\ 0 & 0 & 1 \end{pmatrix}$

7. $\begin{pmatrix} 1 & 0 & 0 & 0 \\ 0 & 1 & (2 - 3\sqrt{3})/41 & (2 + 3\sqrt{5})/41 \\ 0 & 0 & (-1 + \sqrt{5})/2 & (-1 - \sqrt{5})/2 \\ 0 & 0 & 1 & 1 \end{pmatrix}$

11. $\begin{pmatrix} 1 & 0 \\ (1 - 5^{18})/4 & 5^{18} \end{pmatrix}$

Section 8.3

1. None of these;

$$2, 2, \begin{pmatrix} i \\ 1 \end{pmatrix}$$

not diagonalizable

3. Skew-hermitian;

$$0, \begin{pmatrix} 1 \\ 0 \\ (1+i)/2 \end{pmatrix}; \sqrt{3}i, \begin{pmatrix} 1 \\ \sqrt{3}i \\ -1-i \end{pmatrix}; -\sqrt{3}i, \begin{pmatrix} 1 \\ -\sqrt{3}i \\ -1-i \end{pmatrix}, \begin{pmatrix} 1 & 1 & 1 \\ 0 & \sqrt{3}i & -\sqrt{3}i \\ (1+i)/2 & -1-i & -1-i \end{pmatrix}$$

5. Hermitian; eigenvalues satisfy $\lambda^3 - 3\lambda^2 - 5\lambda + 3 = 0$. Eigenvalues and eigenvectors are approximately

$$4.051374, \begin{pmatrix} 1 \\ 0.525687 \\ -0.129755i \end{pmatrix}; 0.482696, \begin{pmatrix} 1 \\ -1.258652 \\ 2.607546i \end{pmatrix}; -1.53407, \begin{pmatrix} 1 \\ -2.267035 \\ -1.477791i \end{pmatrix}$$

$$\begin{pmatrix} 1 & 1 & 1 \\ 0.525687 & -1.258652 & -2.267035 \\ -0.129755i & 2.607546i & -1.477791i \end{pmatrix}$$

7. Skew-hermitian; eigenvalues satisfy $\lambda^3 - i\lambda^2 + 5\lambda - 4i = 0$. Eigenvalues and eigenvectors are approximately

$$-2.164248i, \begin{pmatrix} -i \\ -3.164248 \\ 2.924109 \end{pmatrix}; 0.772866i, \begin{pmatrix} i \\ 0.227134 \\ 0.587772 \end{pmatrix};$$

$$2.391382i, \begin{pmatrix} i \\ -1.391382 \\ -1.163664 \end{pmatrix}, \begin{pmatrix} -i & i & i \\ -3.164248 & 0.227134 & -1.391382 \\ 2.924109 & 0.587772 & -1.163664 \end{pmatrix}$$

9. Hermitian;

$$0, \begin{pmatrix} 0 \\ i \\ 1 \end{pmatrix}; 4 + 3\sqrt{2}, \begin{pmatrix} 4 + 3\sqrt{2} \\ -1 \\ -i \end{pmatrix}; 4 - 3\sqrt{2}, \begin{pmatrix} 4 - 3\sqrt{2} \\ -i \\ -i \end{pmatrix}; \begin{pmatrix} 0 & 4 + 3\sqrt{2} & 4 - 3\sqrt{2} \\ i & -1 & -1 \\ 1 & -i & -i \end{pmatrix}$$

CHAPTER 9

Section 9.1

The following solutions give the fundamental matrix $\Omega(t)$ and the solution of the initial value problem.

1. $\Omega(t) = \begin{pmatrix} -e^{2t} & 3e^{6t} \\ e^{2t} & e^{6t} \end{pmatrix}$; $\mathbf{X}(t) = \begin{pmatrix} -3e^{2t} + 3e^{6t} \\ 3e^{2t} + e^{6t} \end{pmatrix}$

3. $\Omega(t) = \begin{pmatrix} 4e^{(1+2\sqrt{3})t} & 4e^{(1-2\sqrt{3})t} \\ (-1+\sqrt{3})e^{(1+2\sqrt{3})t} & (-1-\sqrt{3})e^{(1-2\sqrt{3})t} \end{pmatrix}$; $\mathbf{X}(t) = \begin{pmatrix} (1+5\sqrt{3}/3)e^{(1+2\sqrt{3})t} + (1-5\sqrt{3}/3)e^{(1-2\sqrt{3})t} \\ (-1+\sqrt{3}/6)e^{(1+2\sqrt{3})t} + (1+\sqrt{3}/6)e^{(1-2\sqrt{3})t} \end{pmatrix}$

5. $\Omega(t) = \begin{pmatrix} e^{t} & 0 & e^{-3t} \\ 0 & e^{t} & 3e^{-3t} \\ e^{-t} & e^{t} & e^{-3t} \end{pmatrix}$; $\mathbf{X}(t) = \begin{pmatrix} 10e^{t} - 9e^{-3t} \\ 24e^{t} - 27e^{-3t} \\ 14e^{t} - 9e^{-3t} \end{pmatrix}$

Section 9.2

1. $\Omega(t) = \begin{pmatrix} 7e^{3t} & 0 \\ 5e^{3t} & e^{-4t} \end{pmatrix}$; $\mathbf{X}(t) = \Omega(t)\mathbf{C} = \begin{pmatrix} 7c_1e^{3t} \\ 5c_1e^{3t} + c_2e^{-4t} \end{pmatrix}$

3. $\Omega(t) = \begin{pmatrix} 1 & e^{2t} \\ -1 & e^{2t} \end{pmatrix}$; $\mathbf{X}(t) = \begin{pmatrix} c_1 + c_2e^{2t} \\ -c_1 + c_2e^{2t} \end{pmatrix}$

5. $\Omega(t) = \begin{pmatrix} 1 & 2e^{3t} & -e^{-4t} \\ 6 & 3e^{3t} & 2e^{-4t} \\ -13 & -2e^{3t} & e^{-4t} \end{pmatrix}$; $\mathbf{X}(t) = \begin{pmatrix} c_1 + 2c_2e^{3t} - c_3e^{-4t} \\ 6c_1 + 3c_2e^{3t} + 2c_2e^{-4t} \\ -13c_1 - 2c_2e^{3t} + c_3e^{-4t} \end{pmatrix}$

7. $\Omega(t) = \begin{pmatrix} 2e^{4t} & e^{-3t} \\ -3e^{4t} & 2e^{-3t} \end{pmatrix}$; $\mathbf{X}(t) = \begin{pmatrix} 6e^{4t} - 5e^{-3t} \\ -9e^{4t} - 10e^{-3t} \end{pmatrix}$

9. $\Omega(t) = \begin{pmatrix} 0 & e^{2t} & 3e^{3t} \\ 1 & e^{2t} & e^{3t} \\ 1 & 0 & e^{3t} \end{pmatrix}$; $\mathbf{X}(t) = \begin{pmatrix} 4e^{2t} - 3e^{3t} \\ 2 + 4e^{2t} - e^{3t} \\ 2 - e^{3t} \end{pmatrix}$

11. $\Omega(t) = \begin{pmatrix} 2e^{2t}\cos(2t) & 2e^{2t}\sin(2t) \\ e^{2t}\sin(2t) & -e^{2t}\cos(2t) \end{pmatrix}$

13. $\Omega(t) = \begin{pmatrix} 5e^{t}\cos(t) & 5e^{t}\sin(t) \\ e^{t}[2\cos(t)+\sin(t)] & e^{t}[2\sin(t)-\cos(t)] \end{pmatrix}$

15. $\Omega(t) = \begin{pmatrix} 0 & e^{-t}\cos(2t) & e^{-t}\sin(2t) \\ 0 & e^{-t}[\cos(2t)-2\sin(2t)] & e^{-t}[\sin(2t)+2\cos(2t)] \\ e^{-2t} & 3e^{-t}\cos(2t) & 3e^{-t}\sin(2t) \end{pmatrix}$

17. $\Omega(t) = \begin{pmatrix} e^{3t} & 2te^{3t} \\ 0 & e^{3t} \end{pmatrix}$

19. $\Omega(t) = \begin{pmatrix} e^{2t} & 3e^{5t} & 27te^{5t} \\ 0 & 3e^{5t} & (3+27t)e^{5t} \\ 0 & -e^{5t} & (2-9t)e^{5t} \end{pmatrix}$

21. $\Omega(t) = \begin{pmatrix} 2 & 3e^{3t} & e^{t} & 0 \\ 0 & 2e^{3t} & 0 & -2e^{t} \\ 1 & 2e^{3t} & 0 & -2e^{t} \\ 0 & 0 & 0 & e^{t} \end{pmatrix}$

Section 9.3

1. $\mathbf{X}(t) = \begin{pmatrix} [c_1(1+2t) + 2c_2t + t^2]e^{3t} \\ [1 - 2c_1t + c_2(1-2t) + t - t^2]e^{3t} + 3e^{t}/2 \end{pmatrix}$

3. $\mathbf{X}(t) = \begin{pmatrix} [c_1 + c_2(1+t) + 2t + t^2 - t^3]e^{6t} \\ [c_1 + c_2t + 4t^2 - t^3]e^{6t} \end{pmatrix}$

5. $\mathbf{X}(t) = \begin{pmatrix} c_2 e^t \\ -2c_2 e^t + (c_3 - 9c_4)e^{3t} + e^t \\ 2c_4 e^{3t} \\ (c_1 - 5c_2)e^t + c_3 e^{3t} + (1 + 3t)e^t \end{pmatrix}$

7. $\mathbf{X}(t) = \begin{pmatrix} (-1 - 14t)e^t \\ (3 - 14t)e^t \end{pmatrix}$

9. $\mathbf{X}(t) = \begin{pmatrix} (6 + 12t + t^2/2)e^{-2t} \\ (2 + 12t + t^2/2)e^{-2t} \\ (3 + 38t + 66t^2 + 13t^3/6)e^{-2t} \end{pmatrix}$

11. $\mathbf{X}(t) = \begin{pmatrix} 3c_1 e^{2t} + c_2 e^{6t} - 4e^{3t} - 10/3 \\ -c_1 e^{2t} + c_2 e^{6t} + 2/3 \end{pmatrix}$

13. $\mathbf{X}(t) = \begin{pmatrix} c_1 e^t + 5c_2 e^{7t} + (68/145)\cos(3t) - (54/145)\sin(3t) + 40/7 \\ -c_1 e^t + c_2 e^{7t} + (2/145)\cos(3t) + (24/145)\sin(3t) - 48/7 \end{pmatrix}$

15. $\mathbf{X}(t) = \begin{pmatrix} 2 + 4(1 + t)e^{2t} \\ -2 + 2(1 + 2t)e^{2t} \end{pmatrix}$

CHAPTER 10

Section 10.1

1. $(f(t)\mathbf{F}(t))' = -12\sin(3t)\mathbf{i} + 12t[2\cos(3t) - 3t\sin(3t)]\mathbf{j} + 8[\cos(3t) - 3t\sin(3t)]\mathbf{k}$
3. $(\mathbf{F} \times \mathbf{G})' = (1 - 4\sin(t))\mathbf{i} - 2t\mathbf{j} - (\cos(t) - t\sin(t))\mathbf{k}$
5. $(f(t)\mathbf{F}(t))' = (1 - 8t^3)\mathbf{i} + (6t^2\cosh(t) - (1 - 2t^3)\sinh(t))\mathbf{j} + (-6t^2 e^t + e^t(1 - 2t^3))\mathbf{k}$
7.

$$\text{position } \mathbf{F}(t) = 2t^2\mathbf{i} + 3t^2\mathbf{j} + 4t^2\mathbf{k}$$

$$\text{tangent } \mathbf{F}'(t) = 2t(2\mathbf{i} + 3\mathbf{j} + 4\mathbf{k})$$

$$\text{length } s(t) = \sqrt{29}(t^2 - 1)$$

$$\text{position } \mathbf{G}(s) = \mathbf{F}(t(s)) = \left(1 + \frac{s}{\sqrt{29}}\right)(2\mathbf{i} + 3\mathbf{j} + 4\mathbf{k})$$

Section 10.2

1. $\mathbf{v}(t) = 3\mathbf{i} + 2t\mathbf{k}, v(t) = \sqrt{9 + 4t^2}, \mathbf{a}(t) = 2\mathbf{k}$,
   $\kappa = \frac{6}{(9 + 4t^2)^{3/2}}, \mathbf{T}(t) = \frac{1}{\sqrt{9 + 4t^2}}(3\mathbf{i} + 2t\mathbf{k})$

3. $\mathbf{v}(t) = 2\mathbf{i} - 2\mathbf{j} + \mathbf{k}, v(t) = 3, \mathbf{a}(t) = \mathbf{O}$
   $\kappa = 0$,
   $\mathbf{T}(t) = \frac{1}{3}(2\mathbf{i} - 2\mathbf{j} + \mathbf{k})$

5. $\mathbf{v}(t) = -3e^{-t}(\mathbf{i} + \mathbf{j} - 2\mathbf{k}), v(t) = 3\sqrt{6}e^{-t}, \mathbf{a}(t) = 3e^{-t}(\mathbf{i} + \mathbf{j} - 2\mathbf{k})$
   $\kappa = 0, \mathbf{T}(t) = \frac{1}{\sqrt{6}}(-\mathbf{i} - \mathbf{j} + 2\mathbf{k})$

7. $\mathbf{v}(t) = 2\cosh(t)\mathbf{j} - 2\sinh(t)\mathbf{k}, v(t) = 2\sqrt{\cosh(2t)}$,
   $\mathbf{a}(t) = 2\sinh(t)\mathbf{j} - 2\cosh(t)\mathbf{k}$
   $\kappa = \frac{1}{2(\cosh(2t))^{3/2}}$,
   $\mathbf{T}(t) = \frac{1}{\sqrt{\cosh(2t)}}(\cosh(t)\mathbf{j} - \sinh(t)\mathbf{k})$

Section 10.3

1. $x = x, y = 1/(x + c), z = e^{x+k}; x = x, y = 1/(x - 1), z = e^{x-2}$
3. $x = x, y = e^x(x - 1) + c, x^2 = -2z + k$;
   $x = x, y = e^x(x - 1) - e^2, z = \frac{1}{2}(12 - x^2)$

5. $x = c, y = y, 2e^z = k - \sin(y);$
    $x = 3, y = y, z = \ln(1 + \sqrt{2}/4 - (1/2)\sin(y))$

7. There are many such vector fields. One whose streamlines are circles about the origin in the plane $z = 0$ is given by

$$\mathbf{R}(x, y, z) = y\mathbf{i} - x\mathbf{j}$$

Section 10.4

1. $yz\mathbf{i} + xz\mathbf{j} + xy\mathbf{k}, \mathbf{i} + \mathbf{j} + \mathbf{k}, \sqrt{3}, -\sqrt{3}$
3. $(2y + e^z)\mathbf{i} + 2x\mathbf{j} + xe^z\mathbf{k}, (2 + e^6)\mathbf{i} - 4\mathbf{j} - 2e^6\mathbf{k},$
    $\sqrt{20 + 4e^6 + 5e^{12}}, -\sqrt{20 + 4e^6 + 5e^{12}}$
5. $(1/\sqrt{3})(8y^2 - z + 16xy - x)$
7. $(1/\sqrt{5})(2x^2z^3 + 3x^2yz^2)$
9. $x + y + \sqrt{2}z = 4; x = y = 1 + 2t, z = \sqrt{2}(1 + 2t)$
11. $x = y; x = 1 + 2t, y = 1 - 2t, z = 0$
13. $x = 1; x = 1 + 2t, y = \pi, z = 1$
15. Level surfaces are planes $x + z = k$.

Section 10.5

In Problems 1, 3, and 5, $\nabla \cdot \mathbf{F}$ is given first, then $\nabla \times \mathbf{F}$.

1. $4, \mathbf{O}$ 3. $2y + xe^y + 2, (e^y - 2x)\mathbf{k}$
5. $2(x + y + z), \mathbf{O}$
7. $\mathbf{i} - \mathbf{j} + 4z\mathbf{k}$
9. $-6x^2yz^2\mathbf{i} - 2x^3z^2\mathbf{j} - 4x^3yz\mathbf{k}$
11. $\nabla \cdot (\varphi\mathbf{F}) = \nabla\varphi \cdot \mathbf{F} + \varphi\nabla \cdot \mathbf{F}$
    $\nabla \times (\varphi\mathbf{F}) = \nabla\varphi \times \mathbf{F} + \varphi\nabla \times \mathbf{F}$

CHAPTER 11

Section 11.1

1. 0   3. $26\sqrt{2}/3$ 5. $\sin(3) - 81/2$
7. 0   9. $-422/5$ 11. $-27/2$

Section 11.2

1. -8 3. -12 5. -40 7. 512π
9. 0   11. $95/4$

Section 11.3

1. 0   3. 2π if C encloses the origin; 0 otherwise   5. 0

Section 11.4

1. conservative, $\varphi(x, y) = xy^3 - 4y$ 3. conservative, $\varphi(x, y) = 8x^2 + 2y - y^3/3$
5. conservative, $\varphi(x, y) = \ln(x^2 + y^2)$
7. $\varphi(x, y, z) = x - 2y + z$ 9. not conservative
11. -27 13. $5 + \ln(3/2)$
15. -5 17. -403 19. $2e^{-2}$

Section 11.5

1. $125\sqrt{2}$ 3. $\pi(29^{3/2}-27)/6$ 5. $28\pi\sqrt{2}/3$
7. $(9/8)(\ln(4+\sqrt{17})+4\sqrt{17})$ 9. $-10\sqrt{3}$

Section 11.6

1. $49/12, (12/35, 33/35, 24/35)$ 3. $9\pi k\sqrt{2}, (0,0,2)$
5. $78\pi, (0,0,27/13)$ 7. $128/3$

Section 11.7

1. $256\pi/3$ 3. 0 5. $8\pi/3$
7. 2π 9. 0 because $\nabla\cdot(\nabla\times\mathbf{F})=0$.

Section 11.8

1. -8π 3. -16π 5. $-32/3$

CHAPTER 12

Section 12.1

1. 4

3. $\frac{1}{\pi}\sinh(\pi)+\frac{2}{\pi}\sinh(\pi)\sum_{n=1}^{\infty}\frac{(-1)^n}{n^2+1}\cos(n\pi x)$

5. $\frac{16}{\pi}\sum_{n=1}^{\infty}\frac{1}{2n-1}\sin((2n-1)x)$

7. $\frac{13}{3}+\sum_{n=1}^{\infty}(-1)^n\left[\frac{16}{n^2\pi^2}\cos\left(\frac{n\pi x}{2}\right)+\frac{4}{n\pi}\sin\left(\frac{n\pi x}{2}\right)\right]$

9. $\frac{3}{2}+\frac{1}{\pi}\sum_{n=1}^{\infty}\frac{1-(-1)^n}{n}\sin(nx)$

11. The series converges to $3/2$ for $x=\pm 3$, to $2x$ if $-3<x<-2$, to -2 if $x=-2$, to 0 if $-2<x<1$, to $1/2$ if $x=1$ and to x^2 if $1<x<3$.

13. The series converges to $(2+\pi^2)/2$ if $x=\pm\pi$, to x^2 if $-\pi<x<0$, to 1 if $x=0$ and to 2 if $0<x<\pi$.

15. The series converges to -1 if $-4<x<0$, to 0 if $x=\pm 4$ or $x=0$, and to 1 if $0<x<4$.

Section 12.2

1. The cosine series is 4, the function itself, for $0\le x\le 3$. The sine series is

$$\frac{16}{\pi}\sum_{n=1}^{\infty}\frac{1}{2n-1}\sin\left(\frac{(2n-1)\pi x}{3}\right),$$

converging to 0 if $x=0$ or $x=3$ and to 4 for $0<x<3$.

3. The cosine series is

$$\frac{1}{2}\cos(x)-\frac{2}{\pi}\sum_{n=1}^{\infty}\frac{(-1)^n(2n-1)}{(2n-3)(2n+1)}\cos\left(\frac{(2n-1)x}{2}\right)$$

converging to 0 if $0\le x<\pi$, to 0 at $x=0$, to $\cos(x)$ for $\pi<x\le 2\pi$, and to $-1/2$ at $x=\pi$.

The sine series is

$$-\frac{2}{3\pi}\sin\left(\frac{x}{2}\right)+\sum_{n\ne 2}\frac{-2n}{\pi(n^2-4)}(\cos(n\pi/2)+(-1)^n)\sin(nx/2)$$

converging to 0 for $0\le x<\pi$ and for $x=2\pi$, and to $\cos(x)$ for $\pi<x<2\pi$.

5. The cosine series is

$$\frac{4}{3} + \frac{16}{\pi^2} \sum_{n=1}^{\infty} \frac{(-1)^n}{n^2} \cos(n\pi x/2),$$

converging to x^2 for $0 \le x \le 2$. The sine series is

$$-\frac{8}{\pi} \sum_{n=1}^{\infty} \left[\frac{(-1)^n}{n} + \frac{2(1-(-1)^n)}{n^3 \pi^2} \right] \sin(n\pi x/2),$$

converging to x^2 if $0 < x < 2$ and to 0 for $x = 0, 2$.

7. The cosine series is

$$\frac{1}{2} + \sum_{n=1}^{\infty} \left[\frac{4}{n\pi} \sin(2n\pi/3) + \frac{12}{n^2\pi^2} \cos(2n\pi/3) - \frac{6}{n^2\pi^2}(1 + (-1)^n) \right] \cos(n\pi x/3),$$

converging to x if $0 \le x < 2$, to 1 if $x = 2$ and to $2 - x$ if $2 < x \le 3$. The sine series is

$$\sum_{n=1}^{\infty} \left[\frac{12}{n^2\pi^2} \sin(2n\pi/3) - \frac{4}{n\pi} \cos(2n\pi/3) + \frac{2}{n\pi}(-1)^n \right] \sin(n\pi x/3),$$

converging to x if $0 \le x < 2$ to 1 if $x = 2$, to $2 - x$ if $2 < x < 3$ and to 0 if $x = 3$.

Section 12.3

1. The Fourier series of f on $[-\pi, \pi]$ is

$$\frac{1}{4}\pi + \sum_{n=1}^{\infty} \left[\frac{(-1)^n - 1}{\pi n^2} \cos(nx) + \frac{(-1)^{n+1}}{n} \sin(nx) \right].$$

This converges to 0 for $-\pi < x < 0$ and to x for $0 < x < \pi$. Because f is continuous, its Fourier series can be integrated term by term, yielding the integral of the sum of the series. Term by term integration yields

$$\frac{1}{4}\pi x + \frac{1}{4}\pi^2$$

$$+ \sum_{n=1}^{\infty} \left[\frac{(-1)^n - 1}{\pi n^2} \frac{1}{n} \sin(nx) + \frac{(-1)^{n+1}}{n} \frac{1}{n}(-\cos(nx)) + (-1)^n \right].$$

3. For $-\pi \le x \le \pi$,

$$x \sin(x) = \pi - \frac{1}{2}\pi \cos(x) + 2\pi \sum_{n=2}^{\infty} \frac{(-1)^{n+1}}{n^2 - 1} \cos(nx).$$

f is continuous with continuous first and second derivatives on $[-\pi, \pi]$, and $f(-\pi) = f(\pi)$, so we can differentiate the series term by term to obtain

$$x \cos(x) + \sin(x) = \frac{1}{2} \sin(x) + 2 \sum_{n=2}^{\infty} \frac{(-1)^n}{n^2 - 1} n \sin(nx)$$

for $-\pi < x < \pi$.

Section 12.4

1. $3 + \frac{3i}{\pi} + \sum_{n=-\infty, n\neq 0}^{\infty} \frac{1}{n} e^{2n\pi ix/3}$

3. $\frac{3}{4} - \frac{1}{2\pi} \sum_{n=-\infty, n\neq 0}^{\infty} \frac{1}{n}[\sin(n\pi/2) + i(\cos(n\pi/2) - 1)]e^{n\pi ix/2}$

5. $\frac{1}{2} + \frac{3i}{\pi} \sum_{n=-\infty, n\neq 0}^{\infty} \frac{1}{2n-1} e^{(2n-1)\pi ix/2}$

CHAPTER 13

Section 13.1

1. The Fourier integral is

$$\int_0^\infty \left[\frac{2\sin(\pi\omega)}{\pi\omega^2} - \frac{2\cos(\pi\omega)}{\omega} \right] \sin(\omega x)\, d\omega,$$

converging to $-\pi/2$ if $x = -\pi$, to x for $-\pi < x < \pi$, to $\pi/2$ for $x = \pi$ and to 0 if $|x| > \pi$.

3.

$$\int_0^\infty \left(\frac{2}{\pi\omega}(1 - \cos(\pi\omega)) \right) \sin(\omega x)\, d\omega,$$

converging to $-1/2$ at $x = -\pi$, to -1 for $-\pi < x < 0$, to 0 if $x = 0$, to 1 if $0 < x < \pi$, to $1/2$ if $x = \pi$ and to 0 if $|x| > \pi$.

5.

$$\int_0^\infty \frac{1}{\pi\omega^3}[400\omega\cos(100\omega) + (20000\omega^2 - 4)\sin(100\omega)]\cos(\omega x)\, d\omega,$$

converging to x^2 if $-100 < x < 100$, to 5000 if $x = \pm 100$, and to 0 if $|x| > 100$.

7.

$$\frac{4}{\pi}\int_0^\infty \left[\frac{\cos(\pi\omega)(\cos^2(\pi\omega) - 1)}{\omega^2 - 1}\cos(\omega x) + \frac{\sin(\pi\omega)\cos^2(\pi\omega)}{1 - \omega^2}\sin(\omega x) \right] d\omega,$$

converging to $\sin(x)$ if $-3\pi < x < \pi$ and to 0 if $x \le 3\pi$ or $x \ge \pi$.

Section 13.2

1. Sine integral:

$$\int_0^\infty \frac{4}{\pi\omega^3}[10\omega\sin(10\omega) - (50\omega^2 - 1)\cos(10\omega) - 1]\sin(\omega x)\, d\omega,$$

Cosine integral:

$$\int_0^\infty \frac{4}{\pi\omega^3}[10\omega\cos(10\omega) - (50\omega^2 - 1)\sin(10\omega)]\cos(\omega x)\, d\omega.$$

Both integrals converge to x^2 for $0 \le x < 10$, to 50 if $x = 10$ and to 0 if $x > 10$.

3. Sine integral:

$$\int_0^\infty \frac{2}{\pi\omega}[1 + \cos(\omega) - 2\cos(4\omega)]\sin(\omega x)\, d\omega,$$

Cosine integral:

$$\int_0^\infty \frac{2}{\pi\omega}[2\sin(4\omega) - \sin(\omega)]\cos(\omega x)\, d\omega.$$

Both integrals converge to 1 for $0 < x < 1$, to $3/2$ for $x = 1$, to 2 for $1 < x < 4$, to 1 for $x = 4$ and to 0 for $x > 4$. The cosine integral converges to 1 at $x = 0$ while the sine integral converges to 0 at $x = 0$.

5. Sine integral:

$$\int_0^\infty \left[\frac{2}{\pi\omega}[1 + (1 - 2\pi)\cos(\pi\omega) - 2\cos(3\pi\omega)] + \frac{4}{\pi\omega^2}\sin(\pi\omega) \right] \sin(\omega x)\, d\omega,$$

Cosine integral:

$$\int_0^\infty \left[\frac{2}{\pi\omega}[(2\pi - 1)\sin(\pi\omega) + 2\sin(3\pi\omega)] + \frac{4}{\pi\omega^2}[\cos(\pi\omega) - 1] \right] \cos(\omega x)\, d\omega.$$

Both integrals converge to $1 + 2x$ for $0 < x < \pi$, to $(3 + 2\pi)/2$ for $x = \pi$, to 2 for $\pi < x < 3\pi$, to 1 for $x = 3\pi$, and to 0 for $x > 3\pi$. The sine integral converges to 0 at $x = 0$, while the cosine integral converges to 1 for $x = 0$.

7. Sine integral:

$$\int_0^\infty \frac{2}{\pi}\left(\frac{\omega^3}{4+\omega^4}\right)\sin(\omega x)\,d\omega,$$

Cosine integral:

$$\int_0^\infty \frac{2}{\pi}\left(\frac{2+\omega^2}{4+\omega^4}\right)\cos(\omega x)\,d\omega.$$

Both integrals converge to $e^{-x}\cos(x)$ for $x > 0$, the cosine integral converges to 1 for $x = 0$, and the sine integral converges to 0 at $x = 0$.

Section 13.3

1. $2i[\cos(\omega) - 1]/\omega$ 3. $[10e^{-7\omega i}\sin(4\omega)]/\omega$

5. $\frac{4}{1+4i\omega}e^{-(1+4i\omega)k/4}$ 7. $\pi e^{-|\omega|}$

9. $\frac{24}{16+\omega^2}e^{2i\omega}$

11. $18\sqrt{\frac{2}{\pi}}e^{-8t^2}e^{-4it}$ 13. $H(t+2)e^{-(10+(5-3i)t)}$

15. $H(t)[2e^{-3t} - e^{-2t}]$ 17. $H(t)te^{-t}$

Section 13.4

1. $\hat{f}_C(\omega) = \frac{1}{1+\omega^2}$, $\hat{f}_S(\omega) = \frac{\omega}{1+\omega^2}$

3. $\hat{f}_C(\omega) = \frac{1}{2}\left[\frac{\sin(K(1-\omega))}{1-\omega} + \frac{\sin(K(1+\omega))}{1+\omega}\right]$,

   $\hat{f}_S(\omega) = \frac{\omega}{\omega^2-1} - \frac{1}{2}\left[\frac{\cos(K(1+\omega))}{1+\omega} - \frac{\cos(K(1-\omega))}{1-\omega}\right]$

5. $\hat{f}_C(\omega) = \frac{1}{2}\left[\frac{1}{1+(1+\omega^2)} + \frac{1}{1+(1-\omega^2)}\right]$,

   $\hat{f}_S(\omega) = \frac{1}{2}\left[\frac{1+\omega}{1+(1+\omega^2)} - \frac{1-\omega}{1+(1-\omega^2)}\right]$

CHAPTER 14

Section 14.1

1. The problem is regular on $[0, L]$ with eigenvalues $((2n-1)\pi/2L)^2$ for $n = 1, 2, \cdots$. The functions $\sin((2n-1)\pi x/2L)$ are eigenfunctions.

3. Regular on $[0, 4]$, $((2n-1)\pi/8)^2$, $\cos((2n-1)\pi x/8)$

5. Periodic on $[-3\pi, 3\pi]$, $n^2/9$ for $n = 0, 1, 2, \cdots$, $a_n\cos(nx/3) + b_n\sin(nx/3)$ with not both a_n and b_n equal to 0

7. Regular on $[0, 1]$, eigenvalues are positive solutions of $\tan(\sqrt{\lambda}) = \frac{1}{2}\sqrt{\lambda}$. If λ is an eigenvalue, an eigenfunction is $2\sqrt{\lambda}\cos(\sqrt{\lambda}x) + \sin(\sqrt{\lambda}x)$.

9. Regular on $[0, \pi]$, $1 + n^2$ and $e^{-x}\sin(nx)$ for $n = 1, 2, \cdots$

11. For $0 < x < 1$,

$$1 - x = \sum_{n=1}^{\infty} \frac{2}{n\pi}\sin(n\pi x).$$

13. The expansion is

$$\sum_{n=1}^{\infty}\frac{4}{\pi}\frac{\sqrt{2}\cos(n\pi/2) - \sqrt{2}\sin(n\pi/2) - (-1)^n}{2n-1}\cos\left(\frac{2n-1}{8}\pi x\right).$$

This converges to -1 for $0 < x < 2$, to 1 for $2 < x < 4$, and to 0 at $x = 2$.

15. For $-3\pi < x < 3\pi$,

$$x^2 = 3\pi^2 + 36\sum_{n=1}^{\infty}\frac{(-1)^n}{n^2}\cos(nx/3).$$

Section 14.2

3. Expand $q(x) = \sum_{j=1}^{m} c_j P_j(x)$. Then

$$\int_{-1}^{1} q(x) P_n(x)\, dx = \sum_{j=0}^{n} c_j \int_{-1}^{1} P_j(x) P_n(x)\, dx = 0$$

because $n \neq j$ for $j = 0, 1, \ldots, m$.

5. For $-1 < x < 1$,

$$\sin(\pi x/2) = \frac{12}{\pi^2} x + 168 \left(\frac{\pi^2 - 10}{\pi^4} \right) P_3(x)$$

$$+ 660 \left(\frac{-112\pi^2 + \pi^4 + 1008}{\pi^6} \right) P_5(x) + \cdots$$

7. For $-1 < x < 1$,

$$\sin^2(x) = -\frac{1}{2} \cos(1) \sin(1) + \frac{1}{2}$$

$$+ \left[-\frac{5}{8} \cos(1) + \frac{15}{8} - \frac{15}{4} \cos^2(1) \right] P_2(x)$$

$$+ \left[\frac{531}{32} \cos(1) \sin(1) - \frac{585}{32} + \frac{585}{16} \cos^2(1) \right] P_4(x) + \cdots$$

9. For $-1 < x < 0$ and $0 < x < 1$,

$$f(x) = \frac{3}{2} P_1(x) - \frac{7}{8} P_3(x) + \frac{11}{16} P_5(x) + \cdots$$

Section 14.3

1. $f(x) = x$

   $2.213145642 J_1(3.831705970x) - 0.5170987826 J_1(7.015586670x)$
   $+1.104611216 J_1(10.17346814x) - 0.4549641786 J_1(13.32369194x)$
   $+0.8113206562 J_1(16.47063005x)$

3. $f(x) = xe^{-x}$

   $1.256395517 J_1(3.831705970x) + 0.08237394412 J_1(7.015586670x)$
   $+0.5976577270 J_1(10.173468144x) - 0.01994105804 J_1(13.32369194x)$
   $+0.4181324338 J_1(16.47063005x)$

5. $f(x) = \sin(\pi x)$

   $3.555896220 J_1(3.831705970x) + 1.670058301 J_1(7.015586670x)$
   $+0.9956101332 J_1(10.173468144x) + 0.7772068876 J_1(13.32369194x)$
   $+0.6036626350 J_1(16.47063005x)$

7. $f(x) = x$

   $7.749400696 J_2(5.135622302x) - 0.1583973994 J_2(8.417244140x)$
   $+1.310726377 J_2(11.61984117x) - 0.2381008476 J_2(14.79595178x)$
   $+0.9524470038 J_2(17.95981949x)$

9. $f(x) = xe^{-x}$

   $1.418532841 J_2(5.135622302x) + 0.2923912667 J_2(8.417244140x)$
   $+0.7581692534 J_2(11.61984117x) + 0.1399888559 J_2(14.79595178x)$
   $+0.5434687461 J_2(17.95981949x)$

11. $f(x) = \sin(\pi x)$

   $3.733991576 J_2(5.135622302x) + 2.468532251 J_2(8.417244140x)$
   $+1.700629359 J_2(11.61984117x) + 1.356527124 J_2(14.79595178x)$
   $+1.099075410 J_2(17.95981949x)$

CHAPTER 15

Section 15.2

1. $y(x,t) = \sum_{n=1}^{\infty} \frac{8}{n^3\pi^3 c} \left(2\sin\left(\frac{n\pi}{2}\right) - n\pi \cos\left(\frac{n\pi}{2}\right)\right) \sin\left(\frac{n\pi x}{2}\right) \sin\left(\frac{n\pi ct}{2}\right)$

3. $y(x,t) = \sum_{n=1}^{\infty} \frac{108}{(2n-1)^4\pi^4} \sin((2n-1)\pi x/3) \sin(2(2n-1)\pi t/3)$

5. $y(x,t) = \sum_{n=1}^{\infty} \frac{24(-1)^{n+1}}{(2n-1)^2\pi} \sin((2n-1)x/2)\cos((2n-1)\sqrt{2}t)$

7. $y(x,t) = \sum_{n=1}^{\infty} \frac{-32}{(2n-1)^3\pi^3} \sin((2n-1)\pi x/2)\cos(3(2n-1)\pi t/2)$
$+ \sum_{n=1}^{\infty} \frac{4}{n^2\pi^2} \left[\cos(n\pi/4) - \cos(n\pi/2)\right] \sin(n\pi x/2)\sin(3n\pi t/2)$

9. Let $Y(x,t) = y(x,t) + h(x)$ and substitute into the problem to choose $h(x) = (x^3 - 4x)/9$. The problem for Y is

$$\frac{\partial^2 Y}{\partial t^2} = 3\frac{\partial^2 Y}{\partial x^2}$$

$$Y(0,t) = Y(2,t) = 0,$$

$$Y(x,0) = \frac{1}{9}(x^3 - 4x), \quad \frac{\partial Y}{\partial t}(x,0) = 0,$$

with solution

$$Y(x,t) = \sum_{n=1}^{\infty} \frac{32}{3n^3\pi^3} \sin(n\pi x/2)\cos(n\pi\sqrt{3}t/2).$$

11. Let $Y(x,t) = y(x,t) + h(x)$ and find that $h(x) = \cos(x) - 1$. The problem for Y is

$$\frac{\partial^2 Y}{\partial t^2} = \frac{\partial^2 Y}{\partial x^2}$$

$$Y(0,t) = Y(2\pi,t) = 0,$$

$$Y(x,0) = \cos(x) - 1, \quad \frac{\partial Y}{\partial t}(x,0) = 0,$$

with solution

$$Y(x,t) = \frac{16}{\pi} \sum_{n=1}^{\infty} \frac{1}{(2n-1)[(2n-1)^2 - 4]} \sin((2n-1)x/2)\cos((2n-1)t/2).$$

13.

$$u(x,t) = e^{-At/2} \sum_{n=1}^{\infty} C_n \sin(n\pi x/L) \left[\cos(r_n t/2L) + \frac{AL}{r_n}\sin(r_n t/2L)\right],$$

where

$$C_n = \frac{2}{L}\int_0^L f(\xi)\sin(n\pi\xi/L)\,d\xi$$

and

$$r_n = \sqrt{4(BL^2 + n^2\pi^2 c^2) - A^2 L^2}.$$

15. (a) With the forcing term the solution is

$$y_f(x,t) = \sum_{n=1}^{\infty} d_n \sin(n\pi x/4)\cos(3(2n-1)\pi t/4)$$

$$+ \frac{1}{9\pi^2}(\cos(\pi x) - 1),$$

where

$$d_n = \frac{64}{\pi^3} \left(\frac{2}{(2n-1)^3} - \frac{1}{9} \frac{1}{(2n-1)((2n-1)^2 - 16)} \right)$$

(b) Without the forcing term the solution is

$$y(x,t) = \sum_{n=1}^{\infty} \frac{128}{\pi^3 (2n-1)^3} \sin((2n-1)\pi x/3) \cos(3(2n-1)\pi t/4).$$

17. $y_{j,1} = 0.25$ for $j = 1, 2, \ldots, 19$
$y_{1,2} = 0.08438, y_{j,2} = 0.05$ for $j = 2, 3, \ldots, 19$
$y_{1,3} = 0.13634, y_{2,3} = 0.077149, y_{j,3} = 0.075$ for $j = 3, 4, \ldots, 19$
$y_{1,4} = 0.17608, y_{2,4} = 0.10786, y_{3,4} = 0.10013, y_{j,4} = 0.1$ for $j = 4, 5, \ldots, 19$
$y_{1,5} = 0.20055, y_{2,5} = 0.14235, y_{3,5} = 0.12574, y_{4,5} = 0.12501,$
$y_{j,5} = 0.125$ for $j = 5, 6, \ldots, 19$.

19.   We give $y_{j,k}$ for $j = 1, 2, \ldots, 9$, first for $k = -1$ and then $k = 0, 1, \ldots, 5$.

$$y_{j,-1} : 0.08075, 0.127, 0.14475, 0.14, 0.11875,$$
$$0.087, 0.05075, 0.016, -0.01125$$
$$y_{j,0} : 0.081, 0.128, 0.147, 0.144, 0.125,$$
$$0.096, 0.063, 0.032, 0.009$$
$$y_{j,1} : 0.079125, 0.1735, 0.14788, 0.147, 0.13063,$$
$$0.10475, 0.075375, 0.0485, 0.030125$$
$$y_{j,2} : 0.0057813, 0.02115, 0.70708, 0.14903, 0.13567,$$
$$0.11328, 0.087906, 0.065531, 0.050516$$
$$y_{j,3} : -0.055066, 0.27160, 1.3199, 0.18908, 0.14015,$$
$$0.12162, 0.10062, 0.083022, 0.068688$$
$$y_{j,4} : -0.092055, 0.3768, 1.7328, 0.29675, 0.14653,$$
$$0.12981, 0.11355, 0.10072, 0.083463$$
$$y_{j,5} : -0.093987, 0.53745, 1.9712, 0.48652, 0.16125,$$
$$0.13803, 0.12669, 0.11814, 0.0941$$

Section 15.3

1. $y(x,t) = \int_0^{\infty} \frac{10}{\pi(25+\omega^2)} \cos(\omega x) \cos(12\omega t) \, d\omega$

3. $y(x,t) = \int_0^{\infty} \frac{1}{2\pi\omega} \frac{\sin(\pi\omega)}{1-\omega^2} \sin(\omega x) \sin(4\omega t) \, d\omega$

5. $y(x,t) = \int_0^{\infty} \left[\left(\frac{e^{-2}}{3\pi\omega} \frac{2\cos(\omega)-\omega\sin(\omega)}{4+\omega^2} \right) \cos(\omega x) \right.$
$$\left. + \left(\frac{e^{-2}}{3\pi\omega} \frac{\omega\cos(\omega)+2\sin(\omega)}{4+\omega^2} \right) \sin(\omega x) \right] \sin(3\omega t) \, d\omega$$

Section 15.4

1. $y(x,t) = \int_0^{\infty} \frac{2}{\pi} \frac{2-\omega\sin(\omega)-2\cos(\omega)}{\omega^3} \sin(\omega x) \cos(3\omega t) \, d\omega$

3. $y(x,t) = \int_0^{\infty} \frac{1}{\pi\omega} \frac{\sin(\pi\omega/2)-\sin(5\pi\omega/2)}{\omega^2-1} \sin(\omega x) \sin(2\omega t) \, d\omega$

5. $y(x,t) = \int_0^{\infty} \frac{3}{7\pi\omega^5} d_\omega \sin(\omega x) \sin(14\omega t) \, d\omega,$

where

$$d_\omega = 2\sin(3\omega) - 4\omega\cos(3\omega) - 3\omega^2\sin(3\omega) - 2\omega$$

Section 15.5

1. $y(x,t) = \frac{1}{2}[(x-t)^2 + (x+t)^2] + \frac{1}{2}\int_{x-t}^{x+t} -\xi \, d\xi = x^2 + t^2 - xt$

3. $y(x,t) = \frac{1}{2}[\cos(\pi(x-7t)) + \cos(\pi(x+7t))] + t - x^2t - \frac{49}{3}t^3$

5. $y(x,t) = \frac{1}{2}[e^{x-14t} + e^{x+14t}] + xt$

7. $y(x,t) = x + \frac{1}{8}(e^{-x+4t} - e^{-x-4t}) + \frac{1}{2}xt^2 + \frac{1}{6}t^3$

9. $y(x,t) = \frac{1}{32}(\sin(2(x+8t)) - \sin(2(x-8t)))$
$$+ \frac{1}{12}xt^4 + x^2 + 64t^2 - x$$

11. $y(x,t) = \frac{1}{2}[\cosh(x-3t) + \cosh(x+3t)] + t + \frac{9}{10}xt^4$

Section 15.6

1. We find that (approximately),

$$a_1 = \frac{2\int_0^1 x(1-x)J_0(2.405x)\,dx}{[J_1(2.405)^2]} \approx 2\frac{0.1057}{0.2695} = 0.78542,$$

$$a_2 \approx 0.06869, a_3 \approx 0.05311, a_4 \approx 0.01736, a_5 \approx 0.01698.$$

The fifth partial sum gives the approximation

$$z(r,t) \approx 0.78442 J_0(2.405r)\cos(2.405t) + 0.06869 J_0(5.520r)\cos(5.520t)$$
$$+ 0.05311 J_0(8.654r)\cos(8.654t) + 0.01736 J_0(11.792r)\cos(11.792t)$$
$$+ 0.01698 J_0(14.931r)\cos(14.931t).$$

3. We find the approximation

$$z(r,t) \approx 1.2534 J_0(2.405r)\cos(2.405t) - 0.80468 J_0(5.520r)\cos(5.520t)$$
$$- 0.11615 J_0(8.654r)\cos(8.654t) - 0.09814 J_0(11.792r)\cos(11.792t)$$
$$- 0.03740 J_0(14.931r)\cos(14.931t)$$

Section 15.7

1. $z(x,y,t) = \frac{1}{\pi}\sum_{n=1}^{\infty}\left[\frac{8(-1)^{n+1}\pi^2}{n} + \frac{16}{n^3}[(-1)^n - 1]\right]\sin(nx/2)\sin(y)\cos(\sqrt{n^2+4}t/2)$

3. $z(x,y,t) = \sum_{n=1}^{\infty}\sum_{m=1}^{\infty} d_{nm}\cos((2n-1)x/2)\sin((2m-1)y/2)\sin(k_{nm}t),$

where

$$d_{nm} = \frac{16}{\pi^2(2n-1)(2m-1)\sqrt{(2n-1)^2 + (2m-1)^2}}$$

and

$$k_{nm} = \sqrt{(2n-1)^2 + (2m-1)^2}.$$

CHAPTER 16

Section 16.1

1. $\partial u/\partial t = k(\partial^2 u/\partial^2 x)$ for $0 < x < L$ and $t > 0$, $u(0,t) = (\partial u/\partial x)(L,t) = 0$ for $t \geq 0$, $u(x,0) = f(x)$.

3. $\partial u/\partial t = k(\partial^2 u/\partial^2 x)$ for $0 < x < L$ and $t > 0$, $(\partial u/\partial x)(0,t) = 0$, $u(L,t) = \beta(t)$ for $t \geq 0$, $u(x,0) = f(x)$.

Section 16.2

In these solutions we sometimes use the notation $\exp(g(t)) = e^{g(t)}$.

1. $u(x,t) = \sum_{n=1}^{\infty} \frac{8L^2}{(2n-1)^3\pi^3}\sin((2n-1)\pi x/L)\exp(-(2n-1)^2\pi^2 kt/L^2)$

3. $u(x,t) = \sum_{n=1}^{\infty} d_n \sin((2n-1)\pi x/L)\exp(-3(2n-1)^2\pi^2 t/L^2),$

where

$$d_n = \frac{-16L}{(2n-1)\pi[(2n-1)^2 - 4]}.$$

5. $u(x,t) = \frac{2}{3}\pi^2 - \sum\limits_{n=1}^{\infty} \frac{4}{n^2}\cos(nx)e^{-4n^2t}$

7. $u(x,t) = \frac{1}{6}(1 - e^{-6}) + 12\sum\limits_{n=1}^{\infty}\left(\frac{1-e^{-6}(-1)^n}{36+n^2\pi^2}\right)\cos(n\pi x/6)e^{-n^2\pi^2t/18}$

9. $u(x,t) = \sum\limits_{n=1}^{\infty} \frac{4B}{(2n-1)\pi}\sin((2n-1)\pi x/2L)\exp(-(2n-1)^2\pi^2 kt/4L^2)$

11. Substitute $e^{\alpha x + \beta t}$ into the partial differential equation and solve for α and β so that $v_t = kv_{xx}$. Obtain $\alpha = -A/2$ and $\beta = k(B - A^2/4)$.

13. Let $u(x,t) = e^{-3x-9t}v(x,t)$. Then $v(0,t) = v(4,t) = 0$ and $v(x,0) = e^{3x}$, so

$$v(x,t) = \sum\limits_{n=1}^{\infty}\left(\frac{2n\pi}{144 + n^2\pi^2}(1 - e^{12}(-1)^n)\right)\sin(n\pi x/4)e^{-n^2\pi^2 t/16}.$$

15. Let $u(x,t) = v(x,t) + f(x)$ and choose $f(x) = 3x + 2$. We obtain $v(0,t) = v(1,t) = 0$ and $u(x,0) = x^2 - f(x)$. The solution for v is

$$v(x,t) = 2\sum\limits_{n=1}^{\infty}\left(\frac{4n^2\pi^2(-1)^n + 2(-1)^n - 2 - 2n^2\pi^2}{n^3\pi^3}\right)\sin(n\pi x)e^{-16n^2\pi^2 t}.$$

17. Let $u(x,t) = e^{-At}w(x,t)$. Then $w(0,t) = w(9,t) = 0$ and $u(x,0) = 3x$. We find that

$$w(x,t) = \sum\limits_{n=1}^{\infty}\frac{54(-1)^{n+1}}{n\pi}\sin(n\pi x/9)e^{-4n^2\pi^2 t/81}.$$

19. $u(x,t) = \sum\limits_{n=1}^{\infty}\left[\frac{1}{8\pi}\frac{1-(-1)^n}{n^5}\left(-1 + 4n^2 t + e^{-4n^2 t}\right)\right]\sin(nx)$
$+ \frac{4}{\pi}\sum\limits_{n=1}^{\infty}\left(\frac{1-(-1)^n}{n^3}\right)\sin(nx)e^{-4n^2 t}$

21. $u(x,t) =$
$\sum\limits_{n=1}^{\infty}\frac{50}{n^3\pi^3}\frac{1-\cos(5)(-1)^n}{n^2\pi^2-25}\left(-25 + n^2\pi^2 t + 25e^{-n^2\pi^2 t/25}\right)\sin(n\pi x/5)$
$+ \sum\limits_{n=1}^{\infty}\frac{500}{n^3\pi^3}\left((-1)^{n+1} - 1\right)\sin(n\pi x/5)e^{-n^2\pi^2 t/25}$

23. $u(x,t) = \sum\limits_{n=1}^{\infty}\frac{27(-1)^n}{128}\left(\frac{16n^2\pi^2 t + 9e^{-16n^2\pi^2 t/9} - 9}{n^5\pi^5}\right)\sin(n\pi x/3)$
$+ 2K\sum\limits_{n=1}^{\infty}\frac{1-(-1)^n}{n\pi}\sin(n\pi x/3)e^{-16n^2\pi^2 t/9}$

25. In the following, j takes on integer values 1 through 9.

$$u_{j,0} : 0.009, 0.032, 0.063, 0.096, 0.125, 0.144,$$
$$0.147, 0.128, 0.081$$
$$u_{j,1} : 0.0125, 0.034, 0.0635, 0.095, 0.1225,$$
$$0.14, 0.1415, 0.121, 0.0725$$
$$u_{j,2} : 0.01475, 0.064125, 0.089, 0.094, 0.1195,$$
$$0.136, 0.136, 0.114, 0.0665$$
$$u_{j,3} : 0.023381, 0.058, 0.084031, 0.099125, 0.11725,$$
$$0.13188, 0.1305, 0.10763, 0.06175$$

Section 16.3

1. $u(x,t) = \frac{1}{\pi} \int_0^\infty \frac{8}{16+\omega^2} \cos(\omega x) e^{-\omega^2 kt} d\omega$

3. $u(x,t) = \int_0^\infty \frac{8\cos(\omega)}{\pi} \frac{\cos^3(\omega) - \cos(\omega)}{\omega^2} \cos(\omega x) e^{-\omega^2 kt} d\omega$

$$+ \int_0^\infty \frac{8}{\pi} \frac{4\omega \sin(\omega)\cos^2(\omega) - 2\omega\sin(\omega)}{\omega^2} \cos(\omega x)e^{-\omega^2 kt} d\omega$$

$$- \int_0^\infty \frac{4}{\pi} \frac{-2\sin(\omega)\cos^3(\omega) + \sin(\omega)\cos(\omega)}{\omega^2} \sin(\omega x)e^{-\omega^2 kt} d\omega$$

$$- \int_0^\infty \frac{4}{\pi} \frac{8\omega\cos^4(\omega) - 8\omega\cos^2(\omega) + \omega}{\omega^2} \sin(\omega x)e^{-\omega^2 kt} d\omega$$

5. $u(x,t) = \frac{2}{\pi} \int_0^\infty \frac{\omega}{\alpha^2+\omega^2} \sin(\omega x) e^{-\omega^2 kt} d\omega$

7. $u(x,t) = \frac{2}{\pi} \int_0^\infty \frac{1-\cos(h\omega)}{\omega} \sin(\omega x) e^{-\omega^2 kt} d\omega$

9. $u(x,t) = \frac{4}{\pi} \int_0^\infty \frac{\omega}{(1+\omega^2)^2} \sin(\omega x) e^{-\omega^2 t} e^{-t^2/2} d\omega$

Section 16.4

1. Write

$$U(r,t) = \sum_{n=1}^\infty \frac{2}{[J_1(j_n)]^2} \left(\int_0^1 \xi^2 J_0(j_n\xi)\, d\xi \right) J_0(j_n r) e^{-j_n^2 t}.$$

Inserting the approximate values, we have

$$U(r,t) \approx 0.8170 J_0(2.405r) e^{-5.785t} - 1.1394 J_0(5/520r) e^{-30.47t}$$

$$+ 0.7983 J_0(8.654r) e^{-74.89t} - 0.747 J_0(11.792r) e^{-139.04t}$$

$$+ 0.6315 J_0(14.931r) e^{-222.93t}.$$

3. $U(r,t) = \sum_{n=1}^\infty \frac{2}{(J_1(j_n))^2} \left(\int_0^1 \xi(9-\xi^2) J_0(j_n\xi)\, d\xi \right) J_0(j_n r/3) e^{-j_n^2 t/18}$

The fifth partial sum approximation is

$$U(r,t) \approx 9.9722 J_0(2.405r/3) e^{-5.78t/18} - 1.258 J_0(5.520r/3) e^{-30.47t/18}$$

$$+ 0.4093 J_0(8.654r/3) e^{-74.89t/18} - 0.1889 J_0(11.792r/3) e^{-139.04t/18}$$

$$+ 0.1048 J_0(14.931r/3) e^{-222.93t/18}$$

Section 16.5

1. $u(x,t,t) = \sum_{n=1}^\infty \sum_{n=1}^\infty b_{nm} \sin(n\pi x/L) \sin(m\pi y/K) e^{-\beta_{nm}kt}$

where

$$\beta_{nm} = \left(\frac{n^2}{L^2} + \frac{m^2}{K^2} \right) \pi^2$$

and

$$b_{nm} = \frac{4}{LK} \int_0^K \int_0^L f(\xi,\eta) \sin(n\pi\xi/L) \sin(m\pi\eta/K)\, d\xi\, d\eta.$$

3. $u(x,y,t) = \frac{8}{\pi} \sum_{m=1}^\infty \frac{m}{4m^2-1} \sin(x) \sin(my) e^{-(1+m^2)t}$

CHAPTER 17

Section 17.2

1. $u(x,y) = \frac{1}{\sinh(\pi^2)} \sin(\pi x) \sinh(\pi(\pi - y))$

3. $u(x,y) = \sum_{n=1}^\infty \frac{32}{\pi^2 \sinh(4n\pi)} \frac{n(-1)^{n+1}}{(2n-1)^2(2n+1)^2} \sin(n\pi x) \sinh(n\pi y)$

5. $u(x, y) = \frac{1}{\sinh(\pi^2)} \sin(\pi x) \sinh(\pi y) + \frac{1}{\sinh(2)} \sin(y) \sinh(x)$

$\qquad + \sum_{n=1, n \neq 2}^{\infty} \frac{16n[(-1)^n - 1]}{\pi^2 (n-2)^2 (n+2)^2 \sinh(n\pi^2/2)} \sin(n\pi x/2) \sinh(n\pi y/2)$

7. $u(x, y) = \sum_{n=1}^{\infty} c_n \sin((2n-1)\pi x/2a) \sinh((2n-1)\pi y/2a)$

where

$$c_n = \frac{2}{a \sinh((2n-1)\pi b/2a)} \int_0^a f(\xi) \sin((2n-1)\pi \xi/2a) \, d\xi$$

9. $u(x, y) = \frac{-1}{\sinh(4\pi)} \sinh(\pi(x-4)) \sin(\pi y)$

$\qquad + \sum_{n=1}^{\infty} \frac{2}{\sinh(4n\pi)} \left(\frac{2(1-(-1)^n)}{\pi^3 n^3} \right) \sinh(n\pi x) \sin(n\pi y)$

Section 17.3

1. $u(r, \theta) = 1$

3. $u(r, \theta) = \frac{1}{3}\pi^2 + \sum_{n=1}^{\infty} \left(\frac{r}{2}\right)^n 2(-1)^n \frac{1}{n^2} [2 \cos(n\theta) + n \sin(n\theta)]$

5. $u(r, \theta) = \frac{1}{\pi} \sinh(\pi)$

$\qquad + \frac{2}{\pi} \sum_{n=1}^{\infty} \left(\frac{r}{4}\right)^n \frac{(-1)^n}{n^2+1} [\cosh(\pi) \cos(n\theta) + n \sinh(\pi) \sin(n\theta)]$

7. $u(r, \theta) = 1 - \frac{1}{3}\pi^2 + \sum_{n=1}^{\infty} \left(\frac{r}{8}\right)^n \frac{4(-1)^{n+1}}{n^2} \cos(n\theta)$

9. In polar coordinates the problem is to solve

$$\nabla^2 U(r, \theta) = 0 \text{ for } r < 4, U(4, \theta) = 16 \cos^2(\theta).$$

This has solution

$$U(r, \theta) = 8 + r^2 \left(\cos^2(\theta) - \frac{1}{2} \right),$$

so

$$u(x, y) = \frac{1}{2}(x^2 - y^2) + 8.$$

11. In polar coordinates, $U(r, \theta) = r^2(2\cos^2(\theta) - 1)$, so $u(x, y) = x^2 - y^2$.

Section 17.4

1. $u(1/2, \pi) = \frac{3}{8\pi} \int_0^{2\pi} \frac{\xi}{5/4 - \cos(\xi - \pi)} \, d\xi \approx 9.8696/\pi$,

$u(3/4, \pi/3) \approx 4.813941646/\pi$,
$u(0, 2, \pi/4) \approx 8.843875590/\pi$

3. $u(4, \pi) \approx 155.25/\pi$, $u(12, 3\pi/2) \approx 302/\pi$,
$\qquad u(8, \pi/4) \approx 111.56/\pi$, $u(7, 0) \approx 248.51/\pi$

Section 17.5

1. $u(x, y) = \frac{1}{\pi} \left[2 \arctan\left(\frac{x}{y}\right) + \arctan\left(\frac{4-x}{y}\right) - \arctan\left(\frac{4+x}{y}\right) \right]$

3. $u(x, y) = \frac{y}{\pi} \int_0^{\infty} \left(\frac{1}{y^2 + (\xi - x)^2} - \frac{1}{y^2 + (\xi + x)^2} \right) e^{-\xi} \cos(\xi) \, d\xi$

5. $u(x, y) = \frac{2}{\pi} \int_0^\infty \left(\int_0^\infty f(\xi) \sin(\omega\xi)\, d\xi \right) \sin(\omega x) e^{-\omega y}\, d\omega$

   $\quad + \frac{2}{\pi} \left(\int_0^\infty g(\xi) \sin(\omega\xi)\, d\xi \right) \sin(\omega y) e^{-\omega x}\, d\omega$

7. $u(x, y) = \frac{2}{\pi} \int_0^\infty \left(\frac{1}{1+\omega^2} \right) \sin(\omega y) e^{-\omega x}\, d\omega$

9. $u(x, y) = \frac{y}{\pi} \int_4^8 \frac{A}{y^2 + (\xi - x)^2}\, d\xi$

   $\quad = \frac{A}{\pi} \left[-\arctan\left(\frac{x-8}{y} \right) + \arctan\left(\frac{x-4}{y} \right) \right]$

11. $u(x, y) = \frac{1}{\pi} \left(\arctan\left(\frac{1-y}{x} \right) + \arctan\left(\frac{1+y}{x} \right) \right)$

Section 17.6

1. $u(x, y, z) = \sum_{n=1}^\infty \sum_{m=1}^\infty \frac{4(-1)^{n+m}}{nm\pi^2 \sinh(\pi\sqrt{n^2+m^2})} \sin(n\pi x) \sin(m\pi y) \sinh(\pi\sqrt{n^2+m^2}z)$

3. $u(x, y, z) = \sum_{n=1}^\infty \sum_{m=1}^\infty c_{nm} \sin((2n-1)\pi x) \sin((2m-1)z) \sinh(\alpha_{nm} y)$

   $\quad + \sum_{n=1}^\infty \sum_{m=1}^\infty d_{nm} \sin((2n-1)\pi x) \sin((2m-1)y/2) \sinh(\beta_{nm} z),$

where

$$\alpha_{nm} = \sqrt{(2m-1)^2 + \pi^2(2n-1)^2},$$

$$c_{nm} = \frac{16}{\pi^2(2n-1)(2m-1)\sinh(2\pi\alpha_{nm})}$$

$$\beta_{nm} = \sqrt{\frac{(2m-1)^2}{4} + \pi^2(2n-1)^2},$$

and

$$d_{nm} = \frac{16}{\pi^2(2n-1)(2m-1)\sinh(\pi\beta_{nm})}.$$

Section 17.7

1. $u(\rho, \varphi) = \sum_{n=0}^\infty \frac{(2n+1)A}{2} \left(\int_{-1}^1 (\arccos(\xi))^2 P_n(\xi) \right) \left(\frac{\rho}{R} \right)^n P_n(\cos(\varphi))$

   $\quad \approx 2.9348A - 3.7011A \left(\frac{\rho}{R} \right) P_1(\cos(\varphi)) + 1.1111A \left(\frac{\rho}{R} \right)^2 P_2(\cos(\varphi))$

   $\quad - 0.5397A \left(\frac{\rho}{R} \right)^3 P_3(\cos(\varphi)) + 0.3200A \left(\frac{\rho}{R} \right)^4 P_4(\cos(\varphi))$

   $\quad - 0.2120A \left(\frac{\rho}{R} \right)^5 P_5(\cos(\varphi)) + \cdots$

3. $u(\rho, \varphi) \approx 6.0784 - 9.8602 \left(\frac{\rho}{R} \right) P_1(\cos(\varphi))$

   $\quad + 5.2360 \left(\frac{\rho}{R} \right)^2 P_2(\cos(\varphi)) - 2.4044 \left(\frac{\rho}{R} \right)^3 P_3(\cos(\varphi))$

   $\quad + 1.5080 \left(\frac{\rho}{R} \right)^4 P_4(\cos(\varphi)) - 0.9783 \left(\frac{\rho}{R} \right)^5 P_5(\cos(\varphi)) + \cdots$

5. $u(\rho, \varphi) = \frac{T_1 R_1}{R_2 - R_1} \left[\frac{1}{\rho} R_2 - 1 \right]$

Index